# Ubiquitin–Proteasome Protocols

# METHODS IN MOLECULAR BIOLOGY™

## *John M. Walker*, SERIES EDITOR

**316. Bioinformatics and Drug Discovery**, edited by *Richard S. Larson, 2005*

**315. Mast Cells:** *Methods and Protocols*, edited by *Guha Krishnaswamy and David S. Chi, 2005*

**314. DNA Repair Protocols:** *Mammalian Systems, Second Edition*, edited by *Daryl S. Henderson, 2005*

**313. Yeast Protocols:** *Second Edition*, edited by *Wei Xiao, 2005*

**312. Calcium Signaling Protocols:** *Second Edition*, edited by *David G. Lambert, 2005*

**311. Pharmacogenomics:** *Methods and Applications*, edited by *Federico Innocenti, 2005*

**310. Chemical Genomics:** *Reviews and Protocols*, edited by *Edward D. Zanders, 2005*

**309. RNA Silencing:** *Methods and Protocols*, edited by *Gordon Carmichael, 2005*

**308. Therapeutic Proteins:** *Methods and Protocols*, edited by *C. Mark Smales and David C. James, 2005*

**307. Phosphodiesterase Methods and Protocols,** edited by *Claire Lugnier, 2005*

**306. Receptor Binding Techniques:** *Second Edition*, edited by *Anthony P. Davenport, 2005*

**305. Protein–Ligand Interactions:** *Methods and Applications*, edited by *G. Ulrich Nienhaus, 2005*

**304. Human Retrovirus Protocols:** *Virology and Molecular Biology*, edited by *Tuofu Zhu, 2005*

**303. NanoBiotechnology Protocols,** edited by *Sandra J. Rosenthal and David W. Wright, 2005*

**302. Handbook of ELISPOT: Methods and Protocols,** edited by *Alexander E. Kalyuzhny, 2005*

**301. Ubiquitin–Proteasome Protocols,** edited by *Cam Patterson and Douglas M. Cyr, 2005*

**300. Protein Nanotechnology:** *Protocols, Instrumentation, and Applications*, edited by *Tuan Vo-Dinh, 2005*

**299. Amyloid Proteins:** *Methods and Protocols*, edited by *Einar M. Sigurdsson, 2005*

**298. Peptide Synthesis and Application,** edited by *John Howl, 2005*

**297. Forensic DNA Typing Protocols,** edited by *Angel Carracedo, 2005*

**296. Cell Cycle Control:** *Mechanisms and Protocols*, edited by *Tim Humphrey and Gavin Brooks, 2005*

**295. Immunochemical Protocols,** *Third Edition*, edited by *Robert Burns, 2005*

**294. Cell Migration:** *Developmental Methods and Protocols*, edited by *Jun-Lin Guan, 2005*

**293. Laser Capture Microdissection:** *Methods and Protocols*, edited by *Graeme I. Murray and Stephanie Curran, 2005*

**292. DNA Viruses:** *Methods and Protocols*, edited by *Paul M. Lieberman, 2005*

**291. Molecular Toxicology Protocols**, edited by *Phouthone Keohavong and Stephen G. Grant, 2005*

**290. Basic Cell Culture Protocols**, *Third Edition*, edited by *Cheryl D. Helgason and Cindy L. Miller, 2005*

**289. Epidermal Cells**, *Methods and Applications*, edited by *Kursad Turksen, 2005*

**288. Oligonucleotide Synthesis**, *Methods and Applications*, edited by *Piet Herdewijn, 2005*

**287. Epigenetics Protocols**, edited by *Trygve O. Tollefsbol, 2004*

**286. Transgenic Plants:** *Methods and Protocols*, edited by *Leandro Peña, 2005*

**285. Cell Cycle Control and Dysregulation Protocols:** *Cyclins, Cyclin-Dependent Kinases, and Other Factors*, edited by *Antonio Giordano and Gaetano Romano, 2004*

**284. Signal Transduction Protocols,** *Second Edition*, edited by *Robert C. Dickson and Michael D. Mendenhall, 2004*

**283. Bioconjugation Protocols,** edited by *Christof M. Niemeyer, 2004*

**282. Apoptosis Methods and Protocols,** edited by *Hugh J. M. Brady, 2004*

**281. Checkpoint Controls and Cancer, Volume 2:** *Activation and Regulation Protocols*, edited by *Axel H. Schönthal, 2004*

**280. Checkpoint Controls and Cancer, Volume 1:** *Reviews and Model Systems*, edited by *Axel H. Schönthal, 2004*

**279. Nitric Oxide Protocols,** *Second Edition*, edited by *Aviv Hassid, 2004*

**278. Protein NMR Techniques,** *Second Edition*, edited by *A. Kristina Downing, 2004*

**277. Trinucleotide Repeat Protocols,** edited by *Yoshinori Kohwi, 2004*

**276. Capillary Electrophoresis of Proteins and Peptides,** edited by *Mark A. Strege and Avinash L. Lagu, 2004*

**275. Chemoinformatics,** edited by *Jürgen Bajorath, 2004*

**274. Photosynthesis Research Protocols,** edited by *Robert Carpentier, 2004*

**273. Platelets and Megakaryocytes, Volume 2:** *Perspectives and Techniques*, edited by *Jonathan M. Gibbins and Martyn P. Mahaut-Smith, 2004*

**272. Platelets and Megakaryocytes, Volume 1:** *Functional Assays*, edited by *Jonathan M. Gibbins and Martyn P. Mahaut-Smith, 2004*

**271. B Cell Protocols,** edited by *Hua Gu and Klaus Rajewsky, 2004*

**270. Parasite Genomics Protocols,** edited by *Sara E. Melville, 2004*

**269. Vaccina Virus and Poxvirology:** *Methods and Protocols*,edited by *Stuart N. Isaacs, 2004*

METHODS IN MOLECULAR BIOLOGY™

# Ubiquitin–Proteasome Protocols

Edited by


## Cam Patterson, MD

*Department of Medicine*
*and the Carolina Cardiovascular Biology Center,*
*University of North Carolina at Chapel Hill,*
*Chapel Hill, NC*

## Douglas M. Cyr, PhD

*Department of Cell and Developmental Biology,*
*University of North Carolina at Chapel Hill,*
*Chapel Hill, NC*



HUMANA PRESS ✷ TOTOWA, NEW JERSEY

Softcover reprint of the hardcover 1st edition 2005
999 Riverview Drive, Suite 208
Totowa, New Jersey 07512

**www.humanapress.com**



This publication is printed on acid-free paper. ∞
ANSI Z39.48-1984 (American Standards Institute)Permanence of Paper for Printed Library Materials.

Cover design by Patricia F. Cleary
Cover illustration: Intracellular localization of misfolded CFTR (Fig. 1, Chapter 21; *see* full caption and discussion on pp. 310, 311).

For additional copies, pricing for bulk purchases, and/or information about other Humana titles, contact Humana at the above address or at any of the following numbers: Tel.: 973-256-1699; Fax: 973-256-8341; E-mail: orders@humanapr.com; or visit our Website: www.humanapress.com



10 9 8 7 6 5 4 3 2 1

eISBN 1-59259-895-1

ISSN 1064-3745

**Library of Congress Cataloging-in-Publication Data**

Ubiquitin/proteasome protocols / edited by Cam Patterson, Douglas M. Cyr.
p. ; cm. — (Methods in molecular biology, ISSN 1064-3745 ; 301)
Includes bibliographical references and index.
ISBN 978-1-61737-453-1 e-ISBN 978-1-59259-895-3
1. Ubiquitin—Laboratory manuals. 2. Proteolytic enzymes—Laboratory manuals.
[DNLM: 1. Ubiquitin—metabolism. 2. Cysteine Endopeptidases. 3. Multienzyme Complexes. QU 56 U1573 2005] I. Patterson, Cam. II. Cyr, Douglas M. III. Methods in molecular biology (Clifton, N.J.) ; v. 301.
QP552.U24U2575 2005
572'.76—dc22 2004017480

# Preface

The year 2003 marked the 25th anniversary of the discovery that the small molecule ubiquitin plays a critical role in the regulated degradation of proteins. Since that time, many other major advances have been made—including the discovery of the enzymes that regulate covalent attachment of ubiquitin to substrates; the purification and characterization of the proteasome; the association of targeted protein degradation with such specific cellular events as cell cycle regulation; and the discovery that defects in protein degradation are linked to specific human pathologies. Although many had once thought of protein degradation as a relatively nonspecific phenomenon, we now know that the ubiquitin–proteasome system is tightly regulated and exhibits precise temporal and substrate specificity.

The central relevance of the ubiquitin–proteasome system was highlighted in 2004 when the Nobel Prize in Chemistry was bestowed on three figures who played crucial roles in the early discoveries in this field: Aaron Ciechanover (a contributor to this book), Avram Hershko, and Irwin Rose. Many others have made great contributions as well, and a debt of gratitude is owed to all the early pioneers by those of us who now work in this growing area of biomedical science.

Because the ubiquitin–proteasome system plays a critical role in so much that goes on within a cell, more and more scientists from different backgrounds—cellular and molecular biology, genetics, pharmacology, pathology, and others—find that they must know how to manipulate this system to address questions in their own areas of research. *Ubiquitin–Proteasome Protocols* is designed for those investigators. We have sought contributions from the scientists who originally developed the protocols that other investigators will find necessary for their work. Because the field of ubiquitin-dependent proteolysis is so broad, the topics in this book cover much ground, from basic biochemistry to cellular assays to discovery techniques using mass spectroscopic analysis. Whether one is new to the field or a long-standing contributor, we hope that *Ubiquitin–Proteasome Protocols* will serve as a useful tool to accelerate discovery and enhance productivity.

We extend our deepest thanks and appreciation to all the contributors to our work. We solicited the leading contributors to this field, and we were thrilled by their willingness to participate, their enthusiasm for this project, and the hard work they invested to make this book as good as it possibly could be. We thank all of our colleagues, coworkers, and trainees for their continued inspiration and devotion to work in this field. We especially thank Edward Dornsmith, Liz Garman, and Chris Horaist for their hard work in helping us prepare the manuscript of this book for submission. We are grateful to Humana Press for the invitation to assemble *Ubiquitin–Proteasome Protocols*, and we thank John Walker for his assistance in making this work a reality.

***Cam Patterson***, MD
***Douglas Cyr***, PhD

# Contents

# Contributors

PAUL S. ANDREWS • *Department of Biochemistry and Biophysics and the Lineberger Comprehensive Cancer Center, University of North Carolina at Chapel Hill, Chapel Hill, NC*

NATHAN BAYS • *Division of Molecular Medicine, Oregon Health Sciences University, Portland, OR*

JACK R. BENNINK • *Laboratory of Viral Diseases, National Institute of Allergy and Infectious Diseases, National Institutes of Health, Bethesda, MD*

CHRISTOPH H. BORCHERS • *Department of Biochemistry and Biophysics and the Lineberger Comprehensive Cancer Center, University of North Carolina at Chapel Hill, Chapel Hill, NC*

JEFFREY L. BRODSKY • *Department of Biological Sciences, University of Pittsburgh, Pittsburgh, PA*

ERIC CARLSON • *Division of Molecular Medicine, Oregon Health Sciences University, Portland, OR*

LILING CHEN • *Department of Cell and Developmental Biology, University of North Carolina at Chapel Hill, Chapel Hill, NC*

AARON CIECHANOVER • *The Center for Vascular Biology and Cancer Research, The Rappaport Faculty of Medicine and Research Institute, Technion–Israel Institute of Technology, Haifa, Israel*

MICHAEL J. CORBOY • *Department of Physiology, University of Texas Southwestern Medical Center at Dallas, Dallas, TX*

DOUGLAS M. CYR • *Department of Cell and Developmental Biology, University of North Carolina at Chapel Hill, Chapel Hill, NC*

LARRY DAVID • *School of Dentistry, Oregon Health Sciences University, Portland, OR*

GEORGE N. DEMARTINO • *Department of Physiology, University of Texas Southwestern Medical Center at Dallas, Dallas, TX*

NEDYALKA N. DICHEVA • *Department of Biochemistry and Biophysics and the Lineberger Comprehensive Cancer Center, University of North Carolina at Chapel Hill, Chapel Hill, NC*

R. JÜRGEN DOHMEN • *Institute for Genetics, University of Cologne, Cologne, Germany*

CHUN-YANG FAN • *Department of Cell and Developmental Biology, University of North Carolina at Chapel Hill, Chapel Hill, NC*

DANIEL FINLEY • *Department of Cell Biology, Harvard Medical School, Boston, MA*

MANABU FURUKAWA • *Lineberger Comprehensive Cancer Center, University of North Carolina at Chapel Hill, Chapel Hill, NC*

MARIA GACZYNSKA • *Institute of Biotechnology, University of Texas Health Science Center at San Antonio, San Antonio, TX*

CHRISTOPH GLANEMANN • *Institute for Genetics, University of Cologne, Cologne, Germany*

MICHAEL H. GLICKMAN • *Department of Biology, The Technion, Haifa, Israel*
SUSANNA F. GREER • *Department of Biochemistry and Biophysics and the Lineberger Comprehensive Cancer Center, University of North Carolina at Chapel Hill, Chapel Hill, NC*
ARTHUR L. HAAS • *Department of Biochemistry and Molecular Biology, Louisiana State University Health Sciences Center, New Orleans, LA*
ROLAND S. HILGARTH • *Department of Molecular and Cellular Biochemistry, University of Kentucky, Lexington, KY*
CHAIM KAHANA • *Department of Molecular Genetics, The Weizmann Institute of Science, Rehovot, Israel*
ROBERT J. LEE • *Department of Biological Sciences, University of Pittsburgh, Pittsburgh, PA*
DAVID S. LEGGETT • *ArQule Biomedical Institute, ArQule Inc., Norwood, MA*
CHANG-WEI LIU • *Department of Physiology, University of Texas Southwestern Medical Center at Dallas, Dallas, TX*
DAVID R. LOISELLE • *Department of Biochemistry and Biophysics and the Lineberger Comprehensive Cancer Center, University of North Carolina at Chapel Hill, Chapel Hill, NC*
MARKUS K. LONDON • *Institute for Genetics, University of Cologne, Cologne, Germany*
ARDYTHE A. MCCRACKEN • *Department of Biology, University of Nevada, Reno, NV*
EDWARD G. MIMNAUGH • *Urologic Oncology Branch, National Cancer Institute, National Institutes of Health, Rockville, MD*
VIOREL MOCANU • *Department of Biochemistry and Biophysics and the Lineberger Comprehensive Cancer Center, University of North Carolina at Chapel Hill, Chapel Hill, NC*
LEONARD M. NECKERS • *Urologic Oncology Branch, National Cancer Institute, National Institutes of Health, Rockville, MD*
ROBERT Z. ORLOWSKI • *The Department of Medicine and the Lineberger Comprehensive Cancer Center, University of North Carolina at Chapel Hill, Chapel Hill, NC*
PAWEL A. OSMULSKI • *Institute of Biotechnology, University of Texas Health Science Center at San Antonio, San Antonio, TX*
CAROL E. PARKER • *Department of Biochemistry and Biophysics and the Lineberger Comprehensive Cancer Center, University of North Carolina at Chapel Hill, Chapel Hill, NC*
CAM PATTERSON • *Department of Medicine and the Carolina Cardiovascular Biology Center, The University of North Carolina at Chapel Hill, Chapel Hill, NC*
CECILE M. PICKART • *Department of Biochemistry and Molecular Biology, Bloomberg School of Public Health, Johns Hopkins University, Baltimore, MD*
SHU-BING QIAN • *Laboratory of Viral Diseases, National Institute of Allergy and Infectious Diseases, National Institutes of Health, Bethesda, MD; Carolina Cardiovascular Center, University of North Carolina, Chapel Hill, NC*

SHAHRI RAASI • *Department of Biochemistry and Molecular Biology, Bloomberg School of Public Health, Johns Hopkins University, Baltimore, MD*
PAULA C. RAMOS • *Departamento de Química e Bioquímica, Faculdade de Ciências e Tecnologia, Universidade do Algarve, Campus de Gambelas, Portugal*
YUVAL REISS • *Proteologics Ltd., Rehovot Science Park, Rehovot, Israel*
MEREDITH F. N. ROSSER • *Department of Cell and Developmental Biology, University of North Carolina at Chapel Hill, Chapel Hill, NC*
NATHANIEL S. RUSSELL • *Graduate Program in Biochemistry, Cell, and Developmental Biology, Emory University School of Medicine, Atlanta, GA*
KEVIN D. SARGE • *Department of Molecular and Cellular Biology, Chandler Medical Center, University of Kentucky, Lexington, KY*
CAMERON O. SCARLETT • *Department of Biochemistry and Biophysics and the Lineberger Comprehensive Cancer Center, University of North Carolina at Chapel Hill, Chapel Hill, NC*
ANTJE SCHÄFER • *Institut für Biochemie, Universität Stuttgart, Stuttgart, Germany*
HANSJÖRG SCHILD • *Institute for Cell Biology, Department of Immunology, University of Tübingen, Tübingen, Germany*
MICHAEL G. SCHLOSSMACHER • *Center for Neurologic Diseases, Brigham and Women's Hospital, Harvard Medical School, Boston, MA*
HIDEKI SHIMURA • *Department of Neurology, Juntendo University School of Medicine, Tokyo, Japan*
WILLIAM R. SKACH • *Division of Molecular Medicine, Oregon Health Sciences University, OR*
ELIZABETH STRICKLAND • *Department of Molecular, Cellular, and Developmental Biology, Yale University, New Haven, CT*
STEFAN TENZER • *Institute of Immunology, University of Mainz, Mainz, Germany*
PHILIP J. THOMAS • *Department of Physiology, University of Texas Southwestern Medical Center at Dallas, Dallas, TX*
MARIA R. WARREN • *Department of Biochemistry and Biophysics, University of North Carolina at Chapel Hill, Chapel Hill, NC*
W. CHRISTIAN WIGLEY • *Reata Discovery, Inc., Dallas, TX*
KEITH D. WILKINSON • *Department of Biochemistry, Emory University School of Medicine, Atlanta, GA*
DIETER H. WOLF • *Institut für Biochemie, Universität Stuttgart, Stuttgart, Germany*
YUE XIONG • *Lineberger Comprehensive Cancer Center and the Department of Biochemistry and Biophysics, University of North Carolina at Chapel Hill, NC*
JONATHAN W. YEWDELL • *Laboratory of Viral Diseases, National Institute of Allergy and Infectious Diseases, National Institutes of Health, Bethesda, MD*
J. MICHAEL YOUNGER • *Department of Cell and Developmental Biology, University of North Carolina at Chapel Hill, Chapel Hill, NC*

# I

# Biochemical Methods to Study the Ubiquitin–Proteasome System

# 1

# Small-Molecule Inhibitors of Proteasome Activity

Maria Gaczynska and Pawel A. Osmulski

**Summary**

The fast-track approval of a proteasome inhibitor, PS-341, to treat multiple myeloma spurred a wave of interest in both the proteasome itself and small-molecule compounds blocking its activities. Besides being candidates for drugs against cancer, autoimmune diseases, inflammation, or stroke, specific proteasome inhibitors are indispensable tools for biochemical and cell biology investigations of the proteasome and proteasome-ubiquitin system. Numerous synthetic peptide derivatives, such as boronates, epoxides, aldehydes, vinyl sulfones, cyclic peptides, and lactones, block the N-terminal threonine-type active centers of the enzyme, halting the cleavage of proteasomal protein substrates both in vitro and in vivo. Because some of the proteasomal inhibitors exhibit a high specificity toward only one particular type of an active center of the proteasome, they constitute valuable probes for testing the mechanism of proteolysis catalyzed by the enzyme. In this chapter we discuss the most common applications of available proteasome inhibitors. In addition to the best-known competitive inhibitors, we also describe the benefits from the use of allosteric inhibitors, which induce distinct but less understood in vitro and in vivo effects on the proteasomal machinery. Finally, we present the application of the basic biochemical procedures to decipher the mechanism of interactions of a novel compound with the proteasome.

**Key Words:** Enzyme kinetics; epoxyketone; inhibitor; lactone; peptidase; peptide; proteasome; proteolysis.

## 1. Introduction

Proteasome holds a unique position among intracellular proteases. First, it is essential for cell physiology, being the major executor of controlled proteolysis *(1)*. Second, it is a giant and modular enzymatic assembly, with multiple activities and multilevel patterns of activity regulation *(2)*. Third, it is an acknowledged drug target *(3)*. The actual or potential benefits of using proteasome inhibitors as drugs range from apoptotic destruction of cancerous cells to confinement of inflammation to inhibition of angiogenesis or promotion of angiogenesis. One of the proteasomal

From: *Methods in Molecular Biology, vol. 301, Ubiquitin–Proteasome Protocols*
Edited by: C. Patterson and D. M. Cyr © Humana Press Inc., Totowa, NJ

inhibitors, a peptide boronic derivative, is already approved as a drug against multiple myeloma. Dozens of other small-molecule compounds, usually peptides or peptide derivatives with an active group pharmacophore, are used to study the proteasome and proteasome-mediated processes, both in vitro and in vivo. Thousands of other compounds are tested for their inhibitory effects on the proteasome, with drug development in mind. In this chapter we describe (1) commonly available proteasome inhibitors and how to use them to the best of their potential and (2) how to characterize biochemically a new proteasome inhibitor in vitro.

Before we describe specific experimental steps, let us briefly present the subject of inhibition, the proteasome. The 700-kDa catalytic core (20S) of this enzymatic assembly is tube-shaped and built from 28 subunits arranged in four stacked, seven-subunit rings *(4)*. The two external α-rings provide a gating mechanism for the proteasomal degradative chamber inside the tube and serve as attachment sites for additional regulatory complexes. Each of the two internal β-rings contains three active centers of distinct specificities, with N-terminal threonines acting as nucleophiles. The three proteasomal peptidase activities can be conveniently probed with model peptide substrates with hydrophobic, basic, or acidic amino acid residues on the carboxyl side of the scissile bond. The peptidases are consequently labeled as chymotrypsin-like (ChT-L), trypsin-like (T-L), and postacidic (caspase-like or peptidylglutamyl peptide hydrolyzing [PGPH)]). Yeast contains one set of three catalytic subunits and the peptidase activities are univocally assigned to them *(5)*. In human and other mammalian 20S proteasomes there are two exchangeable sets of active β-subunits, a "housekeeping" set and an immune response-related set, with subtle differences in specificities within the common three-peptidase frame *(4)*. The ChT-L peptidase activity provided by the β5- (yeast) or β5/β5i (X/LMP7; human)-subunit is considered the most important, and many inhibitors target this active center exclusively or preferentially. Subunits β1/β1i (Y/LMP2) and β2/β2i (Z/MECL1) are responsible for the postacidic and T-L activities, respectively *(5–7)*. One or more of the activities can be blocked permanently or reversibly by competitive inhibitors.

One of the unique features of the proteasome is its modular structure. The 20S core proteasome can act alone or in complex with additional regulatory assemblies. The 20S with 19S caps attached on both sides forms the 26S proteasome, responsible for recognition and degradation of ubiquitinated substrates *(2)*. The activator complex PA28 ("proteasome activator with 28-kDa subunit"; REG, 11S) can be attached to one or both sides of the 20S core to facilitate substrate uptake and to modulate the size of products *(8)*. In general, competitive inhibitors work in the same way on the 20S, 26S, or activated proteasomes. Actions of a noncompetitive inhibitor may differentiate between distinct forms of the giant protease, and the potential presence of a mixed population of proteasomes in preparations should be taken into account.

## 2. Materials

1. Proteasome inhibitors: Calbiochem (San Diego, CA) and Affiniti (Exeter, UK) carry the most extensive collections of the compounds, including all inhibitors (with the exception of PS-341 and ritonavir) mentioned in **Subheading 3**.

2. Fluorogenic peptide substrates (Bachem, Sigma, Calbiochem, Affiniti): succinyl-LeuLeuValTyr-7-amido-4-methylcoumarin (SucLLVY-MCA), butoxycarbonyl-LeuArgArg-MCA (BocLRR-MCA), carbobenzoxy-LeuLeuGlu-MCA (Z-LLE-MCA).
3. Standard of fluorescence: 7-amino-4-methylcoumarin (AMC), 1 m*M* stock solution in dimethyl sulfoxide (DMSO)) (*see* **Note 1**).
4. Pure 20S proteasome preparation from human cultured cells or from yeast (*Saccharomyces cerevisiae*).
5. Reaction buffer: 50 m*M* Tris-HCl, pH 8.0. Store at 4°C.
6. Storage buffer: 50 m*M* Tris-HCl, 20% glycerol, pH 7.0. Store at 4°C.
7. Membrane concentrators: Vivaspin 4-mL concentrators with mol wt cutoff of 100,000 kDa (Vivascience-Sartorius) or other similar concentrators can be used.
8. Centrifuge with 5000*g* capability at 4°C.
9. Fluorometer or a plate reader with capability to measure fluorescence.
10. 96-Well, black, flat-bottom plates (available from several suppliers). The whole plate must be flat to ensure correct measurements.
11. 37°C Incubator.

## 3. Methods

In the first part we provide an overview of commonly available proteasome inhibitors with characteristics of their actions and tips on their usage. The parameters cited refer to human 20S proteasomes or human cell cultures. However, if the appropriate data are available, differences in the responses of human and wild-type yeast proteasomes to various compounds are indicated. In the second part we describe the basic biochemical procedures used to characterize a proteasome inhibitor.

### 3.1. An Overview of Distinct Classes of Small-Molecule Proteasome Inhibitors

Proteasome inhibitors are used to study the ubiquitin–proteasome pathway in vivo and to dissect the proteasomal actions in vitro *(9)*. Obviously, the most desirable for in vivo studies are compounds specifically targeting the proteasome without affecting activities of other serine or cysteine proteases. Fortunately, during the last few years such inhibitors became available and, as a result, dozens of proteins were identified as substrates of the ubiquitin–proteasome pathway. Even more, there are compounds that specifically target only one or two of the three proteasomal active centers. Such inhibitors are excellent tools with which to obtain a precise insight into the mechanism of degradation. The most common uses of proteasome inhibitors include:

1. Interfering with the ubiquitin–proteasome pathway in cultured cells to stop degradation of a particular protein or group of proteins.
2. Inducing apoptosis in cultured cells (*see* **Note 2**).
3. Differentiating between the activity of the proteasome and other proteases in crude cell extracts and specifically abolishing the proteasomal degradation.
4. Studying in vitro the mechanism of proteasomal degradation, including the role of single active centers.

#### 3.1.1. Preparation of Stock Solutions of Proteasome Inhibitors

The peptide derivative proteasome inhibitors described in the following subheadings are soluble in organic solvents such as DMSO (*see* **Note 3**). Stock solutions of millimolar concentrations (up to 100 m*M*) are usually convenient to make and are stable when stored at –20°C or –80°C (*see* **Notes 4** and **5**). The final concentration of DMSO up to 3% (including the substrates dissolved in DMSO; *see* **Subheading 3.2.1.** and **Note 3**) does not affect the performance of the enzyme in the in vitro activity assays. Working dilutions of inhibitors are usually in the range of 50-to 200-fold for in vitro studies and 100- to 1000-fold when used in cell cultures ***(10,11)***.

### 3.1.2. Nonspecific Inhibitors of the Proteasome

The general mechanism of catalytic act of a threonine protease is very similar to that of serine or cysteine proteases. Therefore, it comes to no surprise that many common competitive inhibitors of serine and cysteine proteases, such as chymostatin, leupeptin, calpain inhibitors I and II, antipain, pepstatin, hemin, 3,4-dichloroisocoumarin, *N*-ethylmaleimide, aprotinin, or phenylmethylsulfonyl fluoride affect at least one of the peptidase activities of 20S proteasome ***(12,13)***. Many of these compounds are used in popular inhibitor cocktails added to cell lysates to stop proteolysis during preparation. Of course, cell extracts prepared in the presence of such cocktails are useless for studies on the activity of the proteasome.

The compounds that affect activity of the proteasome in a less obvious manner include ethylenediaminetetraacetate (EDTA) and other chelators of divalent cations such as magnesium. Magnesium ions stabilize the structure of 20S proteasome and, together with ATP, stabilize and maintain the function of 26S proteasome. Glycerol, usually 10% or 20%, is often added to buffers used for preparation and storage of the proteasome as a stabilizing agent ***(14)***. However, the presence of more than 10% of glycerol during testing peptidase activities significantly lowers the velocity of degradation. The mechanism of this inhibition includes, among other possible effects, replacement of water necessary for the catalytic activity of the hydrolase with glycerol molecules.

### 3.1.3. Peptide Aldehydes

Aldehyde derivatives of peptides are commonly used as competitive, reversible inhibitors of serine and cysteine proteases (*see* **Note 6**). Tripeptides with an N-terminus blocked by a bulky group such as carboxybenzyl and derivatized by aldehyde on the C-terminus are straightforward to synthesize and they enter the cell easily. The downside is their less-than-perfect specificity, because they target calpain and cathepsins besides the proteasome. Several inhibitors from this group are commercially available. Two most commonly used are described in **Subheadings 3.1.3.1.** and **3.1.3.2.** and information about others could be found in **ref. *9***.

#### 3.1.3.1. MG132 (Z-LLL-CHO, *N*-Carboxybenzyl-LeuLeuLeu-Aldehyde)

The best known inhibitor from the group is the tripeptide aldehyde MG132. The compound at concentrations of 50–100 μ*M* inhibits all three proteasomal peptidases. Concentrations that are two orders of magnitude lower are sufficient to affect specifically the ChT-L activity in vitro and in cell culture without inhibiting other proteases ***(10,15)***. MG132 is often used in cell cultures when an inexpensive and reversible inhibitor is needed. The ChT-L activity of purified yeast 20S proteasome is completely abolished by 1 μ*M* MG132, similarly to human proteasomes. However, even 100 μ*M* of the compound is not sufficient to inhibit the yeast T-L peptidase and only partially affects the postacidic activity.

#### 3.1.3.2. Proteasome Inhibitor I (*N*-Carboxybenzyl-IleGlu[OtBu]AlaLeu-CHO)

Proteasome inhibitor (PSI) is sometimes used to study proteasome inhibitor–induced apoptosis in cultured cells ***(16,17)***. It targets mostly the ChT-L active center with $IC_{50}$ = 250 n*M* ($IC_{50}$ = inhibitor concentration resulting in 50% inhibition), and activation of the postacidic peptidase by the aldehyde is possible ***(18)***. The $IC_{50}$ for calpain is only 10-fold higher (2.5 μ*M*; [***18***]), which limits the use of PSI.

### *3.1.4. Peptide Vinyl Sulfones*

The vinyl sulfone is a pharmacophore long known for binding the active site thiol of cysteine proteases. When attached to the appropriate peptide core, the vinyl sulfone can specifically and irreversibly react with the active site threonine of the proteasome in vitro and in vivo (*see* **Note 7**) ***(19)***.

#### 3.1.4.1. Z-LLL-VS (*N*-Carboxybenzyl-LeuLeuLeu-Vinyl Sulfone) and NIP-LLL-VS (4-Hydroxy-3-Nitrophenylacetyl-Leu-Leu-Leu-Vinyl Sulfone)

The two inhibitors are derivatives of the blocked trileucine peptide. Z-LLL-VS in micromolar concentrations binds to all active centers of the human proteasome, and 10 μ*M*—20 μ*M* of the inhibitor is sufficient to abolish most of the activities of all three peptidases ***(10)***. The appropriate association constants for the three kinds of active centers differ by less than an order of magnitude, which distinguishes this compound from other proteasome inhibitors ***(10)***. However, the T-L activity of yeast 20S proteasomes is not inhibited by Z-LLL-VS even at concentrations as high as 100 μ*M*.

The advantage of the nitrophenol derivative of the tripeptide vinyl sulfone over the carboxybenzyl derivative is twofold: the former can be radioactively labeled easily by $^{125}I$ and serve to track the proteasome in vivo, and the association constants of NIP-LLL-VS with the active centers are generally one to two orders of magnitude higher than the constants for Z-LLL-VS ***(10,20)***. NIP-LLL-VS is a relatively poor inhibitor of the postacidic peptidase ***(20)***.

Both vinyl sulfones are successfully used in cell culture studies in micromolar concentrations ***(11,21)***. They are not perfectly proteasome specific, because they react with cathepsins B and S. Still, Z-LLL-VS surpasses MG132 by nearly two orders of magnitude when it comes to concentrations significantly affecting nonproteasomal activities in vivo ***(10)***.

#### 3.1.4.2. Aminohexanoic Acid Derivatives of the Trileucine Vinyl Sulfone

Extension of the LLL-VS sequence on the N-terminus by multiple residues of aminohexanoic acid (Ahx) and blocking it with acetate, carboxybenzyl, or adamantanylacetate (Ada) results in a group of inhibitors with an affinity toward the ChT-L activity comparable with that of NIP-LLL-VS and variable affinities toward the other two activities *(22)*. Ada(Ahx)$_3$LLL-VS significantly affects all three proteasomal peptidases at concentration as low as 1 μ*M*. However, the postacidic activity of yeast 20S proteasome is refractory to micromolar concentrations of the inhibitor.

The other commercially available compounds from this group, biotin (Bio)-labeled AdaK(Bio)Ahx$_3$LLL-VS and AdaYAhx$_3$LLL-VS suitable for $^{125}$I labeling of tyrosine, can be used for detection of proteasomal subunits by Western blotting and by autoradiography, respectively *(22)*. Unfortunately, these two are not able to cross the cell membrane and therefore are not suitable for treatment of cultured cells *(22)*.

### *3.1.5. Peptide Boronic Acids*

Peptide boronic acids are slowly binding reversible inhibitors, with very slow dissociation rates (*see* **Note 8**). Unlike other peptide derivative inhibitors that require at least three amino acids with a blocking group, only two amino acid residues are sufficient for boronate derivatives, rendering the compounds more soluble and membrane permeable than larger molecules *(23,24)*.

#### 3.1.5.1. PS-341 (Bortezomib, Velcade, Pyrazylcarbonyl-PheLeu-Boronate)

The anticancer drug PS-341 (Millennium Pharmaceuticals), a dipeptide derivative, is the first proteasome inhibitor approved for treatment of human disease, notably multiple myeloma. PS-341 in nanomolar concentrations affects ChT-L activity in vitro and in vivo in an extremely specific manner, and the cancerous cells are much more susceptible to the cell cycle arrest and apoptosis triggered by PS-341 than noncancerous cells *(3)*. PS-341 is not commercially available, in contrast to the tripeptide boronic acid described in **Subheading 3.1.5.2.**

#### 3.1.5.2. MG262 (Proteasome Inhibitor III [PSIII]; Z-LLL-Boronate, *N*-Carboxybenzyl-LeuLeuLeu-Boronate)

MG262 is another, after aldehyde and vinyl sulfone, derivative of the Z-LLL core peptide. However, its affinity toward the ChT-L active site is at least two orders of magnitude higher than in the case of other trileucine derivatives. Nanomolar concentrations of MG262 are sufficient to affect the ChT-L activity of both human and yeast 20S proteasomes in vitro. The T-L and postacidic activities are not inhibited by the compound even at micromolar concentrations. MG262 at a concentration of less than 100 n*M* significantly inhibits the growth of most cultured cells *(25)*. Although formally a reversible inhibitor, Z-LLL-boronate binds to its target threonyl hydroxyl in a practically irreversible manner.

### *3.1.6. Peptide Epoxyketones*

Inhibitors from this group are covalent and irreversible, with an excellent specificity toward the proteasome (*see* **Note 9**) and a very good specificity toward single active centers. Tetrapeptide epoxyketones are generally more potent than tripeptide derivatives. They are very useful in both in vitro and in vivo studies. Three of them, epoxomicin, YU101, and YU102, are commercially available.

#### 3.1.6.1. Epoxomicin (*N*-Acylo-IleIleThrLeu-Epoxyketone) and YU101 (*N*-Acylo-hPheLeuPheLeu-Epoxyketone)

Both inhibitors target the ChT-L active center, and they work at nanomolar concentrations in vitro, both with human and yeast proteasomes. YU101 is several times more potent toward the ChT-L activity, and less potent toward other activities than epoxomicin. The difference of association constants of YU101 with ChT-L and the other active sites is in the range of five orders of magnitude, making the compound the best available selective inhibitor of the ChT-L activity ***(26)***. Epoxomicin in millimolar concentrations blocks all three active sites, as was proved by a crystal structure of yeast proteasome with the inhibitor ***(27)***. The concentrations used to treat cell cultures range from nanomolar to low micromolar in the case of epoxomicin, and to a fraction of micromolar in the case of YU101 ***(26,28)***.

#### 3.1.6.2. YU102 (*N*-Acylo-GlyProPheLeu-Epoxyketone)

YU102 is the only selective and proteasome-specific inhibitor of the postacidic peptidase available to date. Its affinity toward the T-L active site is negligible, and the association constant for the ChT-L site is 50-fold lower than the constant for the postacidic site ***(29)***. It works well in a low micromolar range of concentrations.

### *3.1.7. Lactone Derivatives*

The compounds target mostly the ChT-L activity by irreversible acylation of the threonyl hydroxyl with the lactone ring as an active pharmacophore (*see* **Note 10**). Some of the compounds require "activation" by rearrangement of their carbon scaffold into the active β-lactone. The lactone ring is a relatively common feature in natural products and many such compounds may have unsuspected antiproteasomal properties (*see* **Notes 11** and **12**).

A lactone derivative, MLN519 (PS-519, LDP-519; Millenium Pharmaceuticals), is a potent antiinflammatory agent. It targets ChT-L activity and is undergoing clinical trials with stroke patients ***(30)***.

#### 3.1.7.1. Lactacystin and Clasto-Lactacystin β-Lactone

Clasto-lactacystin β-lactone, a rearrangement product of lactacystin, is the actual active compound forming an ester adduct with the hydroxyl group of N-terminal threonine ***(31)***. In practice, the rearrangement is achieved by preincubation of a working concentration of lactacystin in aqueous solution, for example for 1 h in the reaction buffer, before monitoring the inhibition of the proteasome. Only clasto-lactacystin β-lactone, and not its precursor lactacystin, is cell permeable ***(32)***. Reducing agents or acidic pH in reaction buffers or in cell culture media will interfere with the efficient rearrangement ***(31)***. Clasto-lactacystin β-lactone, the ready-to-use active compound, should be used whenever administration of a precise concentration of the inhibitor is desired, timing of the inhibition reaction is important, or the "activation" step is impractical to execute (*see* **Note 13**).

β-Lactone preferentially binds to the ChT-L active center in vitro and in vivo and affects proteasome-mediated degradation in cultured cells at micromolar concentrations *(33,34)*. Up to 100-fold higher concentrations of the inhibitor are required to affect all three proteasomal peptidases, with the postacidic active center most resistant to the modification by β-lactone *(10,33)*. Most studies in cultured cells show lactacystin/β-lactone to be specific toward the proteasome *(33,34)* (*see* **Note 14**). The compounds have been successfully used for studies on all aspects of proteasome biology *(35,36)*, including the landmark studies identifying the proteasome as an N-terminal threonine hydrolase *(34)*.

### 3.1.8. Mixed-Action Inhibitors of the Proteasome: Ritonavir

Before turning to true noncompetitive inhibitors, it is worthwhile to mention ritonavir. This peptide analog is used in anti-acquired immune deficiency (AIDS) therapy as an inhibitor of the essential aspartyl protease of HIV virus *(37)*. Nevertheless, proteasome is the target of the drug in therapeutically relevant micromolar concentrations *(38,39)*, which may provide an explanation for some of the side effects *(40)*. The future will show if the bonus activity of ritonavir is useful for treatment of autoimmune diseases *(41,42)*. The "two-site modifier" model of the interactions of ritonavir with the proteasome calls for a mixed-type inhibition, with a competitive blocking of the ChT-L active center and with additional binding to a noncatalytic modifier site *(39)*.

### 3.1.9. Noncompetitive Inhibitors

This group of inhibitors does not target active centers. Instead, they interact with noncatalytic parts of the enzyme to modulate the efficiency of the enzyme. The phenomenon of allostery, or changes in the activity resulting from binding of a compound far from an active center, creates nearly unlimited possibilities of intervention into the proteasomal actions with small-molecule allosteric modulators. The possibilities are poorly explored so far, mostly because of a high level of complication of the proteasome machinery and limited knowledge about the proteasomal allosteric mechanisms *(29,39,43,44)*.

#### 3.1.9.1. PR-39 and PR-11 (PR Peptides)

PR-, a proline- and arginine-rich, 39-residue-long peptide and its 11-residue-long N-terminal fragment apparently interact with α-ring of the 20S proteasome, interfering with efficient movements of the gating mechanism (*see* **Note 15**) *(45)*. The proteasome-related actions of PR peptides are surprisingly substrate specific in vivo. The block of degradation of hypoxia-inducible factor-1α (HIF-1α) by the proteasome is responsible for the PR-39-induced angiogenesis in a mouse model, whereas the block of degradation of IκBα accounts for an important part of the antiinflammatory activity of the compound (*see* **Note 16**) *(46,47)*. There is no accumulation of ubiquitinated proteins and no detectable apoptosis in cells treated with 10 μ*M* of the peptide, whereas the inhibition of IκBα degradation is apparent even with 100 n*M* of PR-39 *(47)*. Less than 100 n*M* of PR-39 are sufficient to inhibit at least 50% of the ChT-L and post-acidic activities in purified human 20S proteasomes (*see* **Note 17**). An order of magnitude higher concentrations of PR-39 inhibit ChT-L activity in yeast

20S proteasome and human 26S complex. Generally, treatment with PR-11 requires about 10-fold higher doses of the peptide than in the case of PR-39 ***(45)***.

#### 3.1.9.2. Gliotoxin

Gliotoxin is produced by several pathogenic fungi and displays strong immunosuppressive and pro-apoptotic activities ***(48,49)***. Some of these actions can be attributed to inhibition of the proteasome. However, relatively high concentrations of gliotoxin, in the range of 40–100 μ*M*, are required for detectable effects in vitro. All three proteasomal peptidases are targeted, with a preference for ChT-L and T-L activities ***(50)***. Reducing agents abrogates the inhibitory effect, apparently by breaking the disulfide bridge in the heterobicyclic core of the toxin ***(50)***.

#### 3.1.9.3. Aclacinomycin A (Aclarubicin)

The anthracycline antibiotic aclacinomycin A (ACM) is used as anticancer drug because of its DNA-intercalative and topoisomerase-inhibiting properties ***(51)***. The drug at high concentrations, at least 100 μ*M*, causes a weak inhibition of the ChT-L activity of 20S proteasome in vitro (*see* **Note 18**) ***(52)***. Interestingly, the sugar moiety of the antibiotic is necessary for the inhibitory effects ***(52)***.

### 3.2. Biochemical Characterization of a Proteasome Inhibitor

In vitro characterization of actions of a potential proteasome inhibitor with purified 20S proteasomes is often the first step in learning about the usefulness of a new compound. Sometimes, this step comes after treatment of cultured cells with a new compound, for example, when general accumulation of ubiquitinated proteins or accumulation of a specific proteasome substrate is observed. Either way, the characterization is necessary to learn about the basic mechanism of action of a compound and to set parameters for the subsequent work. The procedures described in the following subheadings are performed with a pure preparation of human 20S proteasome (*see* **Notes 19–21**). The activity of the proteasome is tested with three fluorogenic peptide substrates. The working concentration of the proteasome preparation should be high enough to produce a signal of about 50% of the linear range of a particular fluorometer. For example, if a fluorometer provides linear readings of up to 1000 arbitrary units (AU), the perfect signal for the control (no inhibitor) proteasome sample incubated for 1 h with a substrate would be about 500 AU (*see* **Subheading 3.2.1., item 1**). Therefore, if a blank signal is about 10 AU then a reasonable signal range from the sample falls between 50 and 500 AU. If electronics and a lamp of a fluorometer are very stable, then the lower limit can be extended to 20 AU.

#### 3.2.1. Which Active Centers of the Proteasome Are Targeted by the Inhibitor?

The range of working concentrations of the compound are established in this step and $IC_{50}$ values (inhibitor concentration resulting in 50% inhibition) for the affected peptidases are calculated. If the compound of interest was already used in cell culture, the concentrations effective in vivo constitute a good starting point for in vitro tests.

1. If the linear range of readings of a fluorometer is not known, this is a good time to gain the knowledge. The range can be easily determined with the AMC standard.
   a. Prepare working dilutions of the AMC in the reaction buffer. A range of 10 concentrations between 10 n*M* and 100 μ*M* should be sufficient to provide readings both within and outside the linear range of the fluorometer. Set the correct values of excitation and emission wavelengths (380 nm and about 460 nm, respectively, for AMC) and read the results.
   b. Plot the readings in arbitrary units as a function of concentration of the fluorescent standard. Assess the range in which the readings are proportional to the concentration of AMC. An identical procedure can be used with other fluorescence standards, for example, β-NA (β-naphthylamide).
2. Prepare stock solutions of the inhibitor in DMSO, for example 1 μ*M*, 10 μ*M*, 100 μ*M*, 1 m*M*, and 10 m*M*.
3. Prepare a working preparation of 20S proteasome in reaction buffer, enough for the control (DMSO added instead of the inhibitor) and four samples with different concentrations of the inhibitor, all in duplicates, and enough for testing of three substrates. Convenient total volume of a single sample is 100 μL, and this will include 98 μL of the diluted proteasome, 1 μL of the inhibitor, and 1 μL of the substrate. Prepare always at least 50 μL more of the preparation than necessary to allow for comfortable pipetting. Pipet the sample into wells of 96-well black plate (*see* **Note 22**). The working concentration of the proteasome is usually in a nanomolar (1–5 n*M*) range.
4. Add 1 μL of DMSO (control) or 1 μL of the stock solutions of the inhibitor to the wells. The final concentrations of the inhibitor will be: 10 n*M*, 100 n*M*, 1 μ*M*, 10 μ*M*, and 100 μ*M*. There should be 6 wells with controls and 30 wells with inhibitor-treated samples, 6 wells with every concentration of the inhibitor (6 concentrations of inhibitor × 3 substrates × duplicates). Mix the samples well by gently pipetting up and down several times. Add one well with a standard of fluorescence: 100 μL of the reaction buffer containing a known amount of AMC (*see* **Note 23**). Cover the plate with a plastic wrap and allow it to preincubate at room temperature for 10 min.
5. Add 1 μL of 10 m*M* stock solutions of the three substrates: SucLLVY-MCA (ChT-L activity), BocLRR-MCA (T-L activity), and Z-LLE-MCA (postacidic activities; *see* **Note 24**). There should be 12 samples for every substrate. In addition, prepare three "blank" samples containing 98 μL of the reaction buffer, 1 μL of DMSO, and 1 μL of a substrate. Mix well. The final concentration of the substrates should be 100 μ*M*.
6. Cover the plate with plastic wrap and incubate for 1 h at 37°C (*see* **Note 25**).
7. Read the results. Calculate the amount of a product released per milligram of the proteasome per hour. Calculate the percentage of proteolytic activity for the samples treated with inhibitor, taking activity in control samples as 100%. Plot the percentage activity as a function of the concentration of the inhibitor ***(10,45)***.
8. Repeat **steps 3–7** with a refined set of concentrations of the inhibitor to determine accurately $IC_{50}$ for the affected peptidase activities ***(45)***.

### 3.2.2. Determination of Inhibitor Reversibility

A similar procedure is described in **ref. *45***. It is usually enough to test reversibility with one of the peptidase activities, for example, with ChT-L (*see* **Note 26**).

1. Prepare a stock solution of the inhibitor in DMSO. The final concentration of the compound should optimally cause between 50% and 90% of inhibition of a chosen peptidase activity.

2. Prepare 450 μL of a working preparation of the proteasome. Divide the preparation into three Eppendorf tubes, 150 μL per tube (*see* **Note 27**).
3. Add to one sample 1.5 μL of DMSO (sample A, control) and to remaining two samples (B and C) 1.5 μL of the inhibitor. Mix well and incubate for 10 min at room temperature.
4. Transfer samples to membrane concentrators and wash with a total of 100 volumes of the reaction buffer.
5. After the last wash concentrate the samples to about 50 μL. Carefully transfer the samples to wells of a 96-well plate. Add small volumes of the reaction buffer to the concentrators and then transfer these "washes" to appropriate wells. Measure the exact volumes of samples with an automatic pipet and adjust to 98 μL with the reaction buffer.
6. Add 1 μL of DMSO to samples A and B and 1 μL of the stock solution of the inhibitor to sample C. Mix, incubate for 10 min and then add 1 μL of the substrate, as earlier. Mix the samples and incubate the plate for 10 min at room temperature..
7. Read the results. If the inhibitor is reversible, samples A and B should display similar activities. If the inhibitor is nonreversible, samples B and C should be very similar, and give much lower readings than sample A.
8. If the inhibitor is nonreversible, the association constant ($k_{assoc}$) can be determined as described in ***(10,34)***. In brief, a set of time-lapse measurements of the activity of 20S proteasome treated with the inhibitor at a concentration close to $IC_{50}$ should be carried out. For example, the time-points can include measurements at 0, 10, 20 min and so on up to 90 min. The association constant $k_{assoc} = \ln(v/v_o)/I$, where $v_o$ is the velocity at time 0, $v$ is the velocity at time *(t)*, and $I$ is the concentration of the inhibitor. The units of $k_{assoc}$ are $M^{-1}$ $s^{-1}$. The higher association constant indicates the higher affinity of an inhibitor toward a target enzyme.
9. If the inhibitor is reversible, the inhibition constant ($K_i$) can be determined (*see* **Subheading 3.2.8., step 8**).

### 3.2.3. Determination of the Type and Kinetic Parameters of Inhibition

The inhibitor may act in a competitive, noncompetitive (mixed), or uncompetitive manner. Inhibitors of the first type bind to active center(s) and are the most common. Noncompetitive inhibitors bind to noncatalytic sites independently of occupation of active center by a substrate. Uncompetitive inhibitors, very rare and sometimes included into the noncompetitive group, bind only to noncatalytic sites of an enzyme–substrate complex. There are no known uncompetitive inhibitors of the proteasome. To learn about the type of inhibition, the Michaelis–Menten kinetic parameters of the control (not inhibited) and inhibited proteasome should be determined, as described in **ref. *45***.

1. Prepare samples of the control proteasome (with DMSO instead of the inhibitor) and of samples treated with the inhibitor. The concentration of the inhibitor should be close to the $IC_{50}$.
2. Prepare stock solutions of a substrate of choice to obtain 10 different final concentrations of the substrate in a range from 5 to 500 μ*M*. Ten concentrations is a reasonable number of points for an accurate kinetic plot. Fewer than six points should never be used.
3. Determine the activities of control and inhibitor-treated enzyme after 1 h of incubation at 37°C. A shorter incubation time can be used, if convenient.
4. The following is a glossary of useful terms used in the discussion that follows:

   $v$ = velocity of the reaction, for example in millimoles of a product released by 1 mg of an enzyme during 1 h;

$S$ = concentration of a substrate;
$I$ = concentration of an inhibitor;
$V_{max}$ = maximal velocity of the reaction;
$V_{max(i)}$ = $V_{max}$ in the presence of an inhibitor;
$K_M$ = Michaelis constant: the concentration of a substrate at which the enzyme reaches half of its maximal velocity;
$K_{M(i)}$ = Michaelis constant in the presence of an inhibitor;
$K_i$ = inhibition constant.

5. Prepare the Lineweaver–Burk plots (L–B plots) of the control and inhibitor-treated enzyme. The $1/v$ should be plotted as a function of $1/S$. The least-squares method should be used to obtain equations for the L–B plots.
6. Determine the $V_{max}$ and $K_M$ parameters from the *L–B* plots. The equation for an L–B plot is: $1/v = a(1/S) + b$, and then $V_{max} = 1/b$, and $K_M = -a/b$.
7. For a competitive inhibitor, $V_{max}$ values of control and inhibitor-treated enzyme are equal, and $K_M$ values are different, with the $K_M$ of the inhibited sample larger than the $K_M$ of the control. For a pure noncompetitive inhibitor, $V_{max}$ values are different whereas $K_M$ values are equal. The uncompetitive inhibition will call for both $V_{max}$ and $K_M$ values significantly different between the control and inhibited samples, with $K_M$ of the inhibited sample smaller than $K_M$ of the control. The competitive/noncompetitive inhibition is characterized by different $V_{max}$ values and different $K_M$ values, with $K_M$ of the inhibited sample larger than the $K_M$ of the control. The type of inhibition can be immediately recognized from the shape of L–B plots ***(45)***.
8. If the inhibitor is reversible, the inhibition constant ($K_i$) can be determined from kinetic plots. In general, $K_i = k_{-1}/k_{+1}$, where $k_{-1}$ is a constant of a dissociation reaction of an enzyme-inhibitor complex, and $k_{+1}$ is a constant of a binding reaction of enzyme and inhibitor. Therefore, the smaller $K_i$ indicates the stronger inhibitor. $K_i$ is given in molar concentration units. For a pure noncompetitive inhibitor $K_i = IC_{50}$, or the constant can be calculated as $K_i = V_{max(i)}I/(V_{max} - V_{max(i)})$ ***(45)***. For a competitive inhibitor $K_i = K_M I/(K_{M(i)} - K_M)$. If the inhibition is of a competitive/noncompetitive type, there are obviously two inhibition constants, $K_i$ and $K_i'$, and they can be calculated as follows:

$$K_i = I/[(K_{M(i)}V_{max}/K_M V_{max(i)}) - 1]$$

$$K_i' = V_{max(i)}I/(V_{max} - V_{max(i)}).$$

## 4. Notes

1. The abbreviation AMC is often used to indicate the free fluorescent tag with a high quantum yield of fluorescence, whereas the abbreviation MCA indicates the peptide-bound tag in its amide form, with the fluorescence effectively quenched. Free AMC is released from substrates as a product of the proteolytic action.
2. Different cell lines have different susceptibilities to the apoptosis induced by the proteasome inhibitors. Therefore, it is necessary to establish on an individual basis the optimal concentration and exposure time of a particular inhibitor to trigger the desired response in particular cells.
3. Use only high-purity DMSO to dissolve inhibitors or substrates. Anhydrous 99.8% DMSO is a good choice. Avoid long-stored open-bottle solvents, because DMSO is hygroscopic and will accumulate water with time. A high water content impairs the capability of DMSO to dissolve hydrophobic compounds and affects the performance of inhibitors and substrates.

4. Inhibitors and fluorogenic peptide substrates are usually sold as lyophilized powders, in amounts of 1 to 5 mg per vial. To make a stock solution, dissolve the whole amount of a compound by adding the solvent directly to the original vial. If the vial was stored in a freezer, bring it to room temperature first. Mix well the contents of the vial. If desired, divide the stock solution into convenient-size, Eppendorf-tube aliquots for storage.
5. If photosensitivity is indicated in the description of a particular compound (all fluorogenic substrates are light sensitive), amber Eppendorf tubes or amber glass vials should be used for stock solutions. Alternatively, the tubes can be tightly wrapped in aluminum foil.
6. The aldehyde group forms a covalent hemiacetal with the active threonine hydroxyl and may form a stable oxazolidine ring by reaction with the N-terminal amine and an accessible side chain hydroxyl ***(24)***. Peptide aldehydes are often listed along with peptide boronic acids as tetrahedral intermediate analogues. However, the tetrahedral complex formed by an aldehyde is too unstable, and the reactivity of the aldehyde group with groups other than N-terminal threonine is too robust to consider peptide aldehydes good models of a tetrahedral intermediate ***(23)***.
7. The vinyl sulfone group is much more inert in the absence of a specific target than the aldehyde group. Peptide vinyl sulfones form ether adducts with the N-terminal threonine of the proteasome ***(10,24)***.
8. Peptide boronic acids form a pseudocovalent adduct with active site threonine. The adduct is stable and mimics well the tetrahedral intermediate complex ***(23,24)***.
9. Peptide epoxyketone inhibitors are now chemically synthesized ***(26,53)***, but they evolved from two fungal antibiotics, epoxomicin and eponemycin ***(54,55)***. They engage both the threonyl hydroxyl and the N-terminal amine in the active center, forming a morpholino adduct. Because the proteasome and its bacterial homolog HSlV are the only known proteases operating via the N-terminal hydroxyl mechanism, the epoxyketones are perfectly proteasome specific. The only other potential candidates for their binding are other N-terminal hydrolases (tnt), which are not a very widespread group of enzymes ***(27)***.
10. The lactone compounds possess peptide-derived backbones, but instead of a linear structure of the former groups of small-molecule inhibitors, their carbon scaffold is closed into a lactone ring. The acylation of active site hydroxyl is formally irreversible; however, the adduct of β-lactone with the proteasome slowly hydrolyzes in water with a $t_{1/2}$ of about 20 h ***(56)***.
11. The most notable examples of unsuspected lactone derivative proteasome inhibitors are statins. These fungal metabolites are commonly used for the treatment of hypercholesterolemia ***(57)***. Three statins—lovastatin (mevinolin), mevastatin, and simvastatin—are administered as prodrugs, converted to active drugs in patient's liver. However, the prodrugs exhibit weak antiproteasomal activity ***(58–60)***. Relatively high concentrations of the three statins, in tens of micromoles, are necessary to cause a detectable inhibition of ChT-L activity in cell extracts ***(58,60)***. It remains to be established how much of the side effects of statins, including anticancer activities, can be attributed to the potential proteasome inhibition. Lovastatin is featured in the Calbiochem catalogue, but it is not available in the United States.
12. Epigallocatechin (EGC), epicatechin (EC), and their derivatives are polyphenols found in tea, and are especially abundant in green tea ***(9)***. Their ester bond carbon may acylate the hydroxyl of the active site threonine in a manner similar to β-lactone. It targets ChT-L and postacidic activities in vitro and in vivo at possibly physiologically relevant ***(61)*** lower than micromolar and micromolar concentrations, respectively ***(62,63)***.
13. Lactacystin and clasto-lactacystin β-lactone are now available as synthetic compounds. However, lactacystin (omuralide) is a natural fungal antibiotic ***(64,65)***. Synthetic β-lac-

tone is more expensive than the synthetic precursor; however, it does not require the time-consuming "activation" procedure with uncertain yield of the product *(33)*.

14. Radioactive lactacystin was used to label specifically the proteasome subunits in live cells or in crude cell extracts, without identifying other targets of the inhibitor *(10,33,34)*. The acylation of active site hydroxyls in serine proteases, if it occurs, probably undergoes rapid hydrolysis and is thus practically insignificant *(56)*. The question about specificity came up recently with the finding that lactacystin/β-lactone may affect the activity of the giant cytosolic serine protease, tripeptidyl peptidase II (TPPII) *(66,67)* and of a serine carboxypeptidase, cathepsin A *(68)*. Fortunately, cathepsin A is a target for β-lactone at much lower pH than that preferred by the proteasome and used for the proteasome-related studies *(68)*. The significance of the potential reaction with TPPII remains to be established.
15. PR-39 is a well known antimicrobial agent from the family of cathelicidins, which are components of innate immunity in mammals *(69)*. PR-39 kills bacteria by a non-pore-forming mechanism, crossing the cell membrane and blocking DNA and protein synthesis *(70)*. There are no known human homologs of PR-39, and the original peptide was first isolated from porcine intestine *(71)*. Synthetic PR-39 based on the porcine sequence and its shortest active fragment, PR-11, is commercially available. Homologous peptides of the size intermediate between 11 and 39 amino acids should be active as well, as long as the three essential N-terminal arginines are conserved *(45)*.
16. The proteasome is not the only intracellular target of PR-39. Other targets include PI3 kinase *(72)*, proteins with Src homology 3 domain *(73)*, and signaling adapter protein p130(Cas) *(74)*.
17. The presented examples of working concentrations of PR peptides in vitro are valid with 20S proteasomes activated with sodium dodecyl sulfate (SDS) *(45)*. One of the rationales of "activation" is to obtain a homogeneous population of the proteasomal particles, presumably with the gate in always open position (*see* **Note 19**). The use of latent 20S complexes may bring many surprises. PR peptides target the gating mechanism, and the gate in latent 20S particles can switch between closed and open conformations *(43–45,75,76)*.
18. Aclacinomycin A is not proteasome specific. High concentrations of the drug inhibit calpain and activate trypsin *(52)*.
19. 20S proteasome can be purified by many different methods, usually a set of chromatographic steps. The enzyme should be electrophoretically pure, without a contamination with 19S cap subunits. To ensure reproducibility of the results, it is wise to use proteasomes always purified by the same method and to avoid using "old" preparations, stored for a long time or frozen and thawed several times. With time such preparations may become spontaneously "activated" and may display properties similar to those of SDS-treated enzymes, however, without the possibility to control the degree of activation. A good rule is to use a preparation stored on ice within a week from purification. Alternatively, the purified enzyme in the storage buffer can be aliquoted and kept at –20°C.
20. The use of SDS-activated 20S proteasomes may be sometimes convenient, especially when a quick assessment of the ChT-L-activity is required. Usually, 0.01–0.03% of SDS is used to activate the proteasome, with the 0.01% concentration being the most reliable.
21. The use of purified yeast 20S proteasomes instead of, or in addition to, human enzyme, have several advantages: (a) preparation of the enzyme from yeast culture is less expensive than from human cultured cells; (b) yeast does not have alternative sets of catalytic subunits, so the yeast proteasome preparation is in that respect homogeneous and with clear-cut three- peptidase activities; (c) if the compound of interest affects in a similar

way both human and yeast enzymes, it will prove that it is a universal modulator of proteasomal activities.

22. Using a fluorometer with a plate reader is the most convenient and allows for automation of both pipetting and readout of the results. The drawback of plate readers is their lower sensitivity than traditional measurements in a quartz cuvette. However, in the case of proteolytic activity measurements, the fluorescence signal is high enough to be reliably measured in a plate. Alternatively, the samples can be prepared in Eppendorf tubes and transferred to a quartz cuvette or to disposable UV-transparent plastic cuvettes for measurements. In the described setting, the enzymatic reaction progresses indefinitely. Alternatively, the reaction can be stopped by addition of ethanol to 70% (v/v) final concentration or with 1% (w/v) SDS and then the fluorescence read. None of these reagents interferes with fluorescence of AMC or βNA. If cuvettes are used to measure fluorescence and kinetic curves are not desired, the reaction is stopped as just described. The released fluorogen is stable for several hours so the readouts can be delayed. The plates can be reused if, immediately after the measurements, they are soaked in a detergent of the Lift-Away (RPI Corp.) strength, and then rinsed several times with distilled water and dried. Otherwise, to avoid spurious results, it is advisable to treat the plates as disposable.
23. The fluorescence standard is necessary to calculate the actual amount of the released products of degradation. The readout of the standard should be within the linear range of a fluorometer. Every plate should include at least one well with the standard.

    Substrates other than the three most common can be used as well, for example, Z-GGL-MCA or SucAAF-MCA for the ChT-L activity, BzVGR-MCA for the T-L activity, and AcYVAD-MCA for the post-acidic activity. Other useful tagging groups include:

    a. 7-Amido-4-trifluoromethylcoumarin (AFC): excitation 400 nm, emission 505 nm, (comparable quantum yield to AMC, but better membrane permeability and retention in cells, good choice if in vivo and in vitro measurements have to be compared directly).
    b. 4-Methoxy-β-naphthylamide (4MβNA): excitation 340 nm, emission 425 nm (much lower quantum yield than AMC; liberated 4MβNA when further derivatized can chelate osmium to localize enzyme reaction in vivo for electron microscopy studies).
    c. β-Naphthylamide (βNA): excitation 326 nm, emission 415 nm (lower quantum yield than AMC; fluorescence not efficiently quenched by peptide, resulting in high background).
    d. *p*-Nitroanilide (pNA): chromogenic 405–410 nm (low sensitivity but one does not need a fluorometer).

24. Alternative to the model peptide substrates are usually fluorescently labeled proteins. The latter perform much better in simulating the reality of proteasome catalysis, but they are much more difficult to use because 20S proteasome is a poor proteinase toward natively folded proteins ***(77)***.
25. If the fluorometer/plate reader can automatically read the results every 5 min or better, it is very convenient to take advantage of this "enzyme kinetic" option. The velocity of the degradation reaction can be then assessed very accurately from the plots of fluorescence as a function of time ***(45)***.
26. Alternative methods of determination of a reversibility of an inhibitor include: (a) dilution of a sample pretreated with the inhibitor prior to incubation with a substrate; an at least 10-fold dilution should be used ***(10)*** and (b) separation of the unbound inhibitor with gel filtration of a sample pretreated with the inhibitor.

27. The addition of extra 50 µL to the standard 100 µL of sample volume is needed to accommodate for inevitable loss of a part of the sample during the washing process, mostly because protein binding to the concentrator's membrane.

## References


1. Zwickl, P., Seemuller, E., Kapelari, B., and Baumeister, W. (2001) The proteasome: a supramolecular assembly designed for controlled proteolysis. *Adv. Protein Chem.* **59,** 187–222.
2. Glickman, M. H. and Maytal, V. (2002) Regulating the 26S proteasome. *Curr. Top. Microbiol. Immunol.* **268,** 43–72.
3. Adams, J. (2002) Proteasome inhibition: a novel approach to cancer therapy. *Trends Mol. Med.* **8**(4 Suppl), S49–54.
4. Bochtler, M., Ditzel, L., Groll, M., Hartmann, C., and Huber, R. (1999) The proteasome. *Annu. Rev. Biophys. Biomol. Struct.* **28,** 295–317.
5. Arendt, C. S. and Hochstrasser, M. (1997) Identification of the yeast 20S proteasome catalytic centers and subunit interactions required for active-site formation. *Proc. Natl. Acad. Sci. USA* **94,** 7156–7161.
6. Gaczynska, M., Rock, K. L., Spies, T., and Goldberg, A. L. (1994) Peptidase activities of proteasomes are differentially regulated by the major histocompatibility complex-encoded genes for LMP2 and LMP7. *Proc. Natl. Acad. Sci. USA* **91,** 9213–9217.
7. Gaczynska, M., Goldberg, A. L., Tanaka, K., Hendil, K. B., and Rock, K. L. (1996) Proteasome subunits X and Y alter peptidase activities in opposite ways to the interferon-γ-induced subunits LMP2 and LMP7. *J. Biol. Chem.* **271,** 17275–17280.
8. Li, J. and Rechsteiner, M. (2001) Molecular dissection of the 11S REG (PA28) proteasome activators. *Biochimie* **83,** 373–383.
9. Gaczynska, M. and Osmulski, P. A. (2002) Inhibitor at the gates, inhibitor in the chamber: allosteric and competitive inhibitors of the proteasome as prospective drugs. *Curr. Med. Chem. Immun. Endocrinol. Metab. Agents* **2,** 279–301.
10. Bogyo, M., McMaster, J. S., Gaczynska, M., Tortorella, D., Goldberg, A. L., and Ploegh, H. (1997) Covalent modification of the active site threonine of proteasomal beta subunits and the *Escherichia coli* homolog HslV by a new class of inhibitors. *Proc. Natl. Acad. Sci. USA* **94,** 6629–6634.
11. Glas, R., Bogyo, M., McMaster, J. S., Gaczynska, M., and Ploegh, H. L. (1998) A proteolytic system that compensates for loss of proteasome function. *Nature* **392,** 618–622.
12. Cardozo, C., Vinitsky, A., Hidalgo, M. C., Michaud, C., and Orlowski, M. (1992) A 3,4-dichloroisocoumarin-resistant component of the multicatalytic proteinase complex. *Biochemistry* **31,** 7373–7380.
13. Matthews, W., Driscoll, J., Tanaka, K., Ichihara, A., and Goldberg, A. L. (1989) Involvement of the proteasome in various degradative processes in mammalian cells. *Proc. Natl. Acad. Sci. USA* **86,** 2597–2601.
14. Gaczynska, M., Rock, K. L., and Goldberg, A. L. (1993) γ-interferon and expression of MHC genes regulate peptide hydrolysis by proteasomes. *Nature* **365,** 264–267.
15. Kisselev, A. F. and Goldberg, A. L. (2001) Proteasome inhibitors: from research tools to drug candidates. *Chem. Biol.* **8,** 739–758.
16. Naito, Y., Handa, O., Takagi, T., et al. (2002) Ubiquitin-proteasome inhibitor enhances tumour necrosis factor-α-induced apoptosis in rat gastric epithelial cells. *Alimen. Pharmacol. Ther.* **16**(52), 59–66.

17. Rideout, H. J., Larsen, K. E., Sulzer, D., and Stefanis, L. (2001) Proteasomal inhibition leads to formation of ubiquitin/alpha-synuclein-immunoreactive inclusions in PC12 cells. *J. Neurochem.* **78,** 899–908.
18. Figueiredo-Pereira, M. E., Berg, K. A., and Wilk, S. (1994) A new inhibitor of the chymotrypsin-like activity of the multicatalytic proteinase complex (20S proteasome) induces accumulation of ubiquitin-protein conjugates in a neuronal cell. *J. Neurochem..* **63,** 1578–1581.
19. Bogyo, M., Gaczynska, M., and Ploegh, H. L. (1997) Proteasome inhibitors and antigen presentation. *Biopolymers* **43,** 269–280.
20. Bogyo, M., Shin, S., McMaster, J. S., and Ploegh, H. L. (1998) Substrate binding and sequence preference of the proteasome revealed by active-site-directed affinity probes. *Chem. Biol.* **5,** 307–320.
21. Wiertz, E. J., Tortorella, D., Bogyo, M., et al. (1996) Sec61-mediated transfer of a membrane protein from the endoplasmic reticulum to the proteasome for destruction. *Nature* **384,** 432–438.
22. Kessler, B. M., Tortorella, D., Altun, M., et al. (2001) Extended peptide-based inhibitors efficiently target the proteasome and reveal overlapping specificities of the catalytic β-subunits. *Chem. Biol.* **8,** 913–929.
23. Adams, J., Behnke, M., and Chen, S., et al. (1998) Potent and selective inhibitors of the proteasome: dipeptidyl boronic acids. *Bioorg. Med. Chem. Lett.* **8,** 333–338.
24. Myung, J., Kim, K. B. and Crews, C. M., et al. (2001) The ubiquitin-proteasome pathway and proteasome inhibitors. *Med. Res. Rev.* **21,** 245–273.
25. Gardner, R. C., Assinder, S. J., Christie, G., et al. (2000) Characterization of peptidyl boronic acid inhibitors of mammalian 20 S and 26 S proteasomes and their inhibition of proteasomes in cultured cells. *Biochem. J.* **346**(Pt 2), 447–454.
26. Elofsson, M., Splittgerber, U., Myung, J., Mohan, R., and Crews, C. M. (1999) Towards subunit-specific proteasome inhibitors: synthesis and evaluation of peptide α',β'-epoxyketones. *Chem. Biol.* **6**, 811–822.
27. Groll, M., Kim, K. B., Kairies, N., Huber, R., and Crews, C. M. (2000) Crystal structure of epoxomicin: 20S proteasome reveals a molecular basis for selectivity of α',β'-epoxyketone proteasome inhibitors. *J. Am. Chem. Soc.* **122,** 1237–1238.
28. Meng, L. H., Mohan, R., Kwok, B. H. B., Elofsson, M., Sin, N., and Crews, C. M. (1999) Epoxomicin, a potent and selective proteasome inhibitor, exhibits in vivo antiinflammatory activity. *Proc. Natl. Acad. Sci. USA* **96,** 10403–10408.
29. Myung, J., Kim, K. B., Lindsten, K., Dantuma, N. P., and Crews, C. M. (2001) Lack of proteasome active site allostery as revealed by subunit-specific inhibitors. *Mol. Cell* **7,** 411–420.
30. Elliott, P. J., Pien, C. S., McCormack, T. A., Chapman, I. D., and Adams, J. (1999) Proteasome inhibition: a novel mechanism to combat asthma. *J. Allergy Clin. Immunol.* **104**(2 Pt 1), 294–300.
31. Dick, L. R., Cruikshank, A. A., Grenier, L., Melandri, F. D., Nunes, S. L., and Stein, R. L. (1996) Mechanistic studies on the inactivation of the proteasome by lactacystin: a central role for clasto-lactacystin β-lactone. *J. Biol. Chem.* **271,** 7273–7276.
32. Dick, L. R., Cruikshank, A. A., Destree, A. T., et al. (1997) Mechanistic studies on the inactivation of the proteasome by lactacystin in cultured cells. *J. Biol. Chem.* **272,** 182–188.
33. Craiu, A., Gaczynska, M., Akopian, T., et al. (1997) Lactacystin and clasto-lactacystin β-lactone modify multiple proteasome beta-subunits and inhibit intracellular protein

degradation and major histocompatibility complex class I antigen presentation. *J. Biol. Chem.* **272,** 13437–13445.
34. Fenteany, G., Standaert, R. F., Lane, W. S., Choi, S., Corey, E. J., and Schreiber, S. L. (1995) Inhibition of proteasome activities and subunit-specific amino-terminal threonine modification by lactacystin. *Science* **268,** 726–731.
35. David, D. C., Layfield, R., Serpell, L., Narain, Y., Goedert, M., and Spillantini, M. G. (2002) Proteasomal degradation of tau protein. *J. Neurochem.* **83,** 176–185.
36. Mo, X. Y., Cascio, P., Lemerise, K., Goldberg, A. L., and Rock, K. (1999) Distinct proteolytic processes generate the C and N termini of MHC class I-binding peptides. *J. Immunol.* **163,** 5851–5859.
37. von der Helm, K. (1996) Retroviral proteases: structure, function and inhibition from a non-anticipated viral enzyme to the target of a most promising HIV therapy. *Biol. Chem.* **377,** 765–774.
38. Schmidtke, G., Holzhutter, H. G., Bogyo, M., et al. (1999) How an inhibitor of the HIV-I protease modulates proteasome activity. *J. Biol. Chem.* **274,** 35734–35740.
39. Schmidtke, G., Emch, S., Groettrup, M., and Holzhutter, H. G. (2000) Evidence for the existence of a non-catalytic modifier site of peptide hydrolysis by the 20 S proteasome. *J. Biol. Chem.* **275,** 22056–22063.
40. Liang, J. S., Distler, O., Cooper, D. A., et al. (2001) HIV protease inhibitors protect apolipoprotein B from degradation by the proteasome: a potential mechanism for protease inhibitor-induced hyperlipidemia. *Nat. Med.* **7,** 1327–1331.
41. Andre, P., Groettrup, M., Klenerman, P., et al. (1998) An inhibitor of HIV-1 protease modulates proteasome activity, antigen presentation, and T cell responses. *Proc. Natl. Acad. Sci. USA* **95,** 131201–3124.
42. Hosseini, H., Andre, P., Lefevre, N., et al. (2001) Protection against experimental autoimmune encephalomyelitis by a proteasome modulator. *J. Neuroimmunol.* **118,** 233–244.
43. Kisselev, A. F., Kaganovich, D., and Goldberg, A. L. (2002) Binding of hydrophobic peptides to several non-catalytic sites promotes peptide hydrolysis by all active sites of 20 S proteasomes—evidence for peptide-induced channel opening in the α-rings. *J. Biol. Chem.* **277,** 22260–22270.
44. Osmulski, P. A. and Gaczynska, M. (2002) Nanoenzymology of the 20S proteasome: proteasomal actions are controlled by the allosteric transition. *Biochemistry* **41,** 7047–7053.
45. Gaczynska, M., Osmulski, P. A., Gao, Y., Post, M. J., and Simons, M. (2003) Proline- and arginine-rich peptides constitute a novel class of allosteric inhibitors of proteasome activity. *Biochemistry* **42,** 8663–8670.
46. Bao, J., Sato, K., Li, M., et al. (2001) PR-39 and PR-11 peptides inhibit ischemia-reperfusion injury by blocking proteasome-mediated I κBα degradation. *Am. J. Physiol. Heart Circ. Physiol.* **281,** H2612–H2618.
47. Gao, Y. H., Lecker, S., Post, M. J., et al. (2000) Inhibition of ubiquitin-proteasome pathway-mediated I κBα degradation by a naturally occurring antibacterial peptide. *J. Clin. Invest.* **106,** 439–448.
48. Ward, C., Chilvers, E. R., Lawson, M. F., et al. (1999) NF-κB activation is a critical regulator of human granulocyte apoptosis in vitro. *J. Biol. Chem.* **274,** 4309–4318.
49. Yamada, A., Kataoka, T., and Nagai, K. (2000) The fungal metabolite gliotoxin: immunosuppressive activity on CTL-mediated cytotoxicity. *Immunol. Lett.* **71,** 27–32.
50. Kroll, M., Arenzana-Seisdedos, F., Bachelerie, F., Thomas, D., Friguet, B., and Conconi, M. (1999) The secondary fungal metabolite gliotoxin targets proteolytic activities of the proteasome. *Chem. Biol.* **6,** 689–698.

51. Wang, J., Maitani, Y., and Takayama, K. (2002) Antitumor effects and pharmacokinetics of aclacinomycin A carried by injectable emulsions composed of vitamin E, cholesterol, and PEG-lipid. *J. Pharmaceut. Sci.* **91,** 1128–1134.
52. Figueiredo-Pereira, M. E., Chen, W. E., Li, J., and Johdo, O. (1996) The antitumor drug aclacinomycin A, which inhibits the degradation of ubiquitinated proteins, shows selectivity for the chymotrypsin-like activity of the bovine pituitary 20 S proteasome. *J. Biol. Chem.* **271,** 16455–16459.
53. Sin, N., Kim, K. B., Elofsson, M., et al. (1999) Total synthesis of the potent proteasome inhibitor epoxomicin: a useful tool for understanding proteasome biology. *Bioorg. Med. Chem. Lett.* **9,** 2283–2288.
54. Hanada, M., Sugawara, K., Kaneta, K., et al. (1992) Epoxomicin, a new antitumor agent of microbial origin. *J. Antibiot.* **45,** 1746–1752.
55. Sugawara, K., Hatori, M., Nishiyama, Y., et al. (1990) Eponemycin, a new antibiotic active against B16 melanoma. I. Production, isolation, structure and biological activity. *J. Antibiot.* **43,** 8–18.
56. Kisselev, A. F., Songyang, Z., and Goldberg, A. L. (2000) Why does threonine, and not serine, function as the active site nucleophile in proteasomes? *J. Biol. Chem.* **275,** 14831–14837.
57. Retterstol, K., Stugaard, M., Gorbitz, C., and Ose, L. (1996) Results of intensive long-term treatment of familial hypercholesterolemia. *Am. J. Cardiol.* **78,** 1369–1374.
58. Kumar, B., Andreatta, C., Koustas, W. T., Cole, W. C., Edwards-Prasad, J., and Prasad, K. N. (2002) Mevastatin induces degeneration and decreases viability of cAMP-induced differentiated neuroblastoma cells in culture by inhibiting proteasome activity, and mevalonic acid lactone prevents these effects. *J. Neurosci. Res.* **68,** 627–635.
59. Murray, S. S., Tu, K. N., Young, K. L., and Murray, E. J. B. (2002) The effects of Lovastatin on proteasome activities in highly purified rabbit 20 S proteasome preparations and mouse MC3T3-E1 osteoblastic cells. *Metab. Clin. Exp.* **51,** 1153–1160.
60. Rao, S., Porter, D. C., Chen, X., Herliczek, T., Lowe, M., and Keyomarsi, K. (1999) Lovastatin-mediated G1 arrest is through inhibition of the proteasome, independent of hydroxymethyl glutaryl-CoA reductase. *Proc. Natl. Acad. Sci. USA* **96,** 7797–7802.
61. Yang, C. S., Lee, M. J., and Chen, L. (1999) Human salivary tea catechin levels and catechin esterase activities: implication in human cancer prevention studies. *Cancer Epidemiol. Biomark. Prev.* **8,** 83–89.
62. Nam, S., Smith, D. M., and Dou, Q. P. (2001) Ester bond-containing tea polyphenols potently inhibit proteasome activity in vitro and in vivo. *J. Biol. Chem.* **276,** 13322–13330.
63. Smith, D. M., Wang, Z. G., Kazi, A., Li, L. H., Chan, T. H., and Dou, Q. P. (2002) Synthetic analogs of green tea polyphenols as proteasome inhibitors. *Mol. Med.* **8,** 382–392.
64. Omura, S., Matsuzaki, K., Fujimoto, T., et al. (1991) Structure of lactacystin, a new microbial metabolite which induces differentiation of neuroblastoma cells. *J. Antibiot.* **44,** 117–118.
65. Omura, S., Fujimoto, T., Otoguro, K., et al. (1991) Lactacystin, a novel microbial metabolite, induces neuritogenesis of neuroblastoma cells. *J. Antibiot.* **44,** 113–116.
66. Geier, E., Pfeifer, G., Wilm, M., et al. (1999) A giant protease with potential to substitute for some functions of the proteasome. *Science* **283,** 978–981.
67. Hilbi, H., Puro, R. J., and Zychlinsky, A. (2000) Tripeptidyl peptidase II promotes maturation of caspase-1 in *Shigella flexneri*-induced macrophage apoptosis. *Infect. Immun.* **68,** 5502–5508.

68. Ostrowska, H., Wojcik, C., Wilk, S., et al. (2000) Separation of cathepsin A-like enzyme and the proteasome: evidence that lactacystin/β-lactone is not a specific inhibitor of the proteasome. *Int. J. Biochem. Cell Biol.* **32,** 747–757.
69. Lehrer, R. I. and Ganz, T. (2002) Cathelicidins: a family of endogenous antimicrobial peptides. *Curr. Opin. Hematol.* **9,** 18–22.
70. Linde, C. M., Hoffner, S. E., Refai, E., and Andersson, M. (2001) In vitro activity of PR-39, a proline-arginine-rich peptide, against susceptible and multi-drug-resistant *Mycobacterium tuberculosis. J. Antimicrob. Chemother.* **47,** 575–580.
71. Agerberth, B., Lee, J. Y., Bergman, T., et al. (1991) Amino acid sequence of PR-39. Isolation from pig intestine of a new member of the family of proline-arginine-rich antibacterial peptides. *Eur. J. Biochem.* **202,** 849–854.
72. Tanaka, K., Fujimoto, Y., Suzuki, M., et al. (2001) PI3-kinase p85α is a target molecule of proline-rich antimicrobial peptide to suppress proliferation of ras-transformed cells. *Jpn. J. Cancer Res.* **92,** 959–967.
73. Shi, J., Ross, C. R., Leto, T. L., and Blecha, F. (1996) PR-39, a proline-rich antibacterial peptide that inhibits phagocyte NADPH oxidase activity by binding to Src homology 3 domains of $p47^{phox}$. *Proc. Natl. Acad. Sci. USA* **93,** 6014–6018.
74. Chan, Y. R. and Gallo, R. L. (1998) PR-39, a syndecan-inducing antimicrobial peptide, binds and affects p130(Cas) *J. Biol. Chem.* **273,** 28978–28985.
75. Liu, C. W., Corboy, M. J., DeMartino, G. N., and Thomas, P. J. (2003) Endoproteolytic activity of the proteasome. *Science* **299,** 408–411.
76. Osmulski, P. A. and Gaczynska, M. (2000) Atomic force microscopy reveals two conformations of the 20 S proteasome from fission yeast. *J. Biol. Chem.* **275,** 13171–13174.
77. Kisselev, A. F., Akopian, T. N., Woo, K. M., and Goldberg, A. L. (1999) The sizes of peptides generated from protein by mammalian 26 and 20 S proteasomes. Implications for understanding the degradative mechanism and antigen presentation. *J. Biol. Chem.* **274,** 3363–3371.

# 2

# Purification of E1 and E1-Like Enzymes

**Arthur L. Haas**

**Summary**

Ubiquitin-activating enzyme is the archetype for a family of enzymes catalyzing the ATP-coupled activation of ubiquitin and other class 1 ubiquitin-like proteins required for their subsequent conjugation to cellular targets. The general physical and mechanistic features of the E1 family appear well conserved. Formation of an obligatory E1–ubiquitin thiol ester intermediate forms the basis of a one-step covalent purification of the enzyme on ubiquitin-linked affinity columns that has been adapted for the isolation of E1 paralogs. We describe the facile purification of active E1 from outdated human red blood cells in yields (2–4 nmol/U of blood) that make this an attractive alternative to expression of the proteolytically labile recombinant protein. In addition, two stoichiometric activity assays are described that rely on formation of the E1 $^{125}$I-ubiquitin thiol ester and ubiquitin [2,8-$^{3}$H]adenylate intermediates.

**Key Words:** Affinity chromatography; AppBp1; E1; Nedd8; purification; Uba3; ubiquitin; ubiquitin-like protein.

## 1. Introduction

The ubiquitin-activating enzyme (E1/Uba1) catalyzes the first step in the conjugation of ubiquitin to protein targets and serves as the archetype for paralogous enzymes catalyzing the activation of other class 1 ubiquitin-like polypeptides including Sumo, Nedd8, ISG15, Hub1, FAT10, and Apg12. The E1 catalytic cycle yields a ternary enzyme complex comprising stoichiometric amounts of a ubiquitin carboxyl terminal thiol ester to an absolutely conserved active site cysteine (Cys$^{632}$, human Uba1a numbering) and a tightly bound ubiquitin adenylate mixed anhydride that serves as the immediate precursor of the thiol ester ***(1,2)*** (*see* **Fig. 1**). Translation from alternative start sites of the E1 mRNA, transcribed from the single gene encoding the enzyme, yields nuclear (Uba1a) and cytoplasmic (Uba1b) isozymes of 117,789 Da and 113,740 Da ***(3,4)***, respectively, that are otherwise functionally indistinguishable with respect to E2 thiol ester formation ***(5)***. The half-reactions of E1-catalyzed ubiquitin activation and E3 ligase-catalyzed isopeptide bond formation are linked through a superfamily of

From: *Methods in Molecular Biology, vol. 301, Ubiquitin–Proteasome Protocols*
Edited by: C. Patterson and D. M. Cyr © Humana Press Inc., Totowa, NJ

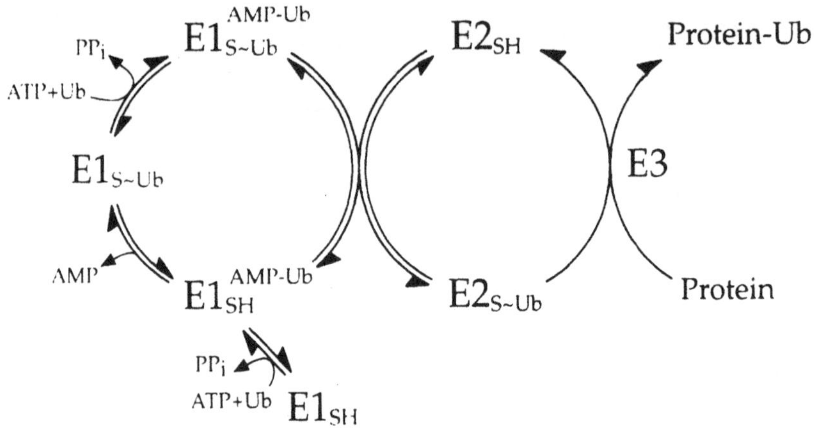


Fig. 1. The mechanism of ubiquitin conjugation. In the presence of ATP and ubiquitin, E1 forms a ternary complex composed of ubiquitin thiol ester and ubiquitin adenylate intermediates. The former intermediate is transferred to E2 to form the corresponding E2 thiol ester. The E3 ligase catalyzes the conjugation of ubiquitin by aminolytic cleavage of the cognate E2–ubiquitin thiol ester.

cognate E2/Ubc isoforms that transfer the polypeptide as an E2–ubiquitin thiol ester *(6,7)* (*see* **Fig. 1**). The hierarchical architecture of ubiquitin conjugation accounts for the broad substrate specificity and evolutionary plasticity of this posttranslational modification that is shared with paralogous pathways for ligation of the ubiquitin-like proteins, reviewed in **refs.** ***6–8***.

In vitro reconstitution of ubiquitin ligation requires the presence of sufficient E1 and E2, determined empirically, to render the overall process rate limiting with respect to E3-catalyzed conjugation in order to yield unambiguous information regarding substrate specificity and function ***(6,9,10)***. The relatively small size of the E2 isoforms (14–35 kDa) favors their expression in high yield within *Escherichia coli*. Small amounts of recombinant human GST–E1 can be similarly expressed ***(11)***; however, the proteolytic instability frequently observed when expressing large recombinant proteins precludes yields sufficient to serve as a practical source of reagent-grade quantities of activating enzyme ***(5)***. Early work demonstrated that ubiquitin-linked affinity columns afford a facile method for isolating E1 from cell extracts from which free ubiquitin has been removed by anion-exchange chromatography ***(1,12)***. In the presence of ATP, $Mg^{2+}$, and a suitable ATP-regenerating system, E1 forms a covalent thiol ester with column-bound ubiquitin that can be specifically eluted by forcing the reaction in reverse on addition of AMP and $PP_i$. Subsequent elution with dithiothreitol (DTT) at alkaline pH yields a mixture of endogenous E2 isoforms, a small fraction of E1 noncovalently bound to E2, and other ubiquitin interacting proteins that can be resolved further by anion-exchange FPLC ***(13,14)***. The relative simplicity of this affinity method makes cell-free extracts an attractive source of E1. However, the presence of a proteolytic activity in cell extracts that inactivates ubiquitin by limited digestion of the C-terminal glycine dipeptide complicates the general use of ubiquitin affinity methods ***(15)***. We have found that human erythrocytes are devoid of this proteolytic activity and contain significant amounts of active E1 ***(9,10)***. We describe a protocol for the facile affinity purification of human E1 from outdated red blood cells that has been refined from earlier methods ***(1,14)***. Because reconstitution of in vitro conjugation requires an accurate knowledge of E1 concentrations, we also provide two alternative methods for quantitating the active enzyme that relies on the stoichiometric formation of the ubiquitin adenylate and thiol ester intermediates ***(1,2)***.

## 2. Materials

### *2.1. Ubiquitin Affinity Column*

1. Affi-Gel 10 activated affinity support (Bio-Rad).
2. Bovine ubiquitin (Sigma).
3. 0.1 *M* Sodium bicarbonate, pH 9.0, at room temperature.
4. Glass-fritted (medium) Büchner funnel and side-arm flask.
5. 0.1 *M* Ethanolamine-HCl, pH 8.0, at room temperature.
6. 50 m*M* Phosphate-buffered saline (PBS), pH 7.4, at room temperature.
7. 0.1 *M* Tris-HCl, pH 9.0, at room temperature.
8. 50 m*M* Tris-HCl, pH 7.5, at room temperature.
9. Bovine serum albumin (BSA) (Sigma).

### *2.2. Preparation of Human Erythrocytes*

1. 5 U of outdated packed human erythrocytes.
2. Surgical gloves, mask, and face shield (for use while harvesting cells from the blood bank bags).
3. 50 m*M* Potassium PBS, pH 7.4, at room temperature.
4. Krebs–Ringer phosphate: 0.1 *M* sodium phosphate buffer, pH 7.4, 120 m*M* NaCl, 5 m*M* KCl, 1.2 m*M* $MgCl_2$ at 37°C.
5. 20 m*M* 2,4-Dinitrophenol.
6. 0.5 *M* 2-Deoxyglucose.
7. 0.1 *M* DTT.
8. DEAE-52 (Whatman) anion-exchange resin (prehydrated).
9. 0.3 *M* Potassium phosphate buffer, pH 7.0, at 4°C.
10. 25 m*M* Potassium phosphate buffer, pH 7.0, containing 1 m*M* DTT at 4°C.
11. 25 m*M* Potassium phosphate buffer, pH 7.0, containing 25 m*M* KCl and 1 m*M* DTT at 4°C.
12. 25 m*M* Tris-HCl, pH 7.2, containing 0.5 *M* KCl and 1 m*M* DTT at 4°C.
13. Solid ammonium sulfate.
14. 50 m*M* Tris-HCl, pH 7.2, containing 1 m*M* DTT at 4°C.

### *2.3. Affinity Isolation of E1*

1. 50 m*M* Tris-HCl, pH 7.5, containing 2 m*M* ATP and 10 m*M* $MgCl_2$ at room temperature.
2. 0.1 *M* ATP.
3. 1.0 *M* $MgCl_2$.
4. 0.5 *M* Creatine phosphate.
5. Creatine phosphokinase (Sigma) at $10^3$ IU/mL in 50 m*M* Tris-HCl, pH 7.5, and 1 m*M* DTT.
6. 50 m*M* Tris-HCl, pH 7.5, at room temperature.

7. 50 m*M* Tris-HCl, pH 7.5, containing 0.2 *M* KCl at room temperature.
8. 25 m*M* Tris-HCl, pH 7.5, containing 2 m*M* AMP and 2 m*M* inorganic pyrophosphate ($PP_i$) at room temperature.
9. 0.1 *M* DTT.
10. 0.1 *M* Tris-HCl, pH 9.0, containing 10 m*M* DTT at room temperature.
11. HR 5/5 Mono Q anion-exchange FPLC column (Pharmacia).
12. Mono Q FPLC buffers: Buffer A: 50 m*M* Tris-HCl, pH 7.5, containing 1 m*M* DTT at 4°C. Buffer B: 50 m*M* Tris-HCl, pH 7.5, containing 1.0 *M* NaCl and 1 m*M* DTT at 4°C.
13. Superose 12 buffer: 50 m*M* Tris-HCl, pH 7.5, containing 50 m*M* NaCl and 1 m*M* DTT at 4°C.

### 2.4. E1 Quantitation

1. Sodium $^{125}$I (carrier-free) or [2,8-$^{3}$H]ATP (30–50 Ci/mmol; Amersham Biosciences).
2. 1 *M* Tris-HCl, pH 7.5, at room temperature.
3. 10 mg/mL of chloramine-T (make fresh).
4. 10 mg/mL of sodium bisulfite (make fresh).
5. 10 m*M* KI.
6. 50 m*M* Tris-HCl, pH 7.5, at room temperature.
7. 10% (w/v) Trichloroacetic acid (TCA) at 4°C.
8. 0.2 *M* Triethylamine-HCl, pH 9.0, at room temperature.

## 3. Methods

### 3.1. Preparation of a Ubiquitin Affinity Column

Any activated affinity support can be used to prepare the ubiquitin affinity column; however, we prefer Affi-Gel 10 (Bio-Rad) because the *N*-hydroxysuccinimide ester chemistry consistently yields a >90% coupling efficiency when used with compatible nonamine buffers and care is taken to minimize hydrolysis of the activating group prior to coupling *(2)*. The latter point requires all reagents to be prepared beforehand. The following protocol assumes a column of 10-cc bed volume and having a ubiquitin concentration of approx 2 mg/mL (234 μ*M*), which is sufficient for processing fraction II cell extract equivalent to 1 U of packed human erythrocytes yielding 2–4 nmol of active E1. The protocol can be readily adapted to prepare similar affinity columns containing recombinant ubiquitin-like proteins *(16)* or E2/Ubc paralogs *(9,10)*.

1. Prepare the coupling solution by dissolving 20 mg of bovine ubiquitin in 10 mL of 0.1 *M* sodium bicarbonate buffer, pH 9.0, equilibrated to room temperature (*see* **Note 1**). Because of the high coupling efficiency, the volume and ligand concentration of the coupling solution should approximate the bed volume and final ligand concentration of the column. Save a 50-μL aliquot of the coupling solution for later determination of the ligand concentration on the column (*see* **Note 2**).
2. Thoroughly suspend the Affi-Gel 10 by vigorous shaking and then pour 1.25 times the desired bed volume into a 15-mL glass-fritted (medium) Büchner funnel attached to a side-arm flask. Using low vacuum, wash the affinity support with five bed volumes each of ice-cold distilled water and 0.1 *M* sodium bicarbonate, pH 9.0. Draw the buffer layer down to the top of the affinity support but DO NOT draw to dryness because the support is difficult to rehydrate. Immediately transfer the desired amount of equilibrated affinity support by spatula to a 50-mL Erlenmeyer flask (*see* **Note 3**).
3. Add the coupling solution to the equilibrated affinity support and incubate at room temperature for 1 h with gentle mixing on a rotary shaker.

4. Block any remaining unreacted activated groups on the affinity support by adding 1 mL of 0.1 *M* ethanolamine-HCl, pH 8.0, and incubating an additional hour at room temperature.
5. Pour the coupled affinity support into a 1.5 × 10 cm column. Collect the coupling solution for determining the amount of bound ligand on the column. Successively wash the column with five bed volumes each of 0.1 *M* sodium bicarbonate, pH 9.0, 50 m*M* PBS, pH 7.4, containing 5 mg/mL of BSA to block nonspecific protein binding sites on the affinity support, 50 m*M* PBS, pH 7.4, then 0.1 *M* Tris-HCl, pH 9.0, to remove noncovalently bound protein that may otherwise elute during the final pH 9 DTT wash of the column (*see* **Subheading 3.2.2.**), and 50 m*M* Tris-HCl, pH 7.5. The column should be stored at 4°C in 50 m*M* Tris-HCl, pH 7.5, containing 0.1% (w/v) sodium azide to retard bacterial growth.

### 3.2. Isolation of Human Ubiquitin E1

Although the majority of cellular proteins present in their nucleated progenitor cells are lost during the terminal differentiation of erythrocytes, low amounts of many enzymes remain in the soluble fraction, making these cells an attractive source from which to purify human E1 and other enzymes of the ubiquitin pathway. Erythrocyte fraction II is prepared from the $10^5 g$ supernatant of red cell lysate by DEAE fractionation at pH 7.0. Free ubiquitin and the ubiquitin-like proteins fail to bind to the anion-exchange matrix and appear in the unadsorbed fraction (fraction I) while E1 is retained in the absorbed fraction, which is subsequently eluted in 0.5 *M* KCl ***(17,18)***. In addition, this step removes hemoglobin, the major protein within erythrocyte cytosol. Erythrocyte fraction II can also be used for the purification of the AppBp1-Uba3 heterodimeric Nedd8 activating enzyme by Nedd8-AffiGel 10 covalent affinity chromatography ***(16)***. Although not tested to date, fraction II presumably also contains the human Aos1-Uba2 heterodimeric Sumo activating enzyme ***(19,20)***.

#### *3.2.1. Preparation of Human Erythrocyte Fraction II*

1. Obtain 5 U of outdated packed human red blood cells (preferably leukocyte free; *see* **Note 4**). Carefully remove the cells from their storage bags by cutting open with scissors and pool.
2. Collect the cells by centrifuging at 4500*g* for 15 min. Remove the plasma by aspiration, and then gently resuspend the cells in twice their volume of 50 m*M* PBS, pH 7.4. Centrifuge again and remove the supernatant by aspiration.
3. Resuspend the cells in an equal volume of Krebs–Ringer phosphate and adjust to 0.2 m*M* 2,4-dinitrophenol and 20 m*M* 2-deoxyglucose. Incubate the cell suspension with gentle shaking for 90 min at 37°C to deplete cellular ATP (*see* **Note 5**).
4. Collect the cells by centrifugation for 15 min at 4500*g*. Remove the supernatant by aspiration and pool with the prior washes for decontamination. Wash the cell pellet three times in an equal volume of 50 m*M* PBS, pH 7.4, removing the supernatant each time by aspiration. Use care during washing and resuspending the cells because ATP depletion makes the cells fragile.
5. Lyse the cells by adding 1.6 pellet volumes of ice-cold distilled water. Immediately add DTT to a final concentration of 1 m*M*. Allow the lysate to stand on ice for 30 min.
6. Centrifuge the lysate at $10^5 g$ for 1 h at 4°C. Collect the supernatant and adjust to pH 7.2 with 1 *M* sodium phosphate buffer, pH 7.2, if necessary. The $10^5 g$ pellet is soft and easily disturbed; therefore, the supernatant should be removed by aspiration. If desired, lysates

can be divided into five equal aliquots and stored at –80°C because they are stable for at least a year without appreciable loss of E1 activity. The subsequent steps can be performed with individual lysate aliquots to avoid processing large volumes. *(The remaining steps assume the processing of a volume of fraction II equal to one unit of packed erythrocytes.)*

The following steps are performed at 4°C.

1. Prepare a DEAE column (Whatman DE-52) having a bed volume equal to 50% of the $10^5g$ supernatant volume to be processed (*see* **Note 6**). Add bulk prehydrated DE-52 to a beaker to give a gently packed approximate volume equaling 25% more than needed in order to allow for loss while removing "fines." The DE-52 should be equilibrated in twice the volume of 0.3 *M* potassium phosphate buffer, pH 7.0, 4°C, and the pH adjusted as necessary. After most of the DE-52 has settled out by gravity, "fines" suspended in the buffer should be removed by aspiration. Resuspend in an equal bed volume of 0.3 *M* potassium phosphate buffer, pH 7.0, 4°C, and allow to settle again. Remove "fines" by aspiration; repeat a third time. Suspend the DE-52 in an equal bed volume of 25 m*M* potassium phosphate buffer, pH 7.0, 4°C, containing 1 m*M* DTT and adjust the pH as necessary. Pour the column to the desired bed volume and then equilibrate the column with three bed volumes of the same buffer by gravity flow. Check the pH of the buffer eluting from the column. If the pH is not 7.0, continue washing the column.
2. Load the $10^5g$ supernatant onto the DE-52 column and begin collecting the unadsorbed fraction immediately as the color begins to elute from the column (*see* **Note 7**). Wash the column with five bed volumes of column buffer or until the $OD_{280nm}$ does not change. Wash the column with three bed volumes of 25 m*M* potassium phosphate buffer, pH 7.0, 4°C, containing 25 m*M* KCl and 1 m*M* DTT to remove residual weakly adsorbed free ubiquitin. Elute the erythrocyte fraction II from the column with three bed volumes of 25 m*M* Tris-HCl, pH 7.2, containing 0.5 *M* KCl and 1 m*M* DTT.
3. Fraction II proteins eluted at 0.5 *M* KCl are concentrated by ammonium sulfate precipitation. With a magnetic stirrer set to a low speed, slowly add solid ammonium sulfate to the 0.5 *M* KCl eluate to achieve 85% saturation at 4°C (610 g/L of eluate). After all of the crystals have dissolved, allow the stirring to continue for at least 1 h at 4°C.
4. Centrifuge at 14,000*g* for 20 min. Resuspend the ammonium sulfate pellet in 5% of the equivalent lysate volume of 50 m*M* Tris-HCl, pH 7.2, containing 1 m*M* DTT (*see* **Note 8**). Using 12-kDa exclusion dialysis tubing, dialyze overnight against 4 L of 50 m*M* Tris-HCl, pH 7.2, containing 1 m*M* DTT.
5. The next morning adjust the pH to 7.2 as necessary. Remove any protein precipitate by centrifuging at 20,000*g* for 10 min. Flash-freeze the fraction II in dry ice–ethanol and store at –80°C.

### 3.2.2. Affinity Isolation of E1

Covalent affinity purification of E1 is adapted from earlier protocols for the isolation of E1 and E2 isoforms from rabbit reticulocyte extract ***(1,14)***. One-step affinity isolation of E1 is based on its ability to form a covalent thiol ester intermediate with column-bound ubiquitin. Because the E1 reaction is highly temperature dependent, all steps should be performed at room temperature and the erythrocyte fraction II should be warmed to 30°C as described prior to applying to the affinity column (*see* **Note 9**).

1. Equilibrate a 2.5 × 2 cm Sephadex G25 precolumn and the 10 cc of ubiquitin-Affi-Gel 10 affinity column prepared earlier with 50 m*M* Tris-HCl, pH 7.5, 20°C, containing 2 m*M* ATP and 10 m*M* $MgCl_2$ (*see* **Note 10**).
2. Flash thaw an aliquot of human erythrocyte fraction II equivalent to 1 U of packed red cells and then place on ice. Adjust the pH to 7.7 with 1 *M* Tris base as necessary and then adjust to 2 m*M* ATP, 10 m*M* $MgCl_2$, 10 m*M* creatine phosphate, and 1 U/mL of creatine phosphokinase (*see* **Note 11**). Warm the fraction II to 30°C in a water bath and then filter through the equilibrated Sephadex G25 precolumn. Save 200 μL as a starting fraction sample.
3. Slowly pass the filtered sample through the ubiquitin-Affi-Gel 10 affinity column at a flow rate of approx 0.5–1 mL/min. Save the unadsorbed flow through fraction as the postcolumn fraction for later determining recovery of E1.
4. Wash the column successively with two bed volumes of 50 m*M* Tris-HCl, pH 7.5, three bed volumes of 50 m*M* Tris-HCl, pH 7.5, containing 0.2 *M* KCl to remove weakly adsorbed proteins, and two bed volumes of 50 m*M* Tris-HCl, pH 7.5.
5. Elute the bound E1 with three bed volumes of 50 m*M* Tris-HCl, pH 7.5, containing 2 m*M* AMP and 2 m*M* $PP_i$. Immediately adjust the eluate to 1 m*M* DTT and place on ice for subsequent FPLC resolution (*see* **Subheading 3.2.3.**).
6. Elute the remaining E1 and E2 isoforms with three bed volumes of 0.1 *M* Tris-HCl, pH 9.0, containing 10 m*M* DTT and then immediately adjust to pH 7.5 with 1 *N* HCl and place on ice for FPLC resolution (*see* **Subheading 3.2.3.**).

### *3.2.3. Fast Protein Liquid Chromatography Purification of E1*

Although the E1 is substantially pure following covalent affinity chromatography, the AMP-$PP_i$ and pH 9-DTT eluates contain variable amounts of E2 isoforms, ubiquitin C-terminal hydrolase, and other ubiquitin-interacting proteins that can interfere with subsequent studies. Fast protein liquid chromatography (FPLC) is used to resolve E1 from these other components ***(14)***. The FPLC steps should be performed at 4°C.

1. Equilibrate a Mono Q HR5/5 anion-exchange column (Pharmacia) with 50 m*M* Tris-HCl, pH 7.5, containing 1 m*M* DTT. Load the entire AMP-$PP_i$ eluate from **step 5** in **Subheading 3.2.2.** onto the column at a flow rate of 1 mL/min and monitor the column eluate at 280 nm with the chart recorder set to 1 absorbance unit full scale. After the unadsorbed fraction has passed through the column (monitored by the marked absorption peak due to AMP), start a linear 0–0.5 *M* NaCl gradient at 12.5 m*M*/mL (0–0.5 *M* in 40 mL) and immediately begin to collect 1-mL fractions. The E1 typically elutes as a sharp peak at 0.23 *M* NaCl (*see* **Fig. 2**). Pool the peak fractions for subsequent assay of E1 activity.
2. Reequilibrate the Mono Q column with 50 m*M* Tris-HCl, pH 7.5, containing 1 m*M* DTT and apply the pH 9 DTT eluate from **step 6** in **Subheading 3.2.2.**. Repeat the NaCl gradient and collect into fresh tubes as before.
3. The Mono Q eluates are sufficiently homogeneous and of sufficient concentration that the NaCl present usually does not interfere because it is later diluted. If necessary, the Mono Q eluates can be resolved further by directly applying as 0.5-mL samples per run to a Superose 12 HR 10/30 gel filtration column (1 mL/min) equilibrated with 50 m*M* Tris-HCl, pH 7.5, containing 50 m*M* NaCl and 1 m*M* DTT. The E1 elutes as a symmetric peak at approx 110 kDa ***(14)***.
4. Because E1 is unstable to repeated freeze–thaw cycles, the enzyme should be divided into appropriate aliquots and flash frozen in liquid nitrogen then stored at –80°C ***(1,14)***.

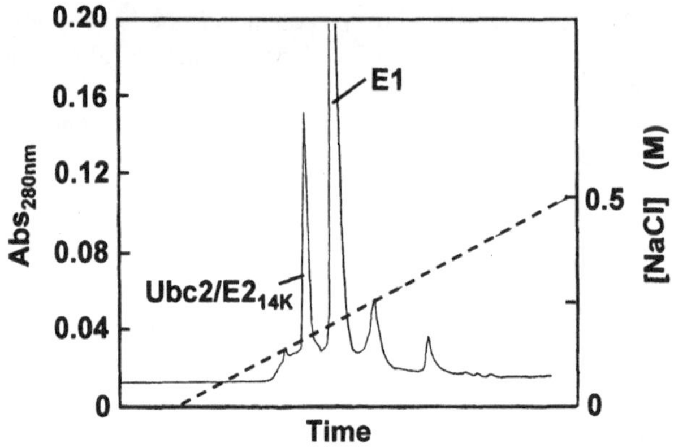


Fig. 2. Representative Mono Q FPLC trace of the AMP-PP$_i$ elution from the ubiquitin affinity column. Eluate recovered from one unit equivalent of human erythrocytes was resolved as described in the text (*see* **Subheading 3.2.2.**). The *dashed line* plots the NaCl gradient in which E1 elutes at 0.23 *M*.

Samples are stable for at least 8 mo at –80°C. Aliquots should be flashed thawed briefly at 37°C and then placed on ice for use. The enzyme typically loses approx 25% of its activity with each thaw. If desired, stability can be enhanced by adding BSA as a carrier protein to a final concentration of 1 mg/mL. The E1 is less stable in 30% glycerol, a typical alternative means of stabilizing proteins.

### 3.3. Stoichiometric Assay of Active E1

The ability of E1 and its paralogs to form a stoichiometric ternary complex containing ubiquitin adenylate and ubiquitin thiol ester serves as the basis for two types of E1 activity assays ***(1,16)***. The *thiol ester assay* measures the amount of E1–$^{125}$I-ubiquitin thiol ester within the ternary complex by direct quantitation of associated radioactivity following nonreducing sodium dodecyl sulfate-polyacrylamide gel electrophoresis (SDS-PAGE) resolution from free radiolabeled polypeptide ***(1,2)***. Because the E1–ubiquitin thiol ester represents the proximal donor of activated ubiquitin in E2–ubiquitin thiol ester formation (*see* **Fig. 1**), this assay best estimates the amount of active E1 present. The *adenylate assay* measures the stoichiometric formation of ubiquitin [$^3$H]adenylate within the E1 ternary complex following fractionation from the [2,8-$^3$H]ATP substrate by TCA precipitation of the enzyme-bound intermediate ***(1,2)***. Analogous assays can be employed to quantitate the AppBp1-Uba3 heterodimeric Nedd8 activating enzyme ***(16)***.

#### 3.3.1. Preparation of $^{125}$I-Ubiquitin

1. Transfer 100 µL of 10 mg/mL of FPLC purified bovine ubiquitin (Sigma) to a 1.5-mL Eppendorf tube (*see* **Note 12**). Add 13 µL of 1 *M* Tris-HCl, pH 7.5. Warm the tube and contents to 37°C.
2. Add 10 µL (1 mCi) of carrier-free Na$^{125}$I (Amersham).
3. Begin the reaction by adding 25 µL of 10 mg/mL of chloramine-T. Incubate at room temperature for 1 min.

4. Quench the reaction by addition of 30 µL of 10 mg/mL of sodium bisulfite. Incubate for 1 min at room temperature.
5. Add 25 µL of 10 m*M* KI as a carrier to dilute the specific activity of the free $^{125}$I.
6. Immediately load the sample onto a 1 × 45 cm column of Sephadex G25 equilibrated with 50 m*M* Tris-HCl, pH 7.5, at room temperature. Resolve $^{125}$I-ubiquitin from free radioiodide at a flow rate of 1 mL/min. Collect 1-mL fractions.
7. To avoid the problem of coincident counts caused by the large amount of radiolabel, determine the $^{125}$I radioactivity present in 5 µL of each fraction by γ-counting. Free $^{125}$I-ubiquitin will elute in the void volume of the column (~fractions 10–12). To avoid excessive dilution of the radiolabeled protein, pool the peak 2 or 3 fractions only.
8. Accurately measure the absorbance of the pooled sample at 280 nm and then subtract the absorbance of a blank determined on a fraction appearing before the void volume (usually fraction 5). Calculate the absolute ubiquitin concentration of the pooled sample using an empirical extinction coefficient of 0.16 $(mg/mL)^{-1}$ ***(21)***.
9. To determine the fixed radioactivity, carefully dilute 10 µL of the pooled sample into 190 µL of 5 mg/mL of BSA. Transfer 10 µL of the latter dilution into 190 µL of 5 mg/mL of BSA and then add 200 µL of 20% (w/v) TCA. Allow the sample to stand on ice for 10 min and then centrifuge 14,000*g* for 10 min. Remove the supernatant by aspiration and determine radioactivity present in the pellet by γ-counting. Calculate the specific radioactivity of the $^{125}$I-ubiquitin (typical values = 6000–12,000 cpm/pmol). The $^{125}$I-ubiquitin should be divided into conveniently sized aliquots and flash frozen in liquid nitrogen and then stored at –20°C. Once thawed, aliquots can be kept at 4°C for several weeks without loss of activity.

### 3.3.2. E1–$^{125}$I-Ubiquitin Thiolester Assay

1. Prepare a standard 10% (w/v) SDS-PAGE gel and running buffer (equilibrated to 4°C).
2. Incubations of 50 µL final volume should contain 50 m*M* Tris-HCl, pH 7.5, 2 m*M* ATP, 10 m*M* $MgCl_2$, 1 m*M* DTT, 1 mg/mL of carrier BSA, 5 µ*M* $^{125}$I-ubiquitin, and E1 sample. Because formation of the E1 ternary complex is rapid ***(2)***, the incubations should be equilibrated for several minutes at 37°C before initiating the assay by the addition of $^{125}$I-ubiquitin (*see* **Note 13**).
3. After 1 min at 37°C, the reaction is quenched by adding 50 µL of standard SDS sample buffer from which 2-mercaptoethanol has been omitted. The sample is allowed to stand on ice for 5 min to allow proteins to unfold. Do not boil the samples because this will destroy the thiol ester linkage.
4. Immediately load the SDS-PAGE gel and resolve under standard conditions at 4°C. To prevent heating of the gel and hydrolysis of the thiol esters during the run, completely immerse the gel in ice cold running buffer for good heat transfer.
5. When the SDS-PAGE is completed, float the gel onto a piece of Whatman filter paper, overlay with Saran Wrap and dry the gel using a standard vacuum gel drier. Mark the filter paper with $^{125}$I-labeled India ink (made by adding 25 µL of $^{125}$I-ubiquitin to 1 mL of India ink) or glow-in-the-dark paint so that the gel can be overlaid on the resulting autoradiogram later. Autoradiograph overnight at –80°C using Kodak X-Omat film and an appropriate intensifying screen. The next day overlay the developed autoradiogram over the dried gel and cut out the corresponding E1 thiol ester bands for quantitation of associated radioactivity by γ-counting. Determine the absolute amount of E1 thiol ester by using the specific radioactivity of the $^{125}$I ubiquitin.

### 3.3.3. E1 Ubiquitin [$^3$H]Adenylate Assay

1. Incubations of 50 μL final volume should contain 50 m*M* Tris-HCl, pH 7.5, 1 μ*M* [2,8-$^3$H]ATP, 10 m*M* $MgCl_2$, 1 m*M* DTT, 1 mg/mL carrier BSA, 5 μ*M* ubiquitin, and E1 sample. Because formation of the E1 ternary complex is rapid *(2)*, the incubations should be equilibrated for several minutes at 37°C before initiating the assay by the addition of ubiquitin.
2. Incubate for 1 min at 37°C then quench the reaction by addition of 150 μL of 10% (w/v) TCA. Set on ice for 10 min and then centrifuge for 10 min at 14,000*g*.
3. Aspirate the supernatant and discard. Gently rinse the surface of the pellet with ice-cold 10% TCA and then aspirate.
4. Dissolve the pellet in 200 μL of 0.2 *M* triethylamine-HCl, pH 9.0, and then quantitatively transfer to a vial containing a suitable scintillation cocktail. Determine the radiolabel present in the solubilized TCA pellet and calculate the absolute E1 content from the specific radioactivity of the [2,8-$^3$H]ATP.

## 4. Notes

1. Commercial bovine ubiquitin is used to prepare the affinity column because it is identical in sequence to human ubiquitin *(6)*.
2. Because the *N*-hydroxysuccinimide released during coupling absorbs at 280 nm and interferes with Lowry protein assays, the ligand concentration on the affinity column can be determined from the difference in unbound ligand protein between the starting and final coupling solutions by the Bradford dye binding assay *(22)* or by using quantitative SDS-PAGE followed by Coomassie staining. Alternatively, a small amount of $^{125}$I-ubiquitin (*see* **Subheading 3.3.1.**) can be added directly to the initial coupling solution to allow one to calculate bound ligand from the radioactivity remaining in the postcoupling solution.
3. The *N*-hydroxysuccinimide ester activated support is sensitive to hydrolysis, which results in a diminished coupling efficiency for the protein ligand. Work quickly while washing the affinity support to avoid excessive hydrolysis. Gentle vacuum filtration is preferably to gravity filtration because the support can be washed more rapidly. Because coupling is through lysyl ε-amino groups on the ligand, never use buffers containing primary amines because they will react with the activated support and diminish coupling efficiency. Care should be taken to avoid introducing water into the unused support, which is stored in anhydrous ethanol; the bottle containing unused support should be carefully resealed with Parafilm and returned to –80°C.
4. The use of leukocyte-free human red blood cells will obviate having to remove the buffy coat by aspiration while washing the cells. Leukocytes do not pose a problem to the protocol but can introduce unwanted enzyme contaminants. Although modern blood bank screening procedures largely eliminate the risk of infectious agents, one should follow normal safety precautions for blood-borne pathogens through the DEAE step: (a) wear a laboratory coat and approved surgical gloves (while cutting open the blood bank bags one should also wear a face shield and surgical mask; (b) dispose of blood bank bags, gloves, and so forth in a biohazard bag and immediately autoclave; (c) decontaminate all glassware with dilute bleach; and (d) collect all disposable supernatants by aspiration into a side-arm flask for subsequent decontamination with bleach before disposal.
5. The ATP depletion step is essential to allow endogenous isopeptidases to disassemble residual ubiquitin conjugates *(23)*. Otherwise, the isopeptidases will disassemble the con-

jugated ubiquitin in later steps and contaminate the erythrocyte fraction II with free ubiquitin. Free ubiquitin efficiently competes with column-bound polypeptide, even though the latter is present at a much higher concentration, and significantly reduces the yield of E1 recovered in the affinity purification step. Depletion of ATP depends on 2,4-dinitrophenol to decouple any remaining mitochondria present in the red cells and endogenous hexokinase to form 2-deoxyglucose-6 phosphate.

6. If fraction II equivalent to one unit of packed erythrocytes is being processed, the bed volume of DEAE will equal 10% of the total $10^5 g$ lysate. Our experience has been that a column of DE-52 is much more efficient at resolving free ubiquitin than bulk adsorption. The conditions described here have been optimized for lysate binding capacity by DE-52 and the resolution of free ubiquitin. Although it is generally bad practice to use an anionic buffer with an anion-exchange matrix, in the present application phosphate buffer aids in blocking ubiquitin adsorption to the DE-52.
7. Free ubiquitin is contained in the unadsorbed fraction of the DE-52 column under the conditions described even though the polypeptide has a pI of 6.7. Equilibrating the column to higher pH than specified results in increased contamination of fraction II with free ubiquitin.
8. Most of the protein will not initially dissolve because of the high concentration of residual ammonium sulfate contained in the pellet; however, protein will resolubilize during dialysis.
9. Covalent affinity purification of E1 requires that the ubiquitin be linked to the activated support so that it is not sterically hindered from proceeding through the catalytic cycle. Although the ubiquitin is probably preferentially linked through $Lys^6$, owing to the inherently greater reactivity of this group *(24,25)*, there will exist a statistical distribution of linkages to all seven lysines present on ubiquitin. We have observed a general increased recovery of E1 with successive uses of the affinity column as sterically preferred sites are ubiquitinated and blocked from further reaction with E1 *(14)*. No more than two bed volumes of fraction II should be processed through the column at one time because later steps in the conjugation reaction tend to elute the E1 from the column-bound thiolester, reducing the overall recovery. Poor recovery of E1 from fraction II usually results from inadequate removal of free ubiquitin in the DEAE step.
10. Protein solutions invariably contain microparticulate denatured proteins that are too small to detect by light scattering when viewed directly. However, these contaminants can collect at the top of affinity columns and, in part, elute with the affinity-bound protein, leading to contamination of the affinity-purified fraction. To prevent this, we use a small Sephadex G25 column as a prefilter to collect the microparticulates. This step proves to be a useful precaution in all affinity methods.
11. Free ATP is a competitive inhibitor of the true substrate, ATP/$Mg^{2+}$ *(2,26)*. This ratio of ATP/$MgCl_2$ guarantees that the ATP is quantitatively present as its $Mg^{2+}$ chelate. The creatine phosphate and creatine phosphokinase are present as an ATP regenerating system. A stock solution of creatine phosphokinase at $10^3$ IU/mL can be made in 50 m*M* Tris-HCl, pH 7.5, containing 1 m*M* DTT. Aliquots should be flash frozen and stored at –20°C. Aliquots should be thawed by hand only once and stored on ice for immediate use.
12. Commercial ubiquitin preparations that are not sufficiently pure for radioiodination can be purified to apparent homogeneity by FPLC *(27)*.
13. Ubiquitin-activating enzyme shows a nonlinear stoichiometry at high concentrations, resulting in an underestimation of the true concentration of active enzyme. Therefore,

one should assay a series of four to five doubling dilutions of the enzyme in 50 m*M* Tris-HCl, pH 7.5, containing 1 m*M* DTT and 1 mg/mL of BSA as a carrier protein. Because the affinity-purified E1 is present at a low concentration, the enzyme should always be diluted in Tris–BSA to prevent loss by nonspecific adsorption to the sides of tubes. The Tris–BSA alone should be used as a negative control. The E1 sample volume within the thiol ester assay should be at least 10 µL for pipetting accuracy.

## Acknowledgments

The author thanks Dr. J. Narasimhan and R. N. Bohnsack for their contributions in adapting the E1 purification protocol. This work was support by USPHS Grant GM34009.

## References

1. Haas, A. L., Warms, J. V., Hershko, A., and Rose, I. A. (1982) Ubiquitin-activating enzyme. Mechanism and role in protein-ubiquitin conjugation. *J. Biol. Chem.* **257**, 2543–2548.
2. Haas, A. L. and Rose, I. A. (1982) The mechanism of ubiquitin activating enzyme. A kinetic and equilibrium analysis. *J. Biol. Chem.* **257**, 10329–10337.
3. Handley-Gearhart, P. M., Stephen, A. G., Trausch-Azar, J. S., Ciechanover, A., and Schwartz, A. L. (1994) Human ubiquitin-activating enzyme, E1. Indication of potential nuclear and cytoplasmic subpopulations using epitope-tagged cDNA constructs. *J. Biol. Chem.* **269**, 33171–33178.
4. Stephen, A. G., Trausch-Azar, J. S., Handley-Gearhart, P. M., Ciechanover, A., and Schwartz, A. L. (1997) Identification of a region within the ubiquitin-activating enzyme required for nuclear targeting and phosphorylation. *J. Biol. Chem.* **272**, 10895–10903.
5. Tokgöz, Z., Bohnsack, R. N., Harder A., and Haas, A. L. (2003) Mutagenesis confirms a structural model for ubiquitin activating enzyme. *J. Biol. Chem*, submitted.
6. Haas, A. L. and Siepmann, T. J. (1997) Pathways of ubiquitin conjugation. *FASEB J.* **11**, 1257–1268.
7. Pickart, C. M. (2001) Mechanism underlying ubiquitination. *Annu. Rev. Biochem.* **70**, 503–533.
8. Larsen, C. N. and Wang, H. (2002) The ubiquitin superfamily: members, features, and phylogenies. *J. Proteome Res.* **1**, 411–419.
9. Baboshina, O. V., Crinelli, R., Siepmann, T. J., and Haas, A. L. (2001) N-end rule specificity within the ubiquitin/proteasome pathway is not an affinity effect. *J. Biol. Chem.* **276**, 39428–39437.
10. Siepmann, T. J., Bohnsack, R. N., Tokgõz, Z., Baboshina, O. V., and Haas, A. L. (2003) Protein interactions within the N-end rule ubiquitin ligation pathway. *J. Biol. Chem.* **278**, 9448–9457.
11. Handley, P. M., Mueckler, M., Siegel, N. R., Ciechanover, A., and Schwartz, A. L. (1991) Molecular cloning, sequence, and tissue distribution of the human ubiquitin-activating enzyme E1 [published erratum appears in *Proc. Natl. Acad. Sci. USA* **88,** 7456]. *Proc. Natl. Acad. Sci. USA* **88**, 258–262.
12. Ciechanover, A., Elias, S., Heller, H., and Hershko, A. (1982) "Covalent affinity" purification of ubiquitin-activating enzyme. *J. Biol. Chem.* **257**, 2537–2542.
13. Rose, I. A. and Warms, J. V. (1983) An enzyme with ubiquitin carboxy-terminal esterase activity from reticulocytes. *Biochemistry* **22**, 4234–4237.

14. Haas, A. L. and Bright, P. M. (1988) The resolution and characterization of putative ubiquitin carrier protein isozymes from rabbit reticulocytes. *J. Biol. Chem.* **263**, 13258–13267.
15. Haas, A. L., Murphy, K. E., and Bright, P. M. (1985) The inactivation of ubiquitin accounts for the inability to demonstrate ATP, ubiquitin-dependent proteolysis in liver extracts. *J. Biol. Chem.* **260**, 4694–4703.
16. Bohnsack, R. N. and Haas, A. L. (2003) Conservation in the mechanism of Nedd8 activation by the human AppBp1-Uba3 heterodimer. *J. Biol. Chem.* **278**, 26823–26830.
17. Ciehanover, A., Hod, Y., and Hershko, A. (1978) A heat-stable polypeptide component of an ATP-dependent proteolytic system from reticulocytes. *Biochem. Biophys. Res. Commun.* **81**, 1100–1105.
18. Ciechanover, A., Heller, H., Elias, S., Haas, A. L., and Hershko, A. (1980) ATP-dependent conjugation of reticulocyte proteins with the polypeptide required for protein degradation. *Proc. Natl. Acad. Sci. USA* **77**, 1365–1368.
19. Okuma, T., Honda, R., Ichikawa, G., Tsumagari, N., and Yasuda, H. (1999) In vitro SUMO-1 modification requires two enzymatic steps, E1 and E2. *Biochem. Biophys. Res. Commun.* **254**, 693–698.
20. Gong, L., Li, B., Millas, S., and Yeh, E. T. (1999) Molecular cloning and characterization of human AOS1 and UBA2, components of the sentrin-activating enzyme complex. *FEBS Lett.* **448**, 185–189.
21. Haas, A. L. and Wilkinson, K. D. (1985) The large scale purification of ubiquitin from human erythrocytes. *Prep. Biochem.* **15**, 49–60.
22. Read, S. M. and Northcote, D. H. (1981) Minimization of variation in the response to different proteins of the Coomassie blue G dye-binding assay for protein. *Analyt. Biochem.* **116**, 53–64.
23. Haas, A. L. and Bright, P. M. (1985) The immunochemical detection and quantitation of intracellular ubiquitin-protein conjugates. *J. Biol. Chem.* **260**, 12464–12473.
24. Jabusch, J. R. and Deutsch, H. F. (1985) Localization of lysines acetylated in ubiquitin reacted with *p*-nitrophenyl acetate. *Arch. Biochem. Biophys.* **238**, 170–177.
25. Macdonald, J. M., Haas, A. L., and London, R. E. (2000) Novel mechanism of surface catalysis of protein adduct formation. NMR studies of the acetylation of ubiquitin. *J. Biol. Chem.* **275**, 31908–31913.
26. Haas, A. L., Warms, J. V., and Rose, I. A. (1983) Ubiquitin adenylate: structure and role in ubiquitin activation. *Biochemistry* **22**, 4388–4394.
27. Baboshina, O. V. and Haas, A. L. (1996) Novel multiubiquitin chain linkages catalyzed by the conjugating enzymes $E2_{epf}$ and Rad6 are recognized by the 26S proteasome subunit 5. *J. Biol. Chem.* **271**, 2823–2831.

# 3

# Assays for RING Family Ubiquitin Ligases

**Manabu Furukawa, Paul S. Andrews, and Yue Xiong**

### Summary

Many eukaryotic proteins are regulated by the covalent attachment of ubiquitin or polyubiquitin chains. These include proteins involved in cell cycle control, tumor suppression, and many signaling pathways. Ubiquitination of proteins occurs through an enzymatic cascade of three discrete steps, which results in covalent attachment of ubiquitin to the substrate. The first two steps in this cascade involve the activating and conjugating enzymes, E1 and E2. The third and final step is performed by the E3 ubiquitin ligase. The ubiquitin ligase is responsible for two distinct activities: targeting specific substrates by bringing the substrate and activated ubiquitin together, as well as catalyzing the ligation of ubiquitin to the substrate. There are two main classes of E3 ligases, the HECT domain and the RING finger-containing ligases. RING finger-based ubiquitination utilizes RING-containing protein subunits, or proteins with intrinsic RING domains, to catalyze the formation of polyubiquitin chains. In this chapter we describe step by step protocols to assay for the activity of the RING finger-type of E3 ligase both in vitro and in vivo.

**Key Words:** RING; in vitro assay; autoubiquitination; cullins; ROC.

## 1. Introduction

The stabilities and the functions of many eukaryotic proteins are regulated by ubiquitin pathways through either 26S proteasome-dependent degradation or conformational changes resulting from covalent ubiquitin conjugation. Protein ubiquitination involves a cascade of enzymes including ubiquitin-activating enzymes (E1), ubiquitin-conjugating enzymes (E2), and ubiquitin ligases (E3) ***(1,2)***. Both ubiquitin-activating and -conjugating enzymes are well characterized and contain highly conserved functional domains. The identity and mechanism of E3 ubiquitin ligases, on the other hand, have been elusive and their activity has been long postulated as responsible for both recognizing substrates and for catalyzing polyubiquitin chain formation. Currently, two mechanistically distinct types of E3s—the HECT and RING families—have been identified. A domain of approx 350 residues located at the C-terminus of E6AP contains an active cysteine residue that can form thioester linkages with ubiquitin ***(3,4)***. Multiple cellular proteins with diverse structures contain a domain *h*omologous to *E*6AP *c*arboxyl *t*erminus (HECT; five in budding yeast and ~30 in human), implying

From: *Methods in Molecular Biology, vol. 301, Ubiquitin–Proteasome Protocols*
Edited by: C. Patterson and D. M. Cyr © Humana Press Inc., Totowa, NJ

that there are many additional substrates and diverse physiological functions for the HECT family of E3 ligases ***(5)***.

The second family of E3s contains either an intrinsic Really Interesting New Gene (RING) finger domain or an associated RING subunit essential for their ubiquitin ligase activity ***(6–8)***. A large number of RING finger-containing proteins exist in all eukaryotes. There are more than 20 in yeast, 100 in *Caenorhabditis elegans*, 100 in *Drosophila melanogaster*, and more than 350 in human genome, implicating a very broad involvement of RING-dependent ubiquitination in vivo. The RING finger comprises eight cysteine and histidine residues that bind two atoms of zinc to form one unique three-dimensional structure referred to as the cross-brace rather than two separate minidomains ***(9)***. Of a dozen well-characterized RING finger proteins, investigations of ROC1 and APC11 have contributed significantly to our realization of RING finger domain–mediated ubiquitination. APC11 is a subunit of the *a*naphase-*p*romoting *c*omplex (APC or cyclosome) that is required for both entry into anaphase as well as exit from mitosis ***(10,11)***. ROC1 (*R*ING *o*f *c*ullins, also known as Rbx1 and Hrt1) is an essential subunit of cullin-dependent ligases ***(12–15)***. Unlike most other RING finger-containing proteins, both ROC1 (108 residues) and APC11 (84 residues) are small proteins with the RING finger taking up most of the coding capacity. Various mutational analyses have demonstrated the requirement of each of the eight conserved Cys or His residues, and thus the integrity of RING finger, for the ubiquitin ligase activity. Unlike the HECT family ligases, the RING-type E3s do not appear to form a ubiquitin thioester intermediate. A surprising finding from studying the RING finger protein c-Cbl ***(16)***, APC11 ***(17,18)***, and ROC1 ***(19)*** was that the RING finger domain alone, in the absence of other flanking sequences, in the case of c-Cbl, or its functional partners such as APC2 or a cullin, can interact with E2 and is sufficient to promote E1- and E2-dependent polyubiquitin chain formation in vitro. Formation of such polyubiquitin chains in the absence of a substrate, often referred to as either substrate-independent ubiquitination or auto-ubiquitination, is achieved via the same chemical reaction as the ubiquitin–substrate ligation, which occurs through a covalent attachment of the C-terminal glycine of ubiquitin to the ε-amino group of a lysine residue of another protein. These findings suggest that the RING finger alone may function as an autonomous component in the E3 to activate allosterically at least some E2s and provide a practical useful mean for initially assaying the ubiquitin ligase activity of any RING finger protein.

This chapter describes detailed step-by-step protocols for in vivo and in vitro ubiquitin ligase assays with RING finger-type E3 ligases. Although these protocols were optimized or developed based on mostly our experience in studying ROC-cullin family of ubiquitin ligases and the MDM2 ligase, they could serve as a simple starting point for initial testing of potential ubiquitin ligase activity of other RING-type ligases. In practice, we recommend dividing the characterization of a RING finger protein for its ubiquitin ligase function into four steps: (1) assaying for in vitro auto-ubiquitination using recombinant protein or the RING finger domain, (2) assaying for in vitro auto-ubiquitination using immunocomplex precipitated from the cells or tissues, (3) determining in vivo ubiquitination of candidate substrate(s), and finally (4) reconstituting in vitro ubiquitination using individual components purified from either bacterial or insect cells. Although assaying for auto-ubiquitination is relative simple, determining substrate-dependent E3 ligase activity requires knowledge of a potential substrate and remains a challenge in most cases because of the lack of a systematic method to identify the substrate of a given ligase. Furthermore, substrate and ligase often interact in a signal-dependent manner. Unlike use of artificial substrates such as myelin basic protein or histone H1 for assaying kinase activity, there is currently no artificial substrate available for assaying ubiquitin ligases. There are only a few successful examples in identifying the substrate of an E3 ligase without any prior knowledge. These include: (1) identification of F-box protein β-TrCP in targeting phosphorylated IκBα to CUL1-mediated $SCF^{b\text{-}TrCP}$ ligase by mass spectrometric analysis of IκBα/NF-κB complex purified from proteasome-inhibited, tumor necrosis factor-α (TNF-α)–stimulated cells ***(20)*** or based on previous genetic analysis in *Drosophila* ***(21)***; (2) F-box protein Fbx2 in targeting glycosylated preintegrin β1 to $SCF^{Fbx}$ ***(2)*** by mass spectrometric screening for proteins bound to various sugar probes in a proteasome inhibition– dependent manner ***(22)***; (3) F-box protein Cdc4/Archipelago/Fbw7 in targeting phosphorylated cyclin E ubiquitination by either a genetic screen for fly mutants with increased proliferation ***(23)*** or by genetic test of individual F-box proteins critical for mammalian cyclin E degradation in yeast ***(24)***; (4) β-arrestin2 in targeting phosphorylated $\beta_2$-adrenergic receptors for ubiquitination by MDM2 by yeast two-hybrid screening with β-arrestin2 as the bait ***(25)***, and (5) RING finger protein HOIL-1 in ubiquitinating iron regulatory protein 2 (IRP2) through differential two-hybrid screens for protein(s) binding to iron-dependent degradation (IDD) domain in yeast cells cultured in either aerobic or anaerobic conditions ***(26)***. These examples could provide a limited guidance for identifying the substrate(s) of other ligases.

## 2. Materials

### *2.1. In Vitro Auto-Ubiquitination Assay*

1. 10X Ubiquitin ligase assay buffer (dilute to 1X prior to use): 500 m*M* Tris-HCl, pH 7.4, 50 m*M* $MgCl_2$, 20 m*M* NaF, 6 m*M* dithiothreitol (DTT), 100 n*M* okadaic acid.
2. 2X SDS–DTT sample buffer: 100 m*M* Tris-HCl, pH 6.8, 4% sodium dodecyl sulfate (SDS), 20% glycerol, 200 m*M* DTT, 0.2% bromophenol blue.
3. Flag-ubiquitin (Sigma-Aldrich, St. Louis, MO).
4. Rabbit E1 (Affiniti Research Products Ltd., Exeter, UK).
5. 0.1 *M* ATP.
6. Bacterially expressed and purified His-UBCH5C (or other E2).
7. Bacterially expressed and purified glutathione-*S*-transferase (GST)–RING finger protein (or other ubiquitin ligase).
8. Anti-Flag M2 antibody (Sigma-Aldrich).

### *2.2. Coupled Immunoprecipitation and In Vitro Auto-Ubiquitination Assay*

1. NP-40 lysis buffer: 150 m*M* NaCl, 0.5% Nonidet P-40 (NP-40), 50 m*M* NaF, 50 m*M* Tris-HCl, pH 7.5, 1 m*M* DTT, 1 m*M* phenylmethylsulfonyl fluoride, 1 m*M* $Na_3VO_4$, 25 μg/mL of leupeptin, 25 μg/mL of aprotinin, 150 μg/mL of benzamidines, 10 μg/mL of trypsin inhibitor.
2. E3-specific antibody of choice.

3. Protein A– agarose.
4. 10X Ubiquitin ligase assay buffer.

### 2.3. In Vitro Substrate-Dependent Ubiquitin Ligase Assay

1. pcDNA3-HA-IKKβ$^{S}$*177*$^{E/S}$*181*$^{E}$.
2. Anti-hemagglutinin (HA) antibody (Covance, Princeton, NJ).
3. Protein A–agarose.
4. Kinase assay buffer: 50 m*M* Tris-HCl, pH 7.4, 0.6 m*M* DTT, 5 m*M* $MgCl_2$, 2 m*M* NaF.
5. 10X Ubiquitin ligase assay buffer.
6. 1 μ*M* ATP.
7. Bacterially expressed and purified GST–Flag-IκBα.
8. pcDNA3-CUL1.
9. pcDNA3-SKP1.
10. pcDNA3-β-TrCP.
11. pcDNA3-HA-ROC1.
12. Bovine-ubiquitin (Sigma-Aldrich).
13. Rabbit E1.
14. His-UBCH5C.
15. 2X SDS–DTT sample buffer.
16. Anti-Flag M2 antibody.
17. 293 T cells (ATCC).

### 2.4. In Vitro Substrate-Dependent Ubiquitin Ligase Assay

1. U2OS cells (ATCC).
2. FuGENE 6 transfection reagent (Roche, Indianapolis, IN).
3. pCMV-(HA-ubiquitin)$_8$.
4. pCMV-HDM2.
5. pcDNA3-p53.
6. MG132 (Peptides International, Louisville, KY).
7. SDS lysis buffer: 50 m*M* Tris-HCl, pH 7.5, 0.5 m*M* EDTA, 1% SDS, 1 m*M* DTT.
8. Anti-p53 antibody (Sigma-Aldrich).
9. Protein A–agarose.
10. Stripping buffer: 2.2 *M* glycine, 0.5 *M* NaCl, 1% SDS, buffer to pH 4.4 with HCl.

## 3. Methods

### 3.1. In Vitro Auto-Ubiquitination Assay Using Purified Recombinant Protein

The following procedures are modified from our studies on the ROC1-cullin ligases ***(19)***. Actual results of auto-ubiquitination assays of RING finger proteins and additional experimental details can also be found in studies of c-Cbl ***(16)*** and APC11 ***(17,18)***.

1. Prepare ubiquitin ligation reaction mixture in a microcentrifuge tube on ice:
   a. 3 μL of 10X ubiquitin ligase assay buffer.
   b. 0.6 μL of 0.1 *M* ATP.
   c. 1 μg of Flag-ubiquitin.
   d. 60 ng of rabbit E1.
   e. 300 ng of bacterially expressed and purified His-UBCH5C as E2 (*see* **Note 1**).
   f. 500 ng of bacterially expressed and purified GST–RING finger protein (*see* **Note 2**).

Bring the final volume to 30 μL.

2. Incubate the reaction mixture at 37°C for 30 min (*see* **Note 3**).
3. Add 30 μL of 2X SDS–DTT sample buffer and boil for 5 min.
4. Separate the reaction mixture by SDS-polyacrylamide gel electrophoresis (SDS-PAGE) on a 10% gel.
5. Transfer to a nitrocellulose filter and perform a standard immunoblotting analysis with anti-Flag M2 antibody (Sigma-Aldrich).

### 3.2. Coupled Immunoprecipitation and In Vitro Auto-Ubiquitination Assay

As an alternative to the use of purified recombinant RING finger protein, immunocomplexes precipitated from cells or tissues can also be used for assaying autoubiquitination activity in vitro by a very similar procedure. Using an immunocomplex as the source of E3 offers two advantages over the use of purified recombinant protein: it avoids often encountered difficulty in purifying large, full-length proteins from bacteria and it allows copurification of a cofactor(s) that might contribute to or is essential for the ligase activity.

1. Perform standard immunoprecipitation of RING finger protein of interest from cultured cells or tissues. If a suitable antibody is not available, construct epitope tagged expression vector and perform standard cell transfection, followed by an immunoprecipitation. There are various detergent-based lysis buffers for cell lysis and subsequent washes of immunoprecipitates. We recommend, as a starting point, to use relative mild NP-40 lysis buffer.
2. After cell lysis, clarify the lysate by centrifugation on a microcentrifuge at maximal speed of 16,000*g* (13,200 rpm) for 10 min at 4°C.
3. Transfer the clarified supernatant to new tube and add appropriate antibody (~1–2 μg for each mL or mg lysate), followed by an incubation for 4 h to overnight at 4°C with rotation.
4. Add 10 μL of Protein A–agarose and rotate for another 1 h at 4°C.
5. Precipitate the agarose beads by centrifugation on a microcentrifuge at a low speed of 800*g* (3000 rpm) for 3 min at 4°C. Carefully aspirate off the supernatant, and add 1 mL of NP-40 lysis buffer to wash the immunocomplex immobilized on the agarose beads. Mix gently but thoroughly by inverting the tubes multiple times.
6. Repeat **step 5** twice.
7. Wash the immunoprecipitate twice, each with 1 mL of 1X ubiquitin ligase assay buffer (diluted freshly from 10X buffer). *Note:* Be careful not to aspirate off the Protein A–agarose beads during the washes.
8. After the final wash, aspirate off the supernatant completely. Add ligase reaction buffer mixture to the immunocomplex immobilized on the agarose beads. Follow **steps 1–5** described above to assay in vitro autoubiquitin ligase activity.

### 3.3. In Vitro Substrate-Dependent Ubiquitin Ligase Assay

Following procedures are modified from our assay for $SCF^{b\text{-}TrCP}$-mediated in vitro IκBα polyubiquitination *(13)*. Additional experimental details can be found in other published literature such as **ref. *14***. $SCF^{b\text{-}TrCP}$-mediated IκBα ubiquitination involves IκBα phosphorylation by IKKβ and subsequent recognition of phosphorylated IκBα by F-box protein β-TrCP *(20)*, providing a good example of signal-dependent in vitro substrate ubiquitination. Assaying for in vitro ubiquitination of other substrates can be modified accordingly.

### 3.3.1. Substrate Preparation

1. Transfect 5 μg of pcDNA3-HA-IKKβ$^{S}$*177*$^{E/S}$*181*$^{E}$ (a constitutively active mutant of IκB kinase) to one 60-mm plate of 293T cells.
2. Perform standard immunoprecipitation using anti-HA antibody as described in **Subheading 3.2., steps 1–4**.
3. Wash the immunocomplex immobilized on the agarose beads three times with NP-40 lysis buffer and twice with kinase assay buffer.
4. Assemble IκBα phosphorylation mixture on ice. For 60 μL IκBα kinase buffer mixture total: 6 μL of 10X ubiquitin ligase assay buffer, 3 μL of 1 m*M* ATP, 10 μg of GST–Flag-IκBα (expressed and purified from bacteria); bring the final volume to total 60 μL with $H_2O$.
5. Add the 60-μL mixture to the IKKβ immobilized on agarose beads. Incubate for 20 min at 37°C with occasional shaking.
6. Terminate the kinase reaction by incubating at 70°C for 5 min.
7. Transfer the supernatant containing phosphorylated substrate IκBα to a fresh tube, make aliquots, and store at –80°C until needed.

### 3.3.2. E3 (SCF $^{b\text{-}TrCP}$) Preparation

1. Transfect following plasmids to 293T cells (per each 60-mm plate): 3 μg of pcDNA3-CUL1; 1 μg of pcDNA3-SKP1; 1 μg of pcDNA3-β-TrCP; 1 μg of pcDNA3-HA-ROC1. *Note:* Individual components can be omitted from the transfection to provide negative controls.
2. Perform standard immunoprecipitation using anti-HA antibody as described in **Subheading 3.2., steps 1–4**.
3. Wash the ROC1-SCF$^{b\text{-}TrCP}$ immunocomplex immobilized on the agarose beads three times with NP-40 lysis buffer and twice with 1X ubiquitin ligase assay buffer (diluted freshly from 10X).

### 3.3.3. Ligase Assay

1. Assemble ligase reaction mixture on ice:
   a. 3 μL of 10X ubiquitin ligase assay buffer.
   b. 0.6 μL of 0.1 *M* ATP.
   c. 12 μg of bovine-ubiquitin.
   d. 60 ng of rabbit E1.
   e. 300 ng of bacterially expressed and purified His-UBCH5C as E2.
   f. 1 μL of phosphorylated substrate IκBα from **step 7.**
   g. Bring the final volume to 30 μL.

   *Note:* assemble a separate reaction that omits E1, E2, and FLAG-ubiquitin as negative controls.
2. Transfer the 30-μL reaction mixture to the tube containing the ROC1-SCF$^{b\text{-}TrCP}$ immunocomplex immobilized on the agarose beads. Incubate at 37°C for 60 min with occasional shaking.
3. Add 30 μL of 2X SDS–DTT sample buffer and boil for 5 min to terminate the reaction.
4. Separate the reaction mixture by SDS-PAGE on a 10% gel.
5. Transfer to a nitrocellulose filter, followed by a standard immunoblotting analysis with anti-Flag M2 antibody (*see* **Note 4**).

### *3.4. In Vitro Substrate-Dependent Ubiquitin Ligase Assay*

MDM2-mediated p53 ubiquitination was first discovered by Honda et al. *(27)* and represents one of most intensively studied built-in RING ubiquitin ligases. The following procedures describe conditions for assaying in vivo p53 ubiquitination and are modified from published literatures as well as our studies on the regulation of MDM2-mediated p53 ubiquitination by ARF and L11, two inhibitors of MDM2 ligase *(28)*. Assaying for in vivo ubiquitination of other substrates can be modified accordingly.

1. Perform standard transfection. One example is shown below for transfection of a 60-mm plate of cultured U2OS cells using FuGENE 6: 0.3 μg of pCMV-(HA–ubiquitin)$_8$ (*see* **Note 5**); 0.3 μg of pCMV-HDM2; 0.3 μg of pcDNA3-p53 (*see* **Note 6**).
2. At 24–36 h after transfection, add proteasome inhibitor, MG132 (final 25 μ*M*) to accumulate polyubiquinated p53 and thus increase the sensitivity of detection. Continue to culture cells for another 4 h.
3. Aspirate off the medium, rinse the plate once with cold phosphate-buffered saline (PBS), and trypsinize the cells. Collect the cells with cold PBS and pellet cells by centrifugation on a tabletop centrifuge at a low speed (e.g., 168*g* [1000 rpm] for 3 min on the Sorvall RT7 centrifuge).
4. Aspirate off PBS. Add 200 μL of preboiled SDS-lysis buffer directly to the cell pellet, resuspend the cell pellet, and boil for 10 min (*see* **Note 7**).
5. Clarify the lysate by centrifugation at 16,000*g* (13,200 rpm) on a microcentrifuge for 10 min at 4°C.
6. Dilute the clarified SDS lysate with 10-fold volume of NP-40 lysis buffer. Measure the total protein concentration using Dc protein Assay Kit (Bio-Rad, Hercules, CA/ *www.bio-rad.com*).
7. Transfer an aliquot of the sample to a separate tube, add an equal volume of 2X SDS–DTT sample buffer, boil for 3 min, and store in a –20°C freezer for subsequent Western blot.
8. Perform standard immunoprecipitation using anti-p53 antibody (~1–2 μg for each milligram of lysate), followed by an incubation for 4 h to overnight at 4°C with rotation.
9. Add 10 μL of Protein A–agarose and rotate for another 1 h at 4°C.
10. Precipitating the agarose beads by centrifugation on a microcentrifuge at a low speed of 800*g* (3000 rpm) for 3 min at 4°C. Carefully aspirate off the supernatant, and add 1 mL of NP-40 lysis buffer to wash three times the immunocomplex immobilized on the agarose beads. Mix gently but thoroughly by inverting the tubes multiple times.
11. After the final wash, aspirate off the supernatant. Add 15 μL of the 2X SDS–DTT sample buffer and boil for 5 min.
12. Separate samples by SDS-PAGE on a 10% gel. Load the lysate stored at –20°C freezer on the same gel, but separated by two blank lanes.
13. Transfer the protein in the gel to a nitrocellulose membrane.
14. Cut the membrane into two pieces: one containing the immunoprecipitation for immunoblotting with HA antibody to determine p53 ubiquitination and one containing the total cell lysate for immunoblotting with p53 and MDM2 antibody to verify the expression of both proteins.
15. After blotting the immunoprecipitation with HA antibody, strip the HA antibody by washing the filter with a stripping buffer three times, 10 min each. Reblot with anti-p53 antibody to examine the p53 on the membrane (*see* **Note 8**).

## 4. Notes

1. There are multiple E2s in different eukaryotic organisms. Many purified E2s are commercially available from Affiniti or BostonBiochem Inc. (www.bostonbiochem.com). A unique feature of E2 ubiquitin-conjugating enzymes is that many of them can be expressed in bacteria as soluble proteins and easily purified as active enzymes as His-tagged recombinant proteins by a simple one-step nickel bead affinity column. This is particularly important if a negative result is obtained when assaying for E2 activation by testing a RING finger protein or an E3 ligase complex. In our experience and many published reports, UBC5 seems to be more readily activated by many different RING finger proteins or E3 including ROC1 and APC11. Confirming the activity of E2 and testing the activation of different E2s is recommended, especially when a negative result is obtained using UBC5.
2. One simple way to purify RING finger protein by many researchers is to express either the full-length RING finger protein or the RING finger domain as a GST fusion protein in bacteria and purify it utilizing a glutathione (GST) affinity column. GST–ROC1 or GST–APC11 would serve as an appropriate positive control for both purification and E2 activation assays.
3. Assemble a separate reaction that omits E1, E2, FLAG-ubiquitin, or GST-fused RING finger protein as negative controls.
4. High molecular weight Flag-IκBα smear characteristic of polyubiquitin chain can be seen in the full reaction mixture but not in the reaction that omits E1, E2, ubiquitin, or transfection of pcDNA3-HA-ROC1. One example can be found in Fig. 5 of Ohta et al., which includes additional controls ***(13)***.
5. Octameric ubiquitin is chosen because it can be processed in vivo by cellular ubiquitin–C-terminal hydrolases and is more efficiently conjugated to the substrate than monomeric ubiquitin ***(29)***.
6. For the negative control, omit pCMV-(HA–ubiquitin)$_8$, pCMV-HDM2, or pcDNA3-p53. Mutation in the RING finger of MDM2 (e.g., MDM2$^{C464A}$) destroys its activity to ubiquitinate and degrade p53 and would provide an additional control for the specificity of MDM2-mediated p53 ubiquitination.
7. Direct lysis of cells by boiling in SDS lysis buffer prevents degradation of polyubiquitinated substrate by proteases after cell lysis. Most protein–protein interactions are dissociated, while polyubiquitin chains remain attached to the substrate after the boiling because they are covalently bound, allowing subsequent detection of ubiquitinated substrate by coupled immunoprecipitation and immunoblotting (IP-Western). The lysate is viscous when you add the SDS lysis buffer but becomes a cleared aqueous solution after 10 min of boiling.
8. Blotting with HA (ubiquitin) and p53 (substrate) antibody would give different pattern of polyubiquitinated p53 ladders, with p53 antibody being able to detect both ubiquitinated and unmodified p53, and HA antibody being more sensitive to detect high molecular weight species because of the link of multiple HA–ubiquitin moieties.

## Acknowledgments

We thank members of the Xiong laboratory for discussion. This study is supported by an NIH grant (GM067113) to Y. X.

## References


1. Hershko, A. and Ciechanover, A. (1998) The ubiquitin system. *Annu. Rev. Biochem.* **67,** 425–479.
2. Hochstrasser, M. (1996) Ubiquitin-dependent protein degradation. *Annu. Rev. Genet.* **30,** 405–439.
3. Huibregtse, J. M., Scheffner, M., Beaudenon, S., and Howley, P. M. (1995) A family of proteins structurally and functionally related to the E6-AP ubiquitin-protein ligase. *Proc. Natl. Acad. Sci. USA* **92,** 2563–2567.
4. Scheffner, M., Nuber, U., and Huibregtse, J. M. (1995) Protein ubiquitination involving an E1-E2-E3 enzymes ubiquitin thioester cascade. *Nature* **373,** 81–83.
5. Pickart, C. M. (2001) Mechanisms underlying ubiquitination. *Annu. Rev. Biochem.* **70,** 503–533.
6. Deshaies, R. J. (1999) SCF and cullin/RING H2-based ubiquitin ligases. *Annu. Rev. Cell Dev. Biol.* **15,** 435–467.
7. Jackson, M. W. and Berberich, S. J. (2000) MdmX protects p53 from Mdm2-mediated degradation. *Mol. Cell. Biol.* **20,** 1001–1007.
8. Zachariae, W. and Nasmyth, K. (1999) Whose end is destruction: cell division and the anaphase-promoting complex. *Genes Dev.* **13,** 2039–2058.
9. Borden, K. L. B., Boddy, M. N., Lally, J., et al. (1995) The solution structure of the RING finger domain from the acute promyelocytic leukaemia proto-oncoprotein PML. *EMBO J.* **14,** 1532–1541.
10. Zachariae, W., Shevchenko, A., Andrews, P. D., et al. (1998) Mass spectrometric analysis of the anaphase-promoting complex from yeast: identification of a subunit related to cullins. *Science* **279,** 1216–1219.
11. Yu, H., Peters, J.-M., King, R. W., Page, A. M., Hieter, P., and Kirschner, M. W. (1998) Identification of a cullin homology region in a subunit of the anaphase-promoting complex. *Science* **279,** 1219–1222.
12. Kamura, T., Koepp, D. M., Conrad, M. N., et al. (1999) Rbx1, a component of the VHL tumor suppressor complex and SCF ubiquitin ligase. *Science* **284,** 657–661.
13. Ohta, T., Michel, J. J., Schottelius, A. J., and Xiong, Y. (1999) ROC1, a homolog of APC11, represents a family of cullin partners with an associated ubiquitin ligase activity. *Mol. Cell* **3,** 535–541.
14. Tan, P., Fuches, S. Y., Angus, A., et al. (1999) Recruitment of a ROC1-CUL1 ubiquitin ligase by Skp1 and HOS to catalyze the ubiquitination of IκBα. *Mol. Cell* **3,** 527–533.
15. Seol, J. H., Feldman, R. M. R., Zachariae, W., et al. (1999) Cdc53/cullin and the essential Hrt1 RING-H2 subunit of SCF define a ubiquitin ligase module that activates the E2 enzyme Cdc34. *Genes Dev.* **13,** 1614–1626.
16. Joazeiro, C. A. P., Wing, S. S., Huang, H.-K., Leverson, J. D., Hunter, T., and Liu, Y.-C. (1999) The tyrosine kinase negative regulator c-Cbl as a RING-type E2-dependent ubiquitin-protein ligase. *Science* **286,** 309–312.
17. Leverson, J. D., Joazeriro, C. A. P., Page, A. M., Huang, H.-K., Hieter, P., and Hunter, T. (2000) The APC11 RING-H2 mediates E2-dependent ubiquitination. *Mol. Biol. Cell* **11,** 2315–2325.
18. Gmachl, M., Gieffers, C., Podtelejnikov, A. V., Mann, M., and Peters, J.-M. (2000) The RING-H2 finger protein APC11 and the E2 enzyme UBC4 are sufficient to ubiquitinate substrates of the anaphase-promoting complex. *Proc. Natl. Acad. Sci. USA* **97,** 8973–8978.

19. Furukawa, M., Ohta, T., and Xiong, Y. (2002) Activation of UBC5 Ubiquitin-conjugating Enzyme by the RING finger of ROC1 and assembly of active ubiquitin ligases by all cullins. *J. Biol. Chem.* **277,** 15758–15765.
20. Yaron, A., Hatzubai, A., Davis, M., et al. (1998) Identification of the receptor component of the IkBα-ubiquitin ligase. *Nature* **396,** 590–594.
21. Spencer, E., Jiang, J., and Chen, Z. J. (1999) Signal-induced ubiquitination of IkBα by the F-box protein Slimb/β-TrCP. *Genes Dev.* **13,** 284–294.
22. Yoshida, Y., Chiba, T., Tokunaga, F., et al. (2002) E3 ubiquitin ligase that recognizes sugar chains. *Nature* **418,** 438–442.
23. Moberg, K. H., Bell, D. W., Wahrer, D. C., Haber, D. A., and Hariharan, I. K. (2001) Archipelago regulates cyclin E levels in *Drosophila* and is mutated in human cancer cell lines. *Nature* **413,** 311–316.
24. Strohmaier, H., Spruck, C. H., Kaiser, P., Won, K. A., Sangfelt, O., and Reed, S. I. (2001) Human F-box protein hCdc4 targets cyclin E for proteolysis and is mutated in a breast cancer cell line. *Nature* **413,** 316–322.
25. Shenoy, S. K., McDonald, P. H., Kohout, T. A., and Lefkowitz, R. J. (2001) Regulation of receptor fate by ubiquitination of activated beta 2-adrenergic receptor and beta-arrestin. *Science* **294,** 1307–1313.
26. Yamanaka, K., Ishikawa, H., Megumi, Y., et al. (2003) Identification of the ubiquitin-protein ligase that recognizes oxidized IRP2. *Nat. Cell Biol.* **5,** 336–340.
27. Honda, R., Tanaka, H., and Yasuda, H. (1997) Oncoprotein MDM2 is a ubiquitin ligase E3 for tumor suppressor p53. *FEBS Lett.* **420,** 25–27.
28. Zhang, Y., Wolf, G. W., Bhat, K., et al. (2003) Ribosomal protein L11 negatively regulates oncoprotein MDM2 and mediates a p53-dependent ribosomal-stress checkpoint pathway. *Mol. Cell Biol.* **23,** 8902–8912.
29. Treier, M., Staszewski, L. M., and Bohmann, D. (1994) Ubiquitin-dependent c-Jun degradation in vivo is mediated by the delta domain. *Cell* **78,** 787–798.

# 4

# Ubiquitin Chain Synthesis

**Shahri Raasi and Cecile M. Pickart**

**Summary**

Several important signaling processes depend on the tagging of cellular proteins with "polyubiquitin chains"—ubiquitin polymers whose building blocks are connected by isopeptide bonds between G76 of one ubiquitin and a specific lysine residue of the next one. Here we describe procedures for the synthesis of polyubiquitin chains of defined lengths that are linked through the K48 or K63 side chains. The method involves a series of enzymatic reactions in which proximally and distally blocked monoubiquitins (or chains) are conjugated to produce a particular chain in high yield. Individual chains are then deblocked and joined in another round of reaction. Successive rounds of deblocking and synthesis can give rise to a chain of any desired length.

**Key Words:** Conjugating enzyme; conjugation; deubiquitinating enzymes; isopeptide; polyubiquitin (chain); ubiquitin.

## 1. Introduction

Polyubiquitin chains linked through ubiquitin-K48 target substrates to 26S proteasomes for degradation, while chains linked through K63 can confer several different nonproteolytic fates on their substrates *(1)*. Other types of chains also exist and in some cases, might represent functionally distinct signals *(2)*. These discoveries, in conjunction with the identification of protein domains that selectively bind polyubiquitin chains *(3–5)*, have created a demand for specific ubiquitin polymers that can be used in biochemical assays or structural studies *(4,6,7)*.

Here we describe methods for the synthesis of K48- and K63-linked chains of defined lengths. The procedure **(Fig. 1)** involves a series of enzymatic reactions catalyzed by ubiquitin conjugation factors that utilize only one of ubiquitin's seven lysine residues as a conjugation site *(8)*. In each round of synthesis, proximally and distally blocked monoubiquitins (or chains) are joined to produce a doubly blocked chain in high yield. The proximal block consists of an extra C-terminal residue (D77) that is labile to ubiquitin C-terminal hydrolase enzymes (UCHs). The distal block consists of a cysteine residue (placed at the normal conjugation site) that can be converted to a

From: *Methods in Molecular Biology, vol. 301, Ubiquitin–Proteasome Protocols*
Edited by: C. Patterson and D. M. Cyr © Humana Press Inc., Totowa, NJ

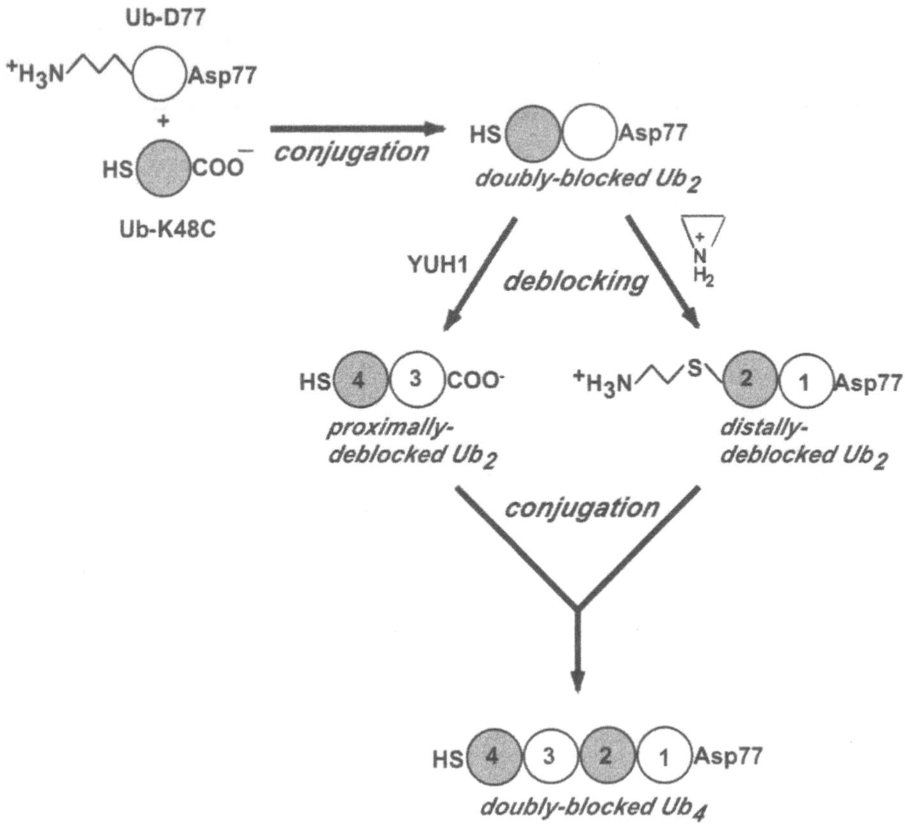


Fig. 1. Synthesis of K48-linked $Ub_4$. This scheme outlines the steps in the synthesis of K48-$Ub_4$ (*see* text). The circles denote Ub molecules; the shading lets the reader keep track of the different ubiquitins in the chain. In the doubly blocked $Ub_4$ molecule, Ub-1 defines the proximal chain terminus and Ub-4 defines the distal terminus.

lysine mimic through alkylation. Successive rounds of deblocking and conjugation can give rise to a chain of any desired length. The method is described in detail for K48-linked chains. The differences that apply during synthesis of K63-linked chains are then outlined briefly.

## 2. Materials

1. Purified recombinant ubiquitin (Ub) proteins: Ub-D77 and Ub-K48C (for K48-linked chains) or Ub-D77 and Ub-K63R for K63-linked chains *(9)* (*see* **Note 1**).
2. Purified recombinant conjugating enzymes: E2-25K for K48-linked chains (9) or Mms2/Ubc13 complex for K63-linked chains ***(10,11)***. E2-25K is available from Affiniti Research Products.
3. Purified ubiquitin-activating enzyme (E1) ***(4,9)***. E2-25K cooperates efficiently only with mammalian E1s but any source of E1 is acceptable for synthesis of K63-linked chains.

Baculoviruses specifying mammalian E1s have been constructed by several laboratories and may be available on request. Alternatively, rabbit E1 is available from Boston Biochem or Affiniti (*see* **Note 2**).
4. Purified yeast ubiquitin C-terminal hydrolase 1 (YUH1) ***(12)***. Commercial UCH-L3 can be substituted.
5. PBDM buffers: 250 m*M* Tris-HCl (50% base, pH 8.0), 25 m*M* $MgCl_2$, 50 m*M* creatine phosphate (Sigma P7396), 3 U/mL of inorganic pyrophosphatase (Sigma I1891), and 3 U/mL of creatine phosphokinase (Sigma C3755). PBDM7.6 is the same except that the buffer is Tris-HCl, pH 7.6 (24% base). Store PBDM buffers at –20°C.
6. ATP, neutralized 0.1 *M* solution (prepared from Sigma A2383; *see* **Note 3**).
7. 2 *N* Acetic acid.
8. Q- and S-Sepharose Fast Flow resins (Amersham-Pharmacia Biotech).
9. Q buffer: 50 m*M* Tris-HCl, pH 7.6, 1 m*M* EDTA, 5 m*M* dithiothreitol (DTT) (freshly prepared DTT).
10. Buffer A: 50 m*M* ammonium acetate pH 4.5, 1 m*M* EDTA, 5 m*M* DTT (freshly prepared DTT).
11. Ethyleneimine: purchase from Chemservice (West Chester, PA), most conveniently in 50-mg aliquots in sealed ampules. Store at 5°C and discard unused portion after ampule is opened (*see* **Note 4**). Neat ethyleneimine is 19.2 *M*.
12. Storage buffer: 20 m*M* Tris-HCl, pH 7.6, 0.5 m*M* EDTA, 2–3 m*M* DTT (freshly prepared DTT).

## 3. Methods

Here we outline (1) the synthesis and partial purification of K48-linked di-ubiquitin (K48-$Ub_2$), (2) the full purification of this molecule, (3) methods for proximal and distal deblocking of K48-$Ub_2$, (4) the synthesis and purification of K48-$Ub_4$, (5) the synthesis and purification of K48-linked chains of other lengths, (6) the synthesis of K63-linked chains, and (7) what to expect in terms of yield and recovery.

### *3.1. Synthesis and Partial Purification of K48-$Ub_2$*

We refer to the end of the chain that would normally carry unconjugated G76 as the proximal end, while the end that would normally carry unconjugated K48 is the distal end. E2-25K is the conjugating enzyme that synthesizes K48-G76 isopeptide bonds exclusively ***(13)***.

#### *3.1.1. Synthesis of $K_{48}$-$Ub_2$*

1. Ub-D77 and Ub-K48C are combined at 7.5 mg/mL each in an incubation containing fivefold diluted PBDM8, 2.5 m*M* ATP, 0.5 m*M* DTT, and 20 μ*M* E2-25K (*see* **Note 5**). Remove 1 μL for sodium dodecyl sulfate-polyacrylamide gel electrophoresis (SDS-PAGE) analysis (*see* **Note 6**).
2. The conjugation reaction, initiated by adding 0.1 μ*M* mammalian E1 (*see* **Note 2**), reaches completion in 4 h at 37°C (**Fig. 2A,** lane 2). To avoid having significant amounts of unreacted $Ub_1$, use precisely equal concentrations of Ub-K48C and Ub-D77, based on measurement of absorption immediately before the reaction ($A_{280}$ = 0.16 for 1 mg/mL of ubiquitin).
3. At the end of the incubation, add DTT (5 m*M*, freshly prepared) and EDTA (1 m*M*) and incubate at room temperature for 20 min. This incubation will reduce any disulfide-linked chains that could precipitate later. Remove 1 μL for SDS-PAGE analysis.

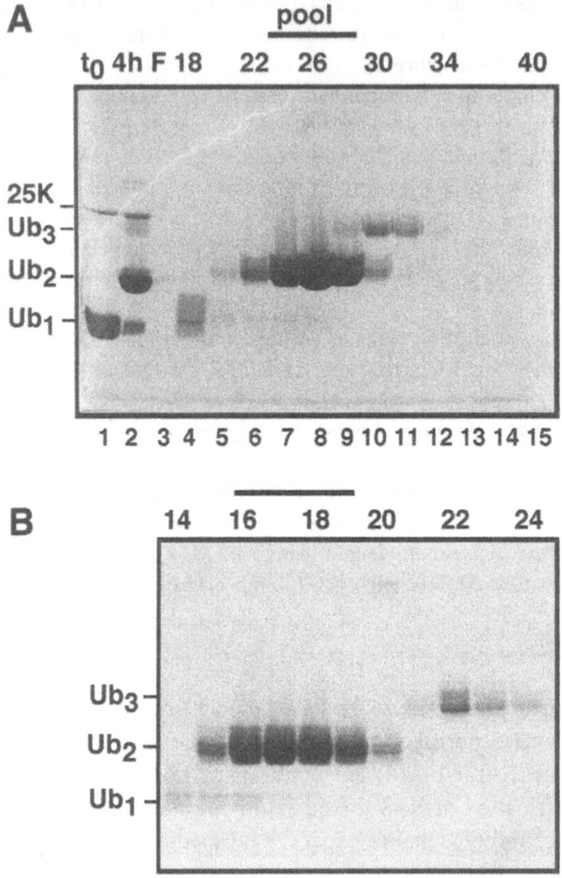


Fig. 2. (**A**) S-Sepharose separation of $Ub_1$ and K48-$Ub_2$. In this particular conjugation reaction, Ub-D77 (20 mg) was reacted with Ub-K48C (20 mg) in a volume of 2.5 mL. *Lanes 1* and *2,* 0.4 μL of the synthetic incubation at time 0 *(lane 1)* and 4 h *(lane 2).* This incubation was acidified and applied directly to 2 mL of S-Sepharose (*see* **Note 8**). *Lane 3,* Unbound fraction. *Lanes 4–15,* Even-numbered fractions from 18 through 40 (2 μL of each). Fractions 23–29 were pooled, concentrated, and buffer-exchanged to yield 36 mg of doubly blocked $Ub_2$. (**B**) FPLC separation of $Ub_1$ and K48-$Ub_2$. The unbound fraction from the Q-Sepharose column was acidified and applied to a MonoS column (*see* **Note 8**). Aliquots of fractions 14–24 were analyzed by SDS-PAGE. Fractions 16–19 were pooled.

### *3.1.2. Partial Purification of K48-$Ub_2$: Removal of E1 and E2-25K Enzymes*

Polyubiquitin carries no net charge at neutral pH, but E1 and E2-25K are anionic. Thus, the enzymes can be removed using an anion-exchange column. Apply the reaction mixture to a 0.5-mL Q-Sepharose column equilibrated with Q buffer. Collect the unbound fraction; wash the column with 4X 0.5 mL of Q buffer, collecting these washes into the unbound fraction (*see* **Note 7**). Remove a volume-normalized aliquot for SDS-PAGE. If $Ub_2$ is the terminal product, you may dispense with further purification if fractional conversion to $Ub_2$ is ≥95% (e.g., **Fig. 2A,** lane 2). If the $Ub_2$ will be used to make longer chains, however, full purification is necessary. Otherwise, the residual $Ub_1$ will produce $Ub_3$ as a side product in the next reaction.

### *3.2. Full Purification of K48-Ub$_2$*

At pH 4.5, the magnitude of the positive charge carried by a given polyubiquitin chain is directly proportional to its length, allowing different chains to be separated by gradient cation-exchange chromatography. However, this method works well only if the chains differ significantly in their lengths. For example, $Ub_2$ can be separated from $Ub_1$ with baseline resolution by fast protein liquid chromatography (FPLC) ***(14)***, but there is little resolution of long chains that differ in length by one ubiquitin unit. This consideration, which motivated us to develop the method described here, is particularly significant when choosing the building blocks for long chains (*see* **Subheading 3.6.**).

1. Add 0.1 volume of 2 *N* acetic acid to the pooled unbound fractions from the Q column; check that the pH is approx 4 by spotting 1 µL onto pH paper.
2. Apply the acidified mixture to an S-Sepharose column preequilibrated with buffer A, using 1 mL of beads per 20 mg of total ubiquitin (*see* **Note 8**).
3. Wash the loaded column with approx 3 vol buffer A (save to verify that $Ub_2$ is absent).
4. Elute the column with a linear gradient of NaCl (0–0.6 *M*) in buffer A, using 20–40 column volumes and collecting 50–60 fractions. $Ub_2$ elutes at approx 0.33 *M* NaCl.

The peak fractions of $Ub_2$ can be located by spotting 0.5-µL aliquots onto filter paper and staining with Coomassie blue. However, the fractions should be examined by SDS-PAGE in order to reject those that contain significant $Ub_1$ **(Fig. 2)**. Often, a small amount $Ub_3$ is formed during the synthetic reaction as a result of carboxypeptidase-mediated removal of D77 from Ub-D77 (a very low level of carboxypeptidase activity contaminates some preparations of E2-25K). The S-Sepharose column separates $Ub_2$ from both $Ub_1$ and $Ub_3$ **(Fig. 2A)**. Better resolution of $Ub_2$ can be obtained using FPLC cation-exchange chromatography (**Fig. 2B** vs **2A**). The peak fractions of $Ub_2$ are pooled and concentrated. This step can also serve to reduce the salt concentration and exchange the sample into storage buffer (*see* **Note 9**). To reduce losses due to nonspecific absorption, chains should be concentrated to 30–80 mg/mL before storing at –80°C.

### *3.3. Deblocking Reactions*

$Ub_2$ resulting from the procedures described in the preceding has D77 at its proximal terminus and a C48 residue at its distal terminus, and thus is doubly blocked **(Fig. 1)**. To synthesize $Ub_4$, half of the material is deblocked at the proximal chain terminus (to expose G76), while the rest of the material is deblocked at the distal chain terminus (by alkylating C48). The two singly blocked dimers are then conjugated to produce $Ub_4$.

#### *3.3.1. Proximal Terminus Deblocking (D77 Removal)*

1. To doubly blocked $Ub_2$ (30–80 mg/mL), add: 50 m*M* Tris-HCl, pH 7.6, 1 m*M* EDTA, and 1 m*M* fresh DTT.

2. Initiate deblocking by adding purified YUH1 to a final concentration of 16 μg/mL. Quantitative removal of D77 occurs in 60 min at 37°C.
3. Add 4 m*M* more DTT and incubate for 10 min at room temperature (to reduce any disulfide bonds).
4. Absorb YUH1 onto Q-Sepharose as described in **Subheading 3.1.2.** Concentrate the unbound fraction and determine the concentration of the proximally deblocked dimer.

#### 3.3.2. Distal Terminus Deblocking (Alkylation of C48)

1. To the remaining doubly blocked dimer (30–80 mg/mL) add: 0.2 *M* Tris-HCl, pH 8.0, and 1 m*M* EDTA.
2. Initiate alkylation by adding ethyleneimine to 55 m*M* (*see* **Note 4**). The reaction proceeds to completion in 60 min at 37°C.
3. Ethyleneimine must be removed to prevent subsequent inactivation of E1 and E2-25K. One can dialyze the incubation against 1 L of 10 m*M* Tris-HCl, pH 8.0 (overnight at 5°C). Or one can repeatedly concentrate and dilute with 10 m*M* Tris-HCl, pH 8.0, 2 m*M* DTT in a centrifugal concentrator until [DTT] = [ethyleneimine]. Concentrate the distally deblocked dimer to 30–80 mg/mL before freezing.

### 3.4. Synthesis and Purification of K48-$Ub_4$

#### 3.4.1. Synthesis of K48-$Ub_4$

Conditions are the same as in the synthesis of K48-$Ub_2$ (**Subheading 3.1.1.**), except: (1) the reactants are the proximally and distally deblocked $Ub_2$ molecules (**Subheading 3.3.**); (2) each chain is added at 10 mg/mL; and (3) the incubation can be shortened to 2 h. E1 and E2-25K are once again removed using Q-Sepharose (**Subheading 3.1.2.**).

#### 3.4.2. Purification of K48-$Ub_4$

The procedure is identical to that used to purify $Ub_2$ (**Subheading 3.3.**) except that $Ub_4$ binds more tightly to S-Sepharose, so a gradient of 0 to 0.7 *M* NaCl is used. Longer chains have a tendency to precipitate, which can be minimized by careful handling (*see* **Note 9**).

### 3.5. Synthesis of K48-Linked Chains of Other Lengths

The principles are as described in **Subheadings 3.1.–3.3.** To maximize separation of reactants from product during cation-exchange chromatography, make long chains by joining two chains of similar lengths. For example, $Ub_{12}$ should be made by linking $Ub_6$ to $Ub_6$ (or $Ub_4$ to $Ub_8$) rather than by linking $Ub_2$ to $Ub_{10}$ (*see* **Subheading 3.2.**). The longer the chain, the higher the salt concentration that is required to elute it, so gradients should be adjusted accordingly. When reactant amounts are low, as is often be the case for long chains, FPLC cation exchange is preferred over an open S-Sepharose column because a smaller gradient volume can be used, thus maximizing recovery. We have successfully made chains up to $n = 12$ (**Fig. 3** and data not shown). Once the chain has reached its final length, the distal C48 residue can be alkylated with ethyleneimine or iodoacetamide, if desired, to reduce the potential for precipitation.

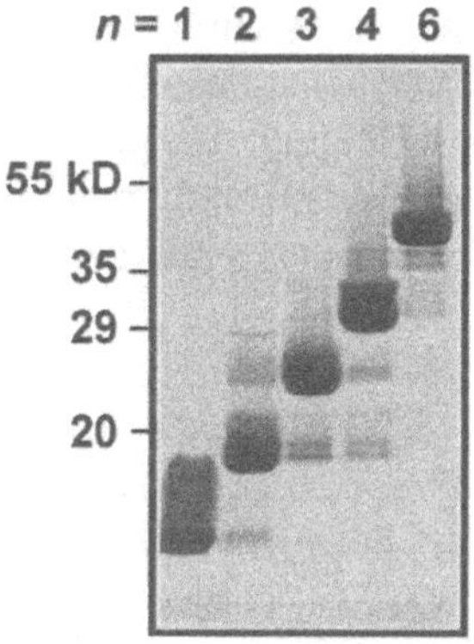


Fig. 3. Defined-length polyubiquitin chains. K48-linked chains consisting of the indicated number of ubiquitin units were synthesized by methods described here (except $n = 1$ is Ub-D77). About 5 µg of material was run in each lane; purity is ≥90% in all cases.

### 3.6. Synthesis of K63-Linked Chains

The principles and procedures are similar to those outlined for K48-linked chains, with three important differences. First, the synthetic reaction contains 8 µ*M* each of yeast Mms2 and Ubc13 (in place of E2-25K *[10]*) and the buffer is fivefold diluted PBDM7.6 (not PBDM8). Second, yeast or mammalian E1 can be used. Finally, the Mms2/Ubc13 complex inefficiently recognizes ethyleneimine-modified C63, which makes it impractical to deblock the distal terminus. Therefore we build K63-linked chains one ubiquitin at a time. K63-$Ub_2$ can be synthesized from Ub-K63R and Ub-D77 at 10 mg/mL each (use of Ub-K63R eliminates precipitation problems), purified (*see* **Subheadings 3.1.2.** and **3.2.**), and deblocked with YUH1 (*see* **Subheading 3.3.1.**). One can then conjugate the proximally deblocked $Ub_2$ to Ub-D77, yielding K63-$Ub_3$. After purification and deblocking, K63-$Ub_3$ is conjugated to Ub-D77 to yield K63-$Ub_4$ *(11)*.

### 3.7. Comments on Yield and Recovery

One important factor in successful chain synthesis is precise normalization of the molar concentrations of the chain reactants. Attention to this factor will maximize fractional conversion, simplify purification, and optimize the yield. For the full purification scheme discussed in the preceding subheading, we routinely recover 65–75% of the input ubiquitins in the specific chain product in 30- to 50-mg scale reactions. For reasons that remain unclear, the Q-Sepharose column is sometimes problematic for recovery. Therefore, researchers who are unconcerned about the presence of trace amounts of (inactive) E1 or E2-25K may dispense with this step, loading the acidified (DTT/EDTA-treated) primary reaction directly onto S-Sepharose (*see* **Note 10**). We do not recommend this alternative for chains of $n > 4$, however, because E2-25K binds to the cation-exchange column and could coelute with the chain. Dispensing with the Q column can increase recovery to 90% during cation-exchange chromatography of large-scale reactions. Recovery during YUH1-dependent deblocking is 80–90%, whereas recovery during ethyleneimine alkylation is 75–90%; both recoveries are better in large-scale reactions. Ubiquitin and polyubiquitin chains are rather sticky; nonspecific absorption to surfaces significantly reduces recovery. Reusing columns, avoiding glass tubes, and maximizing protein concentrations will counteract this problem.

## 4. Notes

1. K48-linked chains can also be synthesized from appropriately mutated ubiquitins tagged at their N-termini with polyhistidine, Flag, or hemagglutinin epitopes. However, because tagged ubiquitins are refractory to conjugation by the Mms2/Ubc13 complex *(11)*, they cannot be used to make K63-linked chains.
2. If the E1 is not highly purified, contaminating deubiquitinating enzymes may catalyze deblocking or chain disassembly during chain synthesis, leading to the formation of undesired products.
3. Adjusting the ATP solution to pH 7 stabilizes it to extended storage (store at –20°C).
4. Ethyleneimine is toxic and should be handled with care. Vials should be opened only in a fume hood and manipulations involving the concentrated stock should be performed there as well. Unused ethyleneimine can be diluted into 10–50 vol of alkaline DTT and allowed to sit for 24 h before disposal.
5. A DTT concentration above 1–2 m*M* can inhibit the conjugation by interfering with E2–ubiquitin thiol ester stability. The recommended reaction vessel is a tightly sealed plastic vial or tube, which should be submerged in a circulating water bath to avoid evaporation and condensation. Repeated use of the same vial (rinsed between uses) will reduce loss of chains caused by nonspecific absorption.
6. Do not boil samples prior to SDS-PAGE, as this may result in nonspecific crosslinking of chains. Also, chains tend to smear on heavily loaded gels.
7. To reduce absorptive losses of chains, the Q- and S-Sepharose columns can be reused. Strip the columns with 1 *M* NaCl and store at 5°C in 1 *M* NaCl containing 0.02% $NaN_3$.
8. Using a larger column than is necessary will reduce recovery because of nonspecific absorption. For a 40-mg-scale reaction, we use a 2-mL S-Sepharose column and elute with an 80-mL gradient, collecting 1.5-mL fractions into uncovered microfuge tubes. Purification can be done at room temperature. One can also do the entire procedure in an automated manner using an FPLC. For a 20-mg scale reaction, we use a 1-mL MonoS column (Amersham-Pharmacia Biotech) and elute with a 40-mL gradient, collecting 1-mL fractions.
9. Polyubiquitin chains show a tendency to precipitate when left at pH 4.5. Therefore we always pool, concentrate, and exchange into storage buffer on the day that the column is run. (We exchange buffer by repeated concentration and dilution.) If precipitation should occur, the chains can be collected by centrifugation and dissolved in a buffer of 10 m*M* Tris-HCl, pH 7.6, 1 m*M* EDTA, 5 m*M* DTT, and 8 *M* urea. Urea is removed through dialysis.
10. In this case, use 0.2 volume of 2 *N* acetic acid to acidify.

## Acknowledgment

We thank R. Cohen for providing the YUH1 expression clone and for a critical reading of the manuscript.

## References


1. Pickart, C. M. (2000) Ubiquitin in chains. *Trends Biochem. Sci.* **25,** 544–548.
2. Peng, J., Schwartz, D., Elias, J. E., et al. (2003) A proteomics approach to understanding protein ubiquitination. *Nat. Biotechnol.* **21,** 921–926.
3. Wilkinson, C. R. M., Seeger, M., Hartmann-Petersen, R., et al. (2001) Proteins containing the UBA domain are able to bind multi-ubiquitin chains. *Nat. Cell Biol.* **3,** 939–943.
4. Raasi, S. and Pickart, C. M. (2003) Rad23 ubiquitin-associated domains (UBA) inhibit 26S proteasome-catalyzed proteolysis by sequestering lysine 48-linked polyubiquitin chains. *J. Biol. Chem.* **278,** 8951–8959.
5. Buchberger, A. (2002) Fro*M* UBA to UBX: new words in the ubiquitin vocabulary. *Trends Cell Biol.* **12,** 216–221.
6. Varadan, R., Walker, O., Pickart, C. M., and Fushman, D. (2002) Structural properties of polyubiquitin chains in solution. *J. Mol. Biol.* **324,** 637–647.
7. Varadan, R., Assfalg, M., Haririnia, A., Raasi, S., Pickart, C. M., and Fushman, D. (2004) Solution conformation of Lys$^{63}$-linked di-ubiquitin chain provides clues to functional diversity of polyubiquitin signaling. *J. Biol. Chem.* **279,** 10.1074/jbc.M309184200.
8. Piotrowski, J., Beal, R., Hoffman, L., Wilkinson, K. D., Cohen, R. E., and Pickart, C. M. (1997) Inhibition of the 26 S proteasome by polyubiquitin chains synthesized to have defined lengths. *J. Biol. Chem.* **272,** 23712–23721.
9. Haldeman, M. T., Xia, G., Kasperek, E. M., and Pickart, C. M. (1997) Structure and function of ubiquitin conjugating enzyme E2-25K: the tail is a core-dependent activity element. *Biochemistry* **36,** 10526–10537.
10. Hofmann, R. M. and Pickart, C. M. (1999) Noncanonical *MMS2*-encoded ubiquitin-conjugating enzyme functions in assembly of novel polyubiquitin chains for DNA repair. *Cell* **96,** 645–653.
11. Hofmann, R. M. and Pickart, C. M. (2001) In vitro assembly and recognition of K63 polyubiquitin chains. *J. Biol. Chem.* **276,** 27936–27943.
12. Johnston, S. C., Riddle, S. M., Cohen, R. E., and Hill, C. P. (1999) Structural basis for the specificity of ubiquitin C-terminal hydrolases. *EMBO J.* **18,** 3877–3887.
13. Chen, Z., Niles, E. G., and Pickart, C. M. (1991) Isolation of a cDNA encoding a mammalian multi-ubiquitinating enzyme (E2-25K), and overexpression of the functional enzyme in *E. coli. J. Biol. Chem.* **266,** 15698–15704.
14. Pickart, C. M., Kasperek, E. M., Beal, R., and Kim, A. (1994) Substrate properties of site-specific mutant ubiquitin protein (G76A) reveal unexpected features in the mechanism of ubiquitin activating enzyme (E1). *J. Biol. Chem.* **269,** 7115–7123.

# 5

# Purification of Proteasomes, Proteasome Subcomplexes, and Proteasome-Associated Proteins From Budding Yeast

David S. Leggett, Michael H. Glickman, and Daniel Finley



## Summary

The proteasome is a highly complex, ATP-dependent protease, consisting of over 30 subunits, and dedicated mainly to the degradation of ubiquitin–protein conjugates. Proteasomes are evolutionarily conserved in the eukaryotic kingdom, and those of yeast are well suited to serve as a general model. We describe techniques for the purification of proteasomes from budding yeast in milligram amounts via conventional and affinity-based strategies. While both approaches yield highly purified material, the affinity method is faster and easier. In addition, the affinity method is more suitable for identifying proteasome-associated proteins. We also describe methods for purifying the major subassemblies of the proteasome, such as the CP, the RP, the lid, and the base. A variety of activity assays and native gel procedures are available to evaluate purified proteasomes functionally. When coupled with the genetic methods available in yeast, these biochemical procedures allow for detailed functional analysis of this unique protein complex.


**Key Words:** Affinity chromatography; base; core particle; lid; native gel electrophoresis; proteasome; proteasome-associated proteins; proteasome purification; proteasome subcomplexes; proteolysis.

## 1. Introduction

The proteasome is a 2.5-megadalton protease, present in all eukaryotes, which degrades proteins conjugated to ubiquitin. The proteasome can be subdivided into two major subcomplexes (*1,2*, and **Fig. 1**) known as the core particle (CP; also called the 20S proteasome) and the 19S regulatory particle (RP; also called PA700 in mammals, or the μ particle in *Drosophila melanogaster*). The CP is composed of four heptameric rings of subunits arranged in a barrel structure *(3)*. The outer rings are composed of seven α-subunits, and the inner rings are composed of seven β-subunits. The peptidase active sites of the CP (encoded by subunits β1, β2, and β5) are sequestered within its lumen *(3)*. Substrate entry into the lumen of the CP is controlled by gated channels, located at the ends of the CP, composed of the N-terminal tails of the α-subunits ***(4–6)***.

From: *Methods in Molecular Biology, vol. 301, Ubiquitin–Proteasome Protocols*
Edited by: C. Patterson and D. M. Cyr © Humana Press Inc., Totowa, NJ

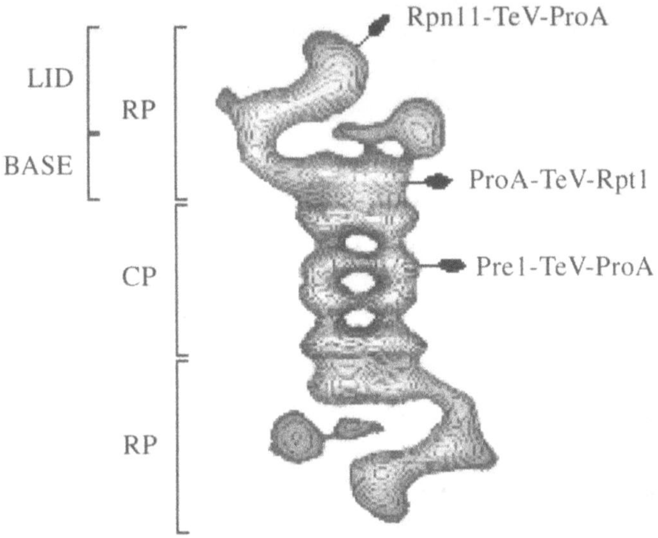


Fig. 1. Proteasome subcomplexes. Schematic of the proteasome indicating the CP, RP, lid, and base subcomplexes. (Modified form from **ref. *28***, with permission from Elsevier.) Also shown are the Protein A-tagged subunits used in the affinity purification protocol.

The RP is thought to be involved in ubiquitin chain binding ***(7,8)***, as well as substrate unfolding and translocation into the CP ***(9)***. The RP contains 19 subunits, which can be dissociated into two subcomplexes called the lid and base ***(10,10a,11)***. The lid contains nine subunits and has been shown to be required for the degradation of ubiquitinated proteins ***(10,11)***. An important component of the lid is Rpn11, a deubiquitinating enzyme ***(11–13)***. The base contains subunits Rpn1, Rpn2, Rpn13 and six proteasomal ATPases. The protein unfolding activity of the proteasome has been shown to reside in the base ***(14–16)***, and the ATPases Rpt2 and Rpt5 have been implicated in opening the CP gate and ubiquitin chain binding, respectively ***(6,7)***.

The yeast proteasome, like those from higher eukaryotes, is traditionally viewed as a stable complex capable of surviving harsh purification conditions. Conventionally purified yeast proteasomes survive exposure to high salt during ion-exchange chromatography to yield 33 apparently stoichiometric subunits (**Table 1**). However, it is likely that affinity-purified proteasomes more closely resemble the proteasome as it is found in vivo. In particular, recent advances in purification of yeast proteasomes using a gentle affinity-based approach have identified three additional major proteasome components: Ecm29, Ubp6, and Hul5 ***(11,17)***. Ubp6 and Hul5 are known components of the ubiquitin system; Ubp6 is a ubiquitin-hydrolase, while Hul5 is a HECT-domain E3 ***(18,19)***. The deubiquitinating activity of Ubp6 is activated upon proteasome binding, suggesting that it may be involved in releasing ubiquitin from proteasome-bound substrates. The role of Hul5 at the proteasome remains unclear. Ecm29 was previously identified during a screen for mutants in cell wall biogenesis, but had no known bio-

**Table 1**
**Proteasome Subunits**

| RP Subunits | | | | CP Subunits | | |
|---|---|---|---|---|---|---|
| Subunit | Previous nomenclature | Mol wt | Subcomplex | Subunit | Previous nomenclature | Mol wt |
| Rpn1 | Hrd2/Nas1 | 109.4 | Base | α1 | Prs2 | 34.6 |
| Rpn2 | Sen3 | 104.3 | Base | α2 | Pre8 | 27 |
| Rpn3 | Sun2 | 60.4 | Lid | α3 | Pre9 | 28.6 |
| Rpn5 | Nas5 | 51.8 | Lid | α4 | Pre6 | 28.3 |
| Rpn6 | Nas4 | 49.8 | Lid | α5 | Pup2 | 28.6 |
| Rpn7 | | 49.0 | Lid | α6 | Pre5 | 25.6 |
| Rpn8 | | 38.3 | Lid | α7 | Pre10 | 31.4 |
| Rpn9 | | 45.9 | Lid | β1 | Pre3 | 16.2 |
| Rpn10 | Mcb1/Sun1 | 29.7 | – | β2 | Pup1 | 25 |
| Rpn11 | Mpr1 | 34.4 | Lid | β3 | Pup3 | 22.6 |
| Rpn12 | Nin1 | 31.9 | Lid | β4 | Pre1 | 22.5 |
| Rpn13 | | 17.9 | Base | | | |
| | Sem1, Dss1 | 10.4 | Lid | | | |
| Rpt1 | Cim5/Yta3 | 52.0 | Base | β5 | Pre2 | 23.3 |
| Rpt2 | Yta5 | 48.8 | Base | β6 | Pre7 | 24.8 |
| Rpt3 | Yta2/Ynt1 | 48.0 | Base | β7 | Pre4 | 25.9 |
| Rpt4 | Crl13/Sug2/Pcs1 | 49.4 | Base | | | |
| Rpt5 | Yta1 | 48.2 | Base | | | |
| Rpt6 | Sug1/Cim3/Crl3 | 45.2 | Base | | | |

| Salt-labile components | | |
|---|---|---|
| Subunit | Mol wt | Location |
| Ecm29 | 210.4 | CP + RP |
| Ubp6 | 57.1 | RP |
| Hul5 | 105.5 | ? |

Conventionally purified proteasomes contain the RP and CP subunits. Affinity purified proteasomes contain the RP and CP and salt-labile components.

chemical function *(20)*. Ecm29 now appears to be involved in maintaining proteasome stability *(17)*.

In this chapter we describe the purification of proteasomes from *Saccharomyces cerevisiae*. Although both conventional and affinity-based proteasome purifications are presented here, there are several reasons for using the affinity-based purification method: (1) The affinity-based purification is much quicker than the conventional method, requiring only 4 h instead of several days. (2) Conventional purification protocols do allow the purification of proteasome and the CP from yeast; however, unlike the affinity-based purification, satisfactory protocols do not exist for conventional purification of the RP, the lid, or base. (3) The affinity method allows the purification of proteasomes containing the additional components Ubp6, Ecm29, and Hul5. Other proteins have been shown to associate with yeast proteasomes ***(8,11,21–24);*** however, the presence or absence of these proteins has not been verified in proteasomes isolated using either the conventional or affinity-based protocols described in this chapter. The purification methods described in this chapter were originally described in abbreviated form ***(2,10,17)***.

## 2. Materials

Unless otherwise stated, all chemicals are purchased from Sigma.

### 2.1. Buffers

#### 2.1.1. Conventional Proteasome Purification

1. Lysis buffer: 10% glycerol (v/v), 25 m*M* Tris-HCl, pH 7.4, 10 m*M* $MgCl_2$, 4 m*M* ATP, 1 m*M* dithiothreitol (DTT).
2. Buffer A: 10% glycerol (v/v), 25 m*M* Tris-HCl, pH 7.4, 10 m*M* $MgCl_2$, 1 m*M* ATP, 1 m*M* DTT.
3. Buffer AN: 10% glycerol (v/v), 25 m*M* Tris-HCl, pH 7.4, 10 m*M* $MgCl_2$, 1 m*M* ATP, 1 m*M* DTT, 100 m*M* NaCl.
4. Buffer B: 25 m*M* Tris-HCl, pH 7.4, 10 m*M* $MgCl_2$, 1 m*M* DTT.
5. Buffer BN: 25 m*M* Tris-HCl, pH 7.4, 10 m*M* $MgCl_2$, 1 m*M* DTT, 100 m*M* NaCl.
6. 1 *M* Tris base.
7. YPD media: 1% yeast extract, 2% Bacto peptone, 2% glucose.

#### 2.1.2. Affinity Proteasome Purification

1. Yeast lysis buffer (buffer 1): 50 m*M* Tris-HCl, pH 8.0, 1 m*M* EDTA, 1 m*M* ATP.
2. Wash buffer (buffer 2): 50 m*M* Tris-HCl, pH 7.5, 1 m*M* EDTA, 1 m*M* ATP, 100 m*M* NaCl.
3. CP/RP disruption buffer (buffer 3): 50 m*M* Tris-HCl, pH 7.5, 1 m*M* EDTA, 1 m*M* ATP, 500 m*M* NaCl.
4. Lid/base disruption buffer (buffer 4): 50 m*M* Tris-HCl, pH 7.5, 1 m*M* EDTA, 1 m*M* ATP, 1 *M* NaCl.
5. TEV protease buffer (buffer 5): 50 m*M* Tris-HCl, pH 7.5, 1 m*M* EDTA, 1 m*M* DTT, 1 m*M* ATP.

### 2.2. Chromatography

#### 2.2.1. Conventional Proteasome Purification

1. DEAE-Affigel Blue resin (Bio-Rad).
2. Resource Q resin (Pharmacia).
3. Superose 6 (Pharmacia).
4. 30-kDa mol wt centrifugal filter (Millipore no. UFV2BTK40).
5. French press.

#### 2.2.2. Affinity Proteasome Purification

1. Antigen Affinity Gel Rabbit IgG resin (MP Biomedical cat. no. 55961).
2. 6His-TEV protease (Invitrogen cat. no. 10127017).
3. French press.
4. 0.5% Glacial acetic acid (v/v).

### *2.3. Native Gel Electrophoresis*

1. 5X Native gel buffer: 450 m*M* Tris-borate, pH 8.35, 25 m*M* $MgCl_2$, 2.5 m*M* EDTA.
2. 5X Loading buffer: 250 m*M* Tris-HCl, pH 7.4, 50% glycerol (v/v), 0.007% (w/v) xylene cyanol.
3. 40% (w/v) Acrylamide–*bis*-acrylamide solution (37.5:1) (Bio-Rad).
4. *N,N,N,',N,'*-Tetramethylethylenediamine (TEMED).
5. 10% Ammonium persulfate (w/v).
6. Adenosine-5,'-triphosphate.
7. DTT.

### *2.4. Sodium Dodecyl Sulfate-Polyacrylamide Gel Electrophoresis (SDS-PAGE)*

1. 40% (w/v) Acrylamide–*bis*-acrylamide solution (37.5:1) (Bio-Rad).
2. 5X Separating buffer: 1.875 *M* Tris-HCl, pH 8.8, 0.5% SDS (w/v).
3. 5X Stacking buffer: 0.625*M* Tris-HCl, pH 6.8, 0.5% SDS (w/v).
4. TEMED.
5. 10% Ammonium persulfate (w/v).
6. Running buffer: 25 m*M* Tris-HCl, 250 m*M* glycine, 0.1% SDS (w/v), pH 8.3.
7. 5X Loading buffer: 50% glycerol (v/v), 250 m*M* Tris-HCl, pH 6.8, 500 m*M* DTT, 10% SDS (w/v), 0.5% bromophenol blue (w/v).
8. Coomassie staining solution: 0.25 mg of Coomassie Brilliant Blue R250 in 100 mL of destain solution.
9. Destain solution: 50% methanol (v/v), 10% glacial acetic acid (v/v) in $H_2O$.

### *2.5. Peptidase Activity Assay*

1. Peptide substrate: 10 m*M* *N*-succinyl-Leu-Leu-Val-Tyr-7-amino-4-methylcoumarin (Suc-LLVY-AMC) (Boston Biochem).
2. 10% SDS (w/v).
3. Fluorimeter.
4. UV transilluminator with 365-nm wavelength.

## 3. Methods

### *3.1. Conventional Proteasome Purification*

The conventional purification protocol described here yields two proteasome species when analyzed by nondenaturing PAGE (*see* **Subheading 3.3.2.** and **Fig. 2**). These two species correspond to proteasome containing either 2 RP (doubly capped proteasome), or 1 RP (singly capped proteasome). The ratio of these two forms of the proteasome varies between purifications, but doubly capped proteasomes usually predominate. Using this protocol, we typically obtain yields of approx 2 mg of proteasome from 10 L of post-logarithmic-phase culture ($OD_{600nm}$ ~15). All buffers used to purify proteasome holoenzyme must contain 10% glycerol and ATP to maintain proteasome stability. The absence of glycerol and ATP in the purification buffers causes the proteasome to dissociate into the RP and CP. Buffers used to purify CP omit glycerol and ATP because this protocol relies on the dissociation of the proteasome.

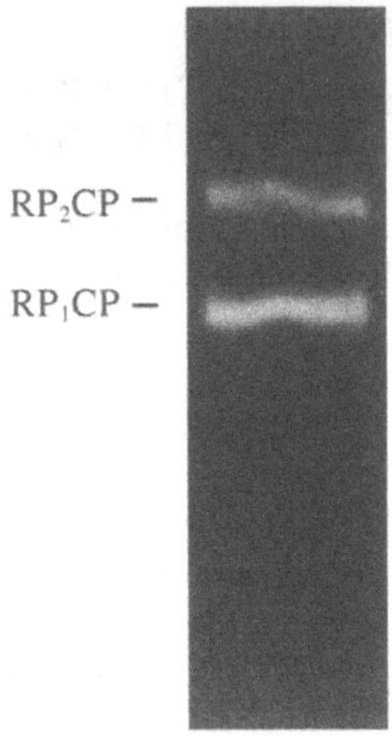


Fig. 2. Native gel electrophoresis of proteasomes. Two micrograms of proteasome were electrophoresed on a 4% nondenaturing polyacrylamide gel and visualized by the peptidase activity assay using Suc-LLVY-AMC. (Modified from **ref. *29***, with permission from Elsevier.) The doubly capped ($RP_2CP$) and singly capped ($RP_1CP$) forms of the proteasome are indicated.

### *3.1.1. Purification of Proteasome Holoenzyme*

1. Grow 10 L of *S. cerevisiae* in YPD media at 30°C (we grow our cultures for 2 d to an $OD_{600nm}$ of ~15 [*see* **Note 1**]). Harvest the cells by centrifugation at 5000*g* for 10 min, wash once with cold lysis buffer, and then resuspend with a twofold volume of cold lysis buffer. From this point on, all steps are performed at 4°C.
2. Lyse the cells by one pass through a French press (*see* **Note 2**) and adjust the pH to 7.4 with 1 *M* Tris base as necessary. Clarify the lysate by centrifugation at 20,000*g* for 30 min.
3. Filter the lysate through cheesecloth to remove any lipids that float on the surface of the lysate after centrifugation.
4. Apply the supernatant to a 100-mL DEAE-affigel blue column equilibrated with buffer A, and wash the resin with one column volume of buffer A, followed by two column volumes of buffer A supplemented with 50 m*M* NaCl. The column is eluted with buffer A supplemented with 150 m*M* NaCl. Fractions containing proteasomes can be identified using the peptidase activity assay (*see* **Subheading 3.3.1.**).
5. Apply the proteasome-containing fractions to a 50-mL Resource Q column equilibrated with buffer AN. Wash the Resource Q column with one column volume of buffer AN, and elute with a 500-mL gradient of 100–500 m*M* NaCl in buffer A, collecting 6-mL fractions. Proteasomes elute at a salt concentration of approx 300–330 m*M* NaCl.
6. It is possible to pause the purification at this point. The proteasomes are relatively pure, and are stable overnight at 4°C.
7. Monitor the elution of proteasome using the peptidase activity assay (*see* **Subheading 3.3.1.**). Pool proteasome-containing fractions, then desalt and concentrate in buffer AN using a 30-kDa mol wt centrifugal filter (Millipore).

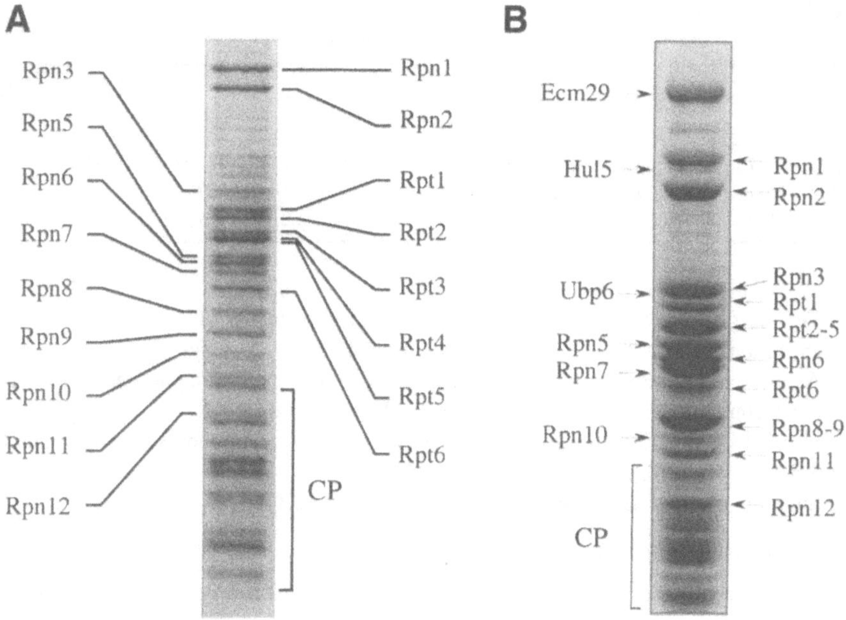


Fig. 3. SDS-PAGE analysis of proteasomes. **(A)** Proteins from conventionally purified proteasomes were separated on a 10–20% SDS-PAGE gradient gel. (Modified form from **ref. *29***, with permission from Elsevier.) **(B)** Proteins from affinity-purified proteasomes were separated on a 10% SDS-PAGE. Both gels were visualized with Coomassie. (Modified form from **ref. *28***, with permission from Elsevier.)

8. Apply the partially purified proteasome onto a Superose 6 column equilibrated with buffer AN. The proteasome is resolved isocratically in the same buffer.
9. Assay the fractions eluted from the Superose 6 column for proteasome using the peptidase activity assay (*see* **Subheading 3.3.1.**) either in the absence or presence of 0.02% SDS to identify proteasome and CP, respectively (*see* **Note 3**). A wide peak of peptidase activity will elute immediately after the void volume, corresponding to a mixture of doubly and singly capped proteasomes followed by CP.
10. Purified proteasome should be assessed for purity and distribution of proteasome and CP by native gel electrophoresis (*see* **Subheading 3.3.2.**) and SDS-PAGE (*see* **Subheading 3.3.3.** and **Fig. 3A**).
11. Store the purified proteasome holoenzyme at –80°C in buffer AN (*see* **Note 4**).

### *3.1.2. Purification of CP*

The purification of CP is performed in essentially the same manner as for the proteasome holoenzyme, except that it is done using buffer B and buffer BN in place of buffer A and buffer AN respectively. The lack of ATP and glycerol in buffer B and buffer BN promotes dissociation of the proteasome into the CP and RP, allowing for the purification of the CP. Using this protocol, we routinely get yields of approx 4 mg of CP from 10 L of stationary-phase culture ($OD_{600nm}$ ~15).

**Table 2**
**Yeast Strains**

| Strain | Genotype | Tag location |
|---|---|---|
| SDL66 | *MAT***a** *lys2-801 leu2-3, 2-112 ura3-52 his3-* Δ *200 trp1-1 rpn11::RPN11-TEVProA (HIS3)* | Lid |
| SDL135 | *MAT***a** *lys2-801 leu2-3, 2-112 ura3-52 his3-* Δ *200 trp1-1 pre1::PRE1-TEVProA (HIS3)* | CP |
| SY36 | *MAT***a** *lys2-801 leu2-3, 2-112 ura3-52 his3-* Δ *200 trp1-1 rpt1::HIS3 pEL36 (TRP1)* | Base |

1. The purification of CP proceeds as described in **Subheading 3.1.1.** until **step 9** with the exception that buffers B and BN are used instead of buffers A and AN, respectively.
2. Assay the fractions eluted from the Superose 6 column for CP using the peptidase activity assay containing 0.02% SDS (*see* **Note 3** and **Subheading 3.3.1.**).
3. Purified CP should be assessed for purity by native gel electrophoresis (*see* **Subheading 3.3.2.**) and SDS-PAGE (*see* **Subheading 3.3.3.**).
4. Store the purified CP at –80°C in buffer BN (*see* **Note 4**).

#### 3.1.3. Purification of RP

At the current time, there is no satisfactory purification protocol for *S. cerevisiae* RP by conventional means. Although it is possible to devise a conventional purification strategy for the RP, the ease of its purification by the affinity-based protocol described in the following subheading makes this unnecessary.

### 3.2. Affinity Proteasome Purification

Affinity purification of the proteasome is based on the tagging of proteasome subunits with a TEV-protease-cleavable Protein A tag **(Fig. 1)**. Protein A binds tightly to IgG, allowing the purification of proteasome on IgG resin. The proteasome can then be eluted from the resin by TEV protease cleavage. We have generated tagged versions of proteasome subunits Rpn11, Rpt1, and Pre1 **(Table 2)** which are located in the lid, base, and CP respectively, and allow the purification of these subcomplexes in addition to proteasome. These strains are available from our laboratory on request. This purification generates proteasome containing the salt-labile components Ubp6, Ecm29, and Hul5, which are absent (or reduced in amount) from proteasomes purified conventionally. It is difficult to isolate proteasomes lacking these components by the affinity method, as the salt concentration required to strip them off is close to the concentration that causes the CP and RP to begin to dissociate. The best way to isolate proteasome lacking one or more of these components is to do the purification in strains with the gene(s) of interest deleted, or use the conventional purification method. The affinity-purification method also allows the preparation of resins loaded with proteasome and its subcomplexes, which are useful for a variety of applications. For example, we have used proteasome and proteasome subcomplex resins to analyze how Ubp6 interacts with the proteasome, as well as to demonstrate competition for proteasome binding between the proteasome-associated proteins Rad23 and Dsk2 *(8,17)*. Unlike in the conventional purification, the stability of proteasomes during affinity purification is not dependent on the presence of 10% glycerol or ATP in the buffers. This is mainly the result of the presence of Ecm29 in the affinity-purified proteasome, which stabilizes the CP/RP interaction *(17)*. In fact, glycerol adversely affects the interaction between the IgG resin and the Protein A affinity tag, significantly reducing proteasome yields. While not required for stability, we include ATP in the affinity purification buffers as it seems to produce more active proteasome. Unless otherwise stated, all purification is performed at 4°C until the TEV cleavage step, which involves a 30°C incubation.

### *3.2.1. Affinity Purification of Proteasome Holoenzyme*

Although it is possible to purify intact proteasome holoenzyme using any of the Protein A-tagged strains described in **Table 1,** we routinely use the Rpn11 strain (sDL66) for this procedure. This technique routinely yields approx 3 mg of proteasome from 10 L of stationary phase culture ($OD_{600nm}$ ~15). Affinity-purified proteasomes are highly enriched for doubly capped proteasome.

1. Grow yeast strains in YPD media to stationary phase (we grow our cultures for 2 d to an $OD_{600nm}$ of ~15 [**Note 1**]). The cells are harvested by centrifugation at 3000*g* for 10 min, washed once with $dH_2O$, and then resuspended in a twofold volume of cold buffer 1 (*see* **Note 5**).
2. Lyse the cells by one pass through a French press, and clarify the lysate by centrifugation at 20,000*g* for 30 min. Filter the clarified lysate through cheesecloth to remove any lipids that float on the surface of the lysate after centrifugation.
3. Incubate the clarified lysate for 1 h at 4°C with IgG resin equilibrated in buffer 1. The optimal ratio is 2 mL of IgG resin for each 2 L of yeast culture grown to an $OD_{600nm}$ of approx 15.
4. After incubation, transfer the resin to a chromatography column and wash with 50 column volumes of buffer 2, followed by 5 column volumes of TEV protease buffer.
5. Incubate the resin for 1 h at 30°C with 1.5 column volumes of TEV protease buffer containing 150 U of 6His-TEV protease for each 2 mL of IgG resin, and elute the holoenzyme with TEV protease buffer. The TEV protease can be removed by incubating the eluate at 4°C for 10 min with NiNTA-resin.
6. Assess the purity of purified proteasome by native gel electrophoresis (*see* **Subheading 3.3.2.**) and SDS-PAGE (*see* **Subheading 3.3.3.** and **Fig. 3B**).
7. Store the purified proteasome at –80°C in TEV protease buffer adjusted to 10% glycerol (v/v) (*see* **Note 4**).
8. Regenerate the IgG resin by washing with 5 column volumes of 0.5% glacial acetic acid (v/v) followed by 5 column volumes of buffer 1 (*see* **Note 6**).

### *3.2.2. Purification of CP and RP*

Purifications of the RP and CP are performed using Rpn11- and Pre1-tagged proteasomes, respectively. The following protocol takes advantage of the fact that incubation of the proteasome with 500 m*M* NaCl causes the CP and RP to dissociate. When the Rpn11 tag is used, the RP is retained on the resin, and the non-RP subunits are eluted in the 500 m*M* NaCl wash step; when the Pre1 tag is used, the CP is retained, and the non-CP subunits are eluted in the 500 m*M* NaCl wash step. Using this protocol, we typically obtain yields of 1 mg of purified CP, or 2 mg of purified RP from 10 L of stationary phase culture ($OD_{600nm}$ ~15).

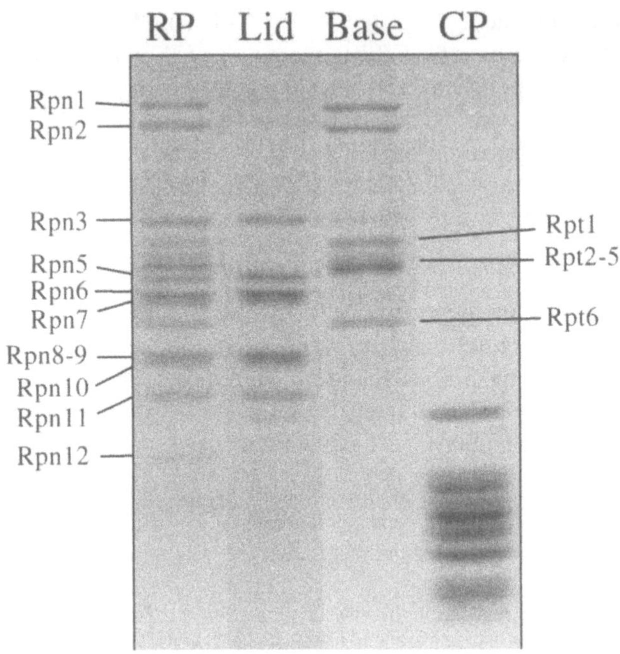


Fig. 4. Proteasome subcomplexes. Proteins from affinity-purified proteasome subcomplexes were separated by 12.5% SDS-PAGE and visualized by Coomassie. (Modified form from **ref.** ***28***, with permission from Elsevier.)

The purification proceeds as described in **Subheading 3.2.1.** until **step 4.** Then:

1. After incubation, transfer the resin to a chromatography column and wash with 10 column volumes of buffer 2. Then wash the resin with 5 column volumes of buffer 3 and incubate for 1 h at 4°C. Wash the resin with 50 column volumes of buffer 3, followed by 5 column volumes of TEV protease buffer.
2. Elute RP or CP from the resin by TEV cleavage as described in **Subheading 3.2.1.**
3. The TEV protease can be removed by incubating the eluate at 4°C for 10 min with NiNTA-resin.
4. Assess the CP for purity by native gel electrophoresis (*see* **Subheading 3.3.2.**) and SDS-PAGE (*see* **Subheading 3.3.3.** and **Fig. 4**). The purity of RP is assessed by SDS-PAGE (*see* **Subheading 3.3.3.** and **Fig. 4**).
5. Store the purified CP or RP at –80°C in TEV protease buffer adjusted to 10% glycerol (v/v) (*see* **Note 4**).
6. Regenerate the IgG resin by washing with 5 column volumes 0.5% glacial acetic acid (v/v) followed by 5 column volumes of buffer 1 (*see* **Note 6**).

### *3.2.3. Purification of Lid and Base*

Purifications of the lid and base are performed using Rpn11- and Rpt1-tagged proteasomes respectively. The following protocol takes advantage of the fact that incubation of the proteasome with 1 *M* NaCl causes the RP to dissociate into the lid and base *(25)*. When the Rpn11 tag is used, the lid is retained on the resin, and all nonlid subunits are eluted in the 1 *M* NaCl wash step; when the Rpt1 tag is used, the base is retained, and all nonbase subunits are eluted in the 1 *M* NaCl wash step. Using this protocol, we typically obtain yields of approx 0.8 mg of purified base or lid from 10 L of logarithmic-phase culture ($OD_{600nm}$ ~15).

The purification proceeds as described in **Subheading 3.2.1.** until **step 4**.

1. After incubation, transfer the resin to a chromatography column and wash with 10 column volumes of buffer 2. Then wash the resin with 5 column volumes of buffer 4 and incubate for 1 h at 23°C. Wash the resin with 50 column volumes of buffer 4, followed by 5 column volumes of TEV protease buffer.
2. Elute lid or base from the resin by TEV cleavage as described in **Subheading 3.2.1.**
3. The TEV protease can be removed by incubating the eluate at 4°C for 10 min with NiNTA-resin.
4. Assess the purity of lid and base should be analyzed by SDS-PAGE (*see* **Subheading 3.3.3.** and **Fig. 4**).
5. The purified lid or base is stored at –80°C in TEV protease buffer adjusted to 10% glycerol (v/v) (*see* **Note 4**).
6. Regenerate the IgG resin by washing with 5 column volumes of 0.5% glacial acetic acid (v/v) followed by 5 column volumes of buffer 1 (*see* **Note 6**).

## 3.3. Analysis of Proteasome

### 3.3.1. Peptidase Activity Assay

Proteasome activity is monitored by the cleavage of the fluorogenic peptide substrate Suc-LLVY-AMC *(26)*. The chymotryptic-like peptidase activity of the proteasome cleaves this substrate, releasing free AMC, which can be monitored by fluorescence. Although we prefer using this peptide substrate, other fluorogenic peptides, which are cleaved by the tryptic and caspase-like peptidase activities of the proteasome, are also available.

1. Combine 10 µL of sample with 40 µL of buffer A containing 100 µ*M* Suc-LLVY-AMC. To assay CP activity, include 0.02% SDS in the reaction (*see* **Note 3**).
2. Incubate reaction for 30 min at 30°C.
3. Stop the reaction by the addition of 1 mL of 1% SDS.
4. Measure the fluorescence at an excitation of 380 nm and an emission of 460 nm.

### 3.3.2. Native Gel Electrophoresis

Native gel electrophoresis of proteasome is performed using 4% nondenaturing polyacrylamide gels. Proteasome holoenzymes and CP can be visualized using an in-gel activity assay. After electrophoresis, gels are immediately immersed in the peptidase activity assay mix described in **Subheading 3.3.1.** This is especially useful in monitoring the amount of doubly capped and singly capped proteasome, as well as CP present in a sample **(Fig. 2)**.

1. Prepare a 4% PAGE gel using native gel buffer containing 1 m*M* ATP and 1 m*M* DTT. We run these gels in a minigel apparatus using 1.5-mm spacers. A 20-mL solution is enough for two gels. 2 mL 40% acrylamide (37.5:1), 4 mL 5X native gel buffer, 40 µL 0.5*M* ATP, 20 µL 1*M* DTT, 200 µL 10% ammonium persulfate, 20 µL TEMED; $H_2O$ to 20 mL.

2. Add 5X native gel loading buffer to each sample, and load the samples on the gel. One microgram of proteasome gives a robust signal.
3. Run gel at 100 V at 4°C until the xylene cyanol dye has migrated to the bottom of the gel (~2 h), or for as long as 3.5 h for greater resolution of the singly and doubly capped proteasomes.
4. Remove the gel from the plates (*see* **Note 7**), and incubate it in buffer A containing 100 µ*M* Suc-LLVY-AMC. To assay CP activity, include 0.02% SDS in the buffer.
5. Incubate the gel for 15 min at 30°C.
6. Visualize the gel on a UV transillumintor with a wavelength of 365nm (*see* **Note 8**).
7. If the signal is weak, you can continue to incubate the gel in buffer A containing 100 µ*M* Suc-LLVY-AMC for a longer period of time.

### 3.3.3. SDS-PAGE

We analyze our proteasome on either 10% or 12.5% SDS-PAGE gels **(Fig. 3)**; however, gradient gels maybe used to achieve better resolution of individual proteasome subunits.

1. Prepare a 10% or 12.5% acrylamide separating gel. Overlay the separating gel with water-saturated butanol to ensure a flat interface. We run these gels in a minigel apparatus using 1.5-mm spacers. A 20-mL solution is enough for two gels.

| | **10% Separating gel** | **12.5% Separating gel** | **Stacking gel** |
|---|---|---|---|
| 40% Acrylamide (37.5:1) | 5 mL | 6.2 mL | 1.25 mL |
| Separating buffer | 4 mL | 4 mL | – |
| Stacking buffer | – | – | 2 mL |
| 10% Ammonium persulfate | 200 µL | 200 µL | 100 µL |
| TEMED | 14 µL | 14 µL | 7 µL |
| $H_2O$ | To 20 mL | To 20 mL | To 10 mL |

2. After the separating gel has set, rinse out the butanol, and pour the stacking gel. Insert the well comb, and allow the stacking gel to polymerize.
3. Add 5X loading buffer to your samples. We find that 10 µg of proteasome gives a good signal when the gel is stained with Coomassie blue dye.
4. Boil the samples and protein markers for 2 min and load onto the gel.
5. Run the gel at 120 V until the bromphenol blue dye has reached the bottom of the gel.
6. Remove the gel from the electrophoresis plates and incubate it in Coomassie staining solution for 30 min at room temperature.
7. Transfer the gel to destain solution until the gel has completely destained, and the protein bands are visible. We find that destaining at 4°C, while slower, results in gels with crisp protein bands and a clear background.

## 4. Notes

1. Although it is possible to purify proteasome from logarithmically growing yeast, we obtain higher yields from post-logarithmic cultures *(27)*.
2. Several factors are crucial for maintaining intact proteasome during lysis. Increases in temperature or dramatic lowering of the pH can cause the holoenzyme to dissociate into the CP and RP. Performing the lysis in the cold with a prechilled French press cell can inhibit temperature increases. After lysis, the pH should be immediately checked and adjusted using 1 *M* Tris base if necessary.

3. Purified CP has very low activity against peptide substrates because the proteasome channels, which allow substrate entry into the CP, are closed *(5)*. The addition of 0.02% SDS to the peptidase activity assay allows the activation of the CP, presumably by selectively denaturing the CP channel, thereby allowing peptide substrate entry. The proteasome holoenzyme does not require the addition of 0.02% SDS to the assay because the association of the RP with the CP opens the channels.
4. 10% Glycerol is necessary for maintaining proteasome activity during freezing. There is no need to snap-freeze the proteasome aliquots. Conventionally purified proteasomes and CP are eluted from the Superose 6 column in buffers containing 10% glycerol and can be frozen directly. Affinity-purified proteasomes, CP, lid, and base are purified in buffers lacking glycerol, and must be adjusted to 10% glycerol prior to freezing. Frozen samples are largely stable to several freeze–thaw cycles; however, repeated freezing and thawing should be avoided.
5. The lysis buffer is designed to maintain the pH during lysis as long as it is used in a 2:1 ratio with the cell pellet.
6. The IgG resin can be reused at least seven times without any decrease in the yield of purified proteasome.
7. 4% Gels are extremely fragile. We find that the best way to remove the gels is to use a water bottle to gently wash the gel from the electrophoresis plates.
8. In order to reduce background from scratched filters, we reserve a transilluminator solely for the visualization of native gels.

## Acknowledgment

The authors thank John Hanna for critical reading of the manuscript.

## References

1. Hoffman, L., Pratt, G., and Rechsteiner, M. (1992) Multiple forms of the 20 S multicatalytic and the 26 S ubiquitin/ATP-dependent proteases from rabbit reticulocyte lysate. *J. Biol. Chem.* **267,** 22362–22368.
2. Glickman, M. H., Rubin, D. M., Fried, V. A., and Finley, D. (1998) The regulatory particle of the Saccharomyces cerevisiae proteasome. *Mol. Cell. Biol.* **18,** 3149–3162.
3. Groll, M., Ditzel, L., Löwe, J., et al. (1997) Structure of 20S proteasome from yeast at 2.4 Å resolution. *Nature* **386,** 463–471.
4. Whitby, F. G., Masters, E. I., Kramer, L., et al. (2000) Structural basis for the activation of 20S proteasomes by 11S regulators. *Nature* **408,** 115–120.
5. Groll, M., Bajorek, M., Köhler, A., et al. (2000) A gated channel into the proteasome core particle. *Nat. Struct. Biol.* **7,** 1062–1067.
6. Köhler, A., Cascio, P., Leggett, D. S., Woo, K.-M., Goldberg, A. L., and Finley, D. (2001) The axial channel of the proteasome core particle is gated by the Rpt2 ATPase and controls both substrate entry and product release. *Mol. Cell* **7,** 1143–1152.
7. Lam, Y. A., Lawson, T. G., Velayutham, M., Zweier, J. L., and Pickart, C. M. (2002) A proteasomal ATPase subunit recognizes the polyubiquitin degradation signal. *Nature* **416,** 763–767.
8. Elsasser, S., Gali, R., Schwickart, M., et al. (2002) Proteasome subunit Rpn1 binds ubiquitin-like protein domains. *Nat. Cell Biol.* **4,** 725–730.
9. Finley, D. (2002) Ubiquitin chained and crosslinked. *Nat. Cell Biol.* **4,** E121–E123.
10. Glickman, M. H., Rubin, D. M., Coux, O., et al. (1998) A subcomplex of the proteasome regulatory particle required for ubiquitin-conjugate degradation and related to the COP9-signalosome and eIF3. *Cell* **94,** 615–623.

10a. Sone, T., Saeki, Y., Toh-e, A., and Yokosawa, H. (2004). Sem1p is a novel subunit of the 26 S proteasome from Saccharomyces cerevisiae. *J. Biol. Chem.* **279**, 28,807–28,816.
11. Verma, R., Chen, S., Feldman, R., et al. (2000) Proteasomal proteomics: identification of nucleotide-sensitive proteasome-interacting proteins by mass spectrometric analysis of affinity-purified proteasomes. *Mol. Biol. Cell* **11,** 3425–3439.
12. Yao, T. and Cohen, R. E. (2002) A cryptic protease couples deubiquitination and degradation by the proteasome. *Nature* **419,** 403–407.
13. Verma, R., Aravind, L., Oania, R., et al. (2002) Role of Rpn11 metalloprotease in deubiquitination and degradation by the 26S proteasome. *Science* **298,** 611–615.
14. Braun, B. C., Kloetzel, P.-M., Kraft, R., et al. (1999) The base of the proteasome regulatory complex exhibits ATP-dependent chaperone-like activity. *Nat. Cell Biol.* **1**, 221–226.
15. Strickland, E., Hakala, K., Thomas, P. J., and DeMartino, G. N. (2000) Recognition of misfolding proteins by PA700, the regulatory subcomplex of the 26 S proteasome. *J. Biol. Chem.* **275,** 5565–5572.
16. Liu, C. W., Millen, L., Roman, T. B., et al. (2002) Conformational remodeling of proteasomal substrates by PA700, the 19 S regulatory complex of the 26 S proteasome. *J. Biol Chem.* **277,** 26815–26820.
17. Leggett, D. S., Hanna, J., Borodovsky, A., et al. (2002) Multiple associated proteins regulate proteasome structure and function. *Mol. Cell* **10,** 495–507.
18. Park, K. C., Woo, S. K., Yoo, Y. J., Wyndham, A. M., Baker, R. T., and Chung, C. H. (1997) Purification and characterization of Ubp6, a new ubiquitin-specific protease in *Saccharomyces cerevisiae. Arch. Biochem. Biophys.* **347,** 78–84.
19. Wang, G., Yang, J., and Huibregtse, J. M. (1999) Functional domains of the Rsp5 ubiquitin-protein ligase. *Mol. Cell. Biol.* **19,** 342–352.
20. Lussier, M., White, A. M., Sheraton, J., et al. (1997) Large-scale identification of genes involved in cell surface biosynthesis and architecture in *Saccharomyces cerevisiae. Genetics* **147,** 435–450.
21. Jager, S., Strayle, J., Heinemeyer, W., and Wolf, D. H. (2001) Cic1, an adaptor protein specifically linking the 26S proteasome to its substrate, the SCF component Cdc4. *EMBO J.* **20,** 4423–4431.
22. Tongaonkar, P., Chen, L., Lambertson, D., Ko, B., and Madura, K. (2000) Evidence for an interaction between ubiquitin-conjugating enzymes and the 26S proteasome. *Mol Cell Biol.* **20,** 4691–4698.
23. Xie, Y. and Varshavsky, A. (2000) Physical association of ubiquitin ligases and the 26S proteasome. *Proc. Natl. Acad. Sci. USA* **97,** 2497–2502.
24. Schauber, C., Chen, L., Tongaonkar, P., et al. (1998) Rad23 links DNA repair to the ubiquitin/proteasome pathway. *Nature* **391,** 715–718.
25. Saeki, Y., Toh-e, A., and Yokosawa, H. (2000) Rapid isolation and characterization of the yeast proteasome regulatory complex. *Biochem. Biophys. Res. Commun.* **273,** 509–515.
26. Hough, R., Pratt, G., and Rechsteiner, M. (1987) Purification of two high molecular weight proteases from rabbit reticulocyte lysate. *J. Biol. Chem.* **262,** 8303–8313.
27. Fujimuro, M., Takada, H., Saeki, Y., Toh-e, A., Tanaka, K., and Yokosawa, H. (1998) Growth-dependent change of the 26S proteasome in budding yeast. *Biochem. Biophys. Res. Commun.* **251,** 818–823.
28. Leggett, D. S., Hanna, J., Borodovsky, A., et al. (2002) Multiple associated proteins regulate proteasome structure and function. *Mol. Cell* **10,** 495–507.
29. Glickman, M. H., Rubin, D. M., Coux, O., et al. (1998) A subcomplex of the proteasome regulatory particle required for ubiquitin-conjugate degradation and related to the COP9-signalosome and eIF3. *Cell* **4,** 615–623.

# 6

# Recognition and Processing of Misfolded Proteins by PA700, the 19S Regulatory Complex of the 26S Proteasome

Chang-Wei Liu, Elizabeth Strickland,
George N. DeMartino, and Philip J. Thomas

**Summary**

The 26S proteasome is composed of the core 20S proteasome in association with the 19S regulatory complex, or PA700. PA700 has multiple activities, including ATPase activity, polyubiquitin-chain binding activity, deubiquitination activity, chaperone-like activity, and substrate remodeling activity. The concerted action of these activities leads to efficient degradation of protein substrates by the 26S proteasome. In this chapter we describe protocols for purifying PA700 and the 20S complexes from bovine red cells and present methods to assay the chaperone-like activity and the substrate remodeling activity of PA700.

**Key Words:** Proteasome; PA700; chaperone activity; remodeling/unfolding.

## 1. Introduction

Proteasome degradation is central to many cellular pathways *(1)*. It is responsible not only for removal of damaged proteins in quality control, but also for precise regulation of the cell cycle, transcription, and antigen presentation. Aberrant action of this protein degradation machinery plays a role in several severe diseases, such as cancer, neurodegenerative diseases, and arthritis *(2)*.

Several forms of the proteasome exist in mammalian cells. Although the core 20S proteasome does exist in a free state in mammalian cells *(1)*, it is frequently found bound by activation complexes. The common 26S proteasome is composed of the 20S proteasome core particle together with the regulatory particle PA700 (also called the 19S cap) *(3)*. Complexes also exist containing the 20S proteasome and PA28 (also called 11S), which is responsible for antigen production when associated with the 20S

From: *Methods in Molecular Biology, vol. 301, Ubiquitin–Proteasome Protocols*
Edited by: C. Patterson and D. M. Cyr © Humana Press Inc., Totowa, NJ

proteasome *(4)*. Furthermore, "hybrid proteasomes" sandwich the 20S proteasome between a PA700 complex on one end and a PA28 particle on the other *(5,6)*. These "hybrid proteasomes" are thought to be important in antigen production. Together these various complexes containing the 20S proteasome and its associated regulatory particles properly identify, prepare for degradation, and proteolyze multiple substrates.

In addition to varying the regulatory particles associated with it, the 20S proteasome can also vary its own subunit composition. Immunoproteasomes have three distinct catalytic subunits (LMP1, MECL-1, and LMP7) compared to the constitutive proteasome (β1, β2, and β5) *(1)*. These subunits are selectively induced by immune stimulation and assembled into newly synthesized proteasomes.

## 1.1. The Structure and Activities of the 20S Proteasome

### 1.1.1. The Structure of the 20S Proteasome

The 20S proteasome is composed of 28 subunits, which are arranged as an α7β7β7α7 stack of four rings to form a cylindrical complex. In *archaea*, the α-ring and the β-ring are each composed of a single gene product *(3)*. In eukaryotes, both the α-ring and the β-ring are composed of different subunits *(4)*. The X-ray crystal structure of the 20S proteasome from *Thermoplasma acidophilum* **(5)** showed that the two α-rings are at the periphery of the cylinder and form central entry ports approx 13 Å in diameter. The annulus at each end of the 20S cylinder controls access through a narrow substrate channel into the proteolytic chamber formed by the two β-rings. In yeast *(6)* and cattle *(7)*, the annulus is completely sealed by interactions between the N-termini of four of the seven α-subunits *(7)*. This sealed annulus can be gated/opened by association with regulatory complexes, such as PA28 *(8)* and PA700 *(9)*. In addition, α-subunits have been reported to bind substrate directly *(10)*, which could open the gate and promote substrate degradation *(11)*.

### 1.1.2. The Peptidases of the 20S Proteasome

The proteasome belongs to the superfamily of the N-terminal nucleophile (Ntn)-hydrolases *(12)*. The catalytic β-subunits of the 20S proteasome have three distinct peptidase activities. The β1-subunit has a "peptidylglutamyl-peptide hydrolyzing" (PGPH) activity, which favors acidic residues at the P1 position. β2 has a "trypsin-like" activity, which favors basic residues at the P1 position. Finally, β5 has a "chymotrypsin-like" activity, which favors hydrophobic residues at the P1 position. The crystal structure of the bovine 20S proteasome *(7)* suggests that β7-subunits may mediate a small neutral amino acid preferring (SNAAP) activity, because the structure around the active β7 N-terminus Thr1 satisfies the requirement for the Ntn-hydrolase active sites. In the immunoproteasome, the β1, β2, and β5 subunits are different than in the constitutive proteasome, namely the LMP2, MECL-1, and LMP7 subunits, respectively. The immunoproteasome has higher measured "chymotrypsin-like" activity, but lower PGPH activity, than the constitutive proteasome *(13,14)*. This alters the standard degradation pattern and favors the production of peptides for presentation by the class I major histocompatibility complexes *(14)*.

The gated annulus and narrow substrate translocation channel prevent folded proteins from entering the central proteolytic chamber. By contrast, some unfolded proteins can open the gate and gain access to the catalytic sites. Unfolded proteins can translocate through the channel either processively from one terminus ***(15)*** or by insertion of a flexible loop ***(11)***. The latter activity is likely responsible for partial degradation by the proteasome to release some transcription factors from inactive precursors ***(16,17)***.

### 1.2. The Functions of PA700

PA700 is an 18-subunit, multifunctional complex ***(18)***. It can be disassociated into two subcomplexes in vitro ***(19)***, called "lid" and "base." The base subcomplex contains all six ATPase subunits (AAA subunits) and the Rpn1 and Rpn2 subunits. The lid subcomplex has all the other subunits. In addition to recognizing and processing substrates for proteasomal degradation, PA700 has also been shown to play a nonproteolytic role in RNA polymerase II transcription ***(20)***. Here, we focus on the functions of PA700 required for efficient proteolysis.

#### *1.2.1. Polyubiquitin Chain Binding*

Many proteasomal substrates are marked for death with ubiquitin. A cascade of enzymatic reactions requiring an ubiquitin-activating enzyme (E1), a ubiquitin-conjugating enzyme (E2), and an ubiquitin ligase (E3) is responsible for attaching a polyubiquitin chain to substrates ***(2)***. PA700 has high affinity toward these polyubiquitin chains (containing at least tetraubiquitin) ***(21)***. The S5a subunit of PA700 specifically binds polyubiquitin chains in vitro ***(22)***. However, in vivo studies in *Saccharomyces cerevisiae* ***(23,24)*** suggest that S5a is not required for the degradation of polyubiquitinated proteins, indicating that additional subunit(s) may be required for ubiquitin chain recognition. A chemical crosslinking study identified the ATPase subunit s6' ***(25)*** as specifically binding polyubiquitin chains in an ATP-dependent manner. This subunit also interacts with ornithine decarboxylase (ODC) ***(26)*** and regulates its antizyme-dependent, ubiquitination-independent degradation ***(27)***. An additional PA700 binding protein, Rad23, has also been shown to interact with polyubiquitin chains, thus assisting delivery of polyubiquitinated substrates to the proteasome ***(28)***.

#### *1.2.2. Isopeptidase Activity*

Polyubiquitin chains are formed by first generating an isopeptide bond between the C-terminal G76 of ubiquitin and the ε-$NH_2$ of a lysine residue of the substrate. The chain is then formed by sequentially attaching additional ubiquitins through their C-termini to K48 of the substrate–ubiquitin conjugate. During proteasomal degradation, only the substrate is degraded, while the ubiquitin chain is detached and released intact as ubiquitin monomers by the isopeptidase activities of either PA700 ***(29–31)*** or a proteasomal associated ubiquitin hydrolase ***(32)***. Two subunits of PA700 exhibit isopeptidase activity. Interestingly, the Uch37 subunit exhibits an ubiquitin chain editing function, which detaches ubiquitin from the chain progressively from the distal end ***(29)***. By contrast, the Rpn11 subunit detaches the whole polyubiquitin chain from the substrate by cutting the bond with the substrate lysine ***(30,31)***. Rpn11 is a

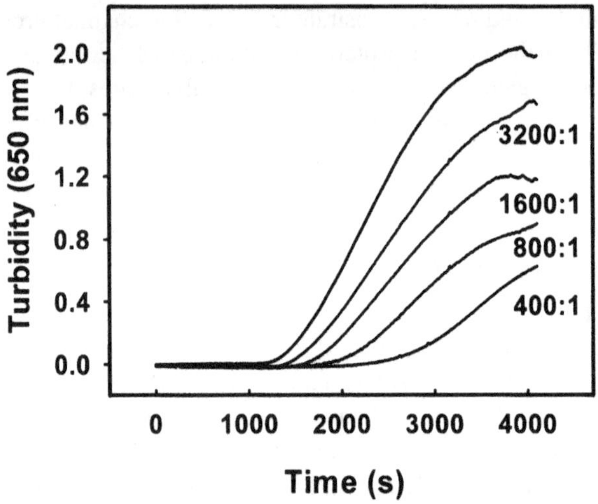


Fig. 1. Inhibition of insulin aggregation by PA700. Aggregation was initiated by dilution of native insulin (160 μ*M*) into reducing buffer (45 m*M* Tris-HCl, pH 7.8 at 37°C, 4 m*M* reduced glutathione, 0.4 m*M* oxidized glutathione) with no PA700, 50 n*M* PA700 (3200:1), 100 n*M* PA700 (1600:1), 200 n*M* PA700 (800:1), or 400 n*M* PA700 (400:1). Reprinted with permission from **ref.** ***35***.

metalloisopeptidase, whose activity requires an intact PA700 complex and is ATP hydrolysis dependent, suggesting that its activity may be coupled to substrate protein unfolding and translocation.

### 1.2.3. ATPase Activity

The six ATPase subunits of PA700 form a hexameric ring, which associates with the apical domain of the α-rings of the 20S proteasome ***(19)***. One of the ATPase subunits, Rpt2, plays a role in regulating the opening of the substrate translocation channel ***(9)***. Another ATPase subunit, s6', binds polyubiquitin chains ***(25)*** and substrates ***(26)***. In summary, ATP hydrolysis is necessary for proteasome assembly ***(33)***, deubiquitination ***(30,31)***, and, probably, unfolding stable substrates and translocating substrates into the proteolytic chamber ***(21,34)***. This suggests the ATPase subunits of PA700 have a critical role in these varied actions.

### 1.2.4. Chaperone-Like Activity

Like the reported chaperone activity of other AAA proteins, the six ATPases contained in the base subcomplex of PA700 exhibit a chaperone-like activity ***(35,36)***. In this regard, the base subcomplex binds denatured proteins, suppresses protein aggregation, and promotes protein folding. In the example shown in **Fig. 1**, PA700 suppresses the aggregation of the insulin B chain. Aggregation of the insulin B chain is a nucleation-dependent process characterized by a lag phase prior to rapid aggregation. Notably, PA700 inhibits aggregation of insulin B chain in a concentration-dependent manner by extending the lag phase. In contrast, reduction of the total

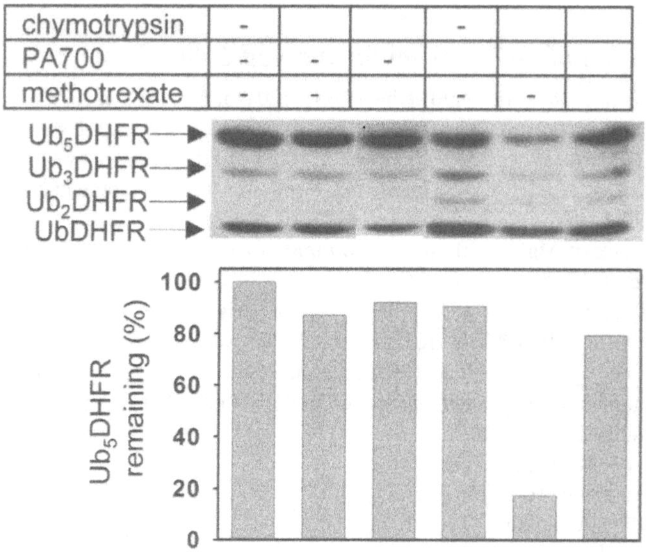


Fig. 2. PA700-dependent exposure of buried chymotryptic sites in $Ub_5DHFR$. Degradation of $Ub_5DHFR$ (80 n*M*) by chymotrypsin (2 n*M*) in presence of PA700 (20 n*M*) and a DHFR ligand, methotrexate (200 μ*M*). $Ub_5DHFR$ was detected by Western blotting with an antibody against a C-terminal HA tag. Reprinted with permission from **ref. *38***.

concentration of insulin did not change the lag phase of the nucleation process ***(35)***, but rather the rate of polymerization. These results suggest that PA700 may act on a species nucleating aggregation ***(35)***. Less clear is whether this chaperone-like activity of PA700 plays a role in vivo. At the very least, it likely contributes to recognition of damaged proteins for degradation.

### *1.2.5. Protein Remodeling Activity*

Prokaryotes contain a protease that is architecturally similar to the eukaryotic proteasome, the ClpP protease. ClpP is regulated by the ATPases ClpA and ClpX, both of which have been demonstrated to globally unfold protein substrates prior to degradation ***(37)***. Although it has been suggested for a long time that PA700 has a similar unfoldase activity ***(21,34)***, this activity has not been demonstrated to date. However, we recently provided evidence that PA700 can promote chymotrypsin digestion of the DHFR moiety of $Ub_5DHFR$, but not methotrexate-stabilized $Ub_5DHFR$ (**Fig. 2**) ***(38)***. The 26S proteasome degrades $Ub_5DHFR$, but not methotrexate-stabilized $Ub_5DHFR$ ***(38)***. These results suggest that PA700 can remodel marginally stable conformers to further promote degradation. Such remodeling activity by PA700 may help the proteasome to selectively degrade damaged proteins while leaving functional, stable proteins intact. In the following subheadings, we introduce assays to monitor both the chaperone-like and substrate remodeling activities of PA700.

## 2. Materials

### 2.1. Purification of PA700 From Bovine Red Cells

1. 4 L of 10X phosphate-buffered saline (PBS): 320 g of NaCl, 8 g of KCl, 46 g of $Na_2HPO_4$, 8 g of $KH_2PO_4$.
2. 4 L of 10X buffer H: 112.32 g of Tris-HCl, 10.88 of Tris-base, 46.4 g of NaCl, 200 mL of 200 m*M* stock of EDTA, pH 7.6, at 4°C, 14.68 mL of β-mercaptoethanol.
3. 4 L of 10X buffer X: 112.32 g of Tris-HCl, 10.88 g of Tris-base, 46.4 g of NaCl, 4.0 mL of 1 *M* stock of $MgCl_2$, 200 mL of 200 m*M* stock of EDTA, pH 7.6, at 4°C, 2.8 g of dithiothreitol (DTT).
4. 4 L of 10X buffer 26: 280.8 g of Tris-HCl, 27.2 g of Tris-base, 14.8 mL of β-mercaptoethanol.
5. Dialysis buffer: 20 m*M* Tris-HCl, pH 7.6, 20 m*M* NaCl, 0.5 m*M* $MgCl_2$, 0.1 m*M* EDTA, 5 m*M* β-mercaptoethanol, and 10% glycerol.
6. 40% Ammonium sulfate–saturated buffer X: 1 L of buffer X + 243 g of ammonium sulfate.
7. Ion-exchange DE52 resin.
8. Sephacryl S-300 (100 × 5 cm).
9. DEAE Fractogel (10 × 2.5 cm).
10. Hydroxylapatite column (7 × 2.5 cm).
11. Fraction collector.

### 2.2. In Vitro Assay: PA700 Association With the 20S Proteasome

1. Standard heatblock.
2. Proteasome peptide substrate, succinyl-Leu-Leu-Val-Trp-7-amino-4-methylcoumarin in 50 m*M* Tris-HCl, pH 8.0, and 5 m*M* β-mercaptoethanol *(33)*.
3. Proteasome assembly buffer: 45 m*M* Tris-HCl, pH 8.0, 5.6 m*M* DTT, 200 μ*M* ATP, and 10 m*M* $MgCl_2$.
4. PA700, 20S proteasome.
5. FL 600 microplate fluorescence reader (Bio-TEK).

### 2.3. Assay of Chaperone-Like Activity

#### 2.3.1. Reductive Aggregation of Insulin

1. Porcine insulin (Calbiochem).
2. Reduced glutathione.
3. Oxidized glutathione.
4. 0.1 *M* Phosphate buffer, pH 7.6.
5. 45 m*M* Tris-HCl buffer, pH 7.8.
6. Aggregation buffer: 45 m*M* Tris-HCl, pH 7.8, 4 m*M* reduced glutathione, 0.4 m*M* oxidized glutathione, 160 μ*M* insulin.
7. Disposable cuvets.
8. UV-Vis scanning spectrophotometer (multichannel) (Shimadzu, Columbia, MD).

#### 2.3.2. Thermal Aggregation of Citrate Synthase

1. Porcine heart citrate synthase (Sigma).
2. 40 mM *N*-(2-hydroxyethyl)piperazine-*N'*-(2-ethanesulfonic acid) (HEPES)–KOH buffer, pH 7.5.
3. Fluorometer (Photon Technology International, Lawrenceville, NJ).
4. Quartz cuvets.
5. Temperature-controlling device.

### *2.4. Conformational Remodeling of $Ub_5$DHFR by PA700*

1. $Ub_5$DHFR.
2. PA700.
3. Chymotrypsin.
4. Methotrexate.
5. Anti-hemagglutinin (HA) antibody (BabCo, Richmond, CA).
6. Enhanced chemiluminescence reagent for detection of horseradish peroxidase-conjugated secondary antibodies.
7. Sodium dodecyl sulfate-polyacrylamide gel electrophoresis (SDS-PAGE) gel box.
8. 10% SDS-polyacrylamide gel.
9. Electrophoretic transfer unit.
10. Immobilon-NC transfer membranes (Millipore).
11. Power supply.
12. Film developer.

## 3. Methods

### *3.1. Purification of PA700 From Bovine Red Cells*

#### *3.1.1. Cells and Preparation of Soluble Extracts*

1. Bovine blood was collected in the presence of heparin from a meat processing plant. Blood cells were collected by centrifugation at 2000*g* for 1 h. The supernatant and the buffy coat were removed by aspiration.
2. The remaining cell pellet was resuspended in four volumes of PBS buffer and recentrifuged. The washing procedure was repeated four times. All of the following steps were carried out at 4°C unless otherwise indicated.
3. Cells were lysed by adding three volumes of lysis buffer H to one volume of the packed cells and stirring for 10 min. The lysate was centrifuged at 13,000*g* for 60 min. The supernatant was removed and saved. The pellet was resuspended in three volumes of buffer H and recentrifuged. The supernatant from the second centrifugation was added to the first.
4. Portions of this crude soluble lysate were dialyzed against dialysis buffer for subsequent gel filtration chromatography or velocity sedimentation centrifugation.
5. The crude soluble lysate was added to DE52 (5 mL of lysate/mL of DE52) that had been equilibrated with buffer H. After gentle mixing for 30 min, the resin was filtered and washed extensively with buffer H until the elute was clear.
6. The resin was then mixed the buffer H containing 0.5 *M* NaCl (1 mL of buffer/mL of DE52), stirred gently for 10 min, filtered, and washed with an additional small volume of the 0.5 *M* NaCl until most of the bound protein has eluted. The eluted fraction was largely free of hemoglobin and was termed "fraction II."
7. Solid ammonium sulfate was added to fraction II to 40% saturation over a period of 30 min with gentle stirring. After an additional 30 min, the precipitated proteins were collected by centrifugation. The pellets were resuspended with a Dounce homogenizer in a large volume of buffer H, which was saturated to 40% with respect to ammonium sulfate. The precipitated proteins were collected by recentrifugation (*see* **Note 1**).
8. The washed pellets were dissolved in a small volume of buffer H that had been supplemented with an additional 100 m*M* NaCl, and were dialyzed for 16 h against large volumes of the same buffer. In some early preparation of PA700, the buffer H used for the dialysis and for the subsequent gel filtration column chromatography (*see* below) also

contained 20% (v/v) glycerol. Removal of the glycerol from the buffer did not have any detectable effect on the purification or properties of PA700. After dialysis, any undisclosed material was removed by centrifugation at 30,000*g* for 30 min.

#### 3.1.2. Purification of PA700

1. The extract from the last centrifugation was loaded unto a column of Sephacryl S-300 (100 × 5 cm) equilibrated with buffer H supplemented with 100 m*M* NaCl and eluted with the same buffer.
2. Fractions of 11 mL were collected and assayed for PA700 activity as described below. A FL 600 microplate fluorescence reader was used for multiple samples.
3. The fractions containing the highest levels of activity were pooled and subjected to ion-exchange chromatography by applying them directly to a column of DEAE Fractogel (EMD Biosciences, 10 × 2.5 cm) equilibrated with the same buffer. The bound proteins were eluted from the column with a linear gradient of NaCl from 100 to 350 m*M* (1000 mL) in buffer H. Samples of the 11-mL fractions were assayed for PA700 activity.
4. The fractions with the highest activity were pooled and dialyzed against a buffer of 20 m*M* potassium phosphate, pH 7.6. The dialyzed sample was applied to a column of hydroxylapatite (7 × 2.5 cm) equilibrated in the same buffer. The bound proteins were eluted from the column with a linear gradient of phosphate buffer (20–200 m*M*, 500 mL total volume). Samples of the 8-mL fractions were assayed for PA700 activity.
5. The fractions containing peak activities were pooled, dialyzed against buffer H, and concentrated to approx 1 mg/mL by ultrafiltration using an Amicon PM10 membrane. The samples could be stored at –70°C with no detectable loss of activity.

### 3.2. In Vitro Assay of PA700 Association With the 20S Proteasome

1. Five microliters of 10X proteasomal assembly buffer were prepared from individual stock components. The 10X mix is: 450 m*M* Tris-HCl, pH 8.0, at 37°C, 56 m*M* DTT, 2 m*M* ATP, 100 m*M* $MgCl_2$.
2. The following were added sequentially: 0.25 μg of purified latent 20S proteasome, and 1.0 μg of purified PA700. The final volume is 50 μL. The tube was tapped gently to mix and incubated at 37°C for 45 min (*see* **Note 2**).
3. One milliliter of preincubated 50 m*M* succinyl-Leu-Leu-Val-Tyr 7-amino-4-methyl-coumarin was mixed in 50 m*M* Tris-HCl, pH 8, at 37°C, 5 m*M* β-mercaptoethanol with the preassembled proteasome from **step 2**. The mixture was transferred to a cuvet for fluorescence determination.
4. Proteasome activity was measured by the hydrolysis of the synthetic peptide Suc-LLVY-AMC. The production of free 7-amino-4-methylcoumarin was monitored continuously at 380 nm (excitation) and 460 nm (emission) for 10 min, and initial steady-state rates were assessed. Generally, the assembled proteasome activity is approx 20-fold higher than the latent 20S proteasome.

### 3.3. Assay of Chaperone-Like Activity

We have examined the antiaggregation activity of PA700 by utilizing several aggregation-prone substrates *(35)*, including ΔF-NBD1, which is the nucleotide binding domain of the cystic fibrosis transmembrane conductance regulator (CFTR) that contains the common cystic fibrosis disease-causing mutation responsible for impaired folding of the full-length CFTR protein *(39)*. Here we provide methods for examining the reductive aggregation of the insulin B chain and the thermal aggregation of citrate synthase, a protein commonly utilized for studying chaperone activity.

#### *3.3.1. Reductive Aggregation of Insulin*

1. A 2 m*M* stock solution of porcine insulin in 0.1 *M* potassium phosphate buffer, pH 7.6, was prepared.
2. Aggregation reactions were performed at 37°C in a final volume of 250 µL containing aggregation buffer and different concentrations of PA700. We tested molar ratios of insulin: PA700 from 400 to 3200.
3. Aggregation was monitored continuously as an increase in turbidity at 650 nm using an UV-vis scanning spectrophotometer (six channels). In independent experiments, the length of the lag phase prior to initial detection of turbidity with insulin alone varied typically from about 300 s to about 1100 s. However, within a given experiment, repetitions varied by less than 10%. A typical result is shown in **Fig. 2**.

#### *3.3.2. Thermal Aggregation of Citrate Synthase*

1. The cuvet was prewarmed in the temperature-controlled cuvet holder of the fluorometer at 43°C.
2. Thermal aggregation of citrate synthase (CS) was carried out by adding 150 n*M* (monomer) CS to a solution of 40 m*M* HEPES–KOH, pH 7.5, heated to 43°C.
3. Aggregation was followed by light scattering at right angles at 500 nm in a fluorometer with 2 nm excitation and 4 nm emission slit widths. PA700 was present before the addition of CS unless stated otherwise. We tested different concentrations of PA700 ranging from 0.9 to 5 n*M*.

### *3.4. Conformational Remodeling of $Ub_5$DHFR by PA700*

1. Thirty microliters of reaction volume contains $Ub_5$DHFR (80 n*M*), chymotrypsin (2 n*M*) with or without PA700 (20 n*M*), and/or methotrexate (200 µ*M*) in 20 m*M* Tris-HCl, pH 7.2, 20 m*M* NaCl, 20 m*M* KCl, 1 m*M* EDTA.
2. For reactions with methotrexate, 25 µL of $Ub_5$DHFR (96 n*M*) were preincubated with methotrexate (240 µ*M*) for 5 min at 37°C, and then other components and buffer were added to a final 30-µL reaction volume.
3. The reactions were stopped by adding 5X SDS sample buffer. The samples were heated at 95°C for 5 min and then subjected to 10% SDS-PAGE and transferred to Immobilon-NC transfer membranes.
4. $Ub_5$DHFR was detected using a monoclonal antihemagglutinin antibody against the hemagglutinin tag at the C-terminus of $Ub_5$DHFR. A typical result is shown in **Fig. 2**.

## 4. Notes

1. Thorough washing of the pellet at this stage is important to eliminate completely the 20S proteasome and PA28, both of which are soluble at this ammonium sulfate concentration, from the precipitated proteins.
2. A four- to sixfold molar excess amount of PA700 was used for the assembling experiment. Under such conditions, more than 85% of the 20S proteasome was assembled in 26S complexes.

## Acknowledgments

This work was supported by grants from the Welch Foundation (to P. J. T.), MDA (to G. N. D.), NIH-DK46818 (to G. N. D.), and NIH-DK49835 (to P. J. T.).

## References


1. DeMartino, G. N. and Slaughter, C. A. (1999) The proteasome, a novel protease regulated by multiple mechanisms. *J. Biol. Chem.* **274,** 22123–22126.
2. Glickman, M. H. and Ciechanover, A. (2002) The ubiquitin–proteasome proteolytic pathway: destruction for the sake of construction. *Physiol Rev.* **82,** 373–428.
3. Puhler, G., Weinkauf, S., Bachmann, L., et al. (1992) Subunit stoichiometry and three-dimensional arrangement in proteasomes from *Thermoplasma acidophilum. EMBO J.* **11,** 1607–1616.
4. Lupas, A., Zwickl, P., and Baumeister, W. (1994) Proteasome sequences in eubacteria. *Trends Biochem. Sci.* **19,** 533–534.
5. Lowe, J., Stock, D., Jap, B., Zwickl, P., Baumeister, W., and Huber, R. (1995) Crystal structure of the 20S proteasome from the archaeon *T. acidophilum* at 3.4 A resolution. *Science* **268,** 533–539.
6. Groll, M., Ditzel, L., Lowe, J., et al. (1997) Structure of 20S proteasome from yeast at 2.4 A resolution. *Nature* **386,** 463–471.
7. Unno, M., Mizushima, T., Morimoto, Y., et al. (2002) The structure of the mammalian 20S proteasome at 2.75 A resolution. *Structure.* (Camb.) **10,** 609–618.
8. Whitby, F. G., Masters, E. I., Kramer, L., et al. (2000) Structural basis for the activation of 20S proteasomes by 11S regulators. *Nature* **408,** 115–120.
9. Kohler, A., Cascio, P., Leggett, D. S., Woo, K. M., Goldberg, A. L., and Finley, D. (2001) The axial channel of the proteasome core particle is gated by the Rpt2 ATPase and controls both substrate entry and product release. *Mol. Cell* **7,** 1143–1152.
10. Touitou, R., Richardson, J., Bose, S., Nakanishi, M., Rivett, J., and Allday, M. J. (2001) A degradation signal located in the C-terminus of p21WAF1/CIP1 is a binding site for the C8 alpha-subunit of the 20S proteasome. *EMBO J.* **20,** 2367–2375.
11. Liu, C. W., Corboy, M. J., DeMartino, G. N., and Thomas, P. J. (2003) Endoproteolytic activity of the proteasome. *Science* **299,** 408–411.
12. Brannigan, J. A., Dodson, G., Duggleby, H. J., et al. (1995) A protein catalytic framework with an N-terminal nucleophile is capable of self-activation. *Nature* **378,** 416–419.
13. Gaczynska, M., Goldberg, A. L., Tanaka, K., Hendil, K. B., and Rock, K. L. (1996) Proteasome subunits X and Y alter peptidase activities in opposite ways to the interferon-gamma-induced subunits LMP2 and LMP7. *J. Biol. Chem.* **271,** 17275–17280.
14. Gaczynska, M., Rock, K. L., and Goldberg, A. L. (1993) Gamma-interferon and expression of MHC genes regulate peptide hydrolysis by proteasomes. *Nature* **365,** 264–267.
15. Navon, A. and Goldberg, A. L. (2001) Proteins are unfolded on the surface of the ATPase ring before transport into the proteasome. *Mol. Cell* **8,** 1339–1349.
16. Rape, M., Hoppe, T., Gorr, I., Kalocay, M., Richly, H., and Jentsch, S. (2001) Mobilization of processed, membrane-tethered SPT23 transcription factor by CDC48(UFD1/NPL4), a ubiquitin-selective chaperone. *Cell* **107,** 667–677.
17. Lin, L., DeMartino, G. N., and Greene, W. C. (1998) Cotranslational biogenesis of NF-kappaB p50 by the 26S proteasome. *Cell* **92,** 819–828.
18. Adams, G. M., Crotchett, B., Slaughter, C. A., DeMartino, G. N., and Gogol, E. P. (1998) Formation of proteasome-PA700 complexes directly correlates with activation of peptidase activity. *Biochemistry* **37,** 12927–12932.
19. Glickman, M. H., Rubin, D. M., Coux, O., et al. (1998) A subcomplex of the proteasome regulatory particle required for ubiquitin-conjugate degradation and related to the COP9-signalosome and eIF3. *Cell* **94,** 615–623.

20. Gonzalez, F., Delahodde, A., Kodadek, T., and Johnston, S. A. (2002) Recruitment of a 19S proteasome subcomplex to an activated promoter. *Science* **296,** 548–550.
21. Thrower, J. S., Hoffman, L., Rechsteiner, M., and Pickart, C. M. (2000) Recognition of the polyubiquitin proteolytic signal. *EMBO J.* **19,** 94–102.
22. Deveraux, Q., Ustrell, V., Pickart, C., and Rechsteiner, M. (1994) A 26 S protease subunit that binds ubiquitin conjugates. *J. Biol. Chem.* **269,** 7059–7061.
23. Fu, H., Sadis, S., Rubin, D. M., Glickman, M., van Nocker, S., Finley, D., and Vierstra, R. D. (1998) Multiubiquitin chain binding and protein degradation are mediated by distinct domains within the 26 S proteasome subunit Mcb1. *J. Biol. Chem.* **273,** 1970–1981.
24. van Nocker, S., Sadis, S., Rubin, D. M., et al. (1996) The multiubiquitin-chain-binding protein Mcb1 is a component of the 26S proteasome in *Saccharomyces cerevisiae* and plays a nonessential, substrate-specific role in protein turnover. *Mol. Cell Biol.* **16,** 6020–6028.
25. Lam, Y. A., Lawson, T. G., Velayutham, M., Zweier, J. L., and Pickart, C. M. (2002) A proteasomal ATPase subunit recognizes the polyubiquitin degradation signal. *Nature* **416,** 763–767.
26. Zhang, M., Pickart, C. M., and Coffino, P. (2003) Determinants of proteasome recognition of ornithine decarboxylase, a ubiquitin-independent substrate. *EMBO J.* **22,** 1488–1496.
27. Murakami, Y., Matsufuji, S., Kameji, T., et al. (1992) Ornithine decarboxylase is degraded by the 26S proteasome without ubiquitination. *Nature* **360,** 597–599.
28. Elsasser, S., Gali, R. R., Schwickart, M., et al. (2002) Proteasome subunit Rpn1 binds ubiquitin-like protein domains. *Nat. Cell Biol.* **4,** 725–730.
29. Lam, Y. A., DeMartino, G. N., Pickart, C. M., and Cohen, R. E. (1997) Specificity of the ubiquitin isopeptidase in the PA700 regulatory complex of 26 S proteasomes. *J. Biol. Chem.* **272,** 28438–28446.
30. Yao, T. and Cohen, R. E. (2002) A cryptic protease couples deubiquitination and degradation by the proteasome. *Nature* **419,** 403–407.
31. Verma, R., Aravind, L., Oania, R., et al. (2002) Role of Rpn11 metalloprotease in deubiquitination and degradation by the 26S proteasome. *Science* **298,** 611–615.
32. Leggett, D. S., Hanna, J., Borodovsky, A., et al. (2002) Multiple associated proteins regulate proteasome structure and function. *Mol. Cell* **10,** 495–507.
33. Ma, C. P., Vu, J. H., Proske, R. J., Slaughter, C. A., and DeMartino, G. N. (1994) Identification, purification, and characterization of a high molecular weight, ATP-dependent activator (PA700) of the 20 S proteasome. *J. Biol. Chem.* **269,** 3539–3547.
34. Johnson, E. S., Ma, P. C., Ota, I. M., and Varshavsky, A. (1995) A proteolytic pathway that recognizes ubiquitin as a degradation signal. *J. Biol. Chem.* **270,** 17442–17456.
35. Strickland, E., Hakala, K., Thomas, P. J., and DeMartino, G. N. (2000) Recognition of misfolding proteins by PA700, the regulatory subcomplex of the 26 S proteasome. *J. Biol. Chem.* **275,** 5565–5572.
36. Braun, B. C., Glickman, M., Kraft, R., et al. (1999) The base of the proteasome regulatory particle exhibits chaperone-like activity. *Nat. Cell Biol.* **1,** 221–226.
37. Weber-Ban, E. U., Reid, B. G., Miranker, A. D., and Horwich, A. L. (1999) Global unfolding of a substrate protein by the Hsp100 chaperone ClpA. *Nature* **401,** 90–93.
38. Liu, C. W., Millen, L., Roman, T. B., et al. (2002) Conformational remodeling of proteasomal substrates by PA700, the 19 S regulatory complex of the 26 S proteasome. *J. Biol. Chem.* **277,** 26815–26820.
39. Qu, B. H. and Thomas, P. J. (1996) Alteration of the cystic fibrosis transmembrane conductance regulator folding pathway. *J. Biol. Chem.* **271,** 7261–7264.

7

# Cell-Free Assay for Ubiquitin-Independent Proteasomal Protein Degradation

**Chaim Kahana and Yuval Reiss**


### Summary

The ATP-dependent degradation of ornithine decarboxylase is an exceptional case whereby a protein is targeted to the 26S proteasome independently of ubiquitin conjugation. Rather, prior association with the polyamine-induced regulatory protein, antizyme, confers susceptibility of ornithine decarboxylase to proteasomal degradation. In this chapter we describe ornithine decarboxylase/antizyme-based in vivo and in vitro systems for the measurement of ATP-dependent, ubiquitin-independent proteasomal degradation, as well as the application of ornithine decarboxylase as a reporter for the targeting of proteins to the 26S proteasome.


**Key Words:** 26S proteasome; ornithine decarboxylase; antizyme; antigen processing.

## 1. Introduction

The majority of cellular short-lived proteins are degraded by the 26S proteasome after their prior conjugation to polyubiquitin chains by the ubiquitin conjugation system *(1)*. There are exceptional cases in which degradation by the proteasome requires ATP hydrolysis but not ubiquitination. The most notable substrate that is degraded by the 26S proteasome without ubiquitination is ornithine decarboxylase (ODC). ODC is a key regulatory enzyme that catalyzes the rate-limiting step in the polyamine biosynthetic pathway. ODC becomes susceptible to degradation by the 26S proteasome through its association with a polyamine-induced regulatory protein termed antizyme (Az). This interaction constitutes an inhibitory feedback mechanism that ensures tight control of polyamine synthesis *(2)*. Recently, Asher et al. described a novel ubiquitin-independent proteasome-dependent degradation pathway regulated by the NQO1-quinone oxidoreductase affecting p53 and p73α 3. Ubiquitin–antigen conjugates are

From: *Methods in Molecular Biology, vol. 301, Ubiquitin–Proteasome Protocols*
Edited by: C. Patterson and D. M. Cyr © Humana Press Inc., Totowa, NJ

extremely difficult to produce and purify in quantities required for studies aimed at the analysis of the 26S proteasome degradation products. Therefore, an ODC-based in vitro degradation system was successfully utilized in the investigation of the function of the 26S proteasome in major histocompatibility class I (MHC class I) antigen processing *(4,5)*. We describe here methods for the measurement of ubiquitin-independent protein degradation by the 26S proteasome through use of ODC and recombinant ODC expressing MHC class I epitopes. The utilization of ODC as a reporter for the targeting of proteins to the 26S proteasome without ubiquitination is also described.

## 2. Materials

1. pGEM-1 expression vector (Promega, Madison, WI).
2. pUC18 plasmid.
3. pET14b, pET19b (Novagen).
4. pMAL (New England Biolabs Inc.).
5. PEFIRES-p, bicistrinic mammalian expression vector.
6. ODC cDNA.
7. A31N, A31N-ts20 mouse fibroblasts.
8. TnT coupled transcription translation system (Promega, Madison, WI).
9. Rabbit reticulocyte lysate, nuclease treated.
10. T7 RNA polymerase.
11. New Zealand white rabbits.
12. Phenylhydrazine-HCl.
13. Heparin.
14. ATP depletion buffer: 20 m*M* 2-deoxyglucose, 0.2 m*M* 2-4 dinitrophenol in phosphate-buffered saline (PBS).
15. Reticulocyte lysis buffer: 10 m*M* Tris-HCl, pH 7.0, 1 m*M* dithiothreitol (DTT), and glycerol is added to 20% (w/v).
16. 8X Reticulocyte buffer: 800 m*M* Tris-HCl, pH 7.8, 40 m*M* $MgCl_2$, 80 m*M* KCl, 4 m*M* DTT, adjusted to pH 7.8.
17. Degradation buffer: 40 m*M* Tris-HCl, pH 7.5, 2 m*M* DTT, 5 m*M* $MgCl_2$, 200 g/mL of ubiquitin, 0.5 m*M* ATP, 10 m*M* phosphocreatine, 1.6 mg/mL of creatine phosphokinase.
18. Buffer P: 10 m*M* Na-phosphate, pH 7.0, 30 m*M* NaCl, 1 m*M* DTT, 1 m*M* EDTA 1-thio-β-D-galactopyranoside (IPTG).
19. Buffer A: 20 m*M* Tris-HCl, pH 7.5, 1 m*M* DTT, 1 m*M* EDTA, 1.5 m*M* ATP, and 0.25 *M* sucrose.
20. Buffer B: 20 m*M* Tris-HCl, pH 7.5, 1 m*m* DTT, 1 m*M* ATP, and 20% (v/v) glycerol
21. Protease inhibitors (Boehringer).
22. DEAE–cellulose (Whatman).
23. Q-Sepharose (Amersham Pharmacia Biotech).
24. Affi-Gel-10 (Bio-Rad).
25. Mono-q 5/5 column (Amersham Pharmacia Biotech).
26. Sepharose 6B (Amersham Pharmacia Biotech).
27. Resource-Q (Amersham Pharmacia Biotech).
28. Enzyme-grade ammonium sulfate.
29. Phosphocreatine.
30. Creatine phosphokinase.

## 3. Methods

### 3.1. Degradation in Reticulocyte Lysate

Because most short-lived proteins are minor constituents in mammalian cells and therefore their purification is not a practical task we use in vitro translated proteins as substrates and regulators in the degradation reaction. The degradation reaction is then performed in a complete reticulocyte lysate based degradation mix or in a fractionated lysate in which ubiquitin is separated from the ubiquitination machinery, therefore constituting a system for testing whether degradation of a given protein requires ubiquitination.

#### 3.1.1. Preparation of Expression Plasmids and Expression of Proteins

ODC and Az cDNAs are cloned into the multiple cloning site of the pGEM1 expression plasmid (Promega) between the *Eco*RI (5) and *BamH*I (3') sites. The corresponding proteins are then expressed in the coupled transcription/translation system, or in a standard rabbit reticulocyte systems programmed with RNA that is synthesized in vitro using T7 RNA polymerase.

#### 3.1.2. Preparation and Fractionation of Reticulocyte Lysate

##### 3.1.2.1. Preparation of Lysate

White New Zealand rabbits are injected with 0.9 mL/kg of 2% phenylhydrazine-HCl on d 1–3 and 4–7. Fifth milliliters of blood are collected on d 9 and mixed immediately with 5000 U of heparin. The cells are washed with PBS to remove clotting factors and buffy coat. Next reticulocytes are incubated at 37°C for 2 h in ATP depletion buffer. After extensive washes with PBS the cells are lysed in 1.5 volume of reticulocyte lysis buffer. The lysate is clarified by 2-h centrifugation at 80,000*g* at 4°C and adjusted to 1X reticulocyte buffer using a 8X concentrate.

##### 3.1.2.2. Preparation of Fraction II

1. The lysate is applied to a DEAE column equilibrated with 10 m*M* Tris-HCl, pH 7.0, 1 m*M* DTT, 20% glycerol.
2. The column is washed with 10 column volumes of loading buffer and the bound material is eluted with a loading buffer containing 500 m*M* KCl.
3. Ammonium sulfate is added to the eluted material to 90% saturation. After 60 min of incubation at 4°C, the insoluble material is precipitated by 20 min of centrifugation at 10,000*g* at 4°C. The resulting pellet is resuspended to a concentration of 10 mg/mL in 10 m*M* Tris-HCl, pH 7.0, 0.5 m*M* DTT, and 20% glycerol and dialyzed against the same buffer (10 h at 4°C).

At present complete reticulocyte lysate and fraction II can be obtained from commercial sources.

#### 3.1.3. Degradation Reaction

1. In a typical degradation reaction 6 µL of the translation lysate were incubated in 50 µL of degradation buffer.

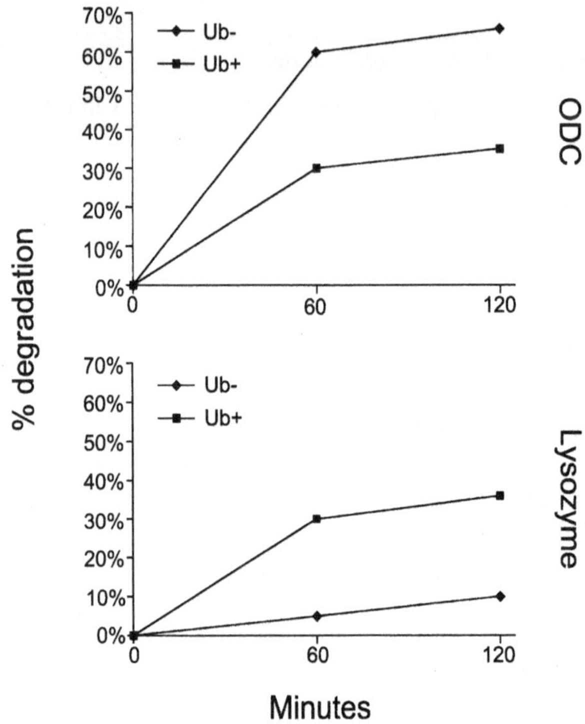


Fig. 1. ODC is degraded efficiently in fraction II of reticulocyte lysate without requiring ubiquitin. $^{32}$S-methionine labeled ODC (translated in vitro in reticulocyte lysate) **(A)** and $^{125}$I-lysozyme (iodinated in vitro) **(B)** were incubated in fraction II of reticulocyte lysate without and with added ubiquitin (5 µg). Aliquots were removed at the indicated times and fractionated by SDS-PAGE. The radioactivity of the ODC and lysozyme bands was determined and the data are presented graphically. (Adapted from **ref. *10***.)

2. Aliquots containing 15 µL are removed at time 0 and after two additional time of incubation at 37°C and fractionated by sodium dodecyl sulfate-polyacrylamide gel electrophoresis (SDS-PAGE).
3. The radioactivity in the monitored band is determined using the Fuji Bus 2500 phosphoimager.
4. Degradation reaction in fraction II is performed as described in **steps 1** and **2** for the complete lysate except that fraction II is used instead of complete reticulocyte lysate with or without added ubiquitin.
5. To minimize the possible introduction of ubiquitin with the translated material, the translated protein is first fractionated on DEAE beads and the bound material eluted as described for the preparation of fraction II.

As shown in **Fig. 1**, ODC is efficiently degraded in fraction II lacking ubiquitin while the degradation of lysozyme is practically inhibited in the absence of ubiquitin and resumed on the addition of ubiquitin, demonstrating that the degradation of ODC does not require ubiquitination.

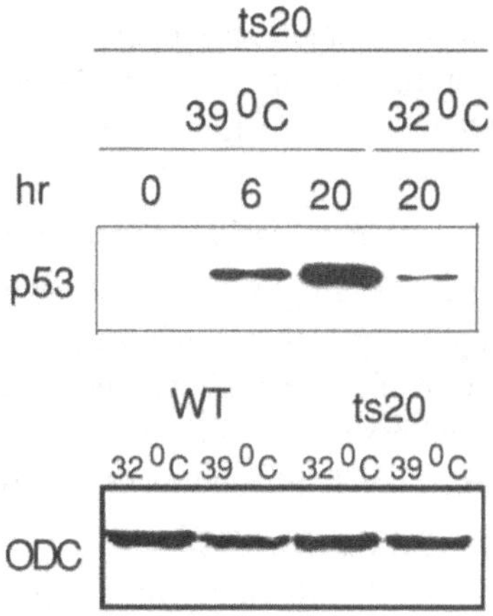


Fig. 2. Degradation of ODC in cells is independent of the integrity of the ubiquitination machinery. Wild-type (A31N) and ts20 cells containing thermosensitive ubiquitin-activating enzyme E1 were transfected with expression constructs encoding ODC. The transfected cells were incubated at 32°C (permissive temperature) and then half of the cells were transferred to 39°C (restrictive temperature) for an additional 20 h. Some cells were then returned to the permissive temperature. Cellular extracts were prepared and fractionated and the proteins of interest were detected by immunoblotting. (Adapted from **ref. *11***.)

### *3.2. Degradation in Mutant Cell Line Harboring Thermosensitive Ubiquitin-Activating Enzyme*

Two main tools are used to determine whether the degradation of a given protein in intact cells requires ubiquitination. (1) The first is expression of a chain-terminating mutant ubiquitin together with the tested protein. The efficiency of this approach is questionable because the expressed mutant ubiquitin have to overcome large amounts of wild-type ubiquitin that are present in the transfected cells. (2) The second method is testing degradation in cell lines containing lesions in specific steps of polyubiquitination. The preferred lesion is that of the ubiquitin-activating enzyme E1 as it is required for the degradation of all proteins that are degraded in a ubiquitin-dependent manner. For this purpose we use A31N-ts20 cells that contain a thermosensitive ubiquitin-activating enzyme E1 and the corresponding parental cells ***(6)***. (*See* **Note 1.**) For this purpose ODC is cloned in the bicistronic expression vector PEFIRES-p *(7)* and the resulting construct is transfected into A31N-ts20 cells and into parental cells. The transfected cells are incubated at 32°C (permissive temperature) and at 24 h posttransfection the cells were shifted to the restrictive temperature (39°C). Endogenous p53, a substrate of the ubiquitin system, is not detected in wild-type cells and in the mutants when grown at the permissive temperature. Massive accumulation of p53 is observed in the mutant cells only on their transfer to the restrictive temperature **(Fig. 2)**. The degradation of p53 is resumed when the mutant cells are returned to the permissive temperature **(Fig. 2)**. In contrast, ODC expressed from the transfected construct is not accumulated in the mutant cells **(Fig. 2)**. In fact the amount of ODC actually declines at the restrictive temperature, probably as a result of enhanced metabolism at this elevated temperature **(Fig. 2)**, demonstrating that ODC is degraded in a ubiquitin-independent manner.

### 3.3. Degradation of ODC in a Purified Reconstituted System

The section outlines (1) construction of plasmids for bacterial expression of Az, ODC, ODC–antigen fusion proteins, and recombinant ODC expressing MHC class I peptide epitope; (2) purification of the recombinant proteins from *E. coli* extracts; and (3) experiments utilizing ODC-based recombinant antigens to measure major histocompatibility complex (MHC) class I antigen processing by purified 26S proteasome.

#### 3.3.1. ODC Expression Plasmids (Fig. 3A)

The cloning method described in this subheading is that described by Ben Shahar et al. ***(4)***. For expression of ODC derivatives in *E. coli,* the full-length mouse ODC cDNA is subcloned in pUC18 at the *Kpn*I (5') and *Bam*HI (3') sites, resulting in the plasmid pUC-ODC. The ODC cDNA is then isolated from pUC-ODC as an *Nco*I/*Bam*HI fragment and cloned in the *Nco*I (5') and *Bam*HI (3') sites of pET19b (Novagen), resulting in the expression plasmid pET-ODC.

The construction of a plasmid encoding an ODC expressing the ovalbumin-derived mouse H2-$K^b$-restricted epitope SIINFEKL is achieved by using adaptor oligonucleotides. The adaptor encoding the peptide SIINFEKL with *Bst*XI-compatible ends is generated by annealing the two synthetic oligonucleotides, 5'-ATAGTATAATCA ACTTCGAAAAACTGAGCTC-3' and 5'-TCAGTTTTTCCGAAGTTGATTATA CTATGGC-3'. The adaptor is then inserted in frame at the unique *Bst*XI site in the ODC sequence in pUC-ODC to generate the plasmid pUC-ODCova. The ODC-ova cDNA is then isolated from pUC-ODCova as an *Nco*I/*Bam*HI fragment and cloned in the *Nco*I (5') and *Bam*HI (3') sites of pET14b (Novagen), resulting in the expression plasmid pET-ODCova. The insertion of the adaptor in the correct orientation is confirmed by DNA sequencing. The plasmids pET-ODC expression plasmids are used for ODC expression in *Escherichia coli* and in reticulocyte lysate. Adaptor oligonucleotides were also employed to insert extended SIINFEKL peptides into the same location in the ODC coding sequence 5.

ODC can also be used as a vehicle to target intact proteins to the proteasome. We have fused the entire HIV-1 Nef coding sequence to the N-terminus of ODCova to create a Nef–ODCova fusion protein ***(8)***.

The construction of a plasmid encoding HIV1–Nef–ODC fusion protein is achieved as follows: An HIV1-nef with 5' and 3' *Nco*I restriction sites is generated by a polymerase chain reaction (PCR) reaction using extended oligonucleotides complementary to nucleotides 5–17 and 598–618 respectively encoding the *Nco*I site. The full-length Nef *Nco*I/*Nco*I fragment is then cloned in frame into the *Nco*I site of pET-ODC. The correct orientation is confirmed by DNA sequencing.

#### 3.3.2. Az Expression Plasmid

The plasmid encoding the fusion protein maltose-binding protein-antizyme (MBP-Az) is constructed from rat full-length AZ cDNA in pET8. The Az cDNA is isolated as an

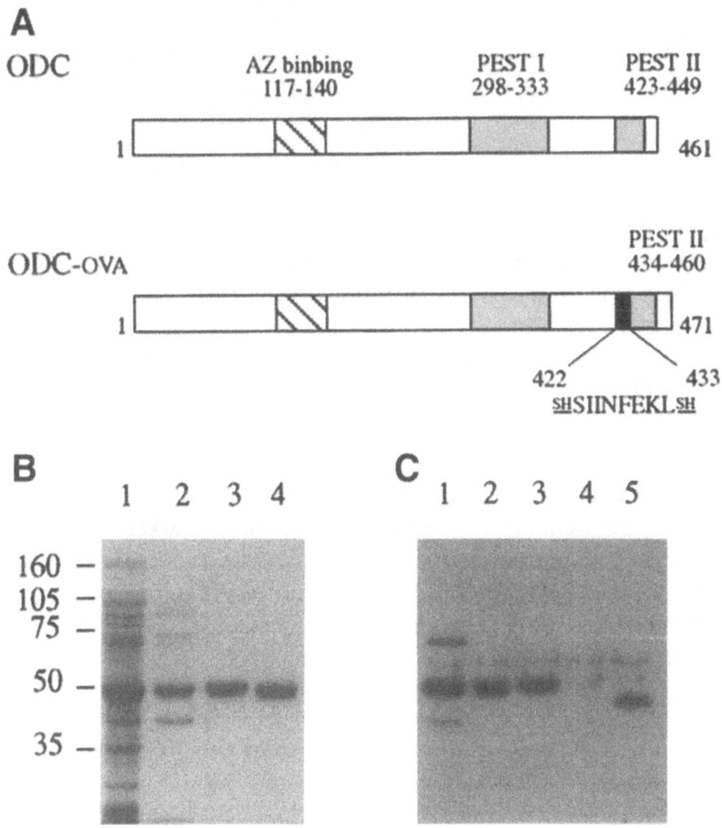


Fig. 3. ODC-ova preparation. **(A)** Structural alignment of ODC and ODC-ova. The $H\text{-}2K^b$ binding peptide SIINFEKL was inserted after amino acid residue 423 of ODC and just preceding the second PEST region. Construction of ODC-ova as described in **Subheading 3.3.1.** described dictated the duplication of histidine and serine on both sides of SIINFEKL (*underlined lowercase letters*). **(B)** Purification of ODC-ova. Samples of bacterially expressed ODC-ova were resolved on 10% SDS-PAGE and stained with Coomassie blue. *Lane 1*, Lysate of induced bacteria (25 µg); *lane 2*, Mono-Q-purified ODC-ova (9 µg); *lane 3*, affinity-purified ODC-ova (2 µg); *lane 4*, affinity-purified ODC (2 µg). **(C)** Immunoblot analysis with SIINFEKL-specific antiserum. *Lane 1*, Lysate of induced bacteria (2.5 µg); *lane 2*, Mono-Q-purified ODC-ova (1 µg); *lane 3*, affinity-purified ODC-ova (0.2 µg); *lane 4*, affinity-purified ODC (0.2 µg); *lane 5*, ovalbumin (0.2 µg). (Reprinted from **ref. *4*.**)

*Nco*I fragment and then treated with DNA polymerase Klenow fragment to produce blunt ends. The blunted *Nco*I fragment is then ligated in frame in pMAL™ (New England Biolabs Inc.). pMAL is initially digested with *Xba*I followed by end-filling of the 5'-overhangs with DNA polymerase Klenow fragment. A plasmid clone with the correct cDNA orientation is then selected by DNA sequencing. The resulting plasmid pMAL-Az is used for expression of Az in *E. coli*. (*See* **Note 2.**)

### 3.3.3. *E. coli* Expression of Recombinant Proteins

1. The ODC and ODC-ova recombinant plasmids are transformed into *E. coli* strain BL21 (DE3).
2. A single colony is grown overnight at 37°C and in the morning is diluted 1:50 into 300 mL of LB medium containing ampicillin (100 μg/mL).
3. The culture is grown until absorbance at 600 nm reaches 0.6–0.8. Expression is then induced with IPTG (1 m*M* final concentration).
4. ODC and ODC-ova are induced for 16 h at 22°C.
5. Bacteria are harvested and washed with ice-cold buffer L (25 m*M* Tris-HCl, pH 7.5, 2.5 m*M* DTT) supplemented with 10 m*M* iodoacetamide and 1:25 (w/v) solution of protease inhibitors (Boehringer).
6. The cells are then resuspended in 20 mL of buffer L and lysed in a French pressure press cell (1260 atm) (Aminco SLM Instruments).
7. After lysis, the extract is supplemented with 5 m*M* DTT, and the insoluble material is removed by centrifugation (10,000*g* for 15 min).

The Nef–ODC expression plasmid is transformed into *E. coli* strain BL21 (DE3lysS). This strain contains an additional resistance marker for the antibiotic chloramphenicol. A single colony is grown overnight. Induction of protein in a 300-mL culture containing 100 μg/mL of ampicillin and 34 μg/mL of chloramphenicol is achieved by addition of IPTG to 1 m*M* and incubation at 20°C for 16 h. Extraction of protein is performed as described in **Subheading 3.3.3., steps 5–7**.

## 3.4. Purification of Recombinant ODC Proteins (Fig. 3B,C)

1. The bacterial lysate (270 mg of protein) is loaded on a 4.5 × 1.6 cm Q-Sepharose column (Amersham Pharmacia Biotech) equilibrated in buffer L.
2. The column is washed with 20 mL of buffer L and then developed with a linear gradient of 0–1 *M* NaCl (in buffer L). ODC recombinant proteins elute from the column between 0.35 and 0.4 *M* NaCl. The peak fractions are combined and subjected to affinity chromatography.
3. The combined protein fraction from the Q-Sepharose column is directly loaded on pyridoxamine 5'-phosphate–agarose column (1 × 7 cm) equilibrated in buffer L containing 0.1 m*M* EDTA and 0.1 m*M* L-ornithine. The sample is applied to the column at a flow rate of 35 μL/min (17 h) at 4°C.
4. The column is then washed with 80 mL of buffer L containing 15 m*M* NaCl. The protein is eluted from the column by successive additions of 7-mL portions of buffer L containing 10 (μ*M* pyridoxal 5'-phosphate.
5. All of the bound protein that eluted in the first five fractions is combined and concentrated to 0.5–2 μg/mL in Centricon 30 concentrator (Millipore) and stored in aliquots at 80°C.
6. Preparation of pyridoxamine 5'-phosphate-agarose column: Solid pyridoxamine 5'-phosphate (50 mg) (custom synthesized as calcium salt) is dissolved in 10 mL of 1 m*M* HCl. To dissolve the material, the suspension is titrated to pH 4.0–5.0 with 16 μL of 6 *N* HCl. Sodium phosphate buffer (0.2 *M*, pH 7.0) is then added to a final concentration of 50 mM. The resulting solution (final volume of ~30 mL) is centrifuged for 5 min at 3000*g* to remove the calcium phosphate salt precipitate. To initiate the coupling to agarose, the clear pyridoxamine 5'-phosphate solution is mixed with 5 mL of Affi-Gel-10 beads (Bio-Rad), prepared according to the manufacturer's instructions. The slurry is gently rotated

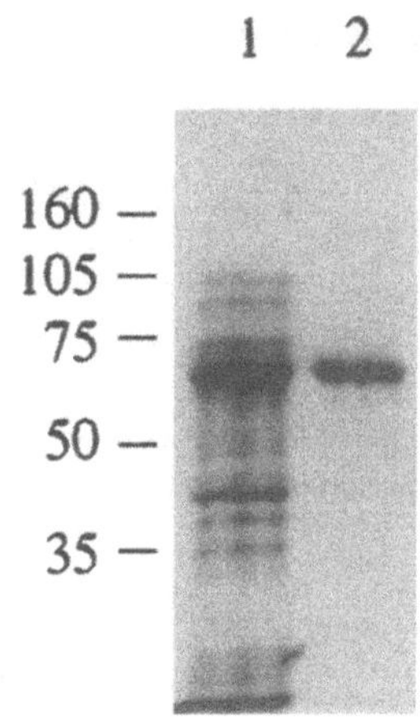


Fig. 4. MBP-Az preparation. Samples of MBP-Az separated on 10% SDS-PAGE and stained with Coomassie blue. *Lane 1*, Lysate of induced bacteria (10 µg); *lane 2*, affinity-purified MPB-Az after ion-exchange chromatography (2 µg). (Reprinted from **ref. *4*.**)

at 4°C for 4 h. Excess unbound pyridoxamine 5'-phosphate is removed by extensive washing with 200 mL of sodium phosphate buffer, pH 7.0, containing 10 m*M* NaCl. The column is stored at 4°C.

### *3.5. Az Production*

1. pMAL-Az is transformed into *E. coli* strain DH10B. A bacterial culture (300 mL) is induced with 0.3 m*M* IPTG. After 2 h of induction at 37°C, the cells are harvested and then washed with ice-cold buffer P supplemented with a 1:25 (v/v) solution of protease inhibitors.
2. The cells are then resuspended in 20 mL of buffer P adjusted to 0.5 *M* NaCl and lysed in a French pressure press cell. The insoluble material is removed by centrifugation (10,000*g*, 15 min).
3. A sample of the bacterial lysate (25 mg of protein) was applied to a 1 mL amylose resin. The column was then washed with buffer P, and MPB-Az was eluted from the column by the sequential addition of 1-mL portions of buffer P containing 10 m*M* maltose. The first 2 mL that contained the bulk of the recombinant protein (~1 mg) were combined and purified further by ion-exchange chromatography (**Fig. 4**).
4. The affinity-purified protein from **step 1** was diluted in 10 mL of buffer L (25 m*M* Tris-HCl, pH 7.5, 2.5 m*M* DTT) and loaded on a Mono-Q 5/5 column (Amersham Pharmacia Biotech) equilibrated in buffer L.
5. The column was washed with buffer L containing 0.1 *M* NaCl and then developed with a linear gradient of 0.1–1.0 *M* NaCl in buffer L. MBP-Az eluted from the column as a sharp protein peak at 0.4 *M* NaCl.
6. The protein was concentrated to 2 mg/mL in Centricon 30 and stored in aliquots at 80°C. The Mono-Q-purified MBP-Az was used in all of the experiments described in this study.

### 3.6. Purification of 26S Proteasome Complex

All purification procedures were performed at 4°C.

1. The 26S proteasome complex is prepared from livers of C57Bl mice (10–13 wk old). A typical preparation is from 10 livers. The livers are thoroughly washed with PBS and then homogenized using a motor-driven Potter–Elvehjem Teflon tissue grinder. Homogenization is in 5 mL/liver of buffer A.
2. The crude extract is then subjected to fractional centrifugation at 1000*g* followed by centrifugation at 10,000*g*. The 10,000*g* supernatant is subjected to ultracentrifugation for 1 h at 70,000*g*. The resulting supernatant (lysate) is then subjected to ammonium sulfate precipitation.
3. The lysate containing approx 250 mg of protein is supplemented with 5 m*M* $MgCl_2$, 10 m*M* phosphocreatine, and 10 µg/mL of creatine phosphokinase and incubated for 1 h at 37°C. The 26S proteasome complex is then precipitated with ammonium sulfate at 38% (w/v) saturation.
4. The 38% ammonium sulfate sediment was dissolved in buffer B and loaded onto a Sepharose 6B fast flow column (2.5 × 40 cm) (Amersham Pharmacia Biotech) equilibrated in buffer B. Fractions of 2 mL were collected, and 26S proteasome activity was assayed in 2-µL samples of column fractions.
5. The proteasome peak from **step 3** is combined and loaded onto a 1 × 4 cm Resource-Q column (Amersham Pharmacia Biotech) equilibrated in buffer B. The column is then washed with 10 mL of buffer B and developed by a linear gradient from 0 to 0.8 *M* NaCl in buffer B over 50 mL. The 26S proteasome is eluted from the column between 0.35 and 0.4 *M* salt.
6. The 26S proteasome complex from the ion-exchange column is concentrated to 250 µL by ultrafiltration in a Centricon 30 (Amicon). The sample is loaded on a 10–40% (v/v) glycerol gradient in buffer B (11.5 mL in a 14 × 95 mm tube). After centrifugation at 28,000 rpm for 18 h at 4°C, fractions of 0.4 mL are collected, and 26S proteasome activity is assayed in 1-µL samples. The proteasome peak is stored in aliquots at 80°C.

### 3.7. Degradation of Bacterially Produced ODC

Metabolically labeled recombinant ODC is incubated with Az and 26S proteasome. The rate of degradation is determined by counting trichloroacetic acid (TCA)-soluble material (**Fig. 5A**). It is noteworthy that the rate of degradation of bacterially expressed ODC proteins is substantially lower than that obtained for the reticulocyte synthesized protein (**Fig. 1**). As increasing the Az concentration does not stimulate the relative rate of degradation, the low degradation rate is likely to result from a relatively large proportion of improperly folded ODC in the bacterial preparation.

1. For production of $^{35}$S-labeled ODC-ova in bacteria, pET-ODCova is transformed into the methionine auxotroph *E. coli* strain B834 (DE3) (Novagen Inc., Madison, WI).
2. A 50-mL culture is grown at 37°C in M9 minimal medium supplemented with thiamine (20 µg/mL) and all 20 amino acids at 0.2 m*M* until absorbance at $A_{600}$ reaches 0.6–0.7.
3. IPTG and Pro-mix L-[$^{35}$S] (Amersham Pharmacia Biotech) (0.5 mCi) are then added for a further incubation at 22°C for 16 h. Purification of $^{35}$S-labeled ODC-ova is then carried out exactly as described in **Subheading 3.3.3., steps 5–7**.

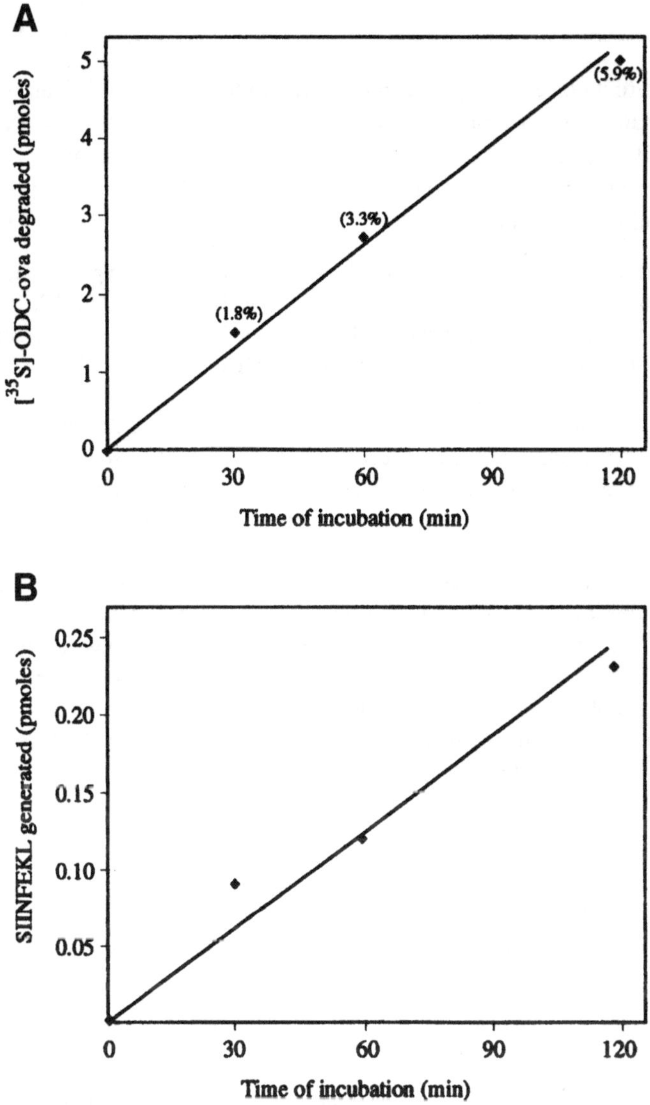


Fig. 5. Processing of $^{35}$S-labeled ODC-ova by purified 26S proteasome. $^{35}$S-Labeled ODC-ova (4.3 μg of protein, 87 pmol, 120,000 cpm) was incubated for the indicated time periods with purified 26S proteasome and MBP-Az in a volume of 350 μL in a standard reaction mixture. At each time point, generation of SIINFEKL (**A**) and degradation of [$^{35}$S]ODC-ova (**B**) were quantified. To determine the percentage of degradation of [$^{35}$S]ODC-ova, duplicate aliquots of 50 μL were withdrawn at each time point. The amount of [$^{35}$S]ODC-ova degraded was then determined by measuring the amount of soluble radioactivity after the addition of TCA. The *numbers* in *parentheses* indicate the percentage of ODC-ova degradation at each time point. To quantify the amount of SIINFEKL, peptides were isolated from the remaining reaction mixture (250 μL) and then incubated with RMA/S cells. The cells were then tested for recognition by mAb 25-D1.16 as described in **Subheading 3.7.**, **steps 3** and **4**. The amount of SIINFEKL produced from ODC-ova was calculated based on the reactivity of RMA/S cells that were incubated in parallel with known amounts of synthetic SIINFEKL. (Reprinted from **ref.** ***4***.)

There is direct correlation between the rate of recombinant ODC antigen degradation and the rate of generation of the MHC class I–restricted epitope (**Fig. 5B**). Therefore, this cell-free ODC degradation system is suitable for the analysis of the 26S proteasome structure–function relationship in MHC class I antigen processing and possibly in other systems as well.

1. Reactions are carried out in a final volume of 250 µL containing the following components: 40 m*M* Tris-HCl, pH 7.5, 5 m*M* $MgCl_2$, 2 m*M* DTT, 1 m*M* ATP, 10 m*M* creatine phosphate, 12.5 U of creatine phosphokinase, 5 µ*M* bestatin, a 1:25 (v/v) solution of protease inhibitors, 4–8 µg of recombinant antigen (ODC–ova or Nef–ODC), 26 µg of MBP-Az, and 20 U of 26S proteasome.
2. Following incubation at 37°C for various time periods, the reaction mixture is adjusted to pH 2 by the addition of trifluoroacetic acid (TFA) and then sonicated for 30 s at full power in a bath sonicator. The acid extract is microcentrifuged in an Amicon Microcon 10 microconcentrator (Millipore Corp., Bedford, MA). The filtrate is collected and lyophilized.
3. The lyophilized low molecular weight material from **step 2** is separated on a 2.1 × 150-mm C18 column (Vydac, Hesperia, CA) (eluant A, 0.1% TFA, 4% acetonitrile; eluant B, 0.085 TFA, 90% acetonitrile; gradient 4–50% B in 45 min; flow rate of 0.2 mL/min). Based on the position of elution of the synthetic SIINFEKL, the material is pooled and tested for biological activity or analyzed directly by mass spectrometry.
4. The combined peptide fraction from **step 2** is incubated with RMA/S cells (expressing empty H2-$K^b$) ***(9)***. The cells are then incubated with mAb 25-D1.16 (that specifically recognizes H2-$K^b$-SIINFEKL complexes) followed by a second incubation with fluorescein isothiocyanate–labeled $F(ab')_2$ goat antimouse IgG and analyzed by flow cytometry (**Figs. 5B** and **6**).
5. For cytotoxicity assays: the peptide fraction from **step 2** is incubated with $^{35}S$-labeled RMA/S cells. Peptide binding makes the cells susceptible to lysis by SIINFEKL-specific cytotoxic T-lymphocytes (CTL). Consequently, the amount of peptide is determined in a standard cytotoxicity assay ***(4)***.

## 4. Notes

1. Although we are using A31N-ts20 as cells harboring thermosensitive E1, there are additional cell lines that can be used for this purpose. It is important to test such cell lines for loss of conjugation activity prior to their use. Also it should be noted that different proteins are differentially affected by specific degree of E1 inactivation.
2. pMAL was used very effectively to produce Az. Other expression vectors were also effective in producing Az in *E. coli*. In each case only part of the produced Az fusion protein was soluble and active. Reducing the culturing temperature can increase the soluble fraction. In all cases it is important to use only soluble material for purification.

## Acknowledgment

This work was supported by a grant from the Israel Science Foundation (ISF) (to C. Kahana)

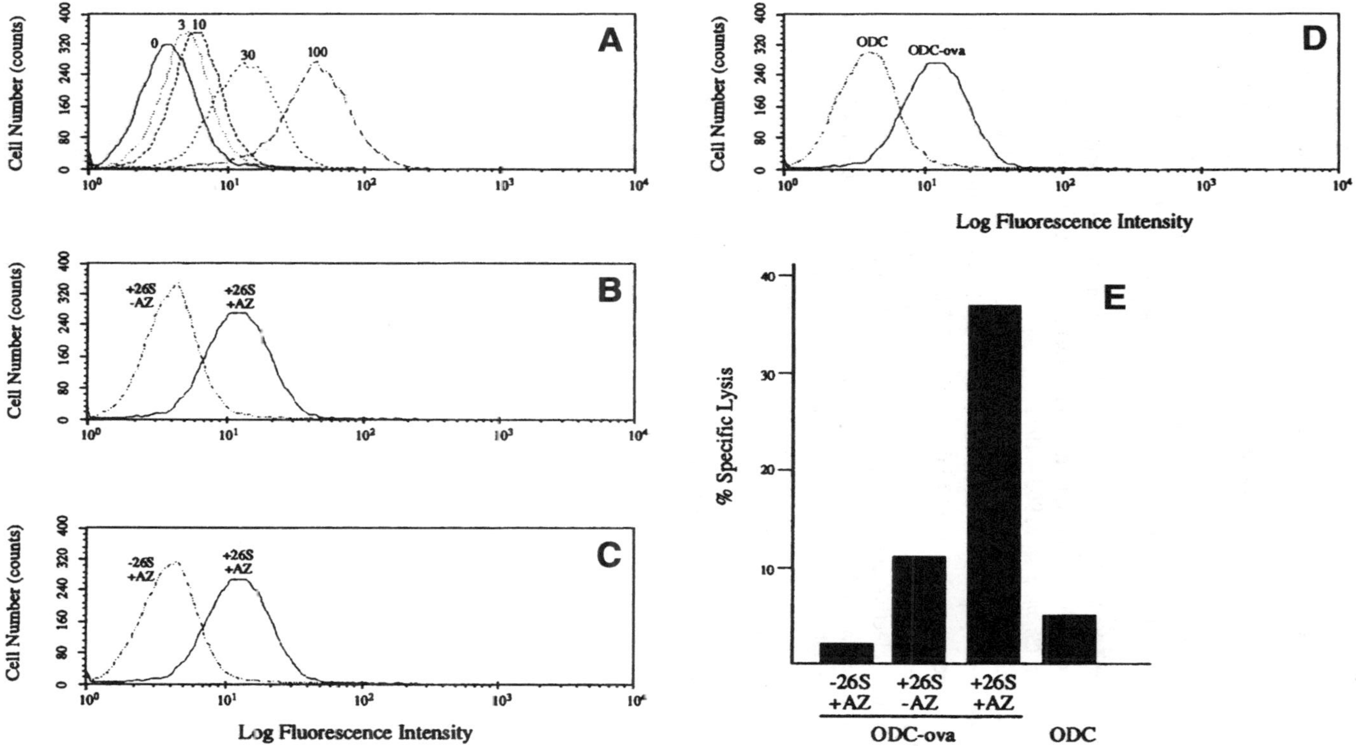


Fig. 6. Specificity of processing of ODC-ova by the 26S proteasome. RMA/S cells were incubated with 0–100 fmol of synthetic SIINFEKL **(A)**, or with peptides isolated from processing reactions **(B–E)**. The cells were then tested for recognition by mAb 25-D1.16 by flow cytometry (A–D) and SIINFEKL-specific CTL in a cytotoxicity assay (E). Processing of ODC-ova in the presence of 26S proteasome and Az was compared with processing with 26S proteasome and without Az (B), with Az and without 26S proteasome (C), or with 26S proteasome and Az but in the presence of ODC instead of ODC-ova (D). (Reprinted from **ref. *5*.**)

## References


1. Hershko, A. and Ciechanover, A. (1998) The ubiquitin system. *Annu. Rev. Biochem.* **67,** 425–479.
2. Coffino, P. (2001) Antizyme, a mediator of ubiquitin-independent proteasomal degradation. *Biochimie* **83,** 319–323.
3. Asher, G., Lotem, J., Sachs, L., Kahana, C., and Shaul, Y. (2002) Mdm-2 and ubiquitin-independent p53 proteasomal degradation regulated by NQO1. *Proc. Natl. Acad. Sci. USA* **99,** 13125–13130.
4. Ben-Shahar, S., Komlosh, A., Nadav, E., et al. (1999) 26 S proteasome-mediated production of an authentic major histocompatibility class I-restricted epitope from an intact protein substrate. *J. Biol. Chem.* **274,** 21963–21972.
5. Komlosh, A., Momburg, F., Weinschenk, T., et al. (2001) A role for a novel luminal endoplasmic reticulum aminopeptidase in final trimming of 26 S proteasome-generated major histocompatability complex class I antigenic peptides. *J. Biol. Chem.* **276,** 30050–30056.
6. Chowdary, D. R., Dermody, J. J., Jha, K. K., and Ozer, H. L. (1994) Accumulation of p53 in a mutant cell line defective in the ubiquitin pathway. *Mol. Cell Biol.* **14,** 1997–2003.
7. Hobbs, S., Jitrapakdee, S., and Wallace, J. C. (1998) Development of a bicistronic vector driven by the human polypeptide chain elongation factor 1alpha promoter for creation of stable mammalian cell lines that express very high levels of recombinant proteins. *Biochem. Biophys. Res. Commun.* **252,** 368–372.
8. Seifert, U., Maranon, C., Shmueli, A., et al. (2003) An essential role for tripeptidyl peptidase in the generation of an MHC class I epitope. *Nat. Immunol.* **4,** 375–379.
9. Ljunggren, H. G., Stam, N. J., Ohlen, C., et al. (1990) Empty MHC class I molecules come out in the cold. *Nature* **346,** 476–480.
10. Bercovich, Z., Rosenberg-Hasson, Y., Ciechanover, A., and Kahana, C. (1989) Degradation of ornithine decarboxylase in reticulocyte lysate is ATP-dependent but ubiquitin-independent. *J. Biol. Chem.* **264,** 15949–15952.
11. Gandre, S., Bercovich, Z., and Kahana, C. (2002) Ornithine decarboxylase-antizyme is rapidly degraded through a mechanism that requires functional ubiquitin-dependent proteolytic activity. *Eur. J. Biochem.* **269,** 1316–1322.

# 8

# Assays of Proteasome-Dependent Cleavage Products

Stefan Tenzer and Hansjörg Schild

**Summary**

The degradation of misfolded, aged, or no longer needed cytosolic proteins depends largely on the ubiquitin–proteasome system. Proteasomes degrade their substrates into fragments of 3–20 amino acids. Human 20S proteasomes can be purified from human erythrocytes by batch adsorption to DEAE-cellulose, ammonium sulfate precipitation, anion-exchange fast protein liquid chromatography (FPLC), and glycerol density gradient ultracentrifugation. 20S proteasomes purified by this method are suitable for the in vitro digestion of synthetic peptides as well as full-length proteins. The degradation products produced by proteasomes are separated by reversed-phase HPLC using an acetonitrile gradient. The obtained fractions are further analyzed by matrix-assisted laser desorption ionization mass spectrometry (MALDI-MS) and Edman degradation, which allows a quantitative analysis of the digestion products.

**Key Words:** Analysis; antigen processing; Edman degradation; in vitro degradation; MALDI-mass spectrometry; peptide; proteasome; purification; quantification.

## 1. Introduction

The degradation of misfolded, aged, or no longer needed cytosolic proteins depends largely on the ubiquitin–proteasome system. Proteasomes degrade their substrates into fragments of 3–20 amino acids *(1)*, which are further broken down by aminopeptidases into single amino acids. A fraction of the proteasomally produced fragments is translocated into the endoplasmatic reticulum by the transporter associated with antigen processing (TAP). There they can associate with major histocompatibility complex (MHC) class I molecules, which are then presented at the cell surface for the inspection by cytotoxic T-lymphocytes *(2)*.

The 20S proteasome is a 700-kDA complex composed of 14 different subunits, which are arranged in four stacked rings with the stoichiometry of $\alpha7\beta7\beta7\alpha7$. The proteolytically active subunits are found in the β-rings. Their active centers face the inner hollow center of the 20S proteasome ***(3,4)***; on stimulation with interferon-γ, the three active β-subunits, Y, Z, and MB1, are exchanged to their immunocounterparts

From: *Methods in Molecular Biology, vol. 301, Ubiquitin–Proteasome Protocols*
Edited by: C. Patterson and D. M. Cyr © Humana Press Inc., Totowa, NJ

LMP2, LMP7, and MECL-1 ***(5)***. This results in a change in proteasomal specificity ***(6–9)***, which influences the generation of CTL epitopes ***(10–14)***.

As the generation of the correct C-terminus by the proteasome is required for the presentation of most CTL-epitopes, the in vitro analysis of proteasomal degradation of peptides has been proven to be an important tool for the identification of CTL ligands ***(15–17)***. The largest number of experimental data on proteasomal cleavage specificity so far stems from the in vitro digestion of whole unmodified proteins and the quantitative analysis of the degradation products by mass spectrometry and Edman sequencing ***(18,19)***. These experimental data have made it possible to develop computer algorithms that are able to predict proteasomal cleavages in a substrate sequence ***(20–23)***.

This chapter focuses on the purification of active human 20S proteasomes and the in vitro digestion of peptides and full-length proteins. The methods utilized for the separation, identification, and quantification of proteasomal cleavage products are illustrated.

## 2. Materials

1. Phosphate-buffered saline (PBS), pH 7.2.
2. Lysis buffer: 30 m*M* Tris-HCl, pH 7.6, 2 m*M* $MgCl_2$, 0.1 m*M* EDTA, 1.6 m*M* dithiothreitol (DTT).
3. TSDG buffers: 20 m*M* Tris-HCl, pH 7.6, 10 m*M* KCl, 2 m*M* $MgCl_2$, 0.1 m*M* EDTA, 1 m*M* DTT, 10% glycerol. TSDG buffers contain variable micromolar amounts of NaCl indicated by the subscript; for example, $TSDG_{100}$ contains 100 m*M* NaCl.
4. DEAE-52 cellulose.
5. Büchner funnel + suction flask + filter.
6. Dialysis tubing (10-kDa cutoff).
7. Fast protein liquid chromatography (FPLC) equipment.
8. Anion-exchange chromatography resin DEAE-Toyopearls 650S (Tosoh Bioscience, Stuttgart, Germany).
9. Column HR16/50 (Amersham Pharmacia, Sweden) or equivalent column for 100 mL gel volume.
10. DEAE wash buffer A: 10 m*M* Tris-HCl, pH 8.0.
11. DEAE wash buffer B: 10 m*M* Tris-HCl, pH 8.0, 0.5 *M* NaCl.
12. Fluorescence sample buffer: 20 m*M* Tris-HCl, pH 7.6, 10 m*M* KCl, 10 m*M* NaCl, 2 m*M* $MgCl_2$, 0.1 m*M* EDTA, 1 m*M* DTT.
13. suc-LLVY-AMC (succinyl-leucyl-leucyl-valyl-tyrosyl-[7-aminomethyl]-coumarine).
14. Fluorimeter (excitation wavelength between 360 nm and 380 nm, emission wavelength between 430 nm and 460 nm).
15. Amicon Ultra-15 (100-kDa cutoff) concentration unit.
16. Gradient buffer 15%: 20 m*M* Tris-HCl, pH 7.6, 100 m*M* NaCl, 10 m*M* KCl, 2 m*M* $MgCl_2$, 0.1 m*M* EDTA, 1 m*M* DTT, 15% glycerol.
17. Gradient buffer 40%: 20 m*M* Tris-HCl, pH 7.6, 100 m*M* NaCl, 10 m*M* KCl, 2 m*M* $MgCl_2$, 0.1 m*M* EDTA, 1 m*M* DTT, 40% glycerol.
18. Gradient mixer.
19. Ultracentrifuge with SW40Ti-Rotor (Beckman) (or equivalent swinging bucket rotor with a capacity of 6 3–14 mL).
20. Synthetic peptide of choice.

21. 10X Peptide digestion buffer: 200 m*M* *N*-(2-hydroxyethyl)piperazine-*N*-(2-ethanesulfonic acid) (HEPES)–NaOH pH 7.6, 20 m*M* $MgAc_2$, 5 m*M* DTT.
22. 25% Trifluoroacetic acid (TFA).
23. Reversed-phase high-performance liquid chromatography (RP-HPLC) buffers A (0.1% TFA in $ddH_2O$) and B (0.1% TFA, 80% acetonitrile, 19.9% $ddH_2O$).
24. μRPC C2/C18 PC2.1/10 Column (Amersham Pharmacia).
25. HPLC equipment.
26. SpeedVac.
27. 50% Methanol–1% formic acid.
28. Matrix-assisted laser desorption ionization time-of-flight (MALDI-TOF) mass spectrometer.
29. DHAP matrix solution (15 mg of 2,5-dihydroxyacetophenon, 5 mg of ammonium citrate, 200 μL of $H_2O$, 800 μL of isopropanol).
30. Biobrene plus (Applied Biosciences, Salt Lake City, UT).
31. Edman sequencing equipment (Applied Biosystems Procise 494A or comparable instrument).

## 3. Methods

The methods described in the following subheadings outline (1) the purification of 20S proteasomes from human erythrocytes, (2) the digestion of synthetic peptides and whole proteins by 20S proteasomes, and (3) the quantitative analysis of the digestion products.

### *3.1. Purification of 20S Proteasomes From Human Erythrocytes*

The purification of active 20S proteasomes from human erythrocytes is described in **Subheadings 3.1.1.–3.1.7.** This includes preparation of the erythrocytes from erythrocyte concentrate, batch adsorption to DEAE-52 cellulose, $(NH_4)_2SO_4$ precipitation, ion-exchange FPLC, glycerol-gradient ultracentrifugation, and the final buffer exchange and concentration. The average yield is 2 mg of >95% pure 20S proteasomes from two erythrocyte concentrate conserves (2X 300 mL).

#### *3.1.1. Preparation of Erythrocytes*

1. Empty the two conserves into a large beaker.
2. Add 1800 mL of ice-cold PBS; mix by stirring.
3. Centrifuge for 20 min, 1500*g* at 4°C.
4. After centrifugation carefully remove the supernatant and the whitish layer (*see* **Note 1**).
5. Repeat **steps 2–4** twice.

#### *3.1.2. Hypotonic Lysis*

1. Add 1.5 volumes of ice-cold lysis buffer (400–600 mL, depending on the loss during the washing steps) to the erythrocytes.
2. Shake (200 rpm) for 45 min at 4°C.
3. Centrifuge (15,000*g*, 20 min, 4°C) to remove cellular debris (*see* **Note 2**).

#### *3.1.3. Batch Adsorption*

This step is required for the removal of hemoglobin, which makes up for >90% of the protein present in the erythrocyte lysate.

1. Preparation of DEAE-52-cellulose: add 600 mL of $dH_2O$ to 80 g of DEAE-52–cellulose, allow to settle for 30 min, then decant to remove fines. Repeat twice. Resuspend in

200 mL of 1 *M* NaCl. Transfer the material to a Büchner funnel. Equilibrate with 200 mL of lysis buffer.
2. Apply erythrocyte lysate to material using vacuum (*see* **Note 3**).
3. Wash with 1000 mL of $TSDG_{10}$ in 100-mL aliquots.
4. Wash with 500 mL of $TSDG_{100}$ in 100-mL aliquots.
5. Elute with 500 mL of $TSDG_{300}$; collect 50-mL fractions.
6. Activity assay: Add 20 µL of each fraction to 200 µL of fluorescence sample buffer containing 100 µ*M* suc-LLVY-AMC. Incubate for 30 min at 37°C. Measure fluorescence (excitation 360 nm, emission: 450 nm) (*see* **Note 4**).
7. Pool fractions containing more than 25% of maximal activity.
8. The DEAE-52–cellulose may be recycled by alternating washes (500 mL each) with 10 m*M* Tris-HCl, pH 8.0, and 10 m*M* TrisHCl, pH 8.0, 0.5 *M* NaCl, until the white color of the resin is restored.

### 3.1.4. $(NH_4)_2SO_4$ Precipitation

1. Measure the volume of pooled fractions.
2. Slowly add 242 mg/mL of $(NH_4)_2SO_4$ while stirring at 4°C.
3. Stir slowly for 1 h at 4°C.
4. Centrifuge at 27,000*g*, 4°C for 30 min.
5. Transfer supernatant to a fresh beaker, slowly add 230 mg/mL (based on the volume measured in **step 1**) $(NH_4)_2SO_4$ while stirring at 4°C.
6. Stir slowly for 1 h at 4°C.
7. Centrifuge at 27,000*g*, 4°C for 30 min.
8. Discard the supernatant.
9. Dissolve the pellet in a minimal volume (25–50 mL) of $TSDG_0$.
10. Transfer the solution to dialysis tubing (100-kDa cutoff).
11. Dialyze for 6–15 h against 2000 mL of $TSDG_{25}$.

### 3.1.5. Ion-Exchange FPLC

For ion-exchange FPLC, a HR16/50 column packed with approx 100 mL of TSK-DEAE 650S Toyopearls resin is used. This resin allows high recovery of active 20S proteasomes while offering very high resolution and therefore higher purity of the final preparates.

The buffers used are: buffer A, $TSDG_0$; buffer B, $TSDG_{1000}$. The flow rate is 2 mL/min.

1. If any precipitate is visible after dialysis of the redissolved $(NH_4)_2SO_4$ precipitate, centrifuge for 30 min at 27,000*g*, 4°C.
2. Filter the dialyzed $(NH_4)_2SO_4$ precipitate through a 0.22-µm filter to remove any precipitates or aggregates, which may clog the column.
3. Equilibrate the column with 8% buffer B.
4. Apply the sample to the column.
5. Wash the column with 300 mL of 8% buffer B.
6. The protein is eluted from the column with the following gradient: 8–25% buffer B in 400 mL; 25–36% buffer B in 20 mL; 36% for 150 mL; 40% for 150 mL. Fractions with a volume of 4 mL are collected (*see* **Fig. 1A**). The 20S proteasome usually elutes at a conductivity of 140 m*M* to 160 m*M* (measured after the column; *see* **Fig. 1B**).
7. Activity assay: add 20 µL of each fraction to 200 µL fluorescence sample buffer containing 100 µ*M* suc-LLVY-AMC. Incubate for 30 min at 37°C. Measure fluorescence (excitation 360 nm, emission: 450 nm) (*see* **Fig. 1C**).

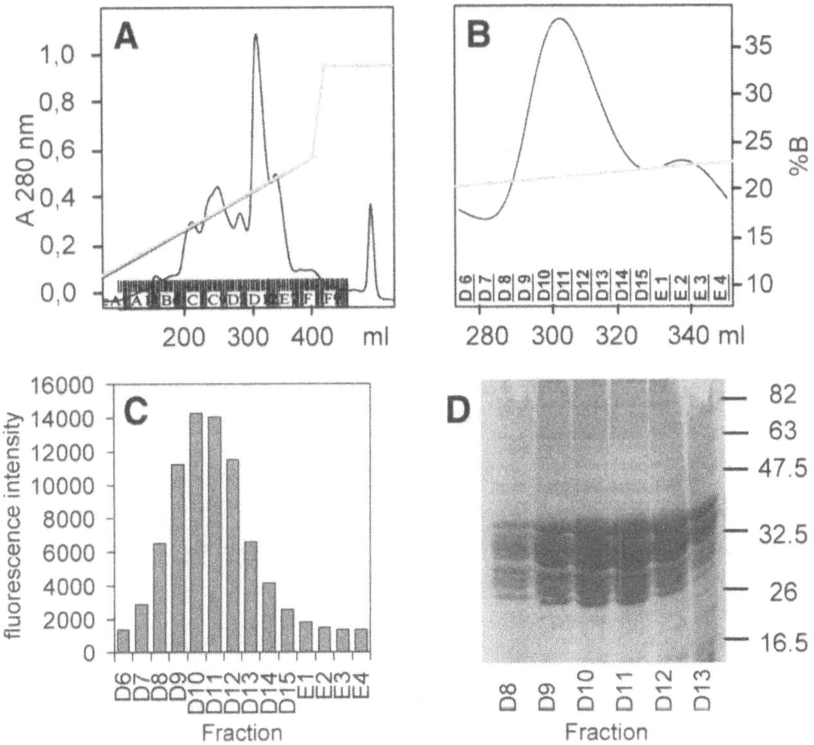


Fig. 1. Purification of 20S proteasomes by anion-exchange chromatography. The dialyzed $(NH_4)_2SO_4$ precipitate was loaded onto a 100-mL TSK-DEAE 650S Toyopearls column. Bound proteins were eluted with a gradient (**A**). Purified proteasomes eluted between 290 mL and 330 mL (**B**). The corresponding fractions were incubated with the fluorogenic substrate suc-LLVY-AMC and measured in a fluorimeter (**C**). Active fractions were analyzed by 12%-SDS-PAGE (**D**).

8. Pool fractions containing more than 50% of maximal activity.
9. Wash column at a flow rate of 1 mL/min with: 500 mL of 10 m*M* Tris-HCl, pH 8.0, 1*M* NaCl; 300 mL of 10 m*M* Tris-HCl, pH 8.0; 300 mL of $dH_2O$; 300 mL of 20% EtOH.

### *3.1.6. Glycerol Gradient Ultracentrifugation*

1. Concentrate the pooled fractions to a volume of 1.5–2 mL using an Amicon Ultra-15 concentration cell (100-kDa cutoff) (*see* **Note 5**).
2. Prepare glycerol density gradients of 15–40% glycerol in $TSDG_{100}$ in 14-mL clear ultracentrifugation tubes using a gradient mixer.
3. Centrifuge at 150,000*g*, 4°C for 18 h (slow acceleration, no brake).
4. Harvest the gradients in 500-µL fractions (*see* **Note 6**).
5. Activity assay: add 10 µL of each fraction to 200 µL of fluorescence sample buffer containing 100 µ*M* suc-LLVY-AMC. Incubate for 30 min at 37°C. Measure fluorescence (excitation 360 nm, emission: 450 nm).
6. Pool fractions containing more than 50% of maximal activity.

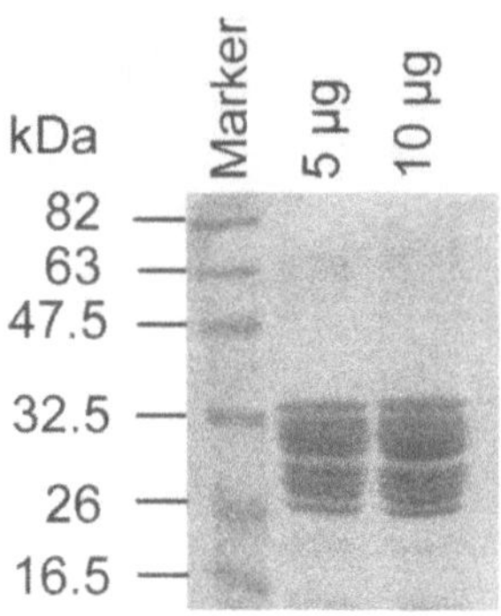


Fig. 2. 12% SDS-PAGE of purified 20S proteasomes. Five and ten micrograms of purified 20S proteasomes were loaded onto 12% SDS-PAGE and stained with Coomassie blue.

### *3.1.7. Buffer Exchange and Concentration*

1. Concentrate the pooled fractions to a volume of 1.5–2 mL using an Amicon Ultra-15 concentration cell (100-kDa cutoff).
2. Add 10 mL of $TSDG_{100}$.
3. Repeat **steps 1** and **2.**
4. Concentrate the pooled fractions to a protein concentration of approx 1 mg/mL using an Amicon Ultra-15 concentration cell (100-kDa cutoff).
5. Aliquot the purified proteasomes and freeze at –80°C.

## *3.2. Characterization of Purified Proteasomes*

The characterization of 20S proteasomes is described in **Subheadings 3.2.1.–3.2.4.** This includes (1) purity analysis by sodium dodecyl sulfate-polyacrylamide gel electrophoresis (SDS-PAGE), (2) characterization of subunits by Western blot, (3) activity assay using fluorogenic substrates, and (4) the use of proteasomal inhibitors to exclude the presence of other proteases.

### *3.2.1. SDS-PAGE*

1. Resolve 10 µg of purified proteasomes on a 12% SDS-PAGE gel.
2. Stain with Coomassie blue.

Proteasomal subunits have a size between 20 kDa and 30 kDa; no other bands should be visible in the stained gel (*see* **Fig. 2**).

### *3.2.2. Western Blot*

1. Resolve 5 µg of purified 20S proteasomes on a 12% SDS-PAGE gel.
2. Transfer to nitrocellulose and analyze by Western blot utilizing antibodies against different proteasomal subunits.

20S proteasomes purified from human erythrocytes are of a constitutive phenotype. In Western blot analysis, bands should be detectable only with antibodies against α-subunits and the proteolytically active constitutive β-subunits Y, Z, and δ.

When staining with antibodies against the immunosubunits LMP2, LMP7, and MECL-1, no bands should be detectable.

### *3.2.3. Fluorogenic Substrates*

The activity of 20S proteasomes can be defined by the ability to cleave fluorogenic substrates. For an exact measurement of the activity, a kinetic analysis is favorable (*see* **Note 7**).

1. Prepare a standard curve for AMC (0.3 μ*M*, 1 μ*M*, 3 μ*M*, 10 μ*M*, 30 μ*M*) in triplicate.
2. Incubate 0.5 μg of purified proteasomes in 200 μL of fluorescence sample buffer containing 200 μ*M* of any of suc-LLVY-AMC, Z-ARR-AMC, or Z-LLE-AMC. Prepare triplicates for each fluorogenic substrate.
3. Measure the fluorescence of each sample every 2 min in a fluorimeter (incubation temperature set to 37°C; excitation at 360 nm, emission at 450 nm) for a period of 100 min.
4. Plot the obtained fluorescence values over time and calculate the fluorescence increase over time for each substrate from the linear part of the curve.
5. Using the AMC standard curve, calculate the activity of the purified proteasomes in units per microgram (one unit is defined as the ability to cleave 1 pmol of substrate in 1 min).

20S Proteasomes purified by this protocol have activities from 20 to 50 U/μg (measured for the substrate suc-LLVY-AMC)

### *3.2.4. Proteasomal Inhibitors*

To exclude the presence of other proteases, the activity of the preparates can be tested in the presence of the proteasomal inhibitors lactacystin and epoxomicin.

1. Preincubate 1 μg of the purified proteasomes in 200 μL of fluorescence sample buffer containing 50 μ*M* lactacystin or 5 μ*M* epoxomicin for 30 min at 37°C
2. Add suc-LLVY-AMC to a final concentration of 200 μ*M*.
3. Measure proteasomal activity as described in **Subheading 3.2.3.**

Proteolytic activities of 20S proteasomes should be inhibited by >98% after preincubation with these inhibitors.

## 3.3. Digestion of Synthetic Peptides and Full-Length Proteins

The in vitro digestion of synthetic peptides is used mainly to assay if a CTL-epitope can be generated by the proteasome from an amino acid sequence carrying the CTL-epitope flanked by the adjacent sequences present in the antigen. This provides important information on the possibility for the CTL-epitope to be generated in vivo. The digestion of full length proteins is the tool of choice for a more general analysis of proteasomal cleavage specificity, as large pools of data can be generated from the analysis of the proteasomal digest of a single substrate protein.

The in vitro digestion of synthetic peptides and full-length proteins is described in **Subheadings 3.3.1.–3.3.6.** This includes (1) choice of solvent for synthetic peptides, (2) purity requirements, (3) digestion conditions for synthetic peptides, (4) preparative digestion of peptides, (5) digestion conditions for full-length proteins, and (6) preparative digestion of full-length proteins.

### 3.3.1. Getting the Peptide Into Solution

The preferred solvent for peptides to be used for digestion experiments is $H_2O$. However, not all peptides are soluble in water at the required concentration. It is therefore advisable to use a small aliquot of the peptide to check the solubility.

Peptides should be dissolved to a concentration of 1 m*M* in either $dH_2O$, 50% dimethyl sulfoxide (DMSO)–$H_2O$, or pure DMSO (*see* **Note 8**).

### 3.3.2. Purity Requirements

As impurities present in the synthetic peptide (e.g., peptides of shorter length or missing amino acids or incomplete removal of protective groups) will severely hamper the detection process and may also possibly lead to incorrect results, synthetic peptides used for in vitro digestion experiments should be of the highest possible purity (at least 90%).

The purity of a synthetic peptide should be checked by RP-HPLC and MALDI-MS before a digestion experiment is set up.

#### 3.3.2.1. Purity Analysis by RP-HPLC

1. Mix 5 nmol of the peptide with 200 µL of RP-buffer A.
2. Load the mixture onto the RP-column.
3. Elute at a flow rate of 150 µL/min with the following gradient:

| | |
|---|---|
| 0% Buffer B | for 10 min |
| 0–60% Buffer B | in 60 min |
| 60–100% Buffer B | in 20 min |
| 100% Buffer B | for 10 min |

4. The peptide should elute as a single sharp peak (usually between 25% buffer B and 60% buffer B, depending on the length and hydrophobicity of the peptide) (*see* **Fig. 3**).

#### 3.3.2.2. Purity Analysis by MALDI-MS

1. Mix 1 nmol of the peptide with 50 µL of RP-buffer A.
2. Pipet 1 µL of DHAP matrix solution onto the MALDI target.
3. Dry the matrix using a vacuum pump. The matrix should form fine crystals and uniformly cover the target area.
4. Apply 1 µL of the diluted peptide to the matrix-covered target. Do not touch the target with the pipet tip.
5. Apply vacuum to dry the sample. If bubbles form during the drying process, release the vacuum for a short time.
6. Measure the sample in the MALDI-TOF mass spectrometer. There should be only very small peak intensities other than the one correlated to the peptide (*see* **Fig. 4**).

### 3.3.3. Digestion Conditions: Peptides

There is considerable variation in the time period required for the digestion of synthetic peptides. The incubation time required to achieve 50% digestion is dependent on several factors, namely purity of the peptide, choice of peptide solvent, sequence of the peptide, and activity of the proteasomes.

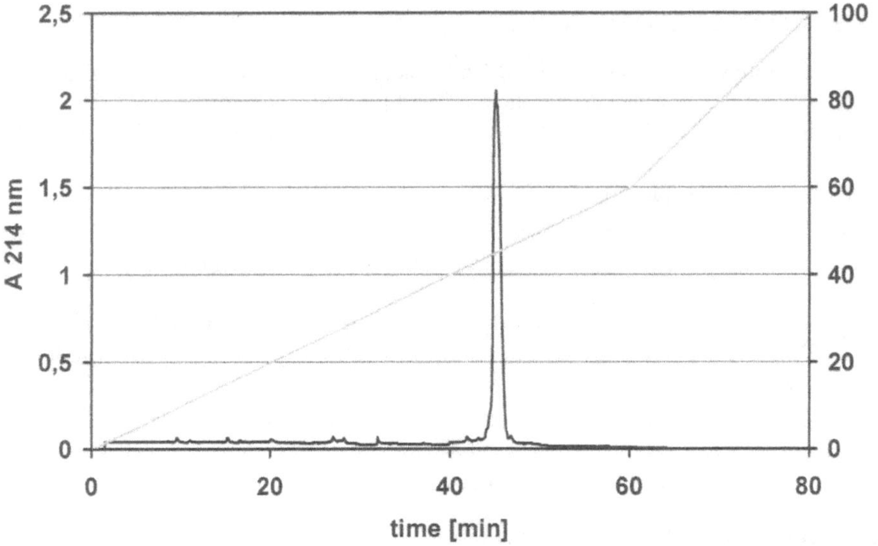


Fig. 3. HPLC profile of the peptide TTIHYKYMCNSSCGGMNRRPILT. Ten nanomoles of the peptide were loaded onto a µRPC-C2-C18 column and eluted with a gradient from 0% buffer B to 60% buffer B in 60 min. The peptide eluted at 46 min.

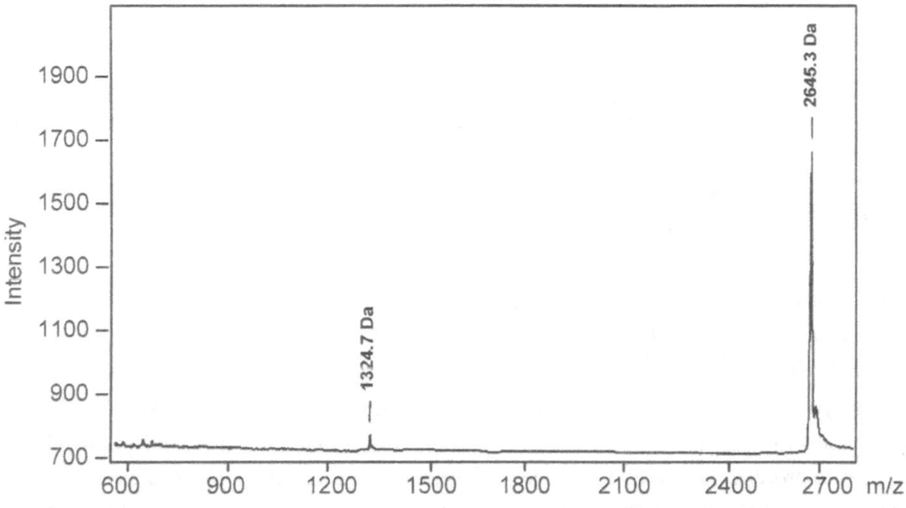


Fig. 4. MALDI-MS spectrum of the peptide TTIHYKYMCNSSCGGMNRRPILT. One picomole of the peptide was loaded on a target coated with DHAP and analyzed by MALDI-MS.

To establish the optimal incubation time:

1. Mix the following components in a 0.5-mL reaction tube: 10 nmol of peptide, 140 μL of $dH_2O$, 15 μL of 10X peptide digestion buffer, 1 μg of 20S proteasome.
2. Incubate for 24 h at 37°C; take 10-μL aliquots after 0 h, 2 h, 4 h, 6 h, 8 h, 12 h, 16 h, 20 h, and 24 h.
3. To prepare the samples for MALDI-MS:
   a. Equilibrate the Zip-Tip with 20 μL of 0.1% TFA –50% acetonitrile (pipet up and down).
   b. Equilibrate the Zip-Tip with 20 μL of 0.1% TFA (pipet up and down).
   c. Apply sample to the Zip-Tip (pipet up and down).
   d. Wash the Zip-Tip three times with 20 μL of 0.1% TFA (pipet up and down)
   e. Elute peptides from the Zip-Tip with 5 μL of 0.1% TFA/50% acetonitrile into a fresh tube.
4. Measure samples in MALDI-MS as described in **Subheading 3.3.1.**
5. The time to achieve 50% digestion is defined by 50% of the time point at which the signal for the full-length peptide is no longer detected.

### *3.3.4. Preparative Digest of Peptides*

1. Mix in a 0.5 mL test tube: 20 μL of 1 m*M* peptide solution, 250 μL of $dH_2O$, 30 μL of 10X peptide digestion buffer, and 2 μg of 20S proteasome.
2. Transfer 150 μL to a fresh tube and freeze at –20°C.
3. Incubate the digestion reaction at 37°C for the digestion time established in **Subheading 3.3.2.**
4. If immediate separation of the digestion products by RP-HPLC is not possible, freeze the sample at –80°C (*see* **Note 9**).

### *3.3.5. Digestion Conditions: Whole Proteins*

Although the digestion conditions for peptides are easily established, the conditions for the digestion of full-length proteins by 20S proteasomes are very difficult to pinpoint. Unfortunately, no general rule can be given for the composition of the reaction buffers required. Some but not all proteins require the addition of 0.01–0.03% SDS to the digestion mixture.

Commonly used buffers are:

1. 30 m*M* Tris-HCl, pH 8.0, 10 m*M* NaCl, 2 m*M* $MgCl_2$, 1 m*M* DTT.
2. 20 m*M* HEPES–NaOH, pH 7.6, 2 m*M* $MgCl_2$, 0.5 m*M* DTT.
3. 20 m*M* HEPES–KOH, pH 7.6, 2 m*M* $MgAc_2$.

These buffers can be used with or without the addition of 0.01–0.03% SDS, which is thought to open the gate of the 20S proteasome, as well as to partially unfold the substrate protein. For an analysis of the fragments generated by 20S proteasomes, the substrate should be of very high purity and absolutely free of other, contaminating, proteases.

To test the digestion conditions:

1. In a 0.5-mL tube, mix 25 μg of protein with 250 μL of the digestion buffer to be tested and 5 μg of purified 20S proteasomes. Prepare a control tube without the addition of 20S proteasomes.
2. Incubate at 37°C.

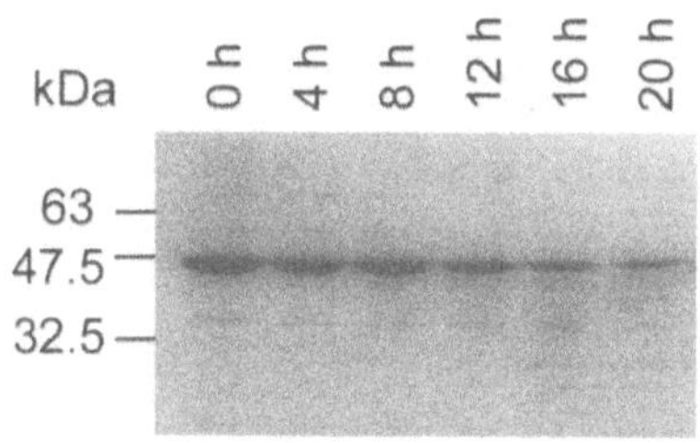


Fig. 5. 10% SDS-PAGE of the digest of enolase with 20S proteasomes. One hundred micrograms of yeast enolase-1 were incubated with 5 µg of purified 20S proteasomes for the indicated periods of time. Aliquots corresponding to 2 µg of protein were separated by 10% SDS-PAGE and stained with Coomassie blue.

3. Take 30-µL aliqouts every 4 h and stop the reaction by the addition of 10 µL of 4X Laemmli loading buffer.
4. Resolve the aliquots on a 12% SDS-PAGE gel and stain by Coomassie (*see* **Fig. 5**).

If buffer conditions allowing the digestion of the protein can be identified, determine the optimal digestion time by quantification of the Coomassie-stained bands using a densitometer (approx 50% of digestion are optimal).

#### 3.3.6. Preparative Digest of Whole Proteins

1. Mix in a 1.5-mL test tube: 150 µg of the protein to be digested, 1500 mL of digestion buffer of choice, and 30 mg of 20S proteasome.
2. Transfer 30 mL to a fresh test tube, add 10 µL of 4X Laemmli loading buffer, and freeze at –20°C.
3. Incubate the digestion reaction at 37°C for the digestion time established in **Subheading 3.3.2.**
4. Take a 30-µL aliquot for SDS-PAGE and add 10 µL of 4X Laemmli loading buffer.
5. Resolve the aliquots (0 h and after digestion) on 12% SDS-PAGE for documentation purposes.
6. If immediate separation of the cleavage products by RP-HPLC is not possible, freeze the sample at –80°C (*see* **Note 9**).

### 3.4. Separation and Analysis of the Cleavage Products

#### 3.4.1. Separation of the Cleavage Products by RP-HPLC

1. Add TFA to the sample to a final concentration of 0.1%.
2. Centrifuge the sample for 30 min at 15,000*g*, 4°C to remove any precipitates.
3. RP-HPLC is performed at a flow rate of 150 µL/min.
4. Equilibrate the RP-column with: 100% Buffer B for 15 min, 0% Buffer B for 15 min.
5. Load the sample to the RP-HPLC column.
6. Wash the column with 0% buffer B for 15 min.
7. Elute the peptides from the column with the following gradient (*see* **Fig. 6**): 0–10% Buffer B in 5 min; 10–50% Buffer B in 60 min; 60–100% Buffer B in 20 min; 100% Buffer B for 15 min. Collect 150-µL fractions by automatic fractionation (*see* **Note 10**).
8. Dry the fractions using a SpeedVac (20°C, ~30 min) and dissolve the pellet in 50% methanol–1% formic acid.
9. Freeze fractions at –80°C until further analysis (*see* **Note 9**).

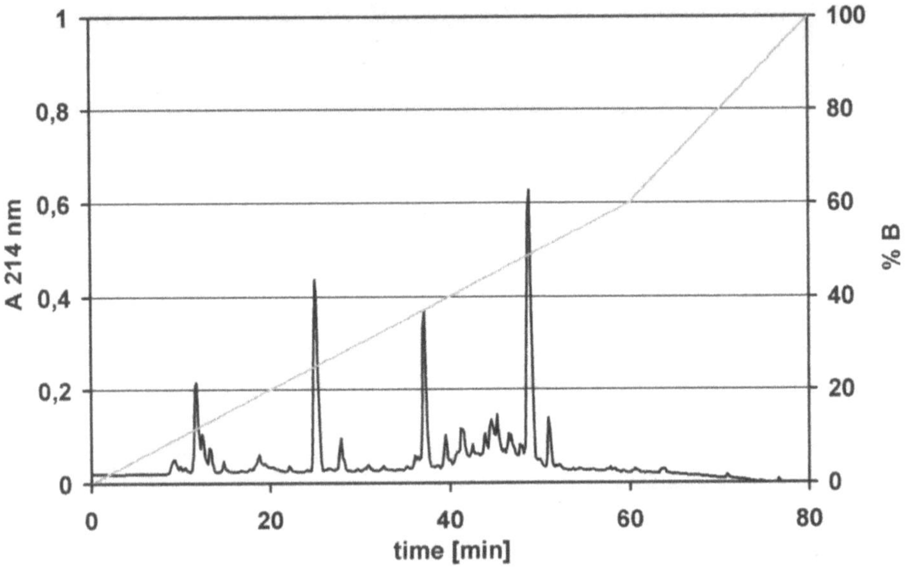


Fig. 6. HPLC elution profile of the digest of the peptide TTIHYKYMCNSS CGGMNRRPILT with purified 20S proteasomes. Ten nanomoles of the peptide were incubated with 2 μg of 20S proteasomes for 6 h at 37°C and the fragments generated were loaded onto a μRPC-C2-C18 column and eluted with a gradient from 0% buffer B to 60% buffer B in 60 min.

### *3.4.2. Analysis of the Cleavage Products by MALDI-MS*

The analysis of the fractions obtained from RP-HPLC by MALDI-MS is the first step in the analysis and identification of proteasomally produced peptide fragment derived either from peptides or whole proteins. However, the analysis by MALDI-MS gives by no means any quantifiable results. This is attributable to the different ionization properties of different peptide fragments. Furthermore, sometimes suppression effects can be observed if more than one peptide is present in a sample.

1. Thaw frozen fractions at 4°C.
2. Pipet 1 μL of DHAP matrix solution onto the MALDI target.
3. Dry the matrix using a vacuum pump. The matrix should form fine crystals and uniformly cover the target area.
4. Apply 1 μL of the diluted peptide to the matrix-covered target. Do not touch the target with the pipet tip.
5. Apply vacuum to dry the sample. If bubbles form during the drying process, release the vacuum for a short time.
6. Measure the sample in the MALDI-TOF mass spectrometer.

If there are detectable peaks in the MALDI-MS spectrum, the peptide can be identified by the corresponding mass. When using a high-resolution MALDI-MS, the direct identification of the fragments is possible in most cases.

Identification of fragments derived from a digest of short peptides is easily accomplished by computing a table with all possible fragments and their corresponding masses. However, for the identification of fragments stemming from the digest of full-length proteins, this is quite laborious and it is preferable to develop computer software for this task.

In some cases, the peptides cannot be directly identified via their mass. This may stem from oxidation of cysteine or methionine residues (which leads to an apparent mass + 16 Da), dimerization via cysteine residues or deamidation of asparagine or glutamine residues.

### *3.4.3. Analysis and Quantification of the Cleavage Products by Edman Degradation*

The only technique allowing a reliable quantification of peptides is analysis by Edman degradation. For this task, automated peptide sequencers have been developed. They allow the reproducible analysis required for the quantification of digestion products; even fractions containing multiple peptides can be analyzed ***(24)***. The operation of an automated peptide sequencer is described in detail in the manual provided by the manufacturer.

The fixation of the sample on the glass fiber filter is accomplished by first coating the filter with Biobrene Plus, which allows hydrophobic and ionic interactions of the peptide fragments with the filter. A covalent fixation is not recommended for peptide samples. The sample is then applied to the filter in 50% methanol–1% formic acid.

Commercially available peptide sequencers have a cycle efficiency of about 96%, allowing the identification of up to 20 N-terminal amino acids. However, this is not necessary for the identification of peptide fragments as five to eight cycles are usually sufficient for the identification of a peptide fragment when used in combination with MALDI-MS data.

Even the identification of multiple sequences present in one sample is possible. To facilitate the analysis, the use of the FINDPATTERN software from GCG is recommended.

For the quantification of the peptide fragments, the PTH-derivates of the amino acids generated in the Edman reaction are separated online by RP-HPLC and detected by UV absorption by the peptide sequencer.

The quantification is based on peak height of the UV-absorption curve. It is recommended to rely on the values obtained for chemically inert amino acids (A, F, G, I, L, M, V).

### *3.4.4. Examples for the Analysis of Digestion Products*

#### 3.4.4.1. Analysis of a Fraction Containing a Single Fragment

The digest of TTIHYKYMCNSSCGGMNRRPILT with 20S proteasomes was separated by RP-HPLC (*see* **Fig. 6**) and fraction 18 (elution volume: 39–40 min) analyzed by MALDI-MS (*see* **Fig. 7**) and Edman-sequencing (*see* **Table 1**).

MALDI-MS analysis shows a single peak with a mass of 925.2 Da, corresponding to amino acids 1–7. In the Edman-sequencing data, the sequence TTIHY… is readable corresponding to the first five amino acids of the peptide.

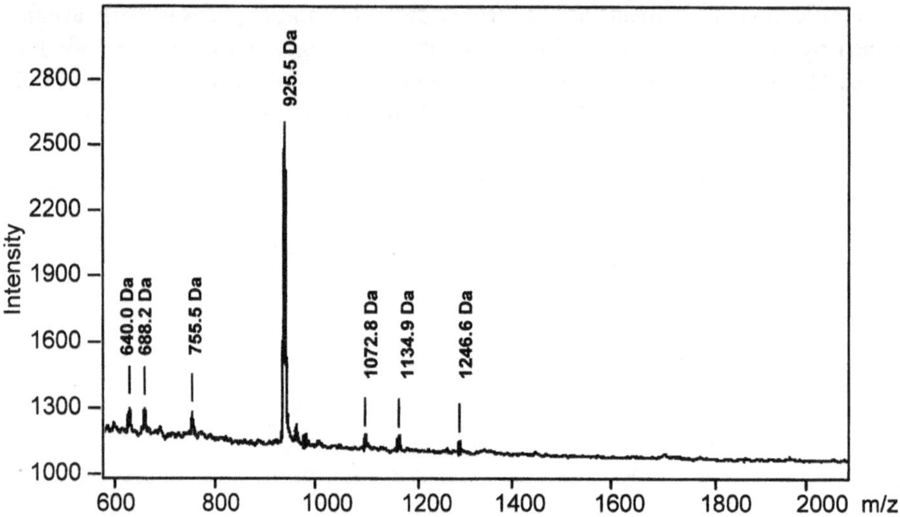


Fig. 7. MALDI-MS spectrum of fraction 18 (retention time: 39–40 min) from the RP-HPLC separation of the digest of the peptide TTIHYKYMCNSSCGGMNRRPILT with purified 20S proteasomes.

**Table 1**
**Edman Sequencing Raw Data From Fraction 18**

| Cycle | A | D | E | F | G | H | I | K | L | M |
|---|---|---|---|---|---|---|---|---|---|---|
| 1 | 3.74 | 3.28 | 2.72 | 0 | 5.09 | 0.72 | 0 | 0.57 | 0.63 | 3.03 |
| 2 | 0.91 | 3.95 | 1.64 | 0 | 5.8 | 0 | 0.44 | 0 | 0 | 1.11 |
| 3 | 0 | 0 | 1.05 | 0 | 4.31 | 0 | 51.91[a] | 0 | 0 | 2.36 |
| 4 | 0 | 0.53 | 0.4 | 0 | 3 | 23.37[a] | 8.72 | 0 | 0 | 1.45 |
| 5 | 0.56 | 0.98 | 0 | 0 | 3.38 | 8.49 | 3.12 | 0 | 0 | 1.71 |
| **Cycle** | **N** | **P** | **Q** | **R** | **S** | **T** | **V** | **W** | **Y** | |
| 1 | 10.89[a] | 0.86 | 0 | 2.47 | 10.66[a] | 39.32[a] | 4.14 | 0 | 1.68 | |
| 2 | 5.23 | 0 | 0 | 1.04 | 4.59 | 30.64[a] | 0 | 2.64 | 3.02 | |
| 3 | 2.09 | 0.91 | 0 | 0 | 4.14 | 3.63 | 0 | 0.37 | 0.9 | |
| 4 | 0.89 | 0.5 | 0 | 0 | 3.79 | 1.68 | 0 | 0 | 2.18 | |
| 5 | 3.85 | 0 | 0 | 0.38 | 1.95 | 0.88 | 0 | 0 | 73.36[a] | |

[a]Amino acids detected in the cycle in an amount > 10 pmol.

Taken together, the peptide TTIHYKY is clearly identified in this fraction. The amount is quantified to 52 pmol using the amino acid isoleucine in the third cycle.

#### 3.4.4.2. Analysis of a Fraction Containing Multiple Fragments

Fraction 20 (elution volume: 41–42 min) was analyzed by MALDI-MS (*see* **Fig. 8**) and Edman sequencing (*see* **Table 2**).

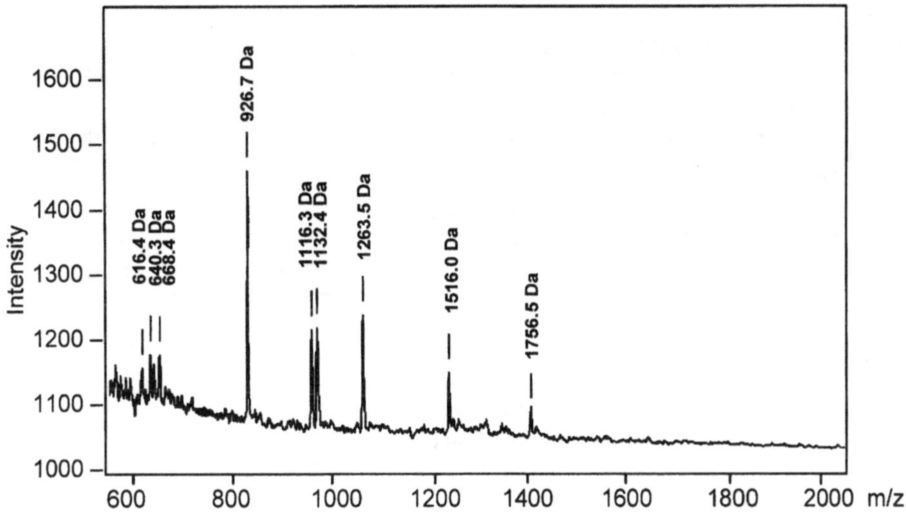


Fig. 8. MALDI-MS spectrum of fraction 20 (retention time: 41–42 min) from the RP-HPLC separation of the digest of the peptide TTIHYKYMCNSSCGGMNRRPILT with purified 20S proteasomes.

**Table 2**
**Edman Sequencing Raw Data From Fraction 20**

| Cycle | A | D | E | F | G | H | I | K | L | M |
|---|---|---|---|---|---|---|---|---|---|---|
| 1 | 3.00 | 7.74 | 3.27 | 0.00 | 28.72[a] | 0.00 | 0.00 | 0.00 | 0.00 | 7.33 |
| 2 | 0.00 | 15.73[a] | 1.21 | 0.00 | 23.05[a] | 0.00 | 0.00 | 0.00 | 0.00 | 0.00 |
| 3 | 0.00 | 1.34 | 0.76 | 0.00 | 16.11[a] | 0.00 | 25.75[a] | 0.00 | 0.00 | 35.04[a] |
| 4 | 0.00 | 4.53 | 1.17 | 0.00 | 9.93 | 10.56[a] | 4.81 | 0.00 | 0.00 | 9.44 |
| 5 | 1.31 | 3.04 | 0.45 | 0.00 | 8.91 | 3.94 | 0.00 | 0.00 | 0.00 | 4.45 |

| Cycle | N | P | Q | R | S | T | V | W | Y |
|---|---|---|---|---|---|---|---|---|---|
| 1 | 3.00 | 0.00 | 0.00 | 0.00 | 4.23 | 16.03[a] | 0.00 | 0.00 | 0.74 |
| 2 | 12.05[a] | 0.00 | 0.00 | 0.00 | 0.00 | 13.53[a] | 0.00 | 0.00 | 0.00 |
| 3 | 8.67 | 0.00 | 0.00 | 2.19 | 10.81[a] | 3.81 | 0.00 | 0.00 | 0.00 |
| 4 | 28.54[a] | 0.00 | 0.00 | 0.00 | 10.48[a] | 0.00 | 1.21 | 0.00 | 0.00 |
| 5 | 10.17 | 0.00 | 0.00 | 16.84[a] | 0.00 | 0.00 | 0.00 | 0.00 | 31.24[a] |

[a]Amino acids detected in the cycle with an amount > 10 pmol.

The MALDI-MS spectrum shows several peaks, four of which can be correlated with fragments from the peptide (*see* **Table 3**).

When analyzing the Edman data, multiple amino acids can be detected in positions 1–5 (*see* **Table 4**).

**Table 3**
**Peptide Fragments Identified in Fraction 20 by MALDI-MS**

| Mass identified by MALDI-MS | Corresponding fragment | Sequence | Calculated mass | Remarks |
|---|---|---|---|---|
| 926.7 Da | 1–7 | TTIHYKY | 925.5 Da | |
| 1116.3 Da | 14–23 | GGMNRRPILT | 1114.6 Da | |
| 1132.2 Da | 14–23 | GGMNRRPILT | 1114.6 Da | +16 Da due to oxidation |
| 1516.0 Da | 9–16 | CNSSCGGM | 758.2 Da | detected as C–C dimer |

**Table 4**
**Evaluation of Edman Sequencing Raw Data From Fraction 20**

| Cycle | Amino acids detected |
|---|---|
| 1 | G,T |
| 2 | D, G, N, T |
| 3 | I, M, S |
| 4 | H, N, S |
| 5 | R,Y |

Using the FINDPATTERN software, the following sequences are found:

1. GGMNR… (Amino acids 14–18) (35 pmol as quantified by M).
2. TTIHY… (Amino acids–5) (25 pmol as quantified by I).
3. NSS… (Amino acids 10–12) (10 pmol as quantified by S).

Together with the MALDI-MS data, the following fragments are identified:

1. TTIHYKY, residues 1–7, 25 pmol.
2. GGMNRRPILT, residues 14–23, 35 pmol.
3. CNSSCGGM, residues 9–16, 10 pmol.

## 4. Notes

1. The whitish layer on top of the erythrocyte pellet contains mostly leukocytes that contain proteasomes harboring also immunosubunits. For a pure preparation of 20S proteasomes from erythrocytes, a complete removal of this layer is necessary.
2. This treatment usually leads to a complete lysis of the erythrocytes. However, if a pellet with a volume of more than 10% of the starting erythrocytes remains, the remaining erythrocytes may be lysed by repeating the process.
3. As the erythrocyte lysate has a very high viscosity because of high protein concentrations, loading by gravity flow is not possible. When applying vacuum, be careful not to let the material run dry. The capacity of the DEAE–cellulose is high enough to bind most of the 20S proteasomes. If the procedure is repeated after regeneration of the material, 10–20% higher yield can be achieved.

4. Standard measuring wavelengths for AMC are excitation at 360 nm and emission at 450 nm. Alternative wavelengths are 380 nm for excitation and 430 nm for emission.
5. Other centrifugation-based concentration units may be used. In some cases, preblocking the membrane with 200 m*M* glycine reduces binding of the proteins to the membrane.
6. Harvesting the gradients is achieved either by inserting a capillary tube from the top and emptying the tube using a peristaltic pump or by drilling a small hole in the base of the tube and emptying by gravity flow, which gives slightly better resolution. A fraction size of 500 µL seems to be optimal for the purification of 20S proteasomes; however, a slightly higher resolution and therefore higher purity may be achieved by collecting smaller fractions.
7. The use of a fluorimeters allowing incubation at 37°C is highly recommended. For kinetic measurements, an interval of 2 min between the measurements is best. For evaluation, a linear part of the curve should be chosen for linear regression, as an initial lag phase caused by cold buffers may affect the results when calculating the activity using an end point assay.
8. It is preferable to dissolve peptides in $dH_2O$, as DMSO may affect proteasomal activity. For testing the best solvent for the peptide, it is recommended to use only small aliquots of the peptide. If the peptide is insoluble in water, test increasing concentrations of DMSO. When a comparison of the digestion of different peptides is planned, all should be dissolved in the same solvent.
9. Peptide digests should be either immediately separated by RP-HPLC or frozen at –80°C, as longer storage at higher temperatures may lead to oxidation or degradation of the digestion products, which may hamper the detection process or give incorrect results.
10. Volume-based fractionation is recommended for the analysis of digests of whole proteins. For the separation of peptide digests, automatic peak fractionation may give better results.

## Acknowledgments

This work was supported by grants from the Deutsche Forschungsgemeinschaft to H. S. (Schi301/2-2, Schi301/2-3 and SFB 510, C1).

## References

1. Kisselev, A. F., Akopian, T. N., Woo, K. M., and Goldberg, A. L. (1999) The sizes of peptides generated from protein by mammalian 26 and 20 S proteasomes. Implications for understanding the degradative mechanism and antigen presentation. *J. Biol. Chem.* **274,** 3363–3371.
2. Rock, K. L. and Goldberg, A. L. (1999) Degradation of cell proteins and the generation of MHC class I-presented peptides. *Annu. Rev. Immunol.* **17,** 739–779.
3. Groll, M., Ditzel, L., Lowe, J., et al. (1997) Structure of 20S proteasome from yeast at 2.4 A resolution. *Nature* **386,** 463–471.
4. Lowe, J., Stock, D., Jap, B., Zwickl, P., Baumeister, W., and Huber, R. (1995) Crystal structure of the 20S proteasome from the archaeon *T. acidophilum* at 3.4 A resolution. *Science* **268,** 533–539.
5. Groettrup, M., Ruppert, T., Kuehn, L., et al. (1995) The interferon-gamma-inducible 11 S regulator (PA28) and the LMP2/LMP7 subunits govern the peptide production by the 20 S proteasome in vitro. *J. Biol. Chem.* **270,** 23808–23815.
6. Eleuteri, A. M., Kohanski, R. A., Cardozo, C., and Orlowski, M. (1997) Bovine spleen multicatalytic proteinase complex (proteasome): replacement of X, Y, and Z subunits by

LMP7, LMP2, and MECL1 and changes in properties and specificity. *J. Biol. Chem.* **272,** 11824–11831.
7. Boes, B., Hengel, H., Ruppert, T., Multhaup, G., Koszinowski, U. H., and Kloetzel, P. M. (1994) Interferon gamma stimulation modulates the proteolytic activity and cleavage site preference of 20S mouse proteasomes. *J. Exp. Med.* **179,** 901–909.
8. Gaczynska, M., Rock, K. L., and Goldberg, A. L. (1993) Gamma-interferon and expression of MHC genes regulate peptide hydrolysis by proteasomes. *Nature* **365,** 264–267.
9. Cardozo, C. and Kohanski, R. A. (1998) Altered properties of the branched chain amino acid-preferring activity contribute to increased cleavages after branched chain residues by the "immunoproteasome." *J. Biol. Chem.* **273,** 16764–16770.
10. Sijts, A. J., Standera, S., Toes, R. E., et al. (2000) MHC class I antigen processing of an adenovirus CTL epitope is linked to the levels of immunoproteasomes in infected cells. *J. Immunol.* **164,** 4500–4506.
11. Sijts, A. J., Ruppert, T., Rehermann, B., Schmidt, M., Koszinowski, U., and Kloetzel, P. M. (2000) Efficient generation of a hepatitis B virus cytotoxic T lymphocyte epitope requires the structural features of immunoproteasomes. *J. Exp. Med.* **191,** 503–514.
12. van Hall, T., Sijts, A., Camps, M., et al. (2000) Differential influence on cytotoxic T lymphocyte epitope presentation by controlled expression of either proteasome immunosubunits or PA28. *J. Exp. Med.* **192,** 483–494.
13. Schwarz, K., van den, B. M., Kostka, S., et al. (2000) Overexpression of the proteasome subunits LMP2, LMP7, and MECL-1, but not PA28 alpha/beta, enhances the presentation of an immunodominant lymphocytic choriomeningitis virus T cell epitope. *J. Immunol.* **165,** 768–778.
14. Morel, S., Levy, F., Burlet-Schiltz, O., et al. (2000) Processing of some antigens by the standard proteasome but not by the immunoproteasome results in poor presentation by dendritic cells. *Immunity* **12,** 107–117.
15. Nussbaum, A. K., Kuttler, C., Tenzer, S., and Schild, H. (2003) Using the World Wide Web for predicting CTL epitopes. *Curr. Opin. Immunol.* **15,** 69–74.
16. Ayyoub, M., Stevanovic, S., Sahin, U., et al. (2002) Proteasome-assisted identification of a SSX-2-derived epitope recognized by tumor-reactive CTL infiltrating metastatic melanoma. *J. Immunol.* **168,** 1717–1722.
17. Kessler, J. H., Beekman, N. J., Bres-Vloemans, S. A., et al. (2001) Efficient identification of novel HLA-A(*)0201-presented cytotoxic T lymphocyte epitopes in the widely expressed tumor antigen PRAME by proteasome-mediated digestion analysis. *J. Exp. Med.* **193,** 73–88.
18. Emmerich, N. P., Nussbaum, A. K., Stevanovic, S., et al. (2000) The human 26 S and 20 S proteasomes generate overlapping but different sets of peptide fragments from a model protein substrate. *J. Biol. Chem.* **275,** 21140–21148.
19. Toes, R. E., Nussbaum, A. K., Degermann, S., et al. (2001) Discrete cleavage motifs of constitutive and immunoproteasomes revealed by quantitative analysis of cleavage products. *J. Exp. Med.* **194,** 1–12.
20. Kuttler, C., Nussbaum, A. K., Dick, T. P., Rammensee, H. G., Schild, H., and Hadeler, K. P. (2000) An algorithm for the prediction of proteasomal cleavages. *J. Mol. Biol.* **298,** 417–429.
21. Nussbaum, A. K., Kuttler, C., Hadeler, K. P., Rammensee, H. G., and Schild, H. (2001) PAProC: a prediction algorithm for proteasomal cleavages available on the WWW. *Immunogenetics* **53,** 87–94.

22. Kesmir, C., Nussbaum, A. K., Schild, H., Detours, V., and Brunak, S. (2002) Prediction of proteasome cleavage motifs by neural networks. *Prot. Eng.* **15,** 287–296.
23. Holzhutter, H. G., Frommel, C., and Kloetzel, P. M. (1999) A theoretical approach towards the identification of cleavage-determining amino acid motifs of the 20 S proteasome. *J. Mol. Biol.* **286,** 1251–1265.
24. Stevanovic, S. and Jung, G. (1993) Multiple sequence analysis: pool sequencing of synthetic and natural peptide libraries. *Anal. Biochem.* **212,** 212–220.

9

# Identification of Components of Protein Complexes

**Carol E. Parker, Maria R. Warren, David R. Loiselle, Nedyalka N. Dicheva, Cameron O. Scarlett, and Christoph H. Borchers**

**Summary**

Protocols are given for a variety of techniques used in protein identification of complexes, including identification of in-gel separated proteins and LC–MS/MS. Gels, staining procedures, and peptide extraction protocols that are compatible with mass spectrometry are described. The detection limits of the various staining procedures and their compatibility with mass spectrometry are discussed. The various mass spectrometric techniques used (MALDI–MS, MALDI–MS/MS, nanospray, and ESI/LC–MS/MS) are also described, along with an indication of the advantages and disadvantages of each, and when they would most appropriately be used. Common pitfalls associated with database searching are also discussed.

**Key Words:** Database searching; in-gel digestion procedures; LC–MS/MS; MALDI–MS; MALDI–MS/MS; peptide mass fingerprinting, protein complexes; protein identification, sequence tag.

## 1. Introduction

The methods used for identifying proteins in protein complexes are essentially the same as those used for "normal" protein identification. In proteomics, what is commonly termed "protein identification" is actually mass spectrometric identification of *peptides* obtained from an unknown protein (*see* **Note 1**). Two common mass spectrometric techniques are used to accomplish this peptide identification *(1,2)*: peptide mass fingerprinting *(3)* and the "sequence-tag approach" *(4)*, which is based on tandem mass spectrometric sequencing (MS/MS) of a peptide through collision-induced dissociation (CID). Which of these techniques is selected depends on what instrumentation is available and the complexity of the protein mixture.

## 2. Materials

### 2.1. Gel Separations

1. Novex precast gels (Invitrogen; Carlsbad, CA).
2. Gel loading buffer (Invitrogen).

From: *Methods in Molecular Biology, vol. 301, Ubiquitin–Proteasome Protocols*
Edited by: C. Patterson and D. M. Cyr © Humana Press Inc., Totowa, NJ

3. Sodium dodecyl sulfate (SDS) running buffer (Invitrogen).
4. Unstained molecular weight markers (Bio-Rad; Hercules, CA).
5. One- and/or two-dimensional (1D and/or 2D) gel electrophoresis equipment (Amersham Biosciences; Piscataway, NJ, Invitrogen).

## 2.2. Gel Staining

### 2.2.1. Coomassie R-250

1. Fixing solution: 25% isopropanol: 10% acetic acid: 65% Milli-Q water (or equivalent).
2. Staining solution: Coomassie R-250 stain (Bio-Rad), 0.01% in 10% acetic acid.
3. Destaining solution: 10% acetic acid.
4. Storage solution: 3% acetic acid.
5. Rocking platform or rotary shaker (Bellco; Vineland, NJ).

### 2.2.2. Sypro Ruby

1. Sypro Ruby protein stain (Molecular Probes; Eugene, OR).
2. Fixing solution: 40% methanol and 10% acetic acid in water.
3. Washing solution: 10% methanol and 7% acetic acid in water.
4. Rocking platform or rotary shaker (Bellco).

### 2.2.3. Silver Stain

1. Silver Quest stain kit (Invitrogen).
2. Fixing solution: 40% methanol and 10% acetic acid.
3. Washing solution: 100 mL containing 30 mL of ethanol and 10 mL of acetic acid, in water.
4. Sensitizing solution: 100 mL containing 30 mL of ethanol and 10 mL of Invitrogen "sensitizer," in water.
5. Staining solution: 100 mL containing 1 mL of Invitrogen "stainer," in water.
6. Rocking platform or rotary shaker (Bellco).

## 2.3. Gel Imaging

1. UV/Vis imager for silver or Coomassie (ProXpress; Perkin Elmer Life and Analytical Sciences, Boston, MA; BioMachines 2DiD; Leap Technologies; Carrboro, NC, etc.).
2. Fluorescent imager for Sypro Ruby (ProXpress, BioMachines 2DiD, etc.).

## 2.4. Gel Cutting

### 2.4.1. Manual Gel Cutting

1. Scalpel blades (Cincinnati Surgical; Cincinnati, OH).
2. Light box (Laboratory Supplies Co.; Hicksville, NY).
3. Laminar flow hood (Labconco; Kansas City, MO).
4. Methanol, HPLC grade (Burdick & Jackson; Muskegon, MI).

### 2.4.2. Automated Gel Cutting

1. Biomachines 2DiD, or Genomic Solutions ProPic, and so forth.

## 2.5. Enzymatic Digestion

### 2.5.1. Manual In-Gel Digestion

1. Water, high-performance liquid chromatography (HPLC) grade (Fisher).
2. Trypsin, sequencing grade (Promega; Madison, WI).

3. Ammonium bicarbonate (Fluka; Milwaukee, WI).
4. Acetonitrile (KSE, Durham, NC).
5. Low-retention Eppendorf tubes (Axygen; Union City, CA).
6. Parafilm (American National Can; Chicago, IL).
7. Thermomixer (Eppendorf; Hamburg, Germany).

#### 2.5.2. Automated In-Gel Digestion

1. Water, HPLC grade.
2. Trypsin, sequencing grade (Promega).
3. Ammonium bicarbonate.
4. Acetonitrile.
5. 88% Formic acid (Fisher).
6. Automated digester (ProGest, Biomachines, etc.).

#### 2.5.3. In-Solution Digestion

1. Water, HPLC grade (Pierce).
2. Trypsin, sequencing-grade (Promega).
3. Ammonium bicarbonate.
4. Low-retention Eppendorf tubes (Axygen).
5. Thermomixer (Eppendorf).

#### 2.5.4. Lyophilization and Reconstitution

1. Freeze dryer (Labconco).
2. Water, HPLC grade.
3. Methanol, HPLC grade.
4. Formic acid (Fisher; Pittsburgh, PA).

### 2.6. Mass Spectrometry

#### 2.6.1. Bruker Reflex III Matrix-Assisted Laser Desorption (MALDI)–MS

1. α-Cyano 4-hydroxycinnamic acid, recrystallized (Aldrich; St. Louis, MO).
2. Acetonitrile (KSE).
3. Water, HPLC-grade.
4. Trifluoroacetic acid, ampules (Pierce).
5. Matrix solvent, 50:50 acetonitrile–water (0.1% TFA)

#### 2.6.2. ABI Q-Star MALDI Q-Time-of-Flight (MALDI Q-TOF)

1. Premixed dihyroxybenzoic acid (DHB) solution (Agilent Technologies; Palo Alto, CA).

#### 2.6.3. ABI 4700 MALDI-TOF/TOF

1. α-Cyano 4-hydroxycinnamic acid, recrystallized (Aldrich).
2. Acetonitrile (KSE).
3. Water, deionized (Milli-Q; Waters or Purelab Plus; US Filter), or HPLC-grade (Fisher).
4. Ammonium citrate (Fluka).
5. Trifluoroacetic acid, ampules (Pierce; Rockford, IL).
6. Matrix solvent: 50:50 acetonitrile: 40 m*M* ammonium citrate in water (0.1% trifluoroacetic acid).

### 2.6.4. ABI Q-Star Nanospray (Nano-Electrospray [NanoESI])

1. Methanol, absolute (Mallinckrodt; Hazelwood, MO).
2. Water, deionized (Milli-Q; Waters or Purelab Plus; US Filter), or HPLC-grade (Fisher).
3. Formic acid, 88% (Fisher).
4. Borosilicate nanospray needles, 0.7-mm id, 2-μm tip (Proxeon [formerly Protana]; Odense, Denmark).

### 2.6.5. Micromass Q-TOF

1. PepMap $C_{18}$ 15 cm × 75 μm id capillary column (Dionex; Sunnyvale, CA).
2. Trapping column 5 mm × 800 μm id $C_{18}$ P3 (Dionex).
3. Water, HPLC grade (Pierce).
4. Acetonitrile, HPLC grade (Pierce).
5. Formic acid (Fisher).

## 2.7. Database Searching

1. *Mascot (Matrix Science; London, England;* www.matrixscience.com*).*
2. Mascot server (IBM; White Plains, NY).
3. *Protein Prospector* (http://prospector.ucsf.edu/).
4. *ProFound* (http://prowl.Rockefeller.edu).
5. *Color printer Tektronix 8600 (Xerox, Stamford, CT).*

# 3. Methods

## 3.1. Mass Spectrometric Methods

### 3.1.1. Peptide Mass Fingerprinting

Peptide mass fingerprinting requires approximately three to five peptides from a given protein *(5)*. If these peptide masses are known to a sufficient degree of accuracy (80 ppm), the protein can be identified by comparison of these masses to those masses predicted by a theoretical digest of all proteins in the database with a specified enzyme. This technique, shown schematically in **Fig. 1**, is usually done by MALDI–MS, because it is a high-throughput technique and is the method of choice for simple mixtures. One problem with this approach can be a somewhat limited dynamic range, owing to the suppression effects encountered in MALDI–MS *(6,7)*. High-sensitivity peptide mass fingerprinting can also be performed by nanoelectrospray *(8)*, although this technique is not high throughput (*see* **Note 2**). Software algorithms have been developed to identify proteins in simple mixtures of proteins (usually no more than two or three proteins). An example of a protein identified by peptide mass fingerprinting from a cullin co-immunoprecipitation experiment is shown in **Fig. 2**, using the Mascot software package *(9)*. **Figure 3** shows a band from a co-immunoprecipitation experiment of the APC complex. This search was done using ProFound *(10)*, and the database search results identified Apc1 as the unknown protein, with an estimated confidence level of 1.5× $10^{-16}$ probability of an incorrect result. For more complex mixtures, separation of the mixture at the protein level by gel electrophoresis or offline HPLC is required prior to MALDI analysis.

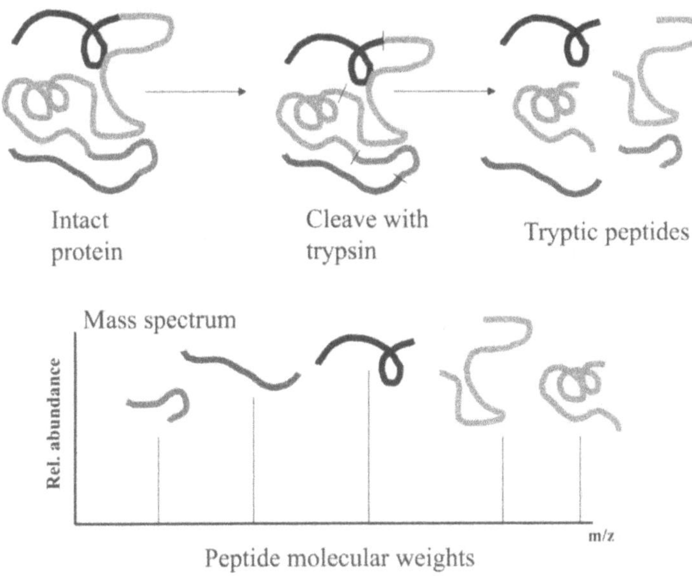


Fig. 1. Protein identification by peptide mass fingerprinting.

### *3.1.2. Sequence-Tag Approach*

The "sequence-tag" approach requires fewer peptides to obtain a similar degree of confidence in the identification, because as the name implies, a "sequence-tag" giving the partial sequence of the peptide is obtained through mass spectrometric dissociation of the peptide (**Fig. 4**). Fortunately, during dissociation, the ionized peptide usually cleaves *between* amino acid residues, giving rise to two series of ions, the *b* series, which contain the N-terminus of the peptide, and the *y* series, which contains the C-terminus of the peptide (*[11,12]*; **Note 3**). Coupled with the peptide molecular weight, this partial sequence tag is often sufficient to identify a peptide, which, in turn, is often sufficient to identify a protein (or a family of homologous proteins) that contains this peptide *(13,14)*. For a higher degree of confidence in the identification, several peptides can be sequenced. Database searching software then compares the molecular masses of these peptides and their sequences to those predicted from a theoretical digest of all of the proteins in the database. Mass accuracy has been shown to be a critical factor for unambiguous protein identification *(14)* (**Table 1** and **Note 4**).

An example of a protein identified by the sequence-tag approach is shown in **Fig. 5**. The sequence-tag approach is particularly useful where there are too few peptides for peptide mass fingerprinting. **Figure 5** shows a low molecular weight protein band from the same experiment as **Fig. 3**. Although the higher molecular weight protein from the same gel was identified by peptide mass fingerprinting **(Fig. 3),** the peptide sequencing approach had to be used on the smaller protein because either (1) fewer peptides were formed or (2) there were fewer peptides in the appropriate mass range for sequencing (*see* **Note 4**).

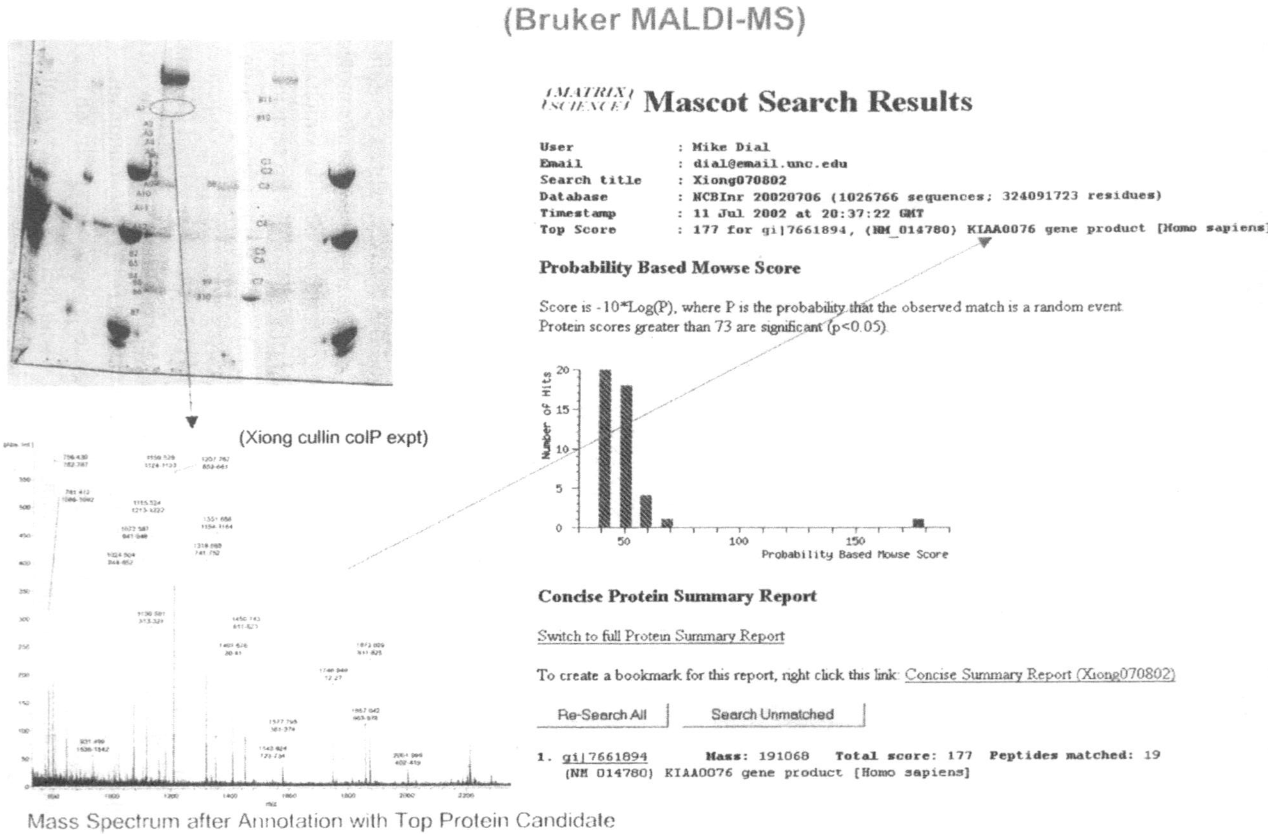


Fig. 2. Example of protein identification by peptide mass fingerprinting using Mascot database searching software (Collaborator: Y. Xiong.)

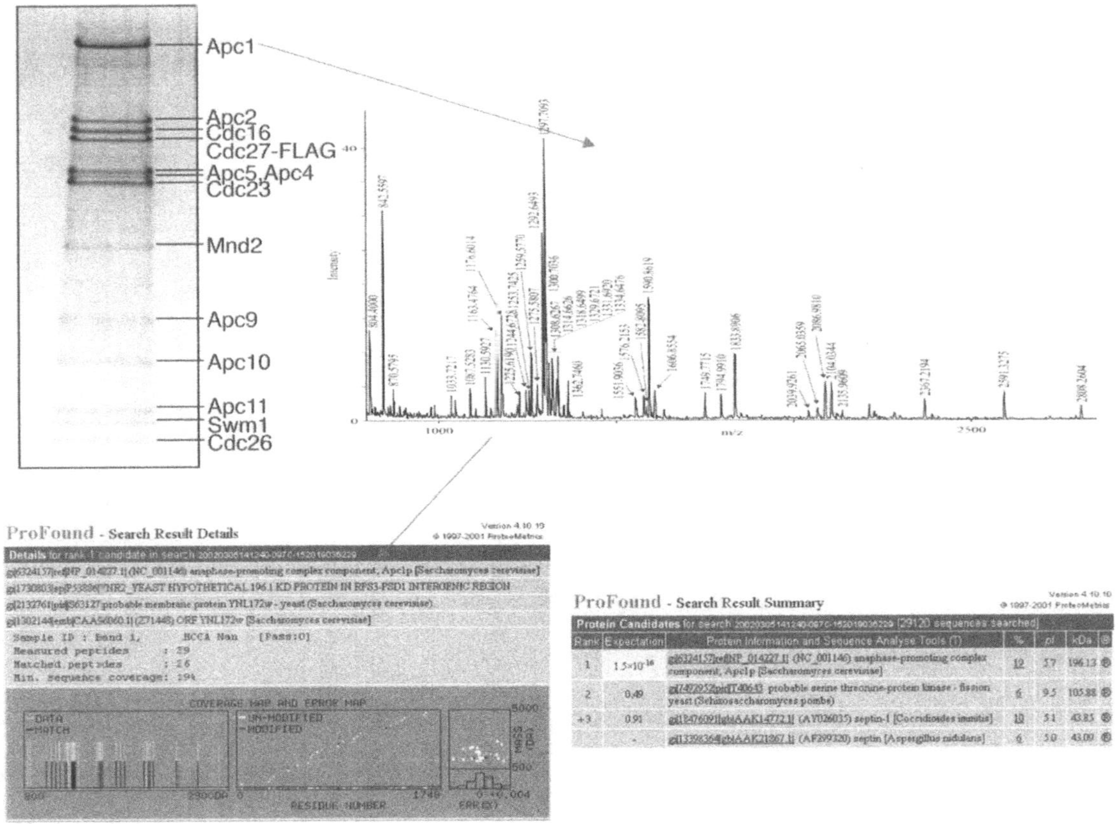


Fig. 3. Example of protein identification by peptide mass fingerprinting using ProFound. Tandem MS identification of APC subunit APC1, one of 13 subunits of the anaphase-promoting complex (APC).

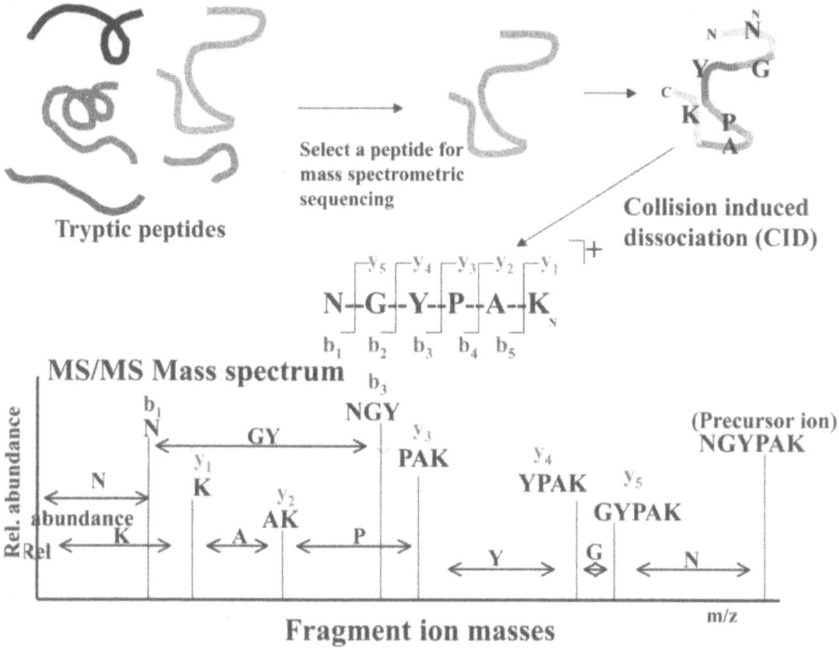


Fig. 4. Protein identification by peptide sequencing.

**Table 1**
**Number of Hits Obtained With Different Resolution (Searching With 2-Amino Acid Tag)**

| Precursor *m/z* | 1000 ppm (QQQ) | 100 ppm | 10 ppm (Q-TOF) | 5 ppm |
|---|---|---|---|---|
| 1171.591 | 573 | 71 | 5 | 0 |
| 1213.207 | 314 | 65 | 1 | 1 |
| 1237.661 | 39 | 9 | 1 | 1 |
| 1488.754 | 202 | 29 | 1 | 1 |
| 1925.837 | 738 | 15 | 1 | 1 |
| 1981.035 | 412 | 38 | 2 | 1 |

Owing to the suppression effects mentioned earlier, this method will not work for peptides from complex mixtures or where proteins are present at widely different concentrations. For mixtures containing a large number of proteins, and thus a large number of peptides, the MALDI-MS spectrum can be too complex for peptide mass fingerprinting. An example of such a mixture of peptides is shown in **Fig. 6**. As can be seen from the inset, the spectrum contains a peak at almost every mass. This means not only that the peptide mass fingerprinting method cannot be used, but also that the sequence-tag approach would most likely fail as well. The reason for this is the width of the mass window used for selecting the precursor ion—the mass window of the

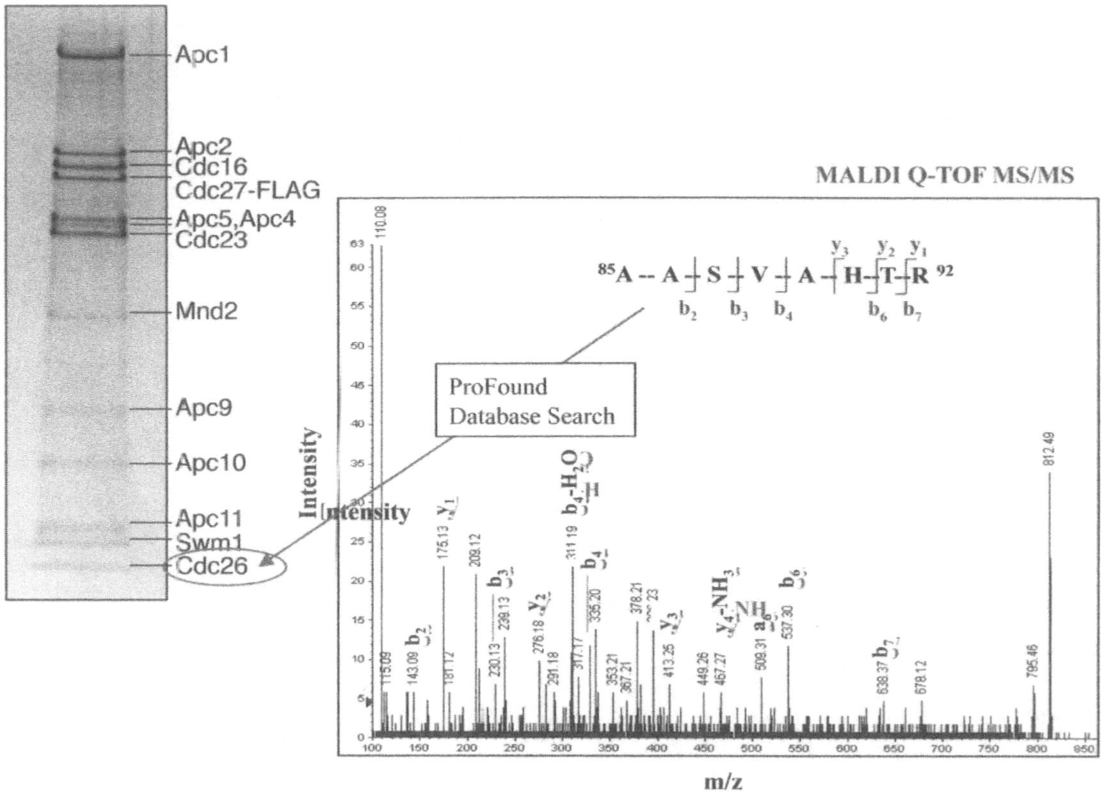


Fig. 5. Example of protein identification requiring peptide sequencing, using ProFound. Tandem MS identification of APC subunit Cdc26, one of 13 subunits of the yeast anaphase-promoting complex (APC). Cdc26 could not be unambiguously identified by peptide mass fingerprinting (too few peptides).

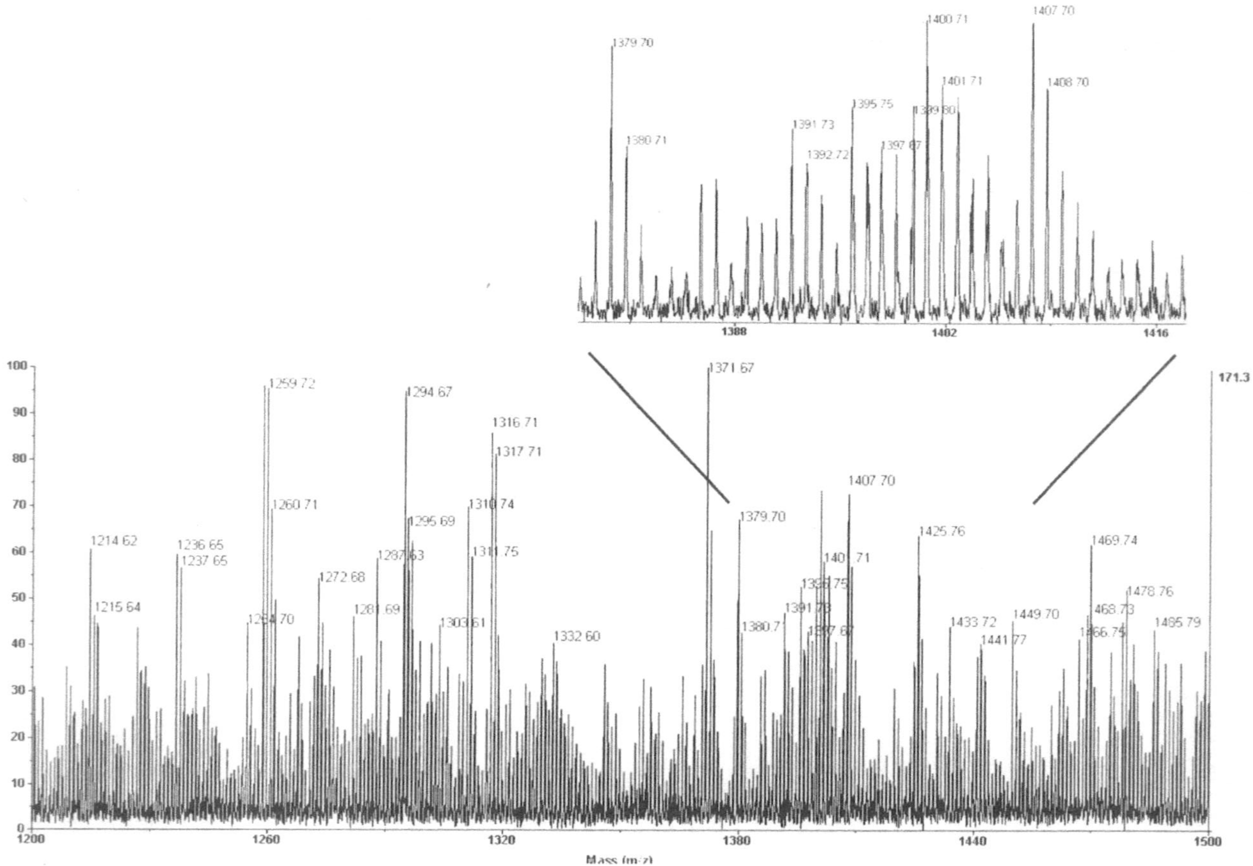


Fig. 6. MALDI–MS spectrum of a mixture too complex for peptide mass fingerprinting.

instrument is likely to include several different peptides, which would be fragmented at the same time. This would lead to a combined MS/MS spectrum that would, most likely, be uninterpretable. As previously, prior separation is needed, either by gel electrophoresis or by off-line HPLC.

#### *3.1.3. LC–MS/MS*

An alternative approach for complex mixtures is on-line LC–MS/MS, where the separation of peptides is done on-line prior to an ESI–MS/MS analysis. This separation has the effect of reducing the suppression effect, as fewer peptides are present in the source at any given time, which reduces the *instantaneous* complexity of the mixture. LC–MS/MS, which still relies on the sequence-tag approach for protein identification, is more suitable than MALDI–MS/MS for complex mixtures, or mixtures in which a high dynamic range is needed. An example of a complex mixture resulting from a co-immunoprecipitation of a FLAG-tagged "bait" protein, analyzed by LC–MS/MS, is shown in **Fig. 7**.

### *3.2. Sample Preparation*

As described earlier, proteins are usually identified by mass spectrometric analysis of their peptides. The challenge in proteomics is to extract these peptides in a way that is compatible with the subsequent mass spectrometric analysis. This key criterion needs to be kept in mind throughout the entire process—from design of the proteomics experiment to the final analysis.

The other challenge is to avoid contamination—keratin is a major problem and can easily swamp out the signal from the target protein. All sample preparation steps must be done in a way to avoid contamination from dust (which contains skin cells, which contain keratin). All solutions and containers used must be scrupulously clean. When transporting gels, cover the containers with plastic wrap. Also, gels should never be stored in a container that, for example, has previously been used for blocking Western blots with a solution of bovine serum albumin (BSA) or milk powder; otherwise BSA or casein peptides will be found all over the gel.

#### *3.2.1. Polyacrylamide Gel Electrophoresis (PAGE)*

The most commonly used separation technique for proteins is one- or two-dimensional (ID- or 2D )PAGE, so most of the samples presented for protein identification are presented as gel-separated proteins **(Fig. 8)**. PAGE is quite an effective clean-up technique, removing many buffers that would interfere with the mass spectrometric analysis. It also allows the separation of highly abundant structural proteins from the target proteins. In pull-down or co-immunoprecipitation experiments, it also allows the separation of the antibody and/or the "bait" protein, and facilitates the visualization and detection of less abundant co-immunoprecipitated proteins.

It is difficult to extract intact proteins from 1D or 2D gels—peptides are easier to extract. Thus, most proteomics laboratories currently perform the enzymatic digestion of a protein while it is still inside the gel, and then elute the resulting peptides, rather than trying to elute the protein and performing the enzymatic digestion *after* extraction. Trypsin is the most common enzyme used because it is stable, works in various

{MATRIX}{SCIENCE} Mascot Search Results

| | | | |
|---|---|---|---|
| User | : | Carol Parker | |
| Email | : | | |
| Search title | : | 111203_Greer_FLAG#2_TRY | |
| MS data file | : | D:\MethodD3_5.PRO\pkl_files\111203_Greer_06.pkl | |
| Database | : | NCBInr (1027784 sequences; 324441803 residues) | |
| Taxonomy | : | Homo sapiens (human) (95238 sequences) | |
| Timestamp | : | 13 Nov 2003 at 17:05:54 GMT | |
| Significant hits | : | gi\|4557749 | (NM_000246) MHC class II transactivator [Homo sapiens] |
| | | gi\|4507895 | (NM_003380) vimentin [Homo sapiens] |
| | | gi\|5729877 | (NM_006597) heat shock 70kD protein 8; heat shock 70kD protein 8 (HSP73); heat shock cognate protei |
| | | gi\|1706622 | Kinesin-related motor protein Eg5 (Kinesin-like spindle protein HKSP) (Thyroid receptor interacting |
| | | gi\|425518 | (S65761) anti-colorectal carcinoma heavy chain [Homo sapiens] |
| | | gi\|425520 | (S65921) anti-colorectal carcinoma light chain [Homo sapiens] |
| | | gi\|4501885 | (NM_001101) beta actin; beta cytoskeletal actin [Homo sapiens] |
| | | gi\|20555572 | (XM_166348) heat shock 70kD protein 1-like [Homo sapiens] |
| | | gi\|14731484 | (XM_010852) similar to Frizzled-7 protein - human [Homo sapiens] |
| | | gi\|87650 | heterogeneous ribonuclear particle protein A1.beta - human |
| | | gi\|15342049 | (BC013296) Similar to ribosomal protein S6 [Homo sapiens] |
| | | gi\|6755901 | (NM_011653) tubulin, alpha 1; tubulin alpha 1 [Mus musculus] |
| | | gi\|4885049 | (NM_005159) actin, alpha, cardiac muscle precursor [Homo sapiens] |
| | | gi\|17438165 | (XM_068316) similar to ACTIN, CYTOPLASMIC TYPE 5 [Homo sapiens] |
| | | gi\|882316 | (U21040) immunoglobulin light chain variable region [Homo sapiens] |
| | | gi\|20556253 | (XM_166357) hypothetical protein XP_166357 [Homo sapiens] |
| | | gi\|20554709 | (XM_114689) similar to H2B histone family, member H [Homo sapiens] |
| | | gi\|88279 | nucleophosmin (clone hpB1) - human (fragment) |
| | | gi\|462325 | Heat shock 70 kDa protein 1 (HSP70.1) (HSP70-1/HSP70-2) |
| | | gi\|17447928 | (XM_069744) similar to 40S RIBOSOMAL PROTEIN S2 [Homo sapiens] |
| | | gi\|16117794 | (NM_015414) ribosomal protein L36; 60S ribosomal protein L36 [Homo sapiens] |
| | | gi\|4758830 | (NM_004559) nuclease sensitive element binding protein 1; Major histocompatibility complex, class I |
| | | gi\|9506651 | (NM_018992) hypothetical protein FLJ20040 [Homo sapiens] |
| | | gi\|17384415 | (AL137072) bA29B11.2 (similar to 40S ribosomal protein S2 (S4) (LLrep3 protein)) [Homo sapiens] |
| | | gi\|3249735 | (AF004426) microtubule-based motor [Homo sapiens] |
| | | gi\|5174431 | (NM_006013) ribosomal protein L10; QM gene; DNA segment on chromosome X (unique) 648 expressed sequ |
| | | gi\|5031512 | (AF155568) NSAP1 protein [Homo sapiens] |
| | | gi\|12734090 | (AJ308460) immunoglobulin light chain variable region [Homo sapiens] |
| | | gi\|18559649 | (XM_087614) similar to 40S ribosomal protein S8 [Homo sapiens] |
| | | gi\|15990426 | (BC015576) Similar to cadherin 1, type 1, E-cadherin (epithelial) [Homo sapiens] |
| | | gi\|4503659 | (NM_001997) ubiquitin-like protein fubi and ribosomal protein S30 precursor; FBR-MuSV-associated ub |
| | | gi\|4506671 | (NM_001004) ribosomal protein P2; 60S acidic ribosomal protein P2; acidic ribosomal phosphoprotein |
| | | gi\|30313 | (X12882) cytokeratin 8 (279 AA) [Homo sapiens] |
| | | gi\|14044052 | (BC007950) Similar to heterogeneous nuclear ribonucleoprotein U (scaffold attachment factor A) [Hom |
| | | gi\|4506713 | (NM_002954) ubiquitin and ribosomal protein S27a precursor; ubiquitin carboxyl extension protein 80 |
| | | gi\|5821952 | Chain A, Rotamer Strain As A Determinant Of Protein Structural Specificity |
| | | gi\|13528660 | (BC004532) Similar to ribosomal protein L4 [Homo sapiens] |
| | | gi\|1469922 | (U44428) D53 [Homo sapiens] |

Fig. 7. LC–MS/MS identification of a complex mixture of proteins directly from a co-immunoprecipitated sample, using on-bead tryptic digestion. (Collaborator: S. Greer.)

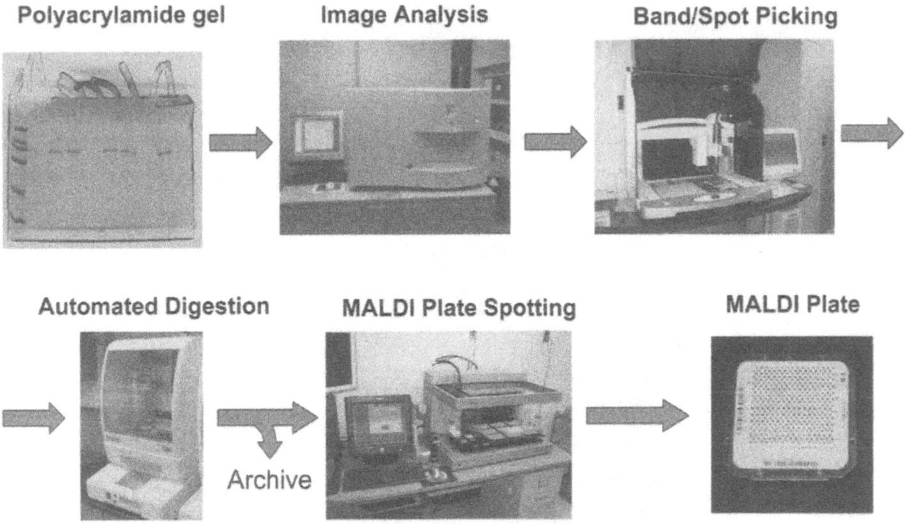


Fig. 8. General scheme for high-throughput protein identification.

buffers, and is quite "aggressive" (*see* **Note 5**). It is tolerant to low levels (<10%) of acetonitrile, in case the acetonitrile used for peptide extraction has not been completely removed. It is "soaked into" the gel, where the digestion takes place. Its cleavage is also "predictable," which is important for database searching. It also cleaves C-terminal to lysine and arginine (except N-terminal to proline), and the frequency of occurrence of lysine and arginine residues in proteins is sufficient to cleave proteins into pieces small enough for sensitive mass spectrometric detection and MS/MS fragmentation (*see* **Note 6**).

Because of their reproducibility, we highly recommend using commercial gels such as Invitrogen Novex Pre-Cast Tris/glycine SDS-PAGE gels to our users, with percent acrylamide >8–10% (*see* **Note 7**). The methods described below for in-gel digestion have been optimized for these gels, and we therefore achieve our highest degree of success for proteins separated on these gels (*see* **Note 8**). We recommend using *unstained* molecular weight markers—we always cut a few of these markers and digest them along with the selected gel bands as part of our quality control procedure. In this way, we can detect a problem in the digestion procedure, and we can distinguish between digestion problems and sample problems (*see* **Note 9**).

While some proteomics facilities recommend reduction and alkylation of the protein prior to gel-based separation, we do not recommend this because the cleanup required after these procedures can lead to sample loss ***(15)***.

### *3.2.2. Staining Procedures*

Coomassie is the least sensitive stain we commonly use, and does not interfere with mass spectrometry. Sypro Ruby is more sensitive than Coomassie, and it is compatible

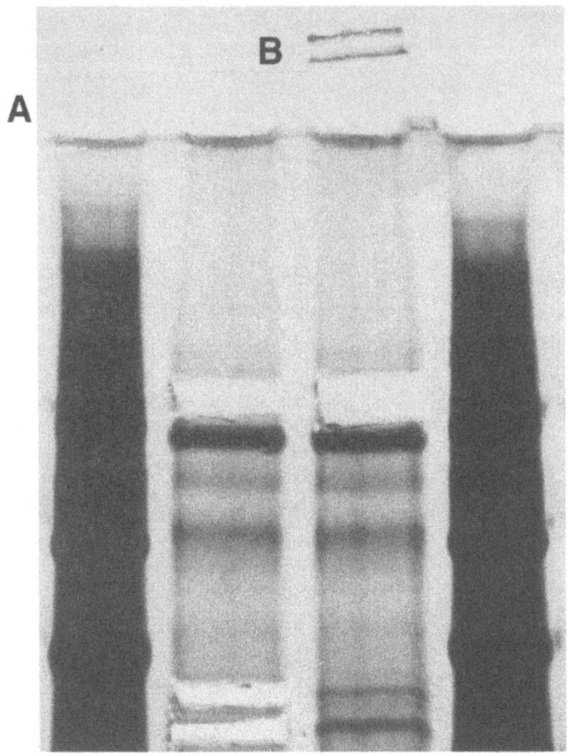


Fig. 9. Overstained silver-stained gel **(A),** and cross-section of overstained gel band **(B).**

with mass spectrometry, but it is a fluorescent stain and requires special visualization techniques. Silver is the most sensitive stain we commonly encounter.

With our current mass spectrometric techniques, we can usually identify protein bands visible with Coomassie or Sypro stain. We are not consistently able to identify protein bands visible only with silver stain. Working with a protein band visible only by silver staining presents two problems. First, we are working at very low levels of protein. Second, silver staining *itself* adversely affects recoveries—the darker the stain, the lower the recoveries. **Figure 9** shows an over-silver-stained gel (A) and a cross-section of gel slice (B). Only peptides from the unstained portion of the gel can be recovered.

All of the staining protocols below can be done at room temperature. All the staining and destaining steps require gentle, continuous agitation, preferably on a rocking platform although a rotary shaker is sufficient.

#### 3.2.2.1. Protocol for Coomassie Staining

1. To fix gel after electrophoresis, soak in Coomassie fixing solution (*see* **Subheading 2.1.1.**) for 20 min for 1-mm thick gels.

2. To stain gel, use Bio-Rad R-250 Coomassie staining solution (*see* **Subheading 2.1.1.**), and stain overnight on an orbital shaker at 40 rpm.
3. To destain, pour off the Coomassie solution from **step 2** and replace destaining solution (**2.1.1.3**), and continue gentle shaking at 40 rpm.
4. As solution turns blue, replace with fresh destaining solution until the gel background is destained (protein bands should remain stained).
5. Leave gel in Coomassie storage solution (*see* **Subheading 2.1.1.**).
6. Destained gel is now ready for imaging and cutting.

#### 3.2.2.2. Protocol for Sypro Ruby

Sypro Ruby Protein Gel Stain is manufactured and sold by Molecular Probes, although several other vendors, including Bio-Rad and Amersham Biosciences, also distribute this stain. Sypro Ruby is a fluorescent, light sensitive stain and therefore must be protected from ambient light during storage, and during the staining and washing steps. The stain should be stored in the dark, or in an amber container, and the trays containing the gels with stain or fix should be covered and wrapped in aluminum foil. Do not use metal or glass trays with this stain. Use only clean plastic (polypropylene, polycarbonate or polyvinyl chloride) trays that have thoroughly rinsed with ethanol prior to use.

1. After electrophoresis, fix the 2D gel in the Sypro Ruby fixing solution (*see* **Subheading 2.2.2.**) for 1 h. Use bottled or Milli-Q water to avoid keratin contamination. Fixing is not required for 1D SDS-PAGE.
2. Incubate the gel in SYPRO Ruby protein stain using enough volume to submerge the gel. For loose or free-floating gels an incubation period of 4 h minimum is required although it is safe to incubate overnight if this is more convenient. Gels that are fixed to a gel plate must be incubated overnight.
3. Remove the gels from the stain and rinse briefly with Milli-Q or bottled water (*see* **Note 10**).
4. Wash gel in Sypro Ruby washing solution (*see* **Subheading 2.2.2.**) for 1 h.
5. The gel is now ready for imaging and cutting.

#### 3.2.2.3. Protocol for Silver Staining With Invitrogen Silver Quest

This protocol can visualize protein amounts down to 3 ng of protein. The protocol listed below is for one minigel, 1.0 mm thick. For large gels, double all solution volumes while keeping the incubation time the same. Do not use the "fast" protocol listed in the Silver Quest manual, and do not use the destainer.

1. After electrophoresis, fix 2D gels in Silver Quest fixing solution (*see* **Subheading 2.2.3.**) for 20 min. Use bottled or Milli-Q water to avoid keratin contamination. 1D gels do not require fixing.
2. Wash with 100 mL of an aqueous solution containing 30 mL methanol, for 10 min.
3. Sensitize with 100 mL of an aqueous solution containing 30 mL of ethanol and 10 mL of Invitrogen "sensitizer" for 10 min.
4. Wash once with 100 mL of an aqueous solution containing 30 mL of ethanol, for 10 min.
5. Wash again with 100 mL of water for 10 min.
6. Stain with 100 mL of an aqueous solution containing 1 mL of Invitrogen "stainer," for 15 min.
7. Wash with 100 mL of water for 1 min.

8. Develop the stain with 100 mL of an aqueous solution containing one drop of Invitrogen "developer/enhancer" until the bands *just start* to become visible. Normally, development times between 5 and 6 min yield the best mass spectrometry results. *Longer development times will reduce peptide recoveries!* (*See* **Fig. 9** and **Note 11**.)
9. Stop the developing by adding 10 mL of Invitrogen "stopper" directly to the developing solution.
10. Wash with 100 mL of water for 10 min.

An example of a stained gel showing peptide recoveries at various levels is shown in **Fig. 10**. As the amount of protein in the gel is reduced, only those peptides with higher relative intensities are detectible. At the lowest levels (3.9 ng), no sample peptides are detectible, only tryptic peptides. At slightly higher levels, if there are even a few peptides detected, the sequence-tag approach may still be successful, even if there are too few peptides for the peptide mass fingerprinting approach. This, of course, depends on the relative sensitivities of the MS and the MS/MS modes on the available instruments.

### *3.2.3. Gel Cutting and In-Gel Extraction*

The excision of the selected gel bands can be done either manually, or with a variety of commercially available robotic systems. For fluorescently-stained gels, there is the added complication of visualizing the stain, which has to be done at wavelengths not visible to the naked eye.

In-gel extraction can be done manually or in an automated robotic system. Manual digestion can provide higher peptide recoveries, but there is more chance of keratin contamination. To avoid contamination of the sample with dust or hair (which contain keratin), the use of a laminar flow hood is desirable, and the use of gloves is mandatory.

#### 3.2.3.1. Manual In-Gel Digestion

##### 3.2.3.1.1. Manual In-Gel Digestion Procedure

1. Excise bands and cut them into 1-mm cubes with a new scalpel blade
2. Transfer cubes from each sample into Axygen low-retention tubes. Wash with 100 μL of HPLC-grade water.
3. Shake in thermomixer for 5 min.
4. Remove water.
5. Destain the gel pieces with several incubations (shaking) of acetonitrile: 50 m*M* ammonium bicarbonate 1:1 (50 μL each) until pieces are clear, discarding the solution in between incubations.
6. Add 100 μL of acetonitrile and incubate for 5 min.
7. Discard the acetonitrile and repeat until pieces are white and hard
8. Puncture sample tube cap, then freeze at –80°C for 30 min.
9. Lyophilize until pieces are completely dry (about 1 h).
10. Prepare trypsin solution just before adding to samples. (Use Promega trypsin and reconstitute one 20 μg aliquot in 1 mL of cold 25 m*M* ammonium bicarbonate). (*See* **Note 12.**)
11. Add 30–50 μL trypsin solution to samples.
12. Incubate for 30 min at 25°C.
13. Remove the enzyme solution.

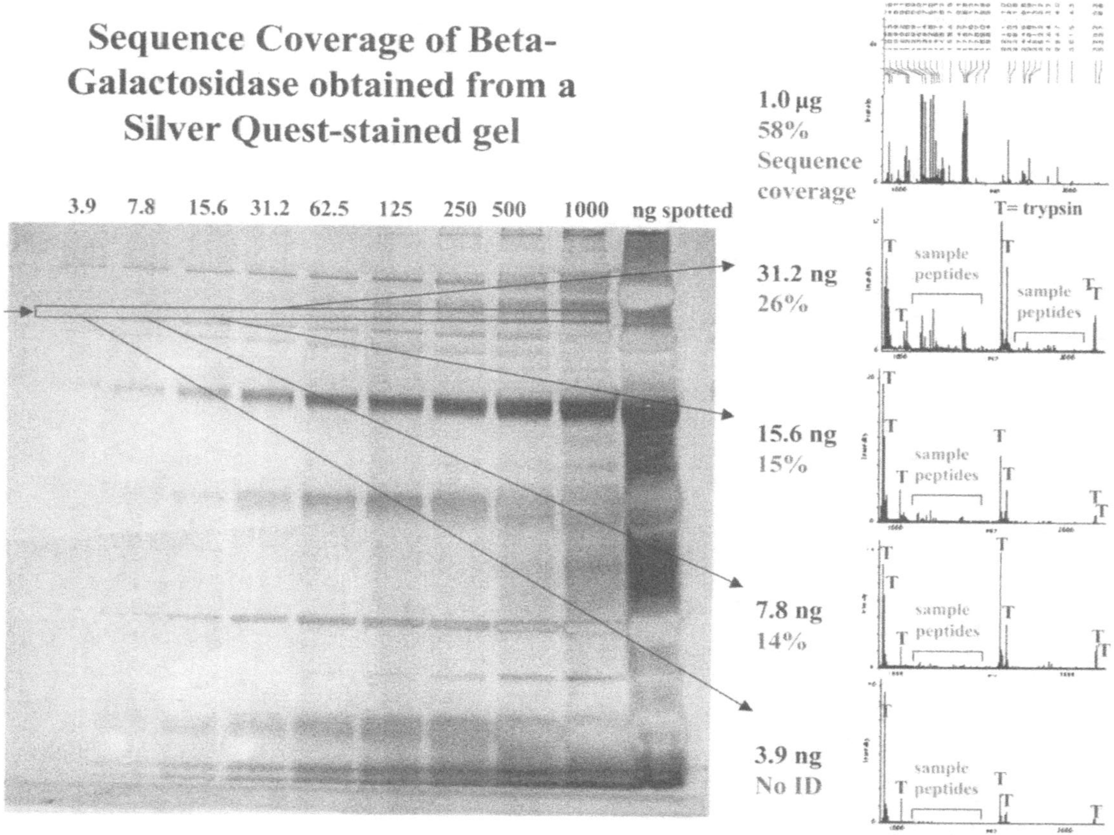


Fig. 10. Effect of the amount of protein loaded on peptide recoveries.

14. Add 100 µL of 50 m*M* ammonium bicarbonate to each sample, wrap tubes in parafilm to seal.
15. Incubate overnight at 35°C in thermomixer at 400 rpm.
16. Remove the trypsin solution and place into a separate tube.
17. Add 25 µL of 10% formic acid in water, and 25 µL of HPLC-grade water and incubate for 10 min.
18. Remove the solution, and combine with solution from **step 16.**
19. Add 25 µL of 10% formic acid in water, and 25 µL of acetonitrile.
20. Incubate for 10 min at room temperature.
21. Remove the solution, and combine with solution from **steps 16** and **18**.
22. Add 50 µL of acetonitrile and incubate for 10 min at room temperature.
23. Remove the solution and combine it with that from **steps 16, 18,** and **21.**
24. Pierce sample tube cap, then freeze at –80°C.
25. Lyophilize the combined solutions.

#### 3.2.3.2. Automated In-Gel Digestion

There are basically two philosophies of automated in-gel digestion. The system we are currently using requires an overnight digestion with trypsin, followed by extraction and lyophilization of the extract. The alternative approach, used by several manufacturers, relies on Zip-Tips (pipet tips prepacked with $C_{18}$ packing material) instead of lyophilization to concentrate the peptides. The Zip-Tip approach has the advantage of speed (the overnight lyophilization is not required) and the removal of salts which might be present in the extraction buffer, but peptide recoveries can be lower and depend on the affinities of the peptides for the $C_{18}$ medium in the Zip-Tip.

Our in-gel digestion procedure following, designed for a Genomics Solutions ProGest robotic system, uses a volatile buffer (ammonium bicarbonate [ABC]) so there is no need for removal of the buffer prior to mass spectrometric analysis. Also, should it become necessary to analyze the remaining extract by LC–MS/MS, the extract can simply be diluted with 10 µL of water prior to injection.

3.2.3.2.1. Automated In-Gel Digestion Procedure: Software Program for ProGest Automated Digestion

```
VERSION,0
TITLE,"Our Digest edit 0304"
VIALMAP,"Solvent A","Acetonitrile"
VIALMAP,"Solvent D","25mM Bicarbonate"
VIALMAP,"Solvent E","10% Formic Acid"
VIALMAP,"Solvent F","Water"
VIALMAP,"Enzyme C","Trypsin, 20 µg in 25 mM ammonium bicarbonate"
; 1
MOVERK,"Moving Rack to run position"," ", 2100,925,110, 447,925, 400, 1
; 2
MOVERK,"Moving Rack to run position"," ", 0,925,110, 93,925, 100, 2
; 3
MOVERK,"Moving Rack to run position"," ", 300,925,80, 419,925, 100, 3
; 4
INCUB,"Pierce Sheet","Reservoir",0, 0.0,0.0,0.0,0.0, 10.000,10.000, 1,0, 0,0,0.0,0.0, 30.000,10.000, 0, 170
; 5
HEAT,"Turn off heaters",0
```

```
; 6
RINSE,"Initial rinse",2000.0,3000.0, 30.000,20.000, 0, 170
; 7
INCUB,"ABC wash","Solvent D",0, 50.0,80.0,10.0,0.0, 10.000,10.000, 1,0, 6,0,1000.0,1000.0,
10.000,10.000, 0, 0
; 8
WAIT,"Soak in ABC",900
; 9
INCUB,"Adding AcN","Solvent A",900, 50.0,80.0,10.0,0.0, 10.000,10.000, 1,0, 0,0,1000.0,1000.0,
10.000,10.000, 0, 0
; 10
INCUB,"Second wash","Solvent D",0, 50.0,80.0,10.0,0.0, 10.000,10.000, 1,0, 6,0,500.0,500.0,
10.000,10.000, 0, 0
; 11
INCUB,"Adding AcN","Solvent A",900, 50.0,80.0,10.0,0.0, 10.000,10.000, 1,0,0,0,1000.0,1000.0,
10.000,10.000, 0, 0
; 12
INCUB,"Third wash","Solvent D",0, 50.0,80.0,10.0,0.0, 10.000,10.000, 1,0, 6,0,500.0,500.0,
10.000,10.000, 0, 0
; 13
INCUB,"Adding ACN","Solvent A",900, 50.0,80.0,10.0,0.0, 10.000,10.000, 1,0,0,0,1000.0,1000.0,
10.000,10.000, 0, 0
; 14
INCUB,"Fourth Wash","Solvent D",0, 50.0,80.0,10.0,0.0, 10.000,10.000, 1,0, 6,0,500.0,500.0,
10.000,10.000, 0, 0
; 15
INCUB,"Adding ACN","Solvent A",900, 50.0,80.0,10.0,0.0, 10.000,10.000, 1,0,0,0,1000.0,1000.0,
10.000,10.000, 0, 0
; 16
INCUB,"Purge Fourth Wash","Solvent A",0, 0.0,0.0,10.0,0.0, 10.000,10.000, 1,0, 6,0,500.0,500.0,
10.000,10.000, 0, 0
; 17
INCUB,"Shrinking in ACN","Solvent A",600, 150.0,80.0,10.0,0.0, 5.000,10.000, 0,0, 0,0,500.0,500.0,
30.000,10.000, 0, 0
; 18
INCUB,"Purge ACN and dry","Solvent A",0, 0.0,0.0,10.0,0.0, 10.000,10.000, 1,0, 30,0,1000.0,1000.0,
10.000,10.000, 0, 0
; 19
INCUB,"Shrinking in Acetonitrile","Solvent A",600, 150.0,80.0,10.0,0.0, 5.000,10.000, 0,0,
0,0,500.0,500.0, 30.000,10.000, 0, 170
; 20
INCUB,"Purge acetonitrile and dry","Solvent A",0, 0.0,0.0,10.0,0.0, 10.000,10.000, 1,0,
30,0,1000.0,1000.0, 10.000,10.000, 0, 0
; 21
PAUSE,"Waiting for trypsin"
; 22
INCUB,"Adding enzyme to samples","Enzyme C",1200, 25.0,20.0,10.0,0.0, 10.000,20.000, 1,0,
0,0,1000.0,1000.0, 30.000,10.000, 0, 170
; 23
MOVERK,"Moving rack for extraction","Sample", 400,925,80, 1625,925, 400, -1
; 24
MOVERK,"Moving rack for extraction"," ", 1610,925,80, 1642,925, 100, 5
; 25
HEAT,"Set heaters to 37C",1
```

```
; 26
WAIT,"Digestion - first stage",7200
; 27
RINSE,"Rinse needles",1000.0,3000.0, 10.000,10.000, 0, 0
; 28
ALIQUOT,"Adding bicarb","Solvent D", 10.0,20.0,10.0, 10.000,20.000, 1
; 29
WAIT,"Digestion - second stage",10800
; 30
ALIQUOT,"Adding water","Solvent F", 10.0,20.0,10.0, 10.000,20.000, 1
; 31
WAIT,"Digestion - third stage",10800
; 32
RINSE,"Rinse needles",2000.0,2000.0, 30.000,10.000, 0, 170
; 33
HEAT,"Turn heater off",0
; 34
INCUB,"Purge supernatant","Solvent A",0, 0.0,0.0,10.0,0.0, 10.000,10.000, 1,0, 6,0,1000.0,1000.0,
10.000,10.000, 1, 0
; 35
ALIQUOT,"Adding water","Solvent F", 20.0,20.0,10.0, 10.000,20.000, 0
; 36
WAIT,"Extracting peptides",1200
; 37
INCUB,"Purge supernatant","Solvent E",0, 0.0,0.0,10.0,0.0, 10.000,10.000, 1,0, 6,0,1000.0,1000.0,
10.000,10.000, 1, 0
; 38
ALIQUOT,"Adding Formic acid","Solvent E", 25.0,20.0,10.0, 10.000,20.000, 0
; 39
WAIT,"Formic acid extraction (I)",1800
; 40
INCUB,"Purge supernatant","Solvent A",0, 0.0,0.0,10.0,0.0, 10.000,10.000, 1,0, 6,0,1000.0,1000.0,
10.000,10.000, 1, 0
; 41
ALIQUOT,"Adding Formic acid","Solvent E", 25.0,20.0,10.0, 10.000,20.000, 0
; 42
ALIQUOT,"Adding CH3CN for extraction","Solvent A", 25.0,20.0,10.0, 5.000,10.000, 0
; 43
WAIT,"Formic Acid/AcN Extraction (II)",1200
; 44
INCUB,"Purge supernatant","Solvent A",0, 0.0,0.0,10.0,0.0, 10.000,10.000, 1,0, 6,0,1000.0,1000.0,
10.000,10.000, 1, 0
; 45
ALIQUOT,"Adding Formic acid","Solvent E", 25.0,20.0,10.0, 10.000,20.000, 0
; 46
ALIQUOT,"Adding CH3CN for extraction","Solvent A", 25.0,20.0,10.0, 5.000,10.000, 1
; 47
WAIT,"Last Extraction (III)",1200
; 48
INCUB,"Purge wells","Solvent A",0, 0.0,0.0,10.0,0.0, 10.000,10.000, 1,0, 15,0,1000.0,1000.0,
30.000,10.000, 0, 170
; 49
RINSE,"Rinse",2000.0,2000.0, 30.000,10.000, 0, 170
```

#### 3.2.3.3. Lyophilization and Reconstitution

All digested samples for MALDI analysis are lyophilized overnight. The lyophilized samples from our automated or manual digestion procedure are resuspended in 5 μL of 50:50 methanol:water (0.1% formic acid) immediately prior to spotting. This solution can be used directly for either nanoESI, or for spotting the MALDI target.

### *3.2.4. Spotting Techniques for MALDI–MS*

Like in-gel digestion, spotting for MALDI analysis can be done manually or robotically. Our normal spotting protocol utilizes the "dried-droplet" approach. For Bruker MALDI–MS, we spot a 0.5-μL aliquot of the extract, followed by a 0.5-μL aliquot of MALDI matrix solution (*see* **Note 13**), a saturated solution of recrystallized (*see* **Note 14**) α-cyano 4-hydroxycinnamic acid (Aldrich) in the 50:50 acetonitrile:water (0.1% trifluoroacetic acid) matrix solvent **(Subheading 2.6.1.5).** For samples to be analyzed on the ABI-TOF/TOF MS/MS analysis, we spot a 0.3-μL aliquot of the extract, followed by a 0.3-μL aliquot of in a MALDI matrix solution which contains ammonium citrate (*see* **Subheading 2.6.3.**). For samples to be analyzed on the ABI Q-Star MALDI-TOF, we use DHB, which is purchased as a premixed solution.

Several robotics systems are also available which spot the target plates for MALDI analysis. These can be "stand-alone" spotting robots (such as the Applied Biosystems Symbiot), or can be combined with the excision step (as in the BioMachines 2DiD system) or the excision and digestion steps (as in the Tecan robotic system).

For high-throughput analyses, as many of the sample preparation steps as possible should be automated **(Fig. 8). Figure 11** shows an example of a Coomassie-stained gel where automated sample preparation and a combination of automated MALDI–MS and MALDI–MS/MS data acquisition and automated database searching were used. This combination of MALDI–MS and MALDI–MS/MS can be quite powerful, as is shown by the successful identification of 25 co-immunoprecipitated proteins in this experiment where an affinity-tagged "bait" protein was used.

### *3.2.5. NanoESI/MS and MS/MS*

Samples, which are simply mixtures of peptides, can be introduced into the mass spectrometer either by infusing the sample from capillary-sized borosilicate needles via "nanospray," that is, nanoESI/MS, or nanoESI/MS/MS *(2,16,17)*. Increasing the number of components results may result in the selection of more than one precursor ion (see **Figs. 4** and **6**). This would result in a combined MS/MS spectrum which would be difficult to interpret. Nanospray also requires clean samples that contain only very low levels of salts and no detergents. Salts or particles can cause plugging of the fine needle orifice (1 μm), and detergents can cause severe suppression of the signal. Nanospray introduction is done at very low flow rates—lower flow rates (1–10 nL/min) than LC–MS/MS (200 nL/min)—and typically requires <5 μL of solvent. The primary advantage of nanospray is the increase in signal-to-noise which can be obtained by summing the spectra during long acquisition times at these low flow rates.

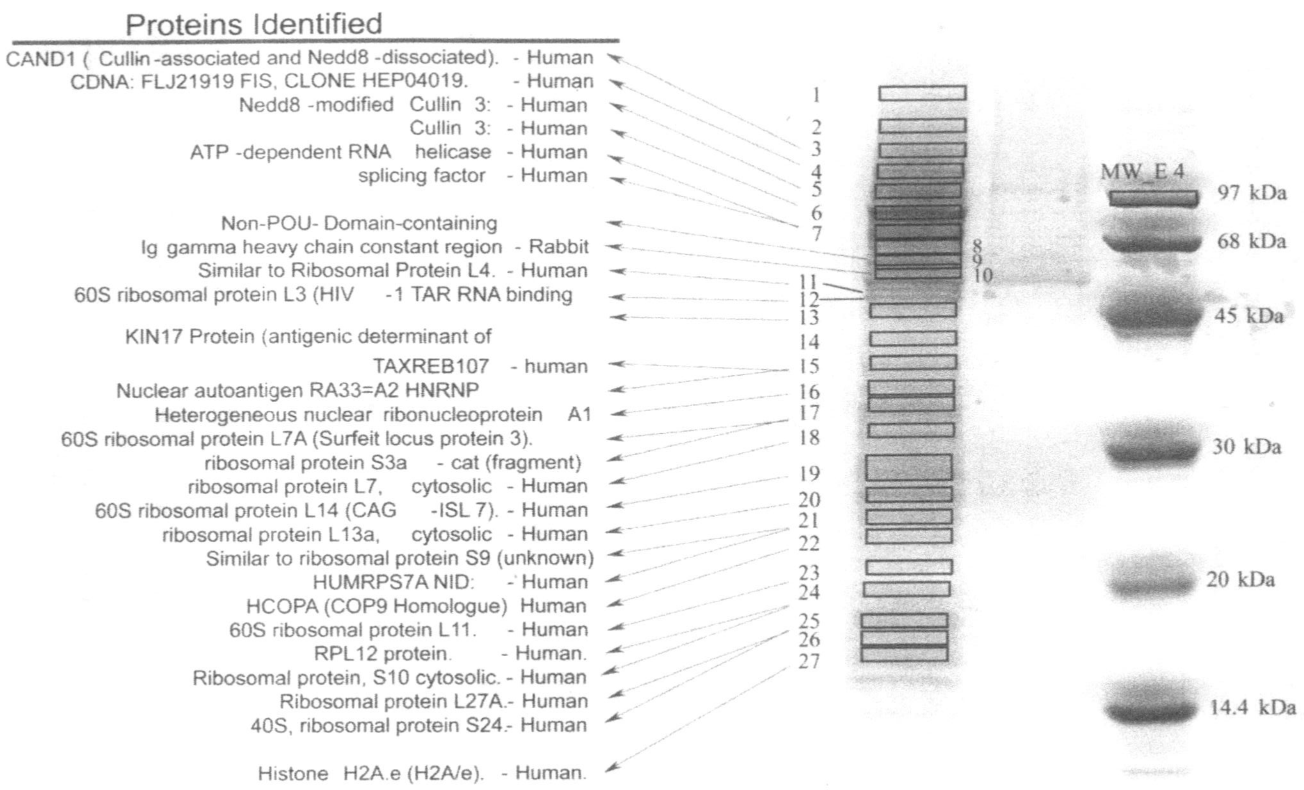


Fig. 11. 1D-PAGE separation and identification of proteins from a cullin co-immunoprecipitation experiment. (Collaborator: Y. Xiong.) The numbered protein bands were excised, digested with trypsin, and then analyzed by MALDI-TOF–MS and MS/MS. The protein identification assignments were made based primarily on the MS/MS fragment-ion matches searched using the Mascot search engine.

#### 3.2.5.1. Protocol for NanoESI

1. Dissolve the lyophilized sample in 50:50 methanol–water (0.1% formic acid), or use the extract from the in-solution digestion protocol directly (*see* **Subheading 3.2.3.3.**).
2. Using a syringe with a fused silica capillary as a needle, load 1 µL of the solution into a Proxeon nanospray needle.

### *3.2.6. LC–MS/MS on In-Gel Digested Samples*

If the high-throughput MALDI-based methods described above fail to provide a protein identification, ESI-LC–MS/MS can be used. The system used in our laboratory is a Micromass Q-tof, with capillary HPLC setup that incorporates a trapping column. As in any gradient LC system, this setup concentrates the sample and provides on-line separation of peptides from the sample. The trapping column, however, allows for more rapid loading of the sample, as well as a "built-in" desalting step, since components not retained by the trapping column do not enter the analytical column and do not enter the MS source (*see* **Note 15**).

Capillary LC–MS/MS, however, is not a high throughput technique–standard runs take a minimum of 60–75 min (the void volume alone is ~30 min) – but it has a major advantage for limited amounts of sample in that all of the sample injected goes through the analytical column and enters the MS source. The splitting of the solvent, which is needed to provide the low flow rate (200 nL/min) for the capillary column, is done prior to the trapping column. The trapping column is loaded at 20 µL/min, and eluted at approx 200 nL/min.

We use a Waters CapLC system that has three pumps and can accept three different solvent solutions. For normal use, buffers A and B form the gradient used for eluting the components from the capillary column, while buffer C is used for loading the trapping column. In our system, buffer A is 5:95 acetonitrile–water (containing 0.1% formic acid), which is the same composition as buffer C, while buffer B consists of 95:5 acetonitrile–water (containing 0.1% formic acid). The trapping column is a Dionex 5 mm × 800 µm id $C_{18}$ P3 cartridge, and the analytical column is a Dionex PepMap $C_{18}$ 15 cm × 75 µm id capillary column. After loading the sample at a flow rate of 20 µL/min, the trapping column continues to be washed with buffer C at 20 µL/min. At 3.5 min, the 10-port valve is switched so that the trapping column is now in line with the analytical column, and the trapped sample is eluted onto the capillary column at a flow rate of approx 200 nL/min. For protein identification work, where there are a limited number of peptides present, a 30-min gradient of 0–80% B is used, with a 5-min hold time at the initial conditions, and a 15-min hold at the "final" conditions, followed by a 5-min wash of the column at 95% B before the gradient returns to the initial conditions.

LC–MS/MS is also the method of choice for mixtures too complex to be analyzed by MALDI–MS, or where a trace amount of a target protein may be present in a large amount of an interfering protein. If the problem with the MALDI–MS or MALDI–MS/MS was contamination—that is, database searches only identified keratin or IgG, a longer linear gradient (2–3 h) can be used to try to separate the peptides and reduce suppression effects, so that a trace amount of a target protein may be identified.

An example of the use of this technique is shown in **Fig. 12**, where successful protein identification was obtained from a gel contaminated with keratin and IgG.

#### *3.2.7. LC–MS/MS on In-Solution Digested Samples*

One way to avoid the difficulties in extracting peptides from gels is to not put the proteins onto gels in the first place. This approach for analyzing mixtures of proteins by digesting them all at once and analyzing the resulting mixture by LC–MS/MS has been championed by Dr. John Yates. There is no sample loss caused by poor recoveries from the gel, but there is also no gel "cleanup" step, so MS-incompatible compounds such as detergents (e.g., SDS) and nonvolatile salts must be avoided (*see* **Note 15**). Also, because there is no gel separation step prior to the LC–MS/MS analysis, there is also no intact protein molecular weight, because the analysis is done entirely on peptides.

For mixtures expected to contain large numbers of proteins, longer gradients are used (200 min) to separate and sequence as many peptides as possible, thus allowing the identification of as many proteins as possible. An example of the use of LC–MS/MS for protein identification of a complex mixture of proteins from an on-bead in-solution digest of FLAG-tagged CIITA was shown in **Fig. 6**.

More than 172 proteins have been identified from a single LC–MS/MS run using this approach, although the additional restrictions on sample preparation must be followed. Also, if the proteins are to be identified subsequent to a pull-down assay, care should be taken to avoid or minimize the use of reducing agents such as β-mercaptoethanol or dithiothreitol (DTT), which will unfold the antibody and allow enzymatic digestion of the antibody *in addition to* the proteins affinity-bound to it. If DTT *must* be used, its concentration should be kept as low as possible. Likewise, any detergent used must be removed by thorough washing of the affinity beads.

If the mixture is too complex for a single LC–MS/MS experiment, multidimensional liquid chromatographic methods ("MUDPIT") can be used ***(18–20)***. In this approach, the mixture is prefractionated, usually by an ion exchange step prior to the LC–MS/MS assays, and each fraction is analyzed by LC–MS/MS. This results in multiple, long, LC–MS runs from one original sample, but can lead to large numbers (>1000) proteins being identified from a given sample ***(21)***.

If LC–MS/MS is to be used without prior gel separation, an in-solution digest procedure must be used. This procedure can be done directly on affinity beads (*see* **Note 16**), or the sample can be eluted with 1:1:8 ethanol:formic acid:water, evaporated and resuspended in the ammonium bicarbonate buffer.

##### 3.2.7.1. In-Solution Protein Digestion Procedure

1. Calculate what a 1:50 enzyme/substrate ratio would be (*see* **Note 17**).
2. Reconstitute Promega trypsin in 20 μL of Promega resuspension buffer (0.015 *M* acetic acid). (Promega trypsin comes in aliquots of 20 μg per vial.)
3. Dissolve sample in approx 20 μL of 100 m*M* ammonium bicarbonate solution.
4. Add calculated amount of trypsin solution to each sample.
5. Vortex mix and centrifuge.
6. Incubate for at least 4 h (or overnight for beads) in sealed Eppendorf tubes, at 35°C with rotation (~400 rpm).

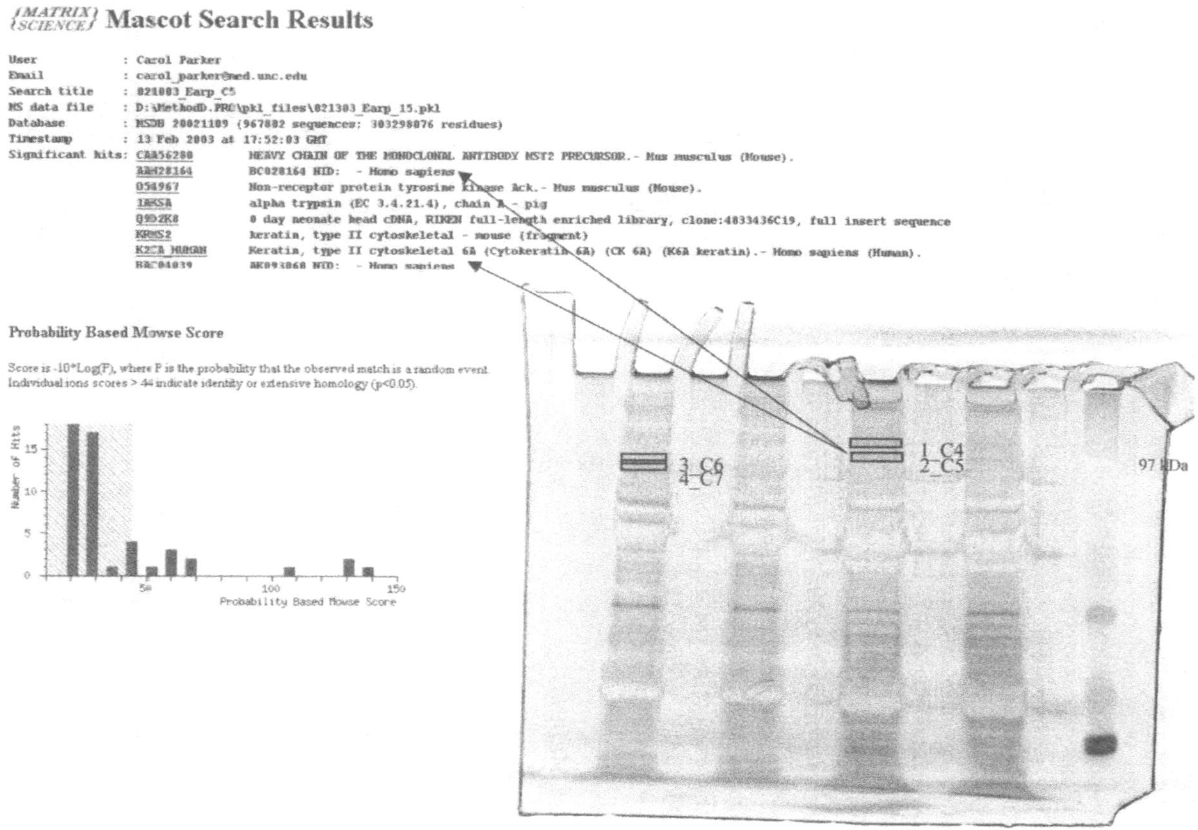


Fig. 12. Example of LC–MS/MS identification of gel-separated proteins in a gel contaminated with a high background of IgG and keratin. (Collaborator: Y. Xiong.)

#### 3.2.7.2. Procedure for On-Bead Digestion

1. Place approx 50–100 µL of beads in an Eppendorf tube.
2. Wash three times with 200 µL of 100 m*M* ammonium bicarbonate
3. Add 100 µL of 100 mM ammonium bicarbonate
4. Reconstitute Promega trypsin in 20 µL of Promega resuspension buffer (0.015 *M* acetic acid). (Promega trypsin comes in aliquots of 20 µg per vial.)
5. Add 2 µL trypsin solution to each sample. For proteins affinity bound to antibody beads, much higher ratios of enzyme to substrate are used than for proteins in solution (e.g., 5:1) *(22)*.
6. Vortex-mix/centrifuge at <1000 rpm to avoid breaking the beads.
7. Incubate overnight in sealed Eppendorf tubes, at 35°C with rotation (~400 rpm) (*see* **Note 16**).
8. Remove the supernatant from the beads.
9. Lyophilize to reduce volume (if concentrated enough, supernatant can be injected directly).
10. Freeze the sample at –80°C until analysis.

### 3.3. Overall Approach to the Identification of In-Gel Separated Samples

More than 80% of gel-separated proteins submitted to our facility are identified by high-throughput MALDI–MS and MALDI–MS/MS. Because of its limited throughput, only those proteins *not* identified by MALDI are analyzed by LC–MS/MS. Of these "difficult" samples, LC–MS/MS can usually identify approximately half, which gives an overall success rate of approx 90% for the combination of all three instruments.

### 3.4. Newer Methods

For extremely critical samples, several additional methods can be used if the more routine methods fail.

#### 3.4.1. MALDI-Directed NanoESI/MS/MS

We have recently developed a procedure that we call MALDI-directed nanoESI-MS/MS *(23)*. This procedure is used if the amount of a particular peptide is too low for sequencing by MALDI–MS. The *m/z* ratios of the singly and doubly-charged forms of a peptide are calculated from its molecular weight as observed in the MALDI spectrum. NanoESI/MS/MS spectra are then collected for these calculated parent ions, even though no ions corresponding to this peptide are observed in the ESI mass spectrum.

The ionization mechanisms of electrospray and MALDI are different, and an analysis of the same protein by MALDI and ESI shows that only approx 50% of the peptides detected will be found in *both* mass spectra. (In other words, ~25% of the peptides will be found only in the ESI/MS spectrum, while another 25% will be found only in the MALDI mass spectrum.) This MALDI-directed nanoESI/MS/MS method targets the 50% of peptides found in both ionization techniques. Using this method, we successfully identified KIAA1309 and/or KIAA1354 from the peptide sequence found by MALDI-directed nanoESI (**Fig. 13**) *(24)*. (The same peptide is found in both proteins.)

#### 3.4.2. Extraction of Proteins From Unstained Gels

As described in **Subheading 3.2.2.3.**, the most sensitive staining technique is silver staining, but this technique can interfere with mass spectrometric protein identification by reducing the recoveries of in-gel-digested peptides just when the recoveries of

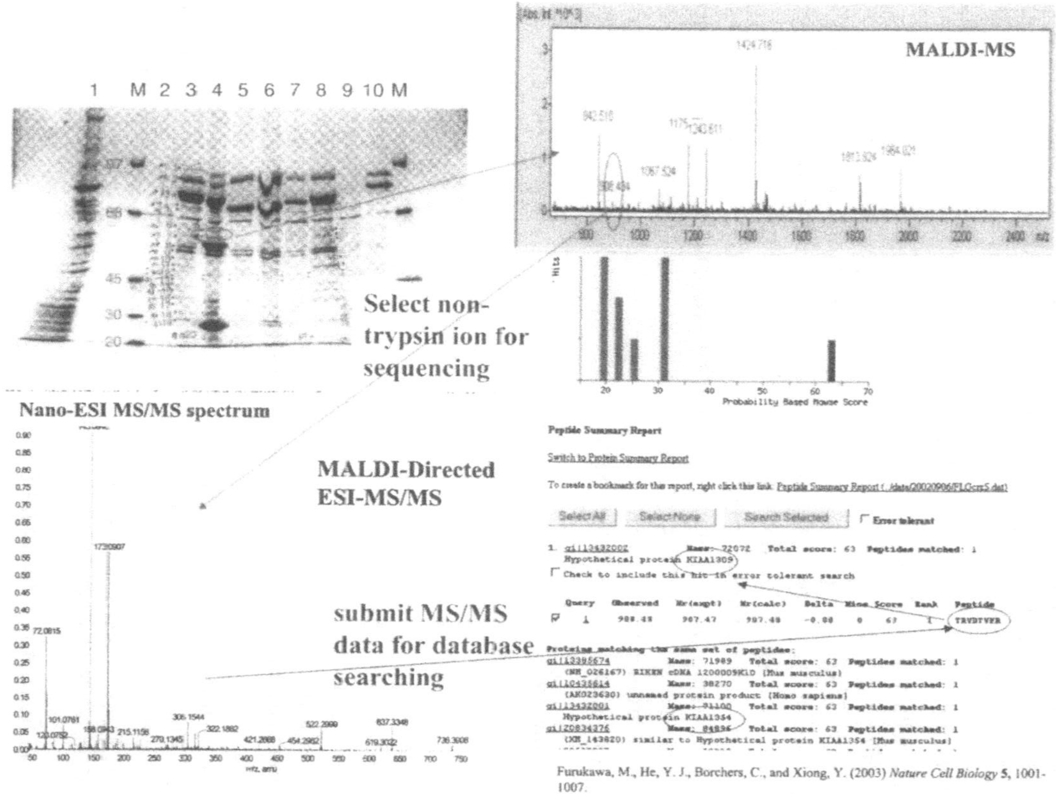


Fig. 13. MALDI-directed nanoESI identification of an unknown in-gel digested protein in a coimmunoprecipitation experiment *(24)*. Note that there are two proteins that match the same peptide, which are from the correct species, and that have approximately the same molecular weight. Which of these two is actually present cannot therefore be distinguished from this experiment.

these low-level peptides are most critical. We are currently developing methods for use where ultimate sensitivities are needed. The methods rely on extraction of peptides from *unstained* gels, and we rely on comparison of stained gels with unstained gels for cutting. Because of the resulting uncertainties of exactly where the protein is in the gel, additional bands or spots are also cut and analyzed. This technique requires high-throughput methods, since a much larger number of bands or spots requiring analysis can easily be generated.

#### 3.4.3. "Top-Down" Protein Identification

Instruments with very high resolution and mass accuracy, such as Fourier-transform ion cyclotron resonance (FTICR)/MS instruments, can perform "top-down" protein identification ***(25–27)***. In "top-down" proteomics, intact *larger* peptides and intact proteins are intensively analyzed in the mass spectrometer. This means no prior enzymatic or chemical protein digestions are required as is the case for "bottom-up" proteomics. Here, nonenzymatic fragmentation leads to nonspecific cleavages of the protein backbone. The "top-down" approach is computationally demanding, because without the known specificity of enzymatic cleavage, there are many more possible peptides from a given protein.

Currently, for larger proteins, these requirements can be fulfilled only by a FTICR mass spectrometer in the ESI mode. FTICR mass spectrometers, however, are very expensive and highly sophisticated instruments. In addition, the top-down approach necessitates proteins to be dissolved in solvents that are compatible with ESI. Nanospray is most often used as the inlet system for this approach, and the same sample restrictions on solvents that were described in **Subheading 3.2.3.** apply here. These requirements are even more stringent for nanospraying proteins than for peptides. Commercial instruments have become available where LC–MS has been coupled on-line to FTICR ***(28,29)***. This will provide a certain level of sample cleanup with the added benefit of separating and concentrating the proteins.

An important advantage of the "top-down" approach is that information about the physical status, for example, the proportion of the protein with the modification, can be obtained. Thus this technique can be used not only to identify but also to *characterize* proteins. "Top-down" proteomics requires both fragmentation of intact proteins in the mass spectrometer and high-resolution mass spectrometry of high molecular weight fragment ions. This is a new and very promising method but, because of the expensive instrumentation involved, it is not yet widely used.

### 3.5. Database Searching and Quality Control

#### 3.5.1. Database Searching

There are several database searching programs for "bottom-up" protein identification, which are available at no cost on the Internet (Protein Prospector, ProFound, Mascot, etc.) ***(9,10,30)***. Advanced versions of these programs can be obtained from these vendors through site licenses, and other commercial database search programs are also available (RADARS, ProteinLynx, etc.). Input data for performing peptide mass fingerprinting using these programs includes the molecular weights of the peptides observed, the mass accuracy of these measured masses, and the species the protein was isolated from. Input data for searching MS/MS data includes tabulated MS/MS spectra including the parent ion mass and charge state, and the mass accuracy of the parent ion and fragment ions, and the species whose genome is to be searched. Relative ion abundances are usually not considered.

Because all of these database-searching programs compare the observed mass spectra to mass spectra predicted from the database, the success of the search depends on the completeness of the database **(Fig. 14).** Databases are continuously being updated, and currently mouse and human databases are fairly complete (*see* **Note 18**). EST databases can also be searched, and human, mouse, and dog EST-database have recently become searchable by the web-based Mascot software. Additional EST-databases can be input and searched against (*see* **Note 19**).

We have already encountered several instances where a protein was not initially in the database, but was re-tried later and was identified because the database was more complete. If you have MS/MS data for a species not in the database, it is sometimes possible to find a homologous protein in a related species. As a "last resort," *de novo* sequencing of the peptide can be done from the *b* and/or *y* ion series, and a BLAST search can be performed on the resulting peptide sequence. An example of a shark apical membrane-associated protein, identified by homology to a known protein from the rat database, is shown in **Fig. 15.** This protein was identified by a BLAST search of the peptide sequence data obtained by nanoESI/MS/MS.

To avoid "false-positives" when searching a database:

1. Do not believe hits requiring "unreasonable" modifications to standard tryptic peptides; for example, do not specify too many missed cleavages; don't specify too many modifications.
2. For MS/MS data, always check the MS/MS spectrum to be certain that noise peaks or $^{13}C$ isotope peaks are not being matched to the theoretical fragment ion masses.
3. As a "first pass," do a search that is *not* restricted to a given species. This will allow the detection of "unexpected" proteins that might otherwise give matches to incorrect proteins from the selected species. Autolysis peptides from the porcine trypsin used for the in-gel digestion, the common contaminant human keratin, and the LC–MS/MS calibration compound human glu-fibrinopeptide B can give "false hits" if, for example, the data is searched against a mouse database.
4. *Then* the data are searched against the database for the target species.
5. If the protein came from an in-gel digest, it is also useful to check the molecular weights for the "hits" against the approximate molecular weight from the gel—often the database contains various truncated or processed forms of a protein with different molecular weights.
6. Beware: the search result *summary* may contain a single protein name, but examination of the search results may show a list of "other proteins matching these peptides." Here again, it is important not to interpret "beyond the data" and to remember that only these few peptides were found, which might match several related proteins.
7. There is no substitute for knowledge of the relevant biology! Hits should "make biological sense"—that is, should be from the appropriate organism, subcellular compartment, and so forth.

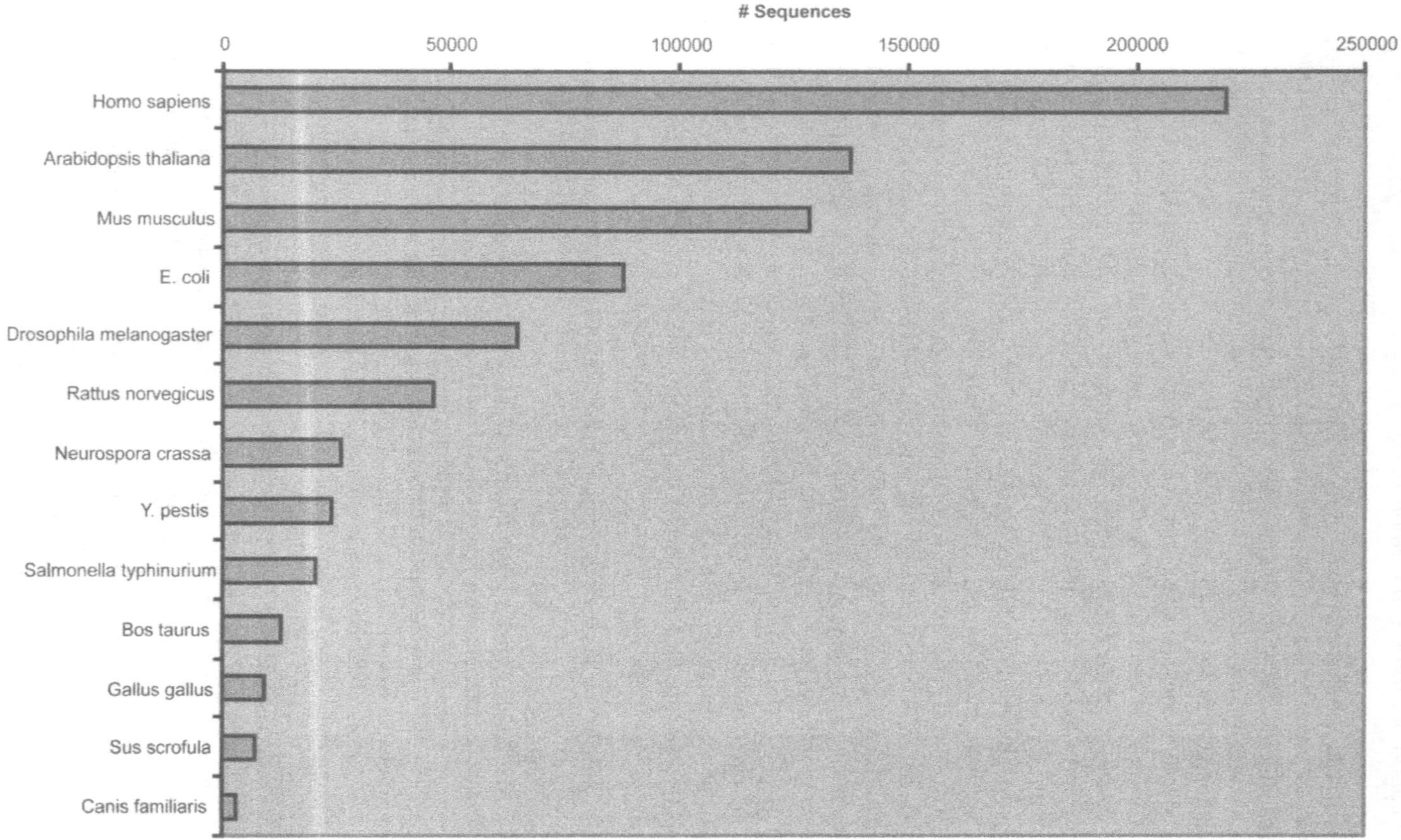


Fig. 14. Comparison of sequences from the NCBI database from different species (not nonredundant).

# Protein ID: Mass Spectrometric *de novo* Sequencing

**NanoESI-MS/MS**

**Peptide sequence from MS/MS spectrum: I/L-I/L-YNNVSNTCR**
**Peptide sequence from BLAST search (Rat clathrin heavy chain): [1216]L-L-YNNVSNFGR[1226]**

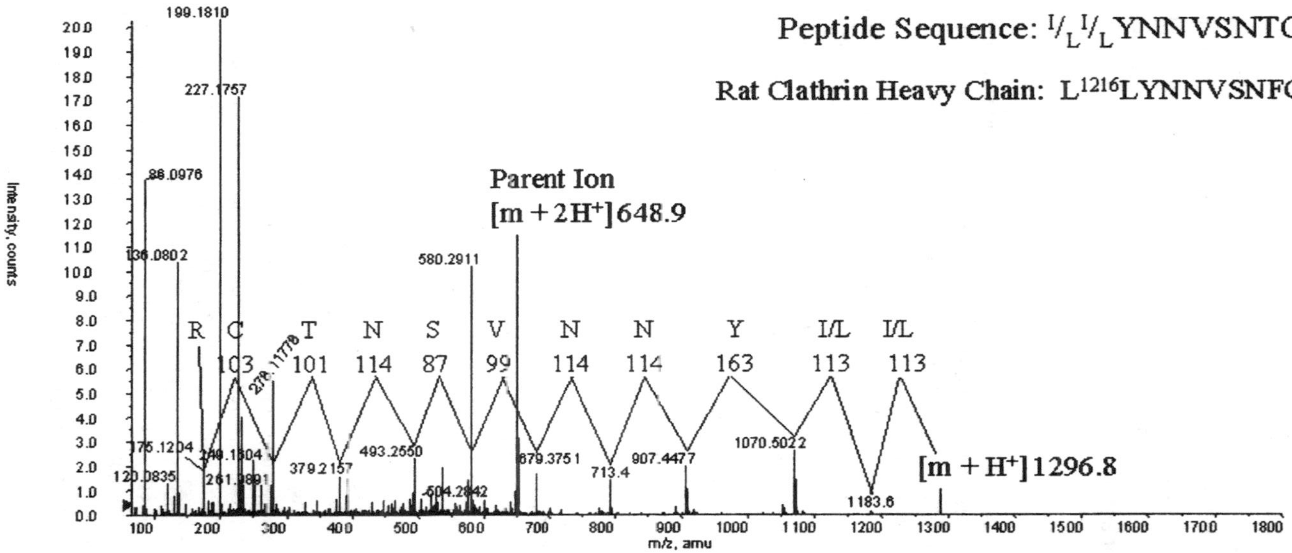


Fig. 15. Identification of proteins by *de novo* sequencing and BLAST searching. Identification of apical membrane-associated proteins in shark which interact with CFTR. (Collaborator: Jack Riordan.)

### 3.5.2. Validation of Search Results

In conclusion, it is always important to remember to remind the submitter that just because a protein is found in a database search, it might not necessarily be a valid identification. Contamination of the sample or the gel can lead to misleading results. Binding in a co-immunoprecipitation experiment may be nonspecific. In addition to the above caveats, a "hit" should not be considered valid unless it makes biological sense, and, ideally, should be confirmed by an independent technique such as a Western blot analysis.

## 4. Notes

1. With the advent of the newer protein identification method, termed the "top-down" approach (described in **Subheading 3.4.3.**), this commonly used peptide-based approach is sometimes called the "bottom-up" protein identification.
2. The recent development of automated robotic nanoelectrospray (or "nanospray") instruments, such as the Advion Nanomate, with microfabricated electrospray nozzles, makes this technique higher throughput than using individually filled nanospray needles. Nanospray is probably the most sensitive MS/MS technique, but it still cannot provide the level of throughput achievable by MALDI–MS, but it does allow ESI/MS/MS sequencing of the peptides detected.
3. Roepstorff nomenclature ***(11,12)*** includes many other types of peptide fragmentation reactions, including *a, c,* and *z* ions. In most commonly used mass spectrometers (MALDI–MS/MS and ESI/MS/MS), *b* and *y* ions are the most abundant sequence ions, so these are the ion types normally used for sequencing.
4. For both approaches, we normally use peptides in the mass range 800–3800 Da, using the trypsin autolysis products at *m/z* 842.5100 and 2211.1046 for internal calibration to ensure the highest possible mass accuracy.
5. The database search programs also support other enzymes, such as GluC, AspN, and LysC, and we have had occasion to use these when the results from tryptic cleavage were not successful, or when we were trying to find a particular peptide containing a modification site.
6. Some databases, such as Mascot, also support combinations of chemical and enzymatic digestions, such as CNBR/trypsin, trypsin/CNBr, and trypsin/chymotrypsin.
7. Very low molecular weight proteins may present a problem for peptide mass fingerprinting, since there may be too few peptides of appropriate molecular weight for peptide mass fingerprinting to identify the protein.
8. It has been our experience that gels with lower percent acrylamide (<8%) tend to turn "gummy" during our extraction procedure, and recovery of peptides from these gels is low. If low-percentage gels must be used, manual in-gel digestion is recommended.
9. If hand-poured gels cannot be avoided, we recommend running a blank gel with unstained molecular weight markers. Our protocols can then be tested on these markers to see if this type of hand-poured gel is compatible with peptide recoveries and mass spectrometric detection. Unfortunately, nonreproducibility of the hand-poured gels can still be a problem.
10. The stain can be reused safely for at least two applications depending on the size of the gel and the amount of protein being stained. The Sypro Ruby dye is slightly orange in color and a clear solution indicates depletion of the dye in the stock or working solutions.

11. Longer development time will result in "burning" of the protein bands and an "overdeveloped" gel, as shown in **Fig. 4**. "Burning" decreases peptide recoveries, which decreases the probability of positive protein identification.
12. Make sure gel pieces are completely submerged.
13. Make the MALDI matrix fresh daily.
14. To make recrystallized α-cyano 4-hydroxycinnamic acid, make a saturated solution of α-cyano 4-hydroxycinnamic acid in boiling methanol. Pour off the solution and discard it. Add more methanol, and again make a saturated solution in boiling methanol. This time, pour off the methanol and save it. Evaporate the methanol to dryness in a hood, while protecting the solution and the crystals from light with aluminum foil. Store in the dark or in a vial wrapped with aluminum foil.
15. Gels are very good for removing interfering compounds such as salts and detergents. We have used the following buffer as an immunoprecipitation steps using FLAG affinity beads *(31)*: 25 m*M* HEPES–NaOH, pH 7.5, 150 m*M* sodium acetate, 10% glycerol, 0.1% Nonidet P-40, and 0.5 m*M* DTT. Because the co-immunoprecipitated proteins were to be separated by SDS-PAGE, these buffer components did not create a problem. For LC–MS/MS analyses *without* prior gel separation, the on-line trapping column *cannot* be relied on to remove these interferences. The on-line trapping column *can* remove low levels of salts (such as a few microliters of PBS, but is not sufficient for removing high levels of salts or detergents such as SDS or Nonidet P-40, which must *not* be used in LC–MS procedures. If possible, use volatile buffers such as ammonium acetate and organic acids such as acetic acid and formic acid.
16. Be sure to add enough digestion buffer so that the beads can "slosh around" in the Eppendorf tube and don't dry out during the overnight digestion.
17. If you do not know the protein concentration, use approx 2 μL of a 1 μg/ μL solution.
18. It is recommended to search the same data against several databases. We routinely search against both MSDB and NCBI. We have found that the same data may give a "hit" against one database and not the other.
19. Mascot, for example, recommends searching the "regular" database first, and *then* searching the EST database.

## Acknowledgments

This study was funded by a gift from an anonymous donor to support research in proteomics and cystic fibrosis, and grants from the Cystic Fibrosis Foundation (CFFTI STUTTS01U0) and from NIH (ES11997, 5U54HD035041-07, and P30 CA 16086-25).

## References

1. Geromanos, S., Freckleton, G., and Tempst, P. (2000) Tuning of an electrospray ionization source for maximum peptide-ion transmission into a mass spectrometer. *Anal. Chem.* **72,** 777–790.
2. Shevchenko, A., Wilm, M., Vorm, O., and Mann, M. (1996) Mass spectrometric sequencing of proteins from silver-stained polyacrylamide gels. *Anal. Chem.* **68,** 850–858.
3. Jungblut, P. and Thiede, B. (1997) Protein identification from 2-DE gels by MALDI mass spectrometry. *Mass Spectrom. Rev.* **16,** 145–162.
4. Mann, M. and Wilm, M. (1994) Error-tolerant identification of peptides in sequence databases by peptide sequence tags. *Anal. Chem.* **66,** 4390–4399.
5. Zhang, W. and Chait, B. T. (2000) ProFound: an expert system for protein identification using mass spectrometric peptide mapping information. *Anal. Chem.* **72,** 2482–2489.

6. Perkins, J. R., Smith, B., Gallagher, R. T., et al. (1993) Application of electrospray mass spectrometry and matrix-assisted laser desorption ionization time-of-flight mass spectrometry for molecular weight assignment of peptides in complex mixtures. *J. Am. Soc. Mass Spectrom.* **4,** 670–684.
7. Billeci, T. M. and Stults, J. T. (1993) Tryptic mapping of recombinant proteins by matrix-assisted laser desorption/ionization mass spectrometry. *Anal. Chem.* **65,** 1709–1716.
8. Griffin, P. R., Coffman, J. A., Hood, L. E., and Yates, J. R., III (1991) Structural analysis of proteins by capillary HPLC electrospray tandem mass spectrometry. *Int. J. Mass Spectrom. Ion Proc.* **111,** 131–149.
9. *www.matrixscience.com.*
10. *http://prowl.rockefeller.edu.*
11. Roepstorff, P. and Fohlman, J. (1984) Proposal for a common nomenclature for sequence ions in mass spectra of peptides. *Biomed. Mass Spectrom.* **11,** 601.
12. Biemann, K. (1988) Contributions of mass spectrometry to peptide and protein structure. *Biomed. Environ. Mass Spectrom.* **16,** 99–111.
13. Griffin, P. R., MacCoss, M. J., Eng , J. K., Blevins, R. A., Aaronson, J. S., and Yates, J. R. (1995) Direct database searching with MALDI-PSD spectra of peptides. *Rapid Commun. Mass Spectrom.* **9,** 1546–1551.
14. Fenyo, D., Qin, J., and Chait, B. T. (1998) Protein identification using mass spectrometric information. *Electrophoresis* **19,** 998–1005.
15. Borchers, C., Peter, J. F., Hall, M. C., Kunkel, T. A., and Tomer, K. B. (2000) Identification of in-gel digested proteins by complementary peptide-mass fingerprinting and tandem mass spectrometry data obtained on an electrospray ionization quadrupole time-of-flight mass spectrometer. *Anal. Chem.* **72,** 1163–1168.
16. Wilm, M. and Mann, M. (1996) Analytical properties of the nanoelectrospray ion source. *Anal. Chem.* **68,** 1–8.
17. Wilm, M., Shevchenko, A., Houthaeve, T., et al. (1996) Femtomole sequencing of proteins from polyacrylamide gels by nano-electrospray mass spectrometry. *Nature* **379,** 466–469.
18. Yates, J. R. I., Link, A. J., and Schieltz, D. (2000) In *Methods in Molecular Biology:* Vol. 146 (Chapman, J. R., ed.), Humana Press, Totowa, NJ, pp. 17–26.
19. Yates, J. R. I. (1998) Mass spectrometry and the age of the proteome. *J. Mass Spectrom.* **33,** 1–19.
20. Yates (III), J. R., Carmack, E., Hays, L., Link, A. J., and Eng, J. K. (1999) In Methods in *Molecular Biology,* Vol. 112 (Link, A. J., ed.), Vol. 112, Humana Press, Totowa, NJ, pp. 553–569.
21. Washburn, M. P., Wolters, D., and Yates, J. R. (2001) Large-scale analysis of the yeast proteome by multidimensional protein identification technology. *Nat. Biotechnol.* **19,** 242–247.
22. Parker, C. E. and Tomer, K. B. (2000) In *Methods in Molecular Biology,* Vol. 146: Mass Spectrometry of Proteins and Peptides (Chapman, J. R., ed.), Humana Press, Totowa, NJ, pp. 185–201.
23. Kast, J., Parker, C. E., van der Drift, K., et al. (2003) MALDI-directed nano-ESI-MS/MS Analysis for Protein Identification. *Rapid Commun. in Mass Spectrom.* **17,** 1825–1834.
24. Furukawa, M., He, Y. J., Borchers, C., and Xiong, Y. (2003) Targeting protein ubiquitination by BTB-cullin 3-Roc1 ubiquitin ligases. *Nat. Cell Biol.* **5,** 1001–1007.
25. Kelleher, N. L., Lin, H. Y., Valaskovic, G. A., Aaserud, D. J., Fridriksson, E. K., and McLafferty, F. W. (1999) Top down versus bottom up protein characterization by tandem high-resolution mass spectrometry. *J. Am. Chem. Soc.* **121,** 806–812.

26. Nemeth-Cawley, J. F., Tangarone, B. S., and Rouse, J. C. (2003) "Top down" characterization is a complementary technique to peptide sequencing for identifying protein species in complex mixtures. *J. Proteome Res.* **2,** 495–505.
27. Ge, Y., ElNaggar, M., Sze, S. K., et al. (2003) Top down characterization of secreted proteins from *Mycobacterium tuberculosis* by electron capture dissociation mass spectrometry. *J. Am. Soc. Mass Spectrom.* **14,** 253–261.
28. Ingendoh, A., Baykut, G., Zhang, T., Speir, P., and Kruppa, G. (1998) HPLC online coupled to Fourier transform ion cyclotron resonance (FT-ICR) MS. *Adv. Mass Spectrom.* **14,** D013240/1–D013240/14.
29. Zubritsky, E. (2000) The best of both worlds with LC/FTMS. *Anal. Chem.* **72,** 633A.
30. *http://prospector.ucsf.edu/*.
31. Hall, M. C., Torres, M. P., Schroeder, G. K., and Borchers, C. H. (2003) Mnd2 and Swm1 are core subunits of the *Saccharomyces cerevisiae* anaphase-promoting complex. *J. Biol. Chem.* **278,** 16698–16705.

# 10

# Mass Spectrometric Determination of Protein Ubiquitination

**Carol E. Parker, Viorel Mocanu, Maria R. Warren, Susanna F. Greer, and Christoph H. Borchers**

**Summary**

Mass spectrometric methods of determining protein ubiquitination are described. Characteristic mass shifts and fragment ions indicating ubiquitinated lysine residues in tryptic and gluC digests are discussed. When a ubiquitinated protein is enzymatically digested, a portion of the ubiquitin side chain remains attached to the modified lysine. The ubiquitinated peptide thus has two N-termini— one from the original peptide and one from the ubiquitin side chain. Thus, it is possible to have two series of b ions and y ions, the additional series is the one that includes fragments containing portions of the ubiquitin side chain. Any diagnostic ions for the modification must include portions of this side chain. Fragment ions involving any part of the "normal" peptide will vary in mass according to the peptide being modified and will therefore not be of general diagnostic use. These diagnostic ions, found through examination of the MS/MS spectra of model ubiquitinated tryptic and gluC peptides, can be used to trigger precursor ion scanning in automated MS/MS data acquisition scanning modes.

**Key Words:** Characteristic ions; mass spectrometry; MS/MS; Ubiquitination.

## 1. Introduction

Protein ubiquitination is a highly regulated process mediated by the activation and serial transfer of ubiquitin to a target protein through a cascade of conserved E1, E2, and E3 enzymes. The covalent linkage of ubiquitin to lysine residues on target proteins has historically been linked to protein destruction by the 26S proteasome, a multisubunit protease that recognizes the polyubiquitin tag. The ubiquitin/proteasome pathway has been shown to regulate a number of critical cellular events including cell division, protein trafficking, cell death, and signal transduction. Conjugation of a single monoubiquitin group has also emerged as an important regulatory pathway independent of degradation, first found in endocytosis and protein localization, and subsequently for gene transcription. The use of mass spectrometry to identify sites of ubiquitination within target proteins in vivo will allow greater understanding of the

From: *Methods in Molecular Biology, vol. 301, Ubiquitin–Proteasome Protocols*
Edited by: C. Patterson and D. M. Cyr © Humana Press Inc., Totowa, NJ

ways in which ubiquitination alters protein function. Therefore, the development of broadly applicable identification approaches is critical.

## 2. Materials

### 2.1. Matrix-Assisted Laser Desorption Ionization (MALDI) Analysis of Intact Proteins

1. Bruker Reflex III MALDI-MS (Bruker Daltonics, Billerica, MA).
2. α-Cyano 4-hydroxycinnamic acid (Aldrich, St. Louis, MO).
3. Ethanol (AAPER Alcohol and Chemical Co., Shelbyville, KY).
4. Water, HPLC grade (Pierce, Rockford, IL).
5. Formic acid (Fisher, Pittsburgh, PA).

### 2.2. Liquid Chromatography–Tandem Mass Spectrometry (LC–MS/MS)

1. Waters/Micromass Q-TOF (Waters/Micromass Corp., Milford, MA).
2. PepMap $C_{18}$15 cm × 75 μm id capillary column (Dionex, Sunnyvale, CA).
3. Trapping column 5 mm × 800 μm id $C_{18}$ P3 (Dionex).
4. Water, HPLC grade (Pierce).
5. Acetonitrile, HPLC grade (Pierce).
6. Formic acid (Fisher).

### 2.3. Affinity Purification

1. Anti-HA antibody beads (Sigma, St. Louis, MO).
2. Anti-FLAG antibody beads (Sigma).
3. Ammonium bicarbonate (Fluka, Milwaukee, WI).
4. Ethanol (AAPER).
5. Water (Pierce).
6. Formic acid (Fisher).

### 2.4. In-Solution Digestion

1. Water, deionized or HPLC grade (Pierce).
2. Trypsin, sequencing grade (Promega, Madison, WI).
3. GluC (Sigma).
4. Ammonium bicarbonate (Fluka).
5. Low-retention Eppendorf tubes (Axygen, Union City, CA).
6. Thermomixer (Eppendorf, Hamburg, Germany).

### 2.5. Lyophilization and Reconstitution

1. Freeze dryer (Labconco, Kansas City, MO).
2. Water (Pierce).

## 3. Methods

### 3.1. Direct Evidence of Ubiquitination

#### 3.1.1. Evidence From Gels

Most direct evidence for ubiquitination comes from gel electrophoresis **(Fig. 1)**, in which a series of higher molecular weight bands are observed above the molecular weight of the protein, or from Western blot analysis using anti-ubiquitin antibodies **(Fig. 2)**.

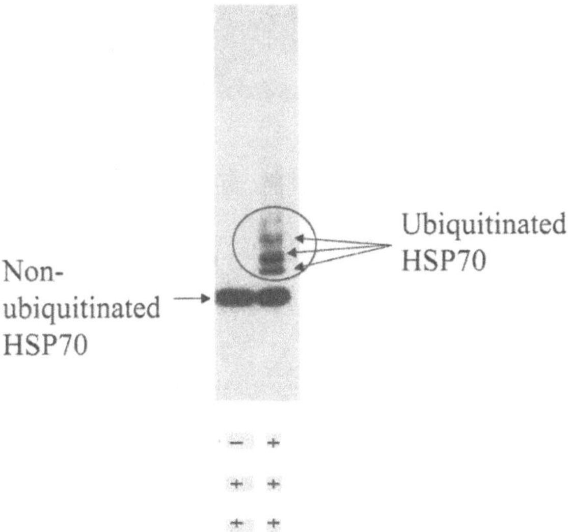


Fig. 1. Ubiquitination evidence from gels. PAGE gel showing a series of bands above the molecular weight of the nonubiquitinated protein. (In collaboration with W. C. Patterson.)

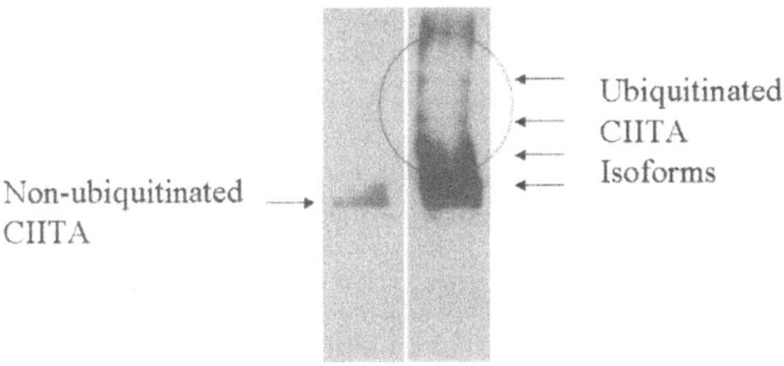


Fig. 2. Ubiquitination evidence from Western blot analyses. PAGE gel showing a series of bands above the molecular weight of the nonubiquitinated protein: Western blot analysis of CIITA showing ubiquitination.

### 3.1.2. Mass Spectrometric Evidence for Protein Ubiquitination

Although ubiquitination is clearly important for protein degradation, most mass spectrometric studies on ubiquitination have focused on protein phosphorylation, rather than being direct mass spectrometric studies of protein ubiquitination. There are three types of mass spectrometric evidence for ubiquitination: the first is ubiquitination of the intact protein, the second is co-electrophoretic migration of the target protein and the ubiquitin (because it is covalently attached), and the third is the mass shift of a ubiquitinated peptide relative to the nonubiquitinated peptide.

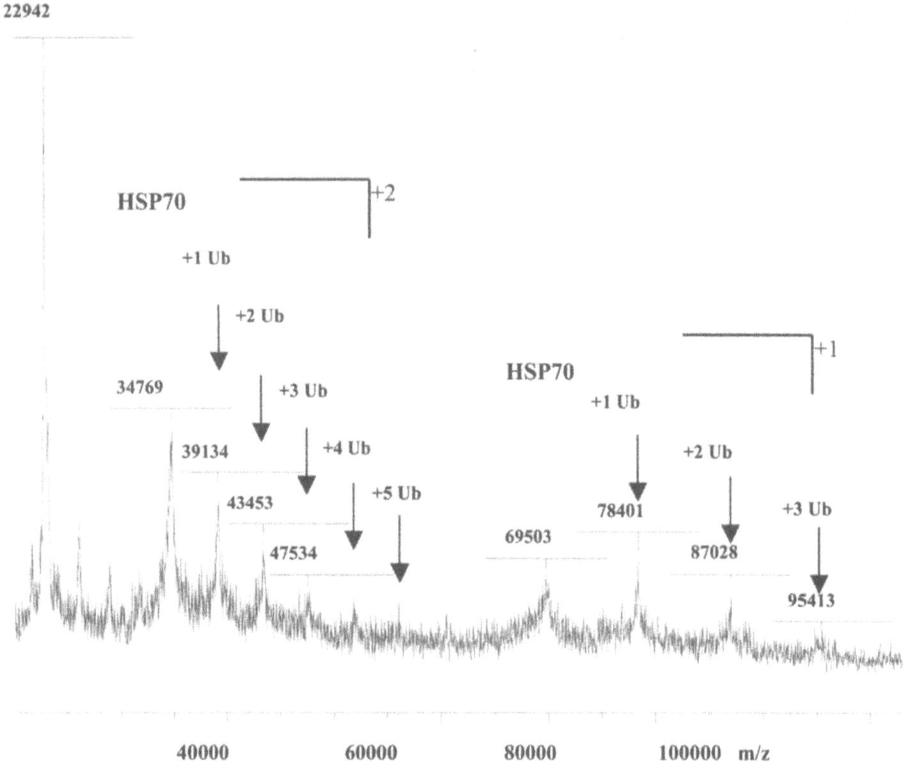


Fig. 3. Ubiquitination evidence from MALDI of intact proteins. MALDI–MS of HSP70 with a series of ubiquitin attached moieties.

### 3.1.2.1. Evidence From MS of the Intact Protein

Direct mass spectrometric evidence of intact ubiquitinated HSP70 is shown in **Fig. 3**. This spectrum was obtained by direct MALDI–MS analysis of ubiquitinated HSP70, affinity-bound to anti-HSP70 affinity beads.

#### 3.1.2.1.1. Direct MALDI Analysis of Proteins Bound to Affinity Beads (*see* **Note 1**)

1. Use antibodies *covalently* bound to the affinity beads (*see* **Note 2**).
2. Bind the antibody according to the bead manufacturer's instructions (*see* **Notes 3** and **4**)
3. Rinse beads three times with two to three bed volumes of 100 m*M* ammonium bicarbonate (*see* **Note 5**).

#### 3.1.2.1.2. Spotting the MALDI Target

1. Prepare the MALDI matrix solution—a saturated solution of recrystallized (see **Note 6**) α-cyano 4-hydroxycinnamic acid in 45:45:10 ethanol:water:formic acid (see **Note 7**)
2. Pipet 0.5 μL of settled beads onto the MALDI target, followed by 0.5 μL of MALDI matrix solution.

#### 3.1.2.2. Evidence From MS and MS/MS of Peptides

The second type of mass spectrometric evidence for ubiquitination comes from in-gel digestion and protein identification studies (*see* Chapter 9). In-gel digestion of the higher molecular weight bands followed by protein identification by MALDI-MS or LC–MS/MS on a gel can sometimes provide evidence for ubiquitination. In this case, the evidence comes from peptides rather than proteins. In the example shown in **Fig. 4**, peptides from the target protein CIITA and peptides from ubiquitin were found in a gel band of approx 200 kDa, clearly higher than the unmodified 150-kDa CIITA or the 8.7- kDa unbound modified monoubiquitin (*see* **Subheading 3.1.2.4.**)

#### 3.1.2.3. MS and MS/MS Spectra of Ubiquitinated Peptides

The third type of mass spectrometric evidence comes from the modified peptide itself, either from a shift in peptide molecular weight, or from MS/MS data. Ubiquitin is a 76-amino acid protein. As discussed in Chapter 9, an E3 ligase attaches this ubiquitin to the ε amino group of a lysine residue in the target protein. This covalent linkage is formed at C-terminal glycine residue of the ubiquitin, with loss of the elements of water. Cleavage with trypsin or gluC (*see* **Note 8**) leaves characteristic "tails" on the modified lysine. These "tails" cause a shift in the molecular weight of the peptide, which can be used to distinguish these modified peptides from the unmodified peptides **(Fig. 5).** The actual site of ubiquitination can then be determined by MS/MS sequencing.

These peptide mass shifts can be used to find and sequence ubiquitinated peptides. Mascot, for example, allows the user to create a modification of a particular mass, which can then be used to search the peptide data from a given digest. Owing to their transient nature and low natural abundances, ubiquitinated peptides are difficult to detect. Surprisingly few ubiquitinated peptides had been found by mass spectrometry until the recent work by Gygi and coworkers *(1)*. In this study, a 6X His-tagged ubiquitin was used, and ubiquitinated proteins from yeast were purified by immobilized metal affinity chromatography (IMAC). The ubiquitin-enriched fraction (0.2 mg out of each original 100 mg of yeast cell lysate) was then digested with trypsin, fractionated into 80 fractions on a SCX column, and the 80 fractions were analyzed by capillary LC–MS/MS. Peptide MS/MS data was searched using Sequest software, allowing for a ubiquitin-modified lysine with a mass shift of +114 Da. In this manner, 110 ubiquitination sites were determined.

3.1.2.3.1. MS/MS Fragmentation of Ubiquitinated Peptides

Prior to the work of Gygi's group *(1)*, there had been only two reports of ubiquitination sites, and these were on specifically targeted proteins of interest. The first was the work of Laub et al. *(2)*, who used gluC for proteolytic digestion of the protein, and the second was the work of Dohlman's group *(3)*, who used trypsin. Both of these articles showed MS/MS spectra of the ubiquitinated peptides, and we decided to use these peptides as the starting point for a detailed study of the fragmentation of ubiquitinated peptides with the goal of finding specific diagnostic fragment ions to aid in detecting low levels of ubiquitinated peptides in biological materials *(4)*.

When a ubiquitinated protein is enzymatically digested, a portion of the ubiquitin side chain remains attached to the modified lysine (**Fig. 6**). A ubiquitinated peptide

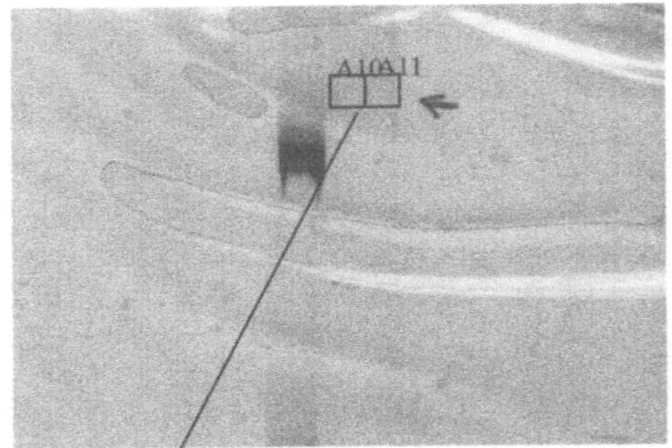


# {MATRIX SCIENCE} Mascot Search Results

| | | |
|---|---|---|
| User | : Carol Parker | |
| Email | : carol_parker@med.unc.edu | |
| Search title | : 070903_Greer_2+_QA50 | |
| MS data file | : D:\MethodD.Pro\pkl_files\071703_Greer_06b.pkl | |
| Database | : MSDB 20030710 (1206286 sequences; 383022796 residues) | |
| Timestamp | : 21 Jul 2003 at 02:47:12 GMT | |
| Significant hits: | A48843 | MHC class II transactivator - human |
| | Q8R3V9 | Hypothetical 52.0 kDa protein.- Mus musculus (Mouse). |
| | CAA56280 | HEAVY CHAIN OF THE MONOCLONAL ANTIBODY MST2 PRECURSOR.- Mus musculus (Mouse). |
| | 2H1PL | 2h1 fab, chain L - mouse |
| | 1CICC | ig heavy chain v regions fab immunoglobulin fragment, chain C - mouse |
| | AAB48774 | Anti-DNA immunoglobulin light chain IgM (Fragment).- Mus musculus (Mouse). |
| | AAB70775 | MAB544 IGG1/K VLJ REGION (FRAGMENT).- Mus musculus (Mouse). |
| | Q8WQK2 | Ubiquitin/ribosomal protein S27a fusion protein.- Branchiostoma belcheri tsingtaunese. |
| | 1UD7A | ubiquitin core mutant 1d7 mutant I3V,V5L,I13V,L15I,I23V,V26F,I67L - synthetic |
| | 1AVWA | trypsin (EC 3.4.21.4), chain A - pig |
| | I61770 | keratin 6e, type II - human |
| | CAA82617 | SINGLE CHAIN FV ANTIBODY (FRAGMENT).- Mus musculus (Mouse). |
| | AAP21111 | AY263789 NID: - synthetic construct |
| | AAC37617 | IMMUNOGLOBULIN HEAVY CHAIN (FRAGMENT).- Mus musculus (Mouse). |
| | G87446 | potassium-transporting ATPase, B subunit CC1592 [imported] - Caulobacter crescentus |
| | Q9EQD7 | Keratin intermediate filament 16b.- Mus musculus (Mouse). |

Fig. 4. Peptides from the target protein CIITA and peptides from ubiquitin from a gel band at mol wt approx 200 kDa.

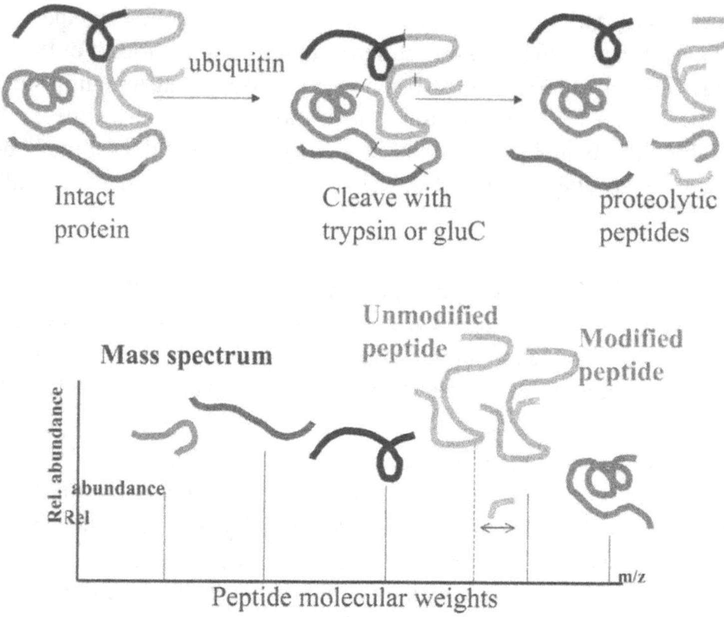


Fig. 5. Schematic showing the expected peptide molecular weight shift caused by ubiquitination.

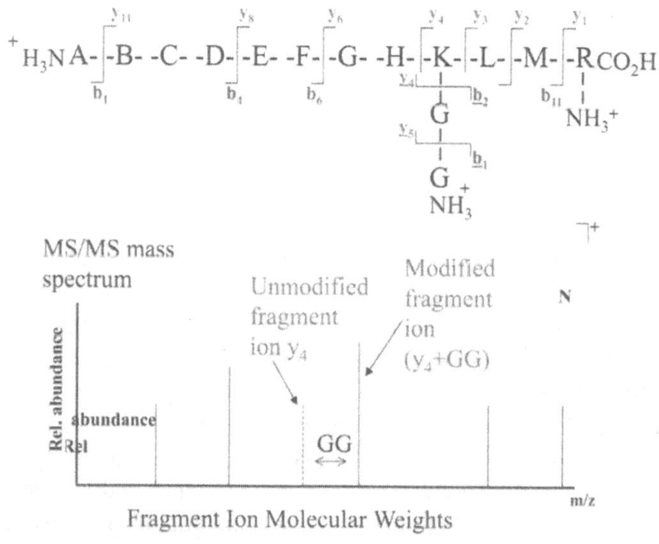


Fig. 6. Schematic showing fragmentation of ubiquitinated peptides.

therefore has two N-termini: one from the original peptide and one from the ubiquitin side chain. Thus, it is possible to have two series of b ions and y ions. For the sake of clarity, we have chosen to refer to those b and y ions involving the ubiquitin side chain as $\underline{b}$ and $\underline{y}$ ions. Obviously, diagnostic ions for the modification must come from fragmentation of this side chain. Fragment ions involving any part of the "normal" peptide will vary in mass according to the peptide being modified and will therefore not be of general diagnostic use.

3.1.2.3.2. Examination of the Literature Tryptic Peptide Spectra

MS/MS spectra of GG-tagged tryptic peptides are shown in both the article from Dohman's group and in that of the Gygi group. Examination of these spectra shows that $\underline{b}$ and $\underline{y}$ ions are produced by dissociation of the peptide from the target protein, but there do not appear to be any $\underline{b}$ ions from the GG side chain or the GGK portion of the peptide.

3.1.2.3.3. Examination of the Literature gluC Peptide Spectra

The MS/MS spectrum of the gluC ubiquitinated peptide from rXL-calmodulin that was shown in the Laub et al. article *(2)* reveals b and y ions from the calmodulin portion of the peptide **(Fig. 7).** Interestingly, it also shows doubly-charged $(b_7\text{-}H_2O)^{2+}$, $(b_{14}\text{-}H_2O)^{2+}$, and $(b_{12}\text{-}H_2O)^{2+}$ ions which contain both of the N-termini (one from the calmodulin peptide and one from the ubiquitin side chain) of the original branched peptide. These two ions (although of low relative abundance) can also be seen in the MS/MS spectrum of the ubiquitinated peptide from natural BT-calmodulin. Careful examination of the both spectra reveals ions that might be from the ubiquitin side chain, and therefore of possible diagnostic utility. The $\underline{y}_{14}$ ion includes ions from both the side chain and the C-terminus of the calmodulin portion of the peptide, so it cannot be used as a ubiquitination marker ion. However, there are also ions from the side chain, $\underline{b}_4$ at *m/z* 439.4 (obs) and, possibly, $\underline{b}_2$, at *m/z* approx 191.2 (obs) which appear to be diagnostic ions of ubiquitination, after digestion with gluC.

3.1.2.3.4. Preparation of Model Ubiquitinated Peptides

To study the fragmentation of ubiquitinated peptides further, we had a model peptide synthesized that should have matched the structure of the gluC ubiquitinated peptide from the Laub article *(2)*. We then planned to digest the peptide with trypsin to generate a model tryptic peptide. These modified peptides should show a characteristic mass shift from their unmodified analogues: in the case of trypsin, a mass shift of 114.0428 Da from the GG "tag" left on the modified lysine. In the case of gluC, a mass shift of 1302.7883 Da, a much larger mass shift resulting from the much longer "tail" (STLHLVLRLRGG-) on the modified lysine, was expected. Unfortunately, the model peptide was synthesized with an amide at the C terminus, and an acetyl group at the N-terminus, so it was not an exact model of the peptide found by Laub et al. *(2)*. Fortunately, however, the acetyl group was in the "calmodulin" portion of the peptide, so diagnostic fragments from the ubiquitin side chain should still be present in the MS/MS spectrum.

ESI-MS/MS spectra of the model peptides were obtained by LC–MS/MS analysis using a Waters/Micromass Q-tof API US, equipped with a Waters capLC system. An aliquot of the sample was injected first onto a Dionex trapping column, which was

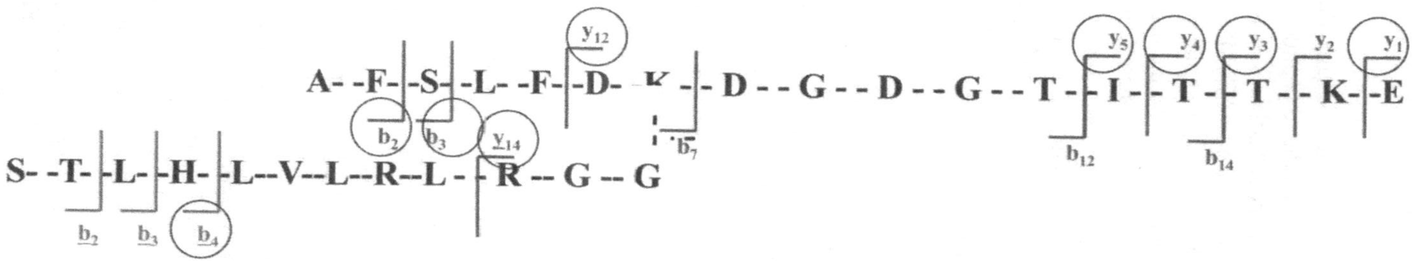


Fig. 7. Calmodulin-ubiquitinated peptide structure after cleavage with gluC. *Circled* peptide fragments were not annotated in the original publication by Laub et al. *(2)*, but appear to have the appropriate masses.

then switched so that it was connected on-line to a 75-μm Dionex Pep-map analytical column and the ESI source. The original synthetic peptide, the model gluC peptide, was dissolved in water and injected without further purification. It proved to be a mixture of acetylated and methylated forms, whose MS/MS spectra could be analyzed separately after separation by LC–MS/MS. The actual methylation sites can be deduced from the MS/MS spectrum: one is on the serine in the calmodulin portion of the peptide, the other is on one of the first two residues (S or T) of the ubiquitin tail.

To create a model tryptic peptide, the tryptic peptide was prepared from the synthetic peptide by means of an in-solution tryptic digest (*see* **Subheading 3.1.2.5.3.**), using 2 μg of trypsin and 1 μg of the original peptide in 100 μL of 100 m*M* ammonium bicarbonate. As stated earlier, an aliquot of the digest was injected into the LC–MS/MS system, and went first onto a Dionex trapping column, which was then switched so that the column was connected on-line to a 75 μm id Dionex Pep-map analytical column and the ESI source.

Two main products were formed: a peptide with the expected "GG" tail on the ubiquitinated lysine, and a second peptide with an "LRGG" tail, resulting from a missed cleavage. Even though this was an in-solution tryptic digest with a large amount of trypsin, a significant amount of peptide with a missed cleavage site on the side chain was produced. Although this missed cleavage was unexpected, it is not unreasonable, as this cleavage site is near the branching point so cleavage at this site is likely to be sterically hindered. MS/MS spectra were obtained for both of these products. The formation of this peptide is of significant analytical interest as it provides a second characteristic molecular weight shift for ubiquitinated peptides after tryptic digestion.

3.1.2.3.5. Fragmentation of Model GluC Ubiquitinated Peptide

Because of the various possible monomethylated isoforms, the first model gluC peptide examined was the dimethylated version. The resulting spectrum is shown in **Fig. 8.** Both b and y fragment ions are found from the "normal" part of the peptide. Most interestingly, several fragments are found that involve only the side chain. As we observed in the literature spectra *(2),* characteristic ions were detected from the "ubiquitin" side chain: these are $\underline{b}_2$, $\underline{b}_3$, $\underline{b}_4$, $\underline{b}_5$, and $\underline{b}_6$. These ions, in their unmethylated forms (*m/z* 189.088, 302.172, 439.231, 555.315, and 651.383), should thus be diagnostic ions for peptides with ubiquitin side chains that have been cleaved with gluC.

3.1.2.3.6. Fragmentation of Model Tryptic Peptides

Cleavage of the synthetic model peptide with trypsin resulted in a GG-tagged peptide whose MS/MS spectrum is shown in **Fig. 9.** Unfortunately, as in the literature spectra, no characteristic fragment ions could be found which were diagnostic for the critical GG(K) portion of the molecule.

The MS/MS spectrum of the model tryptic peptide that had the LRGG tag (resulting from a missed cleavage) was more analytically useful **(Fig. 10).** Diagnostic ions of the ubiquitin tail were found. These include $\underline{b}_2$, $\underline{b}_4$, and an internal fragment ion (LRGGK-28). There is also a LRGGKD ion fragment ion, but because this includes the "D" from the calmodulin peptide, it cannot be used as a diagnostic ion for modification by ubiquitin. The assignment of these peptide fragments is confirmed by the

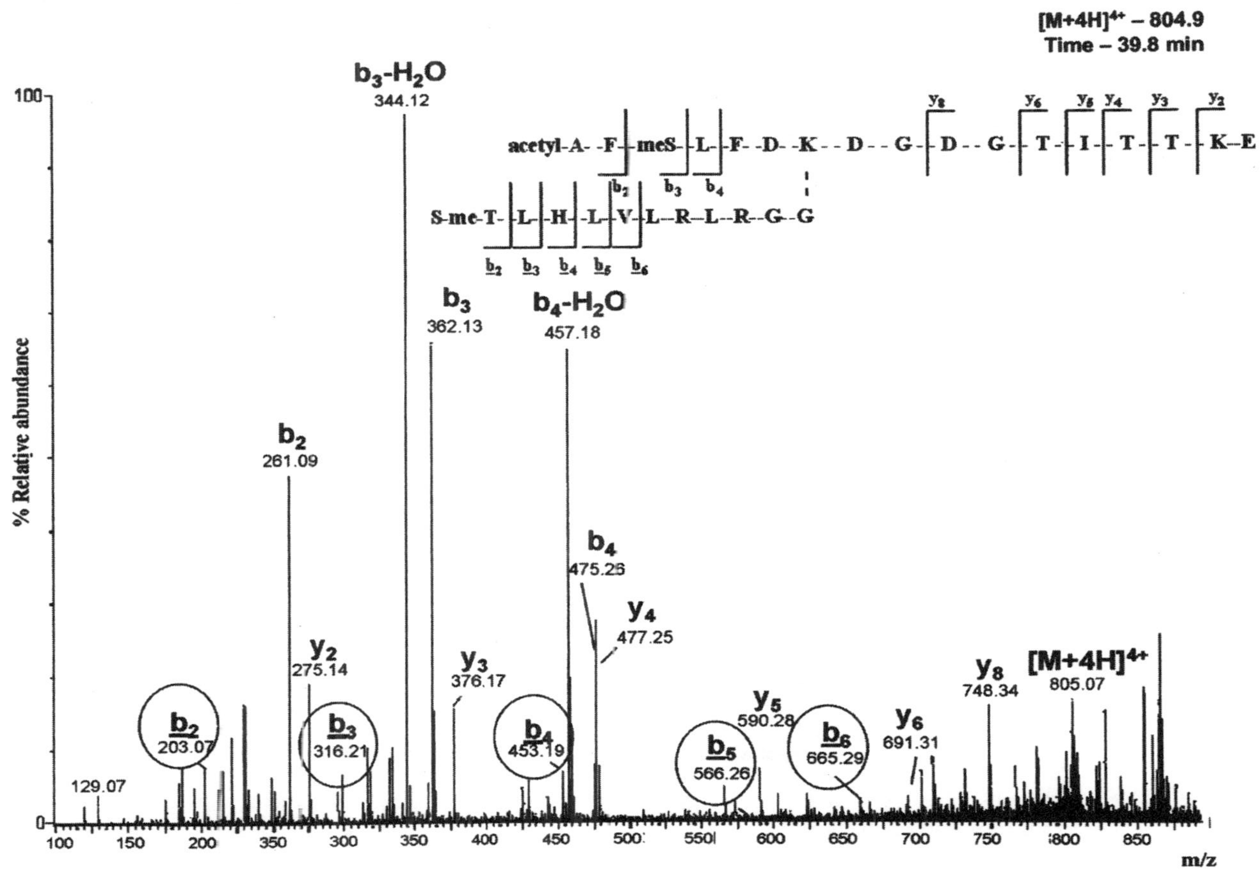


Fig. 8. MS/MS fragmentation of synthetic model calmodulin gluC peptide, dimethyl form. Reproduced with permission from **ref. *4***. Copyright John Wiley & Sons Limited.

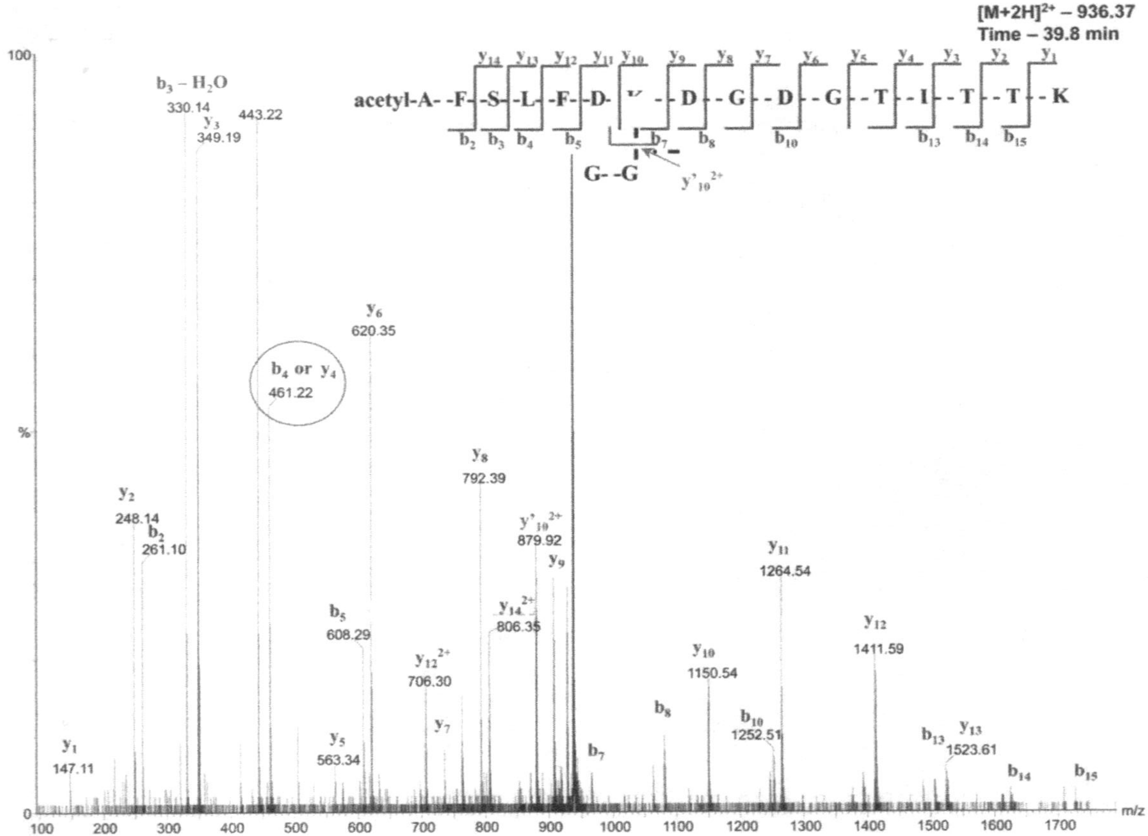


Fig. 9. Fragmentation of model "GG-" ubiquitinated tryptic peptide. Reproduced with permission from **ref. *4***. Copyright John Wiley & Sons Limited.

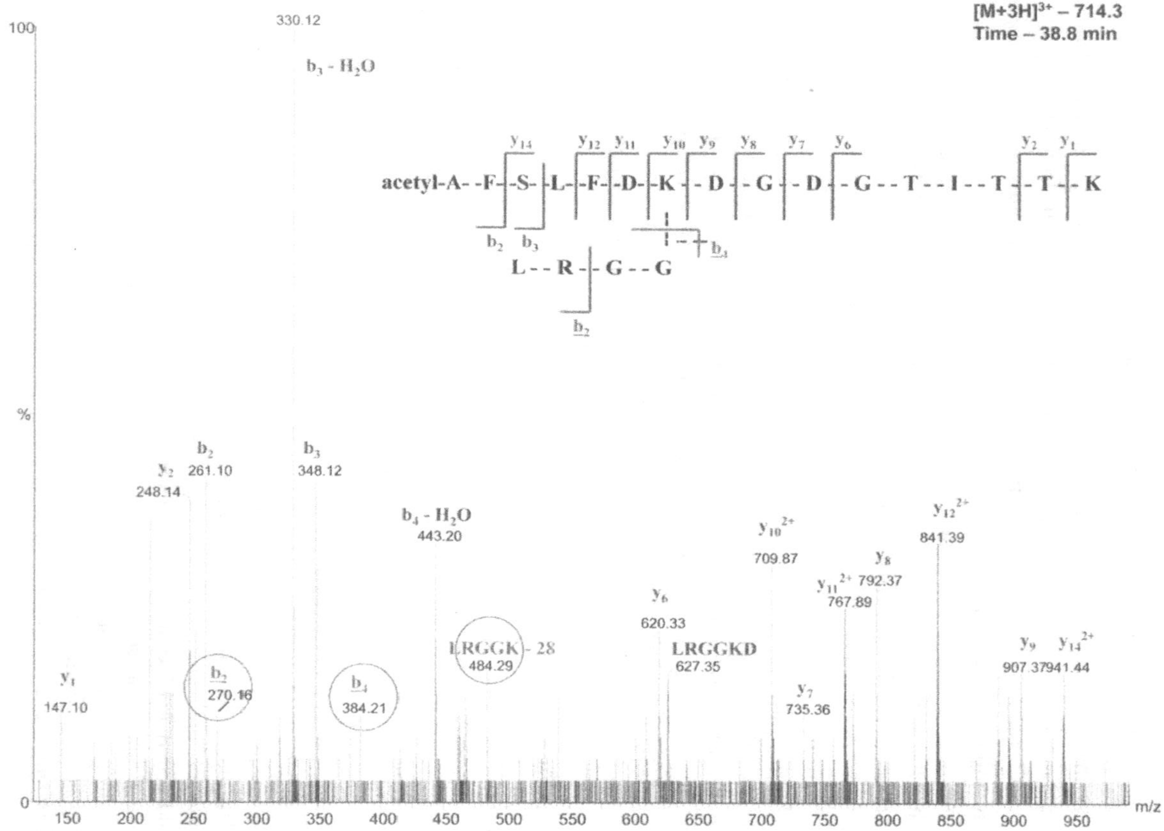


Fig. 10. Fragmentation of model “LRGG-” ubiquitinated tryptic peptide. Reproduced with permission from **ref. *4***. Copyright John Wiley & Sons Limited.

MS/MS spectrum of the *S*-methylated calmodulin peptide is shown in **Fig. 11**. The same diagnostic ions ($\underline{b}_2$, $\underline{b}_4$, and the LRGGK-28 ion) are observed in this spectrum.

#### 3.1.2.4. Enrichment of Samples for Ubiquitin-Modified Proteins.

As described in **Subheading 3.1.2.3.**, the work done by the Gygi group depended on enriching the sample in ubiquitinated peptides. Their goal was to find as many ubiquitination sites as possible in *any* protein, so their approach was to use a 6X His-tagged ubiquitin, and enrich the sample in ubiquitinated proteins through the use of IMAC.

To find ubiquitination sites in a *particular* target protein requires a different approach. One method we have used, if an antibody is available against the target protein, is to first affinity-purify the target protein. If an antibody against the protein of interest is *not* available, an affinity tag can be incorporated into the target protein sequence. In our case, we used a FLAG tag added to the N-terminus of the CIITA target protein so that it could be immunoprecipitated with anti-FLAG antibody beads.

As a second strategy for increasing the proportion of ubiquinated vs. non-ubiquitinated protein, a plasmid was used that coded for a modified ubiquitin that had all of the lysines modified to arginines and an HA tag on the C-terminus. This construct was designed to prevent the formation of polyubiquitin chains and thus to inhibit degradation of the target protein *(5)*. The HA tag allows a second affinity purification step, either before or after proteolysis, this time on anti-HA beads. To avoid proteolysis of the ubiquitin (and loss of the HA tag), LysC was used for the initial digestion of the protein, instead of trypsin.

The affinity-bound protein can then be digested overnight with trypsin or with gluC (*see* **Note 8**) while still attached to the beads, using the protocols described in the next subheading. Alternatively, the protein can be eluted from the affinity beads using eluted with 1:1:8 ethanol:water:formic acid, lyophilized and digested in solution.

#### 3.1.2.5. Elution and Enzymatic Digestion Procedures.

##### 3.1.2.5.1. Elution of Proteins From Beads

1. Place approx 50–100 µL of beads in an Eppendorf tube.
2. Wash beads three times with 200 µL of 100 mM ammonium bicarbonate and discard wash solutions.
3. Add 100 µL 1:1:8 ethanol–formic acid–water.
4. Vortex mix and let settle.
5. Remove and save eluates.
6. Repeat the extraction two more times.
7. Save and combine the eluates.
8. Lyophilize.
9. Store at –80°C.

##### 3.1.2.5.2. On-Bead Digestion Tryptic Procedure *(6)* (*see* **Note 9**)

1. Place approx 50–100 µL of antibody beads (with the attached affinity-bound protein) in an Eppendorf tube.
2. Wash three times with 200 µL of 100 mM ammonium bicarbonate.
3. Add 100 µL of 100 m*M* ammonium bicarbonate.

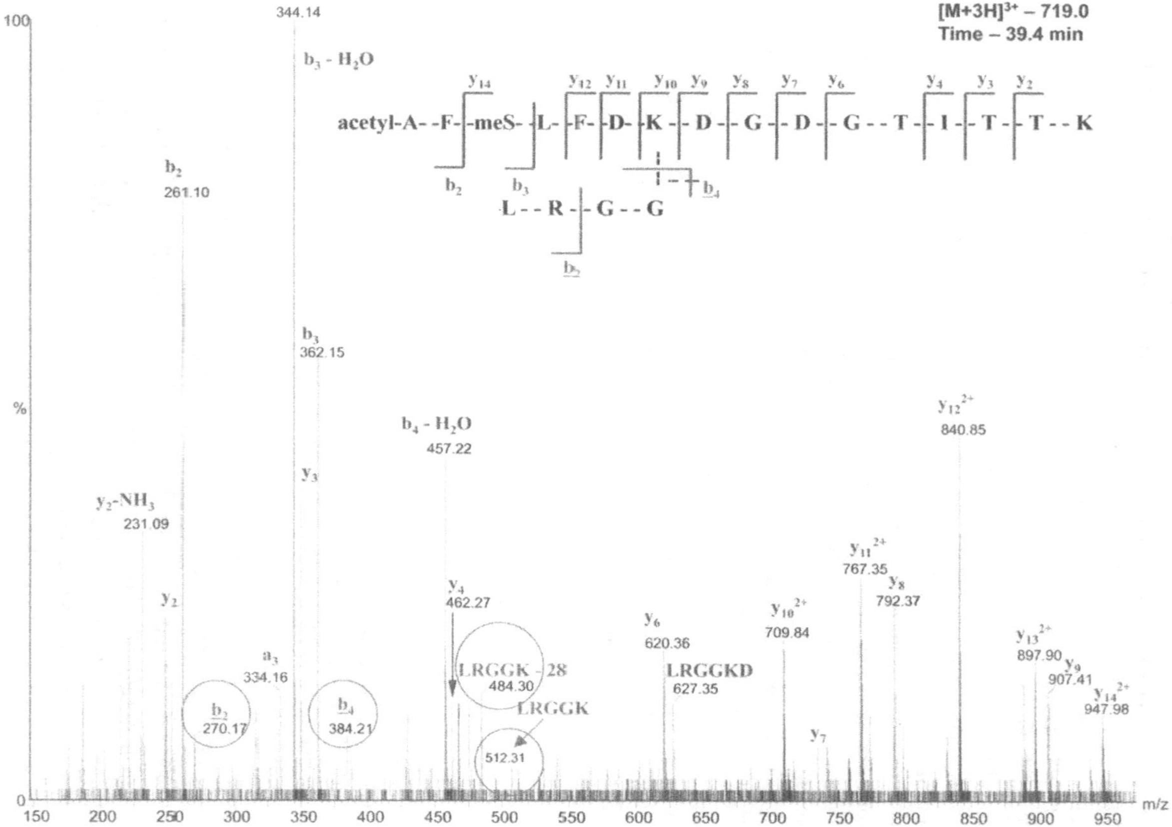


Fig. 11. Fragmentation of model "LRGG-" ubiquitinated tryptic peptide, methylated form. Reproduced with permission from **ref. *4***. Copyright John Wiley & Sons Limited.

4. Reconstitute Promega trypsin in 20 μL Promega resuspension buffer (0.015 *M* acetic acid). Promega trypsin comes in aliquots of 20 μg per vial.
5. Add 2 μL of trypsin solution to each sample (*see* **Note 10**).
6. Vortex mix and centrifuge at <1000 rpm to avoid breaking the beads.
7. Incubate overnight in sealed Eppendorf tubes, at 35°C with rotation (~400 rpm) (*see* **Note 11**).
8. Remove the supernatant from the beads.
9. Lyophilize to reduce volume (if the sample is concentrated enough, the supernatant can be injected directly).
10. Store at –80°C.

3.1.2.5.3. In-Solution Tryptic Digestion Procedure

1. Calculate what a 1:50 enzyme/substrate ratio would be (*see* **Note 12**).
2. Reconstitute Promega trypsin in 20 μL of Promega resuspension buffer (0.015 *M* acetic acid). (Promega trypsin comes in aliquots of 20 μg per vial.)
3. Dissolve sample in approx 20 μL of 100 m*M* ammonium bicarbonate solution (see **Note 13**).
4. Add the calculated amount of trypsin solution to each sample.
5. Vortex mix and centrifuge.
6. Incubate for at least 4 h for a digest in solution, or overnight for beads, in sealed Eppendorf tubes, at 35°C with rotation (~400 rpm).

3.1.2.5.4. In-Solution Digestion With GluC (*see* **Note 14**)

1. Dissolve the enzyme (which comes lyophilized, 25 μg per ampule) in 25 μL of water.
2. Dissolve sample in 100 μL of 100 m*M* ammonium bicarbonate buffer.
3. Add 2 μL of enzyme solution.
4. Incubate overnight at 35°C, with rotation (~400 rpm).
5. Freeze at –80°C.

3.1.2.5.5. On-Bead Digestion With GluC (*see* **Note 14**)

1. Dissolve the enzyme (which comes lyophilized, 25 μg per ampule) in 25 μL of sample in water.
2. Rinse approx 50 μL of affinity beads three times with 100 μL of 100 m*M* ammonium bicarbonate buffer.
3. Add 100 μL of 100 m*M* ammonium bicarbonate to the beads.
4. Add 2 μL of enzyme solution to the beads.
5. Incubate overnight at 35°C, with rotation (~400 rpm).
6. Remove the supernatant, and freeze at –80°C.

3.1.2.6. Mass Spectrometric Approaches for Detection of Ubiquitinated Peptides and Determination of Ubiquitination Sites

The first step is to simply perform LC–MS/MS on the peptide digest. Long runs with long linear gradients (>200 min) are preferred for separating complex mixtures of peptides in order to reduce suppression effects and to try to reduce the number of peptides that coelute, because selection of the various precursor ions is done on the basis of ion abundances. For an unknown modification site in a protein containing many potential sites, where the peptide molecular weight is therefore not known, automatic data-dependent triggering of MS/MS data collection (called "survey scan" mode in the Micromass MassLynx software) is the only feasible automatic scanning option.

The resulting MS/MS spectra are then analyzed by commercially available software packages (such as Mascot or SEQUEST) which can be programmed (*see* **Notes 15** and **16**) to consider a lysine modified with a GG or, as we have learned from the above experiments on our model peptides, LRGG, for a tryptic digest, or with STLHLVLRLRGG for a gluC digest). Ideally, the ubiquitination sites will be found from this automated search routine (*see* **Notes 17** and **18**). **Figure 12** shows an example of a database search of MALDI–MS data from a tryptic digest of a ubiquitinated protein in which a ubiquitinated peptide was identified. Although promising, these results simply mean that there is a peak in the mass spectrum that has the mass of an expected tryptic peptide where a lysine has been modified with a GG tag. Because a peak at this mass could have come from another peptide in the mixture, this ubiquitination site cannot be confirmed without MS/MS sequence data of this peptide.

If MS/MS data has been acquired and searched, as is the case when LC–MS/MS has been used, the identity of the peptide can be confirmed and site of ubiquitination found from the database search results (*see* **Note 19**). This was the method used to find the ubiquitination sites in the Gygi paper. Unfortunately, this approach often fails where lower amounts of biological material are available. Very low levels of the modified peptide mean that there may not be sufficient intensity of the modified peptide to trigger this automatic data-dependent MS/MS sequencing.

In this case, another much more time-consuming option is to examine the MS spectra to search for a peptide shifted by the masses corresponding to these possible modifications. This can be done manually by creating a list of expected "normal" peptide masses, calculating the modified masses, and examining the MS spectra obtained during the LC–MS/MS run. Obviously, at 1 s/scan, thousands of spectra are obtained throughout the course of an extended LC–MS/MS run. Current software systems allow the combination of groups of spectra, and these groups of combined spectra can be examined. Most current software packages also allow the deconvolution of the spectra to singly charged species, which reduces the complexity of the manual data analysis, because multiple charge states of the possible peptides are deconvoluted to singly charged species.

A semi-automated approach to this task is to combine all of the spectra, perform a deconvolution on the MS data to generate a pseudo-singly charged spectrum (*see* **Note 20**), and then submit this data to the database search software for searching as an MS data file. As discussed earlier, the MS data can then be searched for modified peptides (*see* **Note 21**).

If a possible modified peptide is identified in this manner, it is useful to examine the original MS data to see if the calculated +2 and +3 charge states for this peptide cochromatograph. Now that the masses are known, it is possible to perform a different type of MS scanning, where the precursor ion is specified—in Micromass software, it is called "include only" MS/MS. In this scan mode, MS/MS is only performed on the preselected precursor ions. In a previous study *(7)*, we have found that this can lead to an increase in sensitivity of a factor of 50–100 for these ions.

Although this scanning mode is dependent on intensity-based triggering, and thus has the same sensitivity limitations as the "survey scan" mode, knowing these characteristic fragment ions also allows the possibility of "precursor ion scanning." Here, when a preset characteristic ion is detected, the data system switches to the MS mode, detects the precursor, and collects the MS/MS data. Similar approaches are already

## {MATRIX SCIENCE} Mascot Search Results

User : Nely
Email : dicheva@email.unc.edu
Search title : Bead_Susanne_G18_msdb_human
Database : MSDB 20030710 (1206286 sequences; 383022796 residues)
Taxonomy : Homo sapiens (human) (97490 sequences)
Timestamp : 10 Nov 2003 at 15:43:19 GMT
Top Score : 71 for A48843, MHC class II transactivator - human

| | Accession | Mass | Score | Description |
|---|---|---|---|---|
| 1. | A48843 | 123379 | 71 | MHC class II transactivator - human |
| 2. | AAD22767 | 416855 | 66 | AF083037 NID: - Homo sapiens |
| 3. | CAB40713 | 452644 | 65 | HSA131693 NID: - Homo sapiens |
| 4. | CAA09361 | 453387 | 65 | HSA010770 NID: - Homo sapiens |
| 5. | Q29675 | 96496 | 62 | MHC class II transactivator CIITA.- Homo sapiens (Human). |
| 6. | Q9H2G6 | 88383 | 61 | CTCL tumor antigen se2-2 (Fragment).- Homo sapiens (Human). |
| 7. | A55575 | 480107 | 60 | ankyrin 3, long splice form - human |
| 8. | Q96KL4 | 101512 | 60 | MHC2TA.- Homo sapiens (Human). |
| 9. | BAA20828 | 179956 | 59 | AB002371 NID: - Homo sapiens |
| 10. | I53799 | 149519 | 55 | CG1 protein - human |

### Results List

1. A48843 **Mass:** 123379 **Score:** 71
MHC class II transactivator - human

| Observed | Mr(expt) | Mr(calc) | Delta | Start | | End | Miss | Peptide |
|---|---|---|---|---|---|---|---|---|
| 844.54 | 843.53 | 843.46 | 0.07 | 663 | - | 669 | 0 | LAWELGR |
| 850.50 | 849.49 | 849.47 | 0.02 | 406 | - | 411 | 1 | EHRDPR |
| 1007.60 | 1006.59 | 1006.50 | 0.09 | 373 | - | 380 | 0 | SSSKSLER Ubiquitination (K) |
| 1187.71 | 1186.70 | 1186.56 | 0.14 | 381 | - | 390 | 0 | ELATPDWAER |
| 1300.82 | 1299.82 | 1299.72 | 0.10 | 732 | - | 742 | 0 | ELPQYLALTPR |
| 1329.87 | 1328.86 | 1328.76 | 0.10 | 757 | - | 768 | 0 | FLAGLIFQPPAR |
| 1350.77 | 1349.77 | 1349.64 | 0.12 | 334 | - | 343 | 0 | WPEPVEQFYR |
| 1411.86 | 1410.85 | 1410.75 | 0.10 | 145 | - | 156 | 0 | RPFPEELPADLK |
| 1424.89 | 1423.88 | 1423.84 | 0.04 | 1035 | - | 1048 | 0 | LAEALPSLAASLLR |
| 1501.86 | 1500.85 | 1500.75 | 0.11 | 745 | - | 756 | 0 | RPYDNWLEGVPR |
| 1505.89 | 1504.88 | 1504.75 | 0.13 | 851 | - | 864 | 0 | ALEAAGQDFSLDLR |
| 1531.90 | 1530.89 | 1530.80 | 0.08 | 887 | - | 900 | 0 | AALSDTVALWESLR |
| 1560.03 | 1559.02 | 1558.91 | 0.12 | 633 | - | 647 | 0 | LPSTLTGLYVGLLGR |
| 1586.77 | 1585.77 | 1585.65 | 0.12 | 581 | - | 593 | 0 | YFESSGMTEHQDR |
| 1858.02 | 1857.01 | 1856.87 | 0.15 | 671 | - | 686 | 0 | HQSTLQEDQFPSADVR |
| 2215.23 | 2214.22 | 2214.15 | 0.07 | 979 | - | 998 | 0 | ILTAFSSLQHLDLDALSENK |
| 2713.46 | 2712.46 | 2712.36 | 0.10 | 344 | - | 369 | 0 | SLQDTYGAEPAGPDGILVEVDLVQAR |
| 2845.39 | 2844.38 | 2844.35 | 0.04 | 93 | - | 116 | 0 | EAYANIAELDQYVFQDSQLEGLSK |

**No match to:** 913.58, 1039.62, 1109.68, 1243.75, 1386.79, 1403.85, 1446.86, 1515.88, 1545.85, 1550.90, 179

Fig. 12. Mascot database search results from a tryptic digest of ubiquitinated CIITA, showing a "hit" for a potential ubiquitinated peptide. This potential ubiquitination site has not yet been confirmed by MS/MS data.

commonly used, for example, in order to find peptides containing acetylated lysine from the acetylated lysine immonium ion *(8)*, or to find phosphotyrosine-containing peptides from its characteristic immonium ion *(9,10)*.

The identification of the specific and characteristic fragment ions that we have described in the preceding provides a powerful new approach for finding ubiquitinated peptides. Searching the MS/MS chromatograms (*see* **Note 21**) for these characteristic ions (which should cochromatograph because they are fragment ions from the same peptide) should allow one to find peptides containing ubiquitin side chains.

## 4. Notes

1. We call this "direct" analysis because the affinity beads are placed directly on the target, in contrast to methods were the affinity-bound proteins are eluted first, and the eluate is spotted on the target.
2. If binding has to be done through Protein A, or the antibody is dissolved in ascites, the antibody can be crosslinked to the beads *(11)* before the target protein is affinity bound.
3. Because only a small number of beads will be placed on the MALDI target, it is important to obtain much protein as possible on each bead, so use a small amount (~20 μL) of settled affinity bead slurry.
4. To avoid releasing parts of the antibody into the solution, try to avoid reducing agents such as β-mercaptoethanol or dithiothreitol (DTT) in the binding buffer.
5. Most salts will be removed during the wash steps. Other components not compatible with mass spectrometric analysis (such as glycerol or detergents) should be avoided or minimized. Although Zwittergent is thought to be compatible with mass spectrometry, it seems to make agarose beads turn "gummy," so it should not be used.
6. To recrystallize α-cyano 4-hydroxycinnamic acid, make a saturated solution of α-cyano 4-hydroxycinnamic acid in boiling methanol. Pour off the solution and discard it. Add more methanol, and again make a saturated solution in boiling methanol. This time, pour off the methanol and save it. Evaporate the methanol to dryness in a hood, while protecting the solution and the crystals from light with aluminum foil. Store in the dark or in a vial wrapped with aluminum foil.
7. The matrix solvent must contain both organic and acid so that it dissociates the affinity-bound protein from the antibody on the MALDI target.
8. Also known as *Staphylococcus aureus* V 8 protease.
9. The antibody used should be covalently attached to the beads, and DTT should not be used (or used in a very low concentration *[12]*) in the purification step. If the disulfide bridges in the antibody are reduced, the antibody can be enzymatically digested along with the attached protein. This will lead to a high background of IgG peptides along with the peptides from the target protein, and will make finding the modified peptide more difficult.
10. For proteins affinity bound to antibody beads, much higher ratios of enzyme (e.g., 5:1) to substrate should be used than for proteins in solution *(5)*.
11. Be sure to add enough digestion buffer so that the beads can "slosh around" in the Eppendorf tube and do not dry out during the overnight digestion.
12. If you don't know the protein concentration, use approx 2 μL of a 1 μg/μL solution.
13. If the sample to be digested is already dissolved in water or a buffer such as phosphate-buffered saline (PBS), add enough ammonium bicarbonate to the solution to ensure that the pH will be approx 7–8.

14. The specificity of gluC depends on buffer used and pH of the solution. Cleavage can be either C-terminal to glutamic acid (in ammonium acetate at pH 4.0 or ammonium bicarbonate at pH 7.8), or to *both* glutamic and aspartic acid (in PBS, pH 7.8). For a solution digest, you can add a second buffer to the original buffer to adjust the pH, but be sure to take into account both buffers when calculating the expected of the peptide molecular weights.
15. This requires a site-license for Mascot, although the company says that if there is sufficient interest, they will add new modifications to their on-line website *(www.matrixscience.com)*.
16. Programming in these modification means, in effect, telling the software that there are three additional types of lysines, with new masses: 242.1374 for GGK, 511.6294 for LRGGK, and 1430.8824 for STLHLVLRLRGGK.
17. In searching using Mascot, first search with no variable modifications, and *then* select the target protein and search in the error tolerant mode, specifying the appropriate ubiquitination modifications you have previously entered.
18. Be sure to specify at least two missed cleavages. Cleavage is not expected to occur at the modified lysine.
19. The database search can also be "forced" to consider a specific target program. In Mascot, this is done by adding the accession number from the appropriate database as the first line of the peak list being searched (e.g., accession = XXXXX, followed by a blank line)
20. For Micromass MassLynx software, this is MaxEnt 3, under "Tools."
21. If multiple levels of MS/MS spectra are produced, as in the Micromass MassLynx software, *all* of the MS/MS functions must be searched.

## Acknowledgments

This study was funded by a gift from an anonymous donor to support research in proteomics and cystic fibrosis, and grants from the Cystic Fibrosis Foundation (CFFTI STUTTS01U0) and from NIH (ES11997, 5U54HD035041-07, and P30 CA 16086-25).

## References

1. Peng, J., Schwartz, D. R., Elias, J. E., et al. (2003) A proteomics approach to understanding protein ubiquitination. *Nat. Biotechnol.* **21,** 921–926.
2. Laub, M., Steppuhn, J. A., Bluggel, M., Immler, D., Meyer, H. E., and Jennissen, H. P. (1998) Modulation of calmodulin function by ubiquitin-calmodulin ligase and identification of the responsible ubiquitylation site in vertebrate calmodulin. *Eur. J. Biochem.* **255,** 422–431.
3. Marotti, L. A., Jr., Newitt, R., Wang, Y., Aebersold, R., and Dohlman, H. G. (2002) Direct identification of a G protein ubiquitination site by mass spectrometry. *Biochemistry* **41,** 5067–5074.
4. Warren, M. R. E., Parker, C. E., Mocanu, V., Klapper, D., and Borchers, C. H. (2005) Electrospray ionization tandem mass spectrometry of model peptides reveals diagnostic fragment ions for protein ubiquitination. *Rapid Commun. Mass Spectrom.* **19,** 1–9.
5. Greer, S. F., Zika, E., Conti, B., Zhu, X.-S., and Ting, J. P.-Y. (2003) Enhancement of CIITA transcriptional function by ubiquitin. *Nat. Immunol.* **4,** 1074–1082.
6. Parker, C. E. and Tomer, K. B. (2000) In *Methods in Molecular Biology,* Vol. 146: *Mass Spectrometry of Proteins and Peptides* (Chapman, J. R., ed.), Humana Press, Totowa, NJ, pp. 185–201.

7. Kast, J., Parker, C. E., van der Drift, K., et al. (2003) MALDI-directed nano-ESI-MS/MS analysis for protein identification. *Rapid Commun. Mass Spectrom.* **17,** 1825–1834.
8. Borchers, C., Parker, C. E., Deterding, L. J., and Tomer, K. B. (1999) A preliminary comparison of precursor scans and LC/MS/MS on a hybrid quadrupole time-of-flight mass spectrometer. *J. Chromatogr. A* **854,** 119–130.
9. Steen, H., Kuester, B., Fernandez, M., Pandey, A., and Mann, M. (2001) Detection of tyrosine phosphorylated peptides by precursor ion scanning quadrupole TOF mass spectrometry in positive ion mode. *Anal. Chem.* **73,** 1440–1448.
10. Steen, H., Kuster, B., Fernandez, M., Pandey, A., and Mann, M. (2002) Tyrosine phosphorylation mapping of the epidermal growth factor receptor signaling pathway. *J. Biol. Chem.* **277,** 1031–1039.
11. Peter, J. F. and Tomer, K. B. (2001) A general strategy for epitope mapping by direct MALDI-TOF mass spectrometry using secondary antibodies and cross-linking. *Anal. Chem.* **73,** 4012–4019.
12. Hall, M. C., Torres, M. P., Schroeder, G. K., and Borchers, C. H. (2003) Mnd2 and Swm1 are core subunits of the *Saccharomyces cerevisiae* anaphase-promoting complex. *J. Biol. Chem.* **278,** 16698–16705.

# 11

# Reconstitution of Endoplasmic Reticulum-Associated Degradation Using Yeast Membranes and Cytosol

**Robert J. Lee, Ardythe A. McCracken, and Jeffrey L. Brodsky**

### Summary

The first compartment encountered by newly synthesized secreted proteins is the endoplasmic reticulum (ER). Before secreted proteins can traffic beyond the ER they must fold into their final conformations, and components of multiprotein complexes must assemble. Not surprisingly, then, the ER lumen houses a high concentration of molecular chaperones, factors that facilitate protein folding. However, if misfolded secreted proteins arise they may be selected and proteolyzed. This process, which removes potentially toxic proteins from the cell, has been termed ER-associated degradation (ERAD). Surprisingly, ERAD substrates are not degraded within the ER after being selected but are retrotranslocated to the cytoplasm and destroyed by the 26S proteasome. Thus, the ERAD pathway comprises substrate selection, substrate export from the ER, and substrate degradation, and each step in this pathway has been elucidated in part through an in vitro ERAD assay using reagents prepared from a model eukaryote, the yeast *Saccharomyces cerevisiae*. This chapter describes this in vitro ERAD assay in detail, and special considerations when performing the assay are noted.

**Key Words:** Degradation; endoplasmic reticulum; ER-associated degradation; microsome; molecular chaperone; proteasome; translocation; yeast.

## 1. Introduction

Concomitant with their synthesis on endoplasmic reticulum (ER)-tethered ribosomes, soluble secreted proteins are translocated or imported into the lumen of the ER and integral membrane proteins are inserted into the ER membrane. These polypeptides then fold into their final three-dimensional conformations with the aid of ER-associated enzymes that catalyze folding and with the aid of molecular chaperones, factors that prevent the formation of off-pathway aggregates and enhance folding efficiency. However, if protein folding is kinetically retarded or prevented because of mutations, errors, or cellular stress, aberrant polypeptides are targeted for ER-associated degradation, or "ERAD" *(1)*. Therefore, ERAD is one component of an important

From: *Methods in Molecular Biology, vol. 301, Ubiquitin–Proteasome Protocols*
Edited by: C. Patterson and D. M. Cyr © Humana Press Inc., Totowa, NJ

quality control checkpoint, ensuring that only properly folded proteins leave the ER *(2)*. Not surprisingly, many ERAD substrates are proteins whose absence causes a specific human disease, and defects in ERAD and protein folding give rise to several other diseases (reviewed in **ref. *3***).

One surprising feature of ERAD is that aberrant polypeptides are degraded by the cytoplasmic proteasome, an approx 2.5-MDa multicatalytic protease that can be found associated with the ER/nuclear membrane *(4)*. Although integral membrane ERAD substrates display domains that are, in theory, accessible to the proteasome, the degradation of soluble ERAD substrates requires that they are exported, or "retrotranslocated," from the ER to the cytoplasm. Polypeptide retrotranslocation is mediated by a channel that also imports nascent polypeptides into the ER, and several molecular chaperones in the cytoplasm and ER lumen facilitate retrotranslocation and degradation. Therefore, the ERAD pathway can be subdivided into three steps: substrate recognition, polypeptide retrotranslocation, and proteasome-mediated proteolysis *(3,5)*. Detailing the molecular mechanisms of each of these steps represents an active area of ongoing research.

Early evidence for proteasomal degradation of ERAD substrates after their retrotranslocation and the participation of specific lumenal chaperones in the degradation of soluble substrates was obtained by the development and subsequent utilization of an in vitro ERAD assay *(1,6–11)*. The substrate for this assay is a mutated form of pre-pro α-factor (ppαF), a precursor of the secreted α-factor pheromone in yeast. After wild-type ppαF is translocated into ER-derived microsomes, the signal sequence is cleaved and it is core glycosylated at three positions, forming "3GpαF." In contrast, a precursor in which the core glycosylation sites have been mutated (ΔGppαF) is converted into pαF in the ER-derived microsomes. When washed microsomes containing 3GpαF or pαF are resuspended in yeast cytosol and ATP, pαF is degraded but 3GpαF is stable *(1)*. This observation demonstrated that the ERAD machinery selects only an aberrant, but not wild-type polypeptide, for degradation. The purpose of this chapter is to provide a detailed protocol for this assay.

## 2. Materials

1. Yeast microsomes are prepared as described *(12–14)* from cells grown to log phase (optical density at 600 nm [$OD_{600}$] of ~2.0). In brief, the cell walls are removed with lyticase, and the resulting spheroplasts are broken with a Teflon–glass motor-driven homogenizer. Differential sucrose gradient centrifugation is used to obtain a crude microsomal fraction, which is then concentrated and washed by pelleting at approx 15,000*g* for 10 min. Microsomes should be stored in single-use aliquots (~50 μL), which are stable indefinitely at –80°C, and should be thawed on ice immediately before use.
2. Buffer 88 (B88): 20 m*M* *N*-(2-hydroxyethyl)piperazine-*N'*-(2-ethanesulfonic acid) (HEPES)–NaOH, pH 6.8, 150 m*M* potassium acetate, 5 m*M* magnesium acetate, 250 m*M* sorbitol in double-deionized water ($ddH_2O$). The solution should be filter sterilized and stored at 4°C.
3. 10X ATP regenerating system: 10 m*M* ATP, 500 μ*M* creatine phosphate, 2 mg/mL of creatine phosphokinase in B88. The 10X ATP regenerating mix is stable indefinitely at –80°C and should be thawed on ice. Although repeated freeze–thaw cycles do not seem

to be detrimental to the activity of the 10X ATP mix, we typically store this solution in aliquots of only approx 100 µL.

4. [$^{35}$S]Methionine-labeled ΔGppαF and/or wild-type ppαF, prepared by in vitro transcription and translation as described in refs. ***1,14*** at approx 100,000 cpm/µL. The messages for these substrates are prepared using SP6-driven in vitro transcription of plasmids encoding either ΔGppαF or ppαF, the resulting mRNA is translated in yeast lysate, and the proteins are purified by size-exclusion column chromatography. The protein is snap frozen in liquid nitrogen and stored at either –20° or –80°C in 100-µL aliquots, which are thawed on ice. Repeated freeze–thaw cycles do not affect the performance of the substrate in this assay, and it will remain active for approx 9 mo (although older label will result in a weaker signal and a necessity for longer exposure times for autoradiography). The handling of [$^{35}$S]ppαF, as with any radioactive substrate, requires proper training and precautions.
5. Yeast cytosol is prepared as described in **refs. *14,15*** from cells grown to log phase. In brief, 8 L of yeast grown to log phase ($OD_{600}$ = ~2.0) are resuspended in a minimal amount of buffer to form a thick yeast slurry, which is then frozen into pellets by slow addition to liquid nitrogen. The pellets are added to approx 500 mL of liquid nitrogen in a stainless-steel blender, and this volume is maintained by periodic addition of liquid nitrogen during constant, high-speed blending for 8–10 min. After the liquid nitrogen evaporates, approx 20 mL of B88 containing 1 m*M* dithiothreitol (DTT) is added, and the lysed cells are thawed. The supernatant from a 10,000*g* 10 min centrifugation at 4°C is spun at 300,000*g* for 1 h at 4°C to pellet all remaining membranes/aggregated protein. The supernatant from this spin is aliquoted (~100–200 µL), snap-frozen in liquid nitrogen, and stored at –80°C. The concentration of cytosol used in the ERAD assay is between 15 and 30 mg/mL. Cytosol should never be refrozen and thawed more than once because a loss of activity results. A loss of activity may also be observed after storage for >>6 mo.
6. Mammalian 19S and 20S proteasome subunits are purified as described in **refs. *16–18***. In brief, bovine red blood cells are harvested and lysed before dialysis in preparation for a series of chromatographic steps, beginning with gel filtration and followed by ion-exchange chromatography. Ammonium sulfate precipitation is used to obtain independent 19 and 20S particles, and the proteins are resuspended to a final concentration of approx 0.7–1.5 mg/mL. Aliquots should be snap frozen and stored at –80°C, and should not be thawed and refrozen more than once. *N*-Ethyl maleimide (NEM) modification of 19S is accomplished by adding a 1000-fold excess of NEM in 20 m*M* Tris-HCl, pH 7.5, 20 m*M* NaCl, 1 m*M* EDTA, and 10% glycerol at 4°C for 6 h. NEM is removed by extensive dialysis at 4°C against the same buffer. Yeast 26S proteasomes are purified as described from yeast cytosol prepared from a strain expressing a FLAG-epitope-tagged subunit of the 20S core ***(19)***. In brief, cytosol is incubated with anti-FLAG-M2 agarose for 4 h at 4°C, the resin is washed and the proteasomes are eluted by incubation with buffer containing FLAG-peptide for 3 h at 4°C. The resulting eluent is concentrated in Centricon YM-100 cartridges (Amicon) to approx 1 mg/mL and snap-frozen before storage at –80°C. Aliquots should not be thawed and refrozen more than once.
7. Proteasome reassembly buffer (PRB): 45 m*M* Tris-HCl, pH 8.0, 5 m*M* DTT, 10 m*M* $MgCl_2$. This solution should be filter sterilized and stored at 4°C.
8. ATPγS: prepared at a final concentration of 50 m*M* in B88, stored at –80°C, thawed on ice, and used at a final concentration of 5 m*M*.
9. Proteasome inhibitors: lactacystin (100 µ*M* in dd$H_2O$) and MG132 (25 m*M* in dimethyl sulfoxide [DMSO]) are stable indefinitely when stored at –20°C and are used at final

concentrations of 10 μ*M* and 100 μ*M*, respectively. Lactacystin should be thawed on ice and added to B88 or PRB on ice. MG132 in DMSO should be thawed at room temperature (because DMSO freezes on ice) and should be added to buffer only at room temperature, followed by vigorous mixing for 10 s. Tubes can then be placed on ice and chilled before the addition of other reaction components.

10. Serine protease inhibitors: leupeptin (1 mg/mL in dd$H_2O$, or ~2 m*M*) and pepstatin A (5 mg/mL in DMSO, or ~7 m*M*). Both of these are used where indicated at a 1:1000 dilution, and are stable indefinitely at –20°C.
11. Trichloroacetic acid (TCA): 100% (w/v) in dd$H_2O$. TCA is severely corrosive and caustic, so caution is advised when handling or pipetting this solution. The solution should be stored at 4°C, and kept on ice when used. It is advisable to work with a 5- to 10-mL aliquot in a 50-mL disposable conical tube that is resistant to corrosion.
12. Acetone, stored at –20°C. Acetone is highly flammable and can be corrosive.
13. Sodium dodecyl sulfate-polyacrylamide gel electrophoresis (SDS-PAGE) sample buffer: 65 m*M* Tris-HCl, pH 6.8, 1% (v/v) β-mercaptoethanol, 2% (w/v) SDS, 0.5 mg/mL of bromophenol blue, 10% (v/v) glycerol in dd$H_2O$. This solution should be made as a 5X stock and stored at 4°C. Aliquots are stable at room temperature for several months.
14. SDS-PAGE running buffer: 25 m*M* Tris, 192 m*M* glycine, 0.1% (w/v) SDS, pH 8.3.
15. SDS-PAGE fixative: 25% (v/v) isopropanol, 10% (v/v) glacial acetic acid in dd$H_2O$.
16. SDS-PAGE protein gel reagents and equipment. Typically, best resolution of p/pαF is achieved on a discontinuous 18% polyacrylamide gel containing 4*M* urea. The gels we use are 1.5 mm thick and 15 cm high, made up of approx 11 cm of resolving gel and approx 4 cm of stacking gel. Stacking gel: 125 m*M* Tris-HCl, pH 6.8, 0.1% (w/v) SDS, 4.02% (w/v) acrylamide/0.107% (w/v) *bis*-acrylamide (37.5:1 ratio), 4*M* urea, 0.05% (w/v) ammonium persulfate, 0.1% (v/v) *N,N,N',N'*-tetramethylethylenediamine (TEMED). Resolving gel: 375 m*M* Tris-HCl, pH 8.8, 0.1% (w/v) SDS, 18.0% acrylamide/0.48% *bis*-acrylamide (37.5:1 ratio), 4*M* urea, 0.03% (w/v) ammonium persulfate, 0.07% (v/v) TEMED.

## 3. Methods

The methods described in this subheading outline the posttranslational translocation of ppαF into yeast microsomes, and assays used to study the degradation and export of pαF in vitro, as well as data collection and analysis.

### 3.1. *Translocation of ΔGppαF into Yeast Microsomes*

1. A single "master" translocation reaction should be set up in a microcentrifuge tube on ice using multiples of the following values (*see* **Note 1**): 45 μL of ice-cold B88, 6 μL of 10X ATP regenerating system, 5 μL of yeast microsomes, and 4 μL of radiolabeled ΔGppαF (*see* **Note 2**), to a final reaction volume of 60 μL for individual ERAD reactions. Mix by pipetting gently.
2. Incubate the translocation reaction for 1 h in a 20°C water bath.
3. Following the incubation, centrifuge the translocation reaction at top speed (~16,000*g*) in a refrigerated microcentrifuge (4°C) for 3 min to pellet the microsomes, which contain the translocated pαF.
4. Return the tubes to ice, then carefully aspirate the supernatant (*see* **Note 3**), which contains untranslocated ppαF, and dispose of it properly as liquid radioactive waste.
5. Gently resuspend the pellet by pipetting in an equal volume of ice-cold B88 and recentrifuge as described in **step 3** (*see* **Note 4**). After centrifugation, return the tubes to ice.

6. Aspirate the supernatant again and resuspend the pellet to 1/12th the original reaction volume in ice-cold B88 (e.g., 5 µL per 60 µL of the translocation reaction).
7. Aliquot 5 µL into separate microfuge tubes on ice for each subsequent reaction.

### 3.2. PαF Degradation Assay

The following section describes the in vitro assay in which the degradation of pαF is assessed using microsomes containing pαF, an ATP-regenerating system, and cytosol. Negative controls can include the use of the wild type ppαF substrate (*see* **Note 2**), or reactions utilizing the pαF substrate but lacking cytosol and/or ATP (but containing ATPγS). Importantly, microsomes or cytosol from either wild type or mutant yeast strains can be used in order to assess the contribution of specific components on ERAD.

#### *3.2.1. PαF Degradation Reconstituted With Yeast Cytosol*

1. Set up a master translocation reaction as described in **Subheading 3.1.**, and add 5 µL of microsomes containing translocated pαF to a microcentrifuge tube on ice for each reaction. Keep the tubes on ice at all times.
2. To control reactions, add 55 µL of B88 and mix by gently pipetting.
3. To reactions in which ERAD is desired, add 6 µL of 10X ATP regenerating system.
4. Add inhibitors (e.g., ATPγS or proteasome inhibitors) if desired.
5. To reactions in which ERAD is to proceed, add cytosol to the appropriate final concentration (typically 1–3 mg/mL; *see* **Note 5**).
6. If required, add purified cytosolic components (e.g., molecular chaperones or proteasomes or proteasomal subunits), which can be used in addition to or instead of yeast cytosol (*see* **Subheadings 3.2.2.** and **3.3.2.**).
7. Adjust the volume of the reactions to 60 µL with B88.
8. Incubate the reactions at 30°C for 20 min (higher temperatures can be used if temperature-sensitive mutations are to be monitored; for example, *see* **refs. *7,8,10***).
9. After incubation, quickly return the reactions to ice and add 12 µL of ice-cold 100% TCA (final concentration ~17%) into each tube to stop the reactions. Agitate vigorously on a vortex mixer for approx 3 s after TCA addition, then return each tube to ice and incubate for 15–20 min.
10. Centrifuge the reactions at top speed (16,000*g*) in a refrigerated microcentrifuge at 4°C, then carefully aspirate the supernatant using a gel-loading pipet tip. Dispose of the supernatant as radioactive liquid waste.
11. Add 80 µL of ice-cold acetone (removed from storage at –20°C immediately before use) to each pellet, agitate for approx 2 s on a vortex mixer, and recentrifuge as described in **Subheading 3.1., step 3** (*see* **Note 6**). Carefully aspirate the supernatant and dispose of it as radioactive liquid waste. Allow the tubes to air-dry for 2–3 min.
12. Resuspend pellets in SDS-PAGE sample buffer (*see* **Note 7**).
13. Incubate the samples for 10 min at 74°C prior to analysis via SDS-PAGE (*see* **Subheading 3.4.**); alternatively, the samples can be frozen at –20°C. If frozen, the samples are stable indefinitely, but they should be thawed at room temperature and incubated at 74°C for 10 min prior to electrophoresis.

#### *3.2.2. PαF Degradation Reconstituted With Purified Proteasomes*

The in vitro ERAD assay can be used to assay the contributions of purified components in addition to or in place of cytosol. For example, we recently found that purified proteasomes can support the export and degradation of pαF *(20)*. In this case, 19S and

20S components of the proteasome are incubated in the presence of ATP to reconstitute 26S (*see* **Note 8**). Purified 26S proteasomes can also be used; however, they seem to exhibit more robust proteolytic activity after preincubation with ATP.

1. Set up a master translocation reaction as described in **Subheading 3.1.**, except use PRB in place of B88.
2. During the 1-h translocation incubation, preassemble control reactions (PRB only) as well as the 26S assembly reactions containing PRB, purified proteasomes, ATP or ATPγS, and inhibitors such as MG132 or lactacystin if desired, to a final volume of 55 μL. This allows room for the addition of 5 μL of microsomes containing pαF. Incubate all preassembled reaction mixtures for 30 min at 37°C, which permits 19S and 20S reassembly and allows for maximal 26S activity (*see* previously in this subheading and **ref. *20***).
3. After centrifugation and resuspension of the translocation reaction, add 5 μL of microsomes containing translocated pαF to the preassembled reaction mixtures without returning them to ice. Mix gently by pipetting.
4. Incubate reactions for 20 min at 30°C.
5. Continue with the procedure outlined in **Subheading 3.2.1., steps 9–13.**

## 3.3. Assaying for pαF Export

By adding a brief centrifugation step before TCA precipitation, this assay can be co-opted to monitor pαF export from the microsomes under various conditions (i.e., using cytosol with added proteasome inhibitors that block degradation but not export of pαF; **ref. *6***). For example, we found that purified 19S cap in the presence of ATP is sufficient to stimulate pαF export without degradation in the absence of the 20S core ***(20)***.

### 3.3.1. Assaying for pαF Export Using Yeast Cytosol

1. Set up reactions as desired, using the procedure outlined in **Subheading 3.2.1., steps 1–7.**
2. Incubate for 20 min at 30°C.
3. Following incubation, pellet the microsomes in a refrigerated microcentrifuge at 16,000*g* for 3 min at 4°C.
4. Return tubes to ice and quickly remove the supernatant (containing exported pαF) using a gel-loading micropipet tip. Be very careful not to disturb the pellet. Add the supernatant to a clean microfuge tube on ice containing 12 μL of 100% TCA. Agitate each tube vigorously on a vortex mixer for approx 3 s after addition, and then return each tube to ice.
5. Resupend the pellet in 60 μL of B88 and add 12 μL of 100% ice-cold TCA with a micropipet. Agitate each tube vigorously on a vortex mixer for approx 3 s after addition, then return each tube to ice.
6. Incubate tubes containing supernatant and pellet fractions on ice for 15–20 min.
7. Spin the reactions at 16,000*g* in a refrigerated microcentrifuge at 4°C, and then carefully aspirate the supernatants using a gel-loading pipet tip. Dispose of the supernatants as radioactive liquid waste.
8. Add 80 μL of ice-cold acetone (removed from storage at –20°C immediately before use) to each pellet, agitate on a vortex mixer for approx 2 s, and then recentrifuge as above (*see* **Note 6**). Return tubes to ice. Carefully aspirate the supernatant and dispose of it as radioactive liquid waste. Allow the tubes to air-dry for 2–3 min.
9. Resuspend pellets in SDS-PAGE sample buffer (*see* **Note 7**).

10. Freeze samples at –20°C or incubate them for 10 min at 74°C prior to SDS-PAGE (*see* **Subheading 3.4.**). If frozen, the samples are stable indefinitely, but they should be thawed at room temperature and incubated at 74°C for 10 min prior to electrophoresis.

#### *3.3.2. PαF Export From Yeast Microsomes Using Purified 19S Proteasome Subunit*

1. Set up a master translocation reaction as described in **Subheading 3.1.** using PRB instead of B88. Microsomes containing translocated pαF are incubated with either the 19S proteasome particle (*see* **Note 8**) or NEM-treated 19S proteasome particle as a negative control (*see* **Note 9**). Leupeptin (final concentration 20 μ*M*), pepstatin A (final concentration 70 μ*M*), and MG132 (final concentration 100 μ*M*) are added to all reactions to inhibit the activity of residual 20S particle and/or contaminating serine proteases. Use PRB to bring reaction volumes to 60 μL.
2. Incubate reactions for 20 min at 30°C, then immediately pellet the microsomes in a refrigerated microcentrifuge at 16,000*g* for 3 min at 4°C.
3. Return the tubes to ice and continue with the procedure outlined in **Subheading 3.3.1., steps 4–10.**

### *3.4. Data Collection and Analysis*

Samples should be incubated for 10 min at 74°C before proteins are resolved by SDS-PAGE (described in **Subheading 2.**). Gels should be run at 40 mA (constant current) until the bromophenol blue dye front is at the bottom of the gel. A typical run time is about 3 h. The p/pαF will reside approximately in the center of the separating phase of the gel. After the glass plates are removed, the dye front and stacking gel are removed with a razor blade. The gel is then gently place in SDS-PAGE fixative for 45 min to 2 h (*see* **Note 10**). Gels are dried on filter paper for approx 1.5 h on vacuum drier with heating (~80°C), and are then cooled to room temperature before the vacuum is broken. Typically, the resulting autoradiograph requires about 2 d of exposure time on a phosphorimage screen, although the use of half reactions or old label may require longer exposure times.

On the imaged gel, two bands should be visible, the ppαF precursor (slower migrating) and pαF (faster migrating due to signal sequence cleaveage during translocation) (*see* **Fig. 1**). Only translocated pαF should be quantified. When assaying for degradation, the pαF remaining under conditions that either promote or inhibit its degradation should be averaged for the duplicate or triplicate reactions and compared to the average of the control reactions that contain only buffer. The data are presented as "%pαF remaining," as the control reactions represent 100% of the initial pαF.

When assaying pαF export, the amount of pαF in the supernatant and pellet should be summed to determine the total pαF. The amount of pαF in the supernatant should be divided by the total pαF to calculate the % pαF exported to the supernatant. These values can then be averaged among duplicate or triplicate sets, and then compared to background export, which is typically approx 10–15% of control reactions.

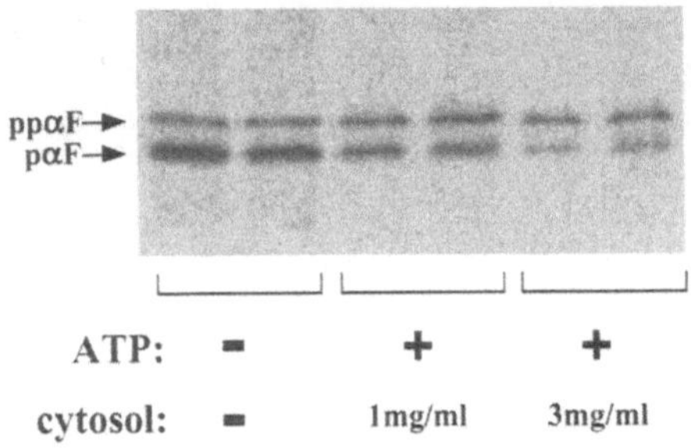


Fig. 1. The in vitro ERAD assay. Duplicate ERAD reactions were performed as described in **Subheading 3.2.1.** using ER-derived microsomes from wild-type yeast containing pαF. The microsomes were subsequently incubated in the presence of a final concentration of 0, 1, or 3 mg/mL cytosol prepared from wild-type yeast and in the presence of ATP ("+") or ATPγS ("–"). Note that a small amount of ppαF is present despite washing the microsomes several times following the initial translocation reaction. This material is likely due to its aggregation on the surface of the microsomes, and in some cases can be removed by incubating the microsomes prior to the washes in trypsin at a final concentration of 0.2 mg/mL for 20–30 min at 4°C. The separation of p/pαF shown is achieved on an 18% polyacrylamide/4*M* urea gel (*see* **Subheading 3.4.**).

## 4. Notes

1. Reaction volumes can be halved to conserve material, as long as all other manipulations are adjusted accordingly. However, this will result in a weaker signal, leading to a need for increased exposure time and a lower signal-to-noise ratio. It is desirable to perform separate ERAD reactions in duplicate or triplicate, but beginning all ERAD assays from a single "master" translocation reaction leads to more consistent data.
2. Reactions with wild-type ppαF can also be prepared, which serves as a negative control. Because this substrate is signal-sequence cleaved and triply glycosylated in the yeast microsomes to form 3GpαF, it will not be an ERAD substrate. 3GpαF migrates at a molecular mass of approx 28 kDa, which is significantly different than ppαF (18 kDa) and pαF (16 kDa).
3. Sometimes the pellet in this or subsequent steps may be difficult to see, but by slowly and carefully aspirating the supernatant with a gel-loading tip held against the side of the tube away from the pellet, loss of material can be prevented. Use extreme care when resuspending the microsomal pellet in order to avoid membrane rupture. Avoid bubbles when pipetting, and never agitate reaction mixtures unless specifically indicated in the procedure.
4. If the ΔGppαF preparation being used shows radiolabeled bands on the gel that do not correspond to p/pαF in the final autoradiograph, multiple washes may be performed to remove the contaminants at this step. Also, multiple washes will help to remove ppαF, which are typically present on the gel because of its tendency to stick to the microsomes. However, one wash is usually sufficient.
5. Optimal degradation is noted with cytosol concentrations of 3–5 mg/mL, and degradation plateaus at higher concentrations (*see* **ref. *10***). It may be necessary to titrate new cytosol preparations to find the optimal concentration. Cytosol prepared and used at this concen-

tration also supports the degradation of the mammalian ERAD substrate, apolipoprotein B, in vitro ***(21)***.

6. Pelleted samples are stable on ice after addition of acetone for only 20–30 min, but best results are obtained if the samples are processed immediately.
7. Heat the sample buffer to 74°C before resuspending the acetone-washed pellets. Resuspending the final pellets is tedious, and using preheated buffer facilitates this process, especially in the case of reactions containing large amounts of cytosol. If the sample buffer turns orange-yellow upon addition, this is due to residual TCA. Neutralize the sample with 1–2 μL of 1 *M* Tris HCl, pH 6.8, until the orange color turns back to blue.
8. While the protocol suggests a 60-μL reaction, halving the volume of the reaction conserves material. Typically in a 30-μL "half"-reaction, 5 and/or 10 μg of purified mammalian 20S and/or 19S particles are used, respectively. Also, 30 μg of affinity purified yeast 26S particles have been used successfully ***(20)***. Double these values for a 60-μL reaction.
9. Omit DTT from the PRB when using NEM-treated 19S, because reducing agents will reverse NEM modification.
10. Overnight fixation is not recommended, as this seems to increase significantly the likelihood of cracked gels during drying.

## References

1. McCracken, A. A. and Brodsky, J. L. (1996) Assembly of ER-associated protein degradation in vitro: dependence on cytosol, calnexin, and ATP. *J. Cell Biol.* **132,** 291–298.
2. Ellgaard L. and Helenius, A. (2003) Quality control in the endoplasmic reticulum. *Nat. Rev. Mol. Cell Biol.* **4,** 181–91.
3. McCracken, A. A. and Brodsky, J. L. (2003) Evolving questions and paradigm shifts in endoplasmic reticulum associated degradation (ERAD). *Bioessays* **25,** 868–877.
4. Voges, D., Zwickl, P., and Baumeister, W. (1999) The 26S proteasome: a molecular machine designed for controlled proteolysis. *Annu. Rev. Biochem.* **68,** 1015–1068.
5. Kostova, Z. and Wolf, D. H. (2003). For whom the bell tolls: protein quality control of the endoplasmic reticulum and the ubiquitin–proteasome connection. *EMBO J.* **22,** 2309–2317.
6. Werner, E. D., Brodsky, J. L., and McCracken, A. A. (1996) Proteasome-dependent endoplasmic reticulum–associated protein degradation: an unconventional route to a familiar fate. *Proc. Natl. Acad. Sci. USA* **93,** 13797–13801.
7. Pilon, M., Schekman, R., and Römisch, K. (1997) Sec61p mediates export of a misfolded secretory protein from the endoplasmic reticulum to the cytosol for degradation. *EMBO J.* **16,** 4540–4548.
8. Nishikawa, S., Fewell, S. W., Kato, Y., Brodsky, J. L., and Endo, T. (2001) Molecular chaperones in the yeast endoplasmic reticulum maintain the solubility of proteins from retro-translocation and degradation. *J. Cell Biol.* **153,** 1061–1069.
9. Zhou, M. and Schekman, R. (1999) The engagement of Sec61p in the ER dislocation process. *Mol. Cell* **4,** 925–34.
10. Brodsky, J. L., Werner, E. D., Dubas, M. E., Goeckeler, J. L., Kruse, K. B., and McCracken, A. A. (1999) The requirement for molecular chaperones during ER-associated protein degradation (ERAD) demonstrates that protein import and export are mechanistically distinct. *J. Biol. Chem.* **274,** 3453–3460.
11. Gillece, P., Luz, J. M., Lennarz, W. J., de La Cruz, F. J., and Romisch, K. (1999) Export of a cysteine-free misfolded secretory protein from the endoplasmic reticulum for degradation requires interaction with protein disulfide isomerase. *J. Cell Biol.* **147,** 1443–56.

12. Rothblatt, J. A. and Meyer, D. I. (1986) Secretion in yeast: reconstitution of the translocation and glycosylation of alpha-factor and invertase in a homologous cell-free system. *Cell* **44,** 619–628.
13. Deshaies, R. J. and Schekman, R. (1989) SEC62 encodes a putative membrane protein required for protein translocation into the yeast endoplasmic reticulum. *J. Cell Biol.* **109,** 2653–2664.
14. Brodsky, J. L. and Schekman, R. (1993) A Sec63p–BiP complex from yeast is required for protein translocation in a reconstituted proteoliposome. *J. Cell Biol.* **123,** 1355–63.
15. Sorger, P. K. and Pelham, H. R. (1987) Purification and characterization of a heat-shock element binding protein from yeast. *EMBO J.* **6,** 3035–3041.
16. McGuire, M. J., McCullogh, M. L., Croall, D. E., and DeMartino, G. N. (1989) The high molecular weight multicatalytic proteinase, macropain, exists in a latent form in human erythrocytes. *Biochim. Biophys. Acta* **995,** 181–186.
17. Chu-Ping, M., Vu, J. H., Proske, R. J., Slaughter, C. A., and DeMartino, G. N. (1994) Identification, purification, and characterization of a high molecular weight, ATP-dependent activator (PA700) of the 20S proteasome. *J. Biol. Chem.* **269,** 3539–3547.
18. DeMartino, G. N., Moomaw, C. R., Zagnitko, O. P., et al. (1994) PA**700,** an ATP-dependent activator of the 20S proteasome, is an ATPase containing multiple members of a nucleotide-binding protein family. *J. Biol. Chem.* **269,** 20878–20884.
19. Verma, R., Chen, S., Feldman, R., et al. (2000) Proteasomal proteomics: identification of nucleotide-sensitive proteasome-interacting proteins by mass spectrometric analysis of affinity-purified proteasomes. *Mol. Biol. Cell* **11,** 3425–3439.
20. Lee, R., Liu, C. W., Harry, C., et al. (2004) Retro-translocation and degradation can be uncoupled during the ER Associated Degradation (ERAD) of a soluble protein. *EMBO J.* **23,** 2206–2215.
21. Gusarova, V., Caplan, A. J., Brodsky, J. L., and Fisher, E. A. (2001) Apoprotein B degradation is promoted by the molecular chaperones Hsp90 and Hsp70. *J. Biol. Chem.* **276,** 24891–24900.

# 12

# Reticulocyte Lysate as a Model System to Study Endoplasmic Reticulum Membrane Protein Degradation

**Eric Carlson, Nathan Bays, Larry David, and William R. Skach**

**Summary**

Recent studies have revealed that rabbit reticulocyte lysate (RRL) efficiently reconstitutes endoplasmic reticulum-associated degradation (ERAD) of mutant and misfolded membrane proteins. When supplemented with canine pancreas microsomal membranes, the RRL system faithfully carries out ER targeting, translocation, glycosylation, and membrane integration events and therefore provides a ready source of 35S-labeled protein with defined transmembrane topology. These substrates can be rapidly isolated in native ER membranes which, when incubated in RRL lacking exogenous hemin, are degraded in an ATP-dependent manner by the ubiquitin–proteasome pathway. Because the newly translated protein is the only source of radiolabel, degradation can be followed to its end state by conversion into trichloroacetic acid (TCA)-soluble peptide fragments. A particularly useful aspect of this system is that both membrane-associated and cytosolic components are amenable to biochemical and pharmacological manipulation. Here we describe techniques for preparing translation- and degradation-competent RRL, affinity depletion, identification of cytosolic factors involved in degrading the cystic fibrosis transmembrane conductance regulator (CFTR), and reconstitution of ERAD by add-back of purified recombinant proteins. These techniques provide a powerful tool for dissecting components involved in ubiquitination, degradation, and in particular, extraction of transmembrane ERAD substrates.

**Key Words:** Cystic fibrosis transmembrane conductance regulator; endoplasmic reticulum-associated degradation; in vitro degradation/proteolysis; membrane protein degradation; rabbit reticulocyte lysate.

## 1. Introduction

### *1.1. Endoplasmic Reticulum-Associated Degradation (ERAD)*

Misfolded and unassembled proteins in the endoplasmic reticulum (ER) are degraded through a process referred to as ER-associated degradation (ERAD) ***(1–5)***.

From: *Methods in Molecular Biology, vol. 301, Ubiquitin–Proteasome Protocols*
Edited by: C. Patterson and D. M. Cyr © Humana Press Inc., Totowa, NJ

Although it was initially proposed that this proteolytic machinery resided within the ER itself ***(6)***, it is now clear that most if not all ERAD substrates are degraded through the ubiquitin–proteasome pathway ***(7)***. This finding has driven major efforts to identify the mechanism by which proteins are recognized by quality control machinery ***(8–10)***, the E2 and E3 enzymes responsible for ubiquitin conjugation, and the means of delivering ER proteins to the cytosolic 26S proteasome ***(11,12)***. In this last respect, membrane proteins pose a particular challenge because hydrophobic transmembrane segments must be extracted from the lipid bilayer prior to degradation.

Much of our early knowledge of the ubiquitin proteasome pathway came from biochemical characterization of cytosolic extracts such as rabbit reticulocyte lysate (RRL). Although early studies focused primarily on cytosolic substrates ***(13)***, RRL also contains the necessary factors to carry out ERAD and has recently been used to identify and dissect various components and steps involved in degrading secretory ***(14,15)*** and membrane proteins ***(16–18)***. This approach has proven particularly useful when combined with the capacity of RRL for membrane protein biogenesis. When supplemented with ER-derived microsomal membranes, RRL retains newly synthesized membrane proteins in a functional ER compartment that is readily amenable to biochemical manipulation.

This chapter focuses on the use of RRL for reconstitution, quantitation, and manipulation of the ERAD pathway. The methods described use rough microsomes derived from canine pancreas (CRM) as a source of functional ER membranes, although other sources such as *Xenopus laevis* oocytes ***(19,20)***, semipermeabilized cells ***(18)***, or yeast ***(21)*** could potentially be used. The RRL system has a higher fidelity for reconstituting many mammalian membrane protein biogenesis (particularly polytopic proteins) than bacterial or wheat germ-based systems (our unpublished observations). RRL also has robust degradation activity that likely reflects the terminally differentiated state of reticulocytes in which cellular organelles and most cytosolic proteins are being degraded. RRL degradation machinery also appears to select certain proteins (e.g., cystic fibrosis transmembrane conductance regulator [CFTR]) over other misfolded trafficking mutants (e.g., aquaporins), although the reason for this is unclear.

A major advantage of RRL is that the substrate can be synthesized in vitro, and is thus the only protein that contains significant incorporated radioisotope ***(22)***. It is therefore possible to quantitatively follow substrate entry into the degradation pathway (ubiquitination) and progressive conversion into TCA-soluble peptide fragments. Release of substrate from the ER membranes is also accomplished by simple fractionation techniques, and both microsomes and lysate are directly accessible to pharmacological and/or biochemical modification prior to or during the degradation reaction. Here we will briefly cover basic strategies for reconstituting ERAD in RRL using *de novo* synthesized substrates. Detailed protocols are provided on manipulation of cytosolic and membrane components using proteolysis and alkylation of microsomes, affinity depletion of RRL, and reconstitution of microsomes and lysates with recombinant proteins.

## 2. Materials

### 2.1. Preparation of RRL

#### 2.1.1. Reticulocyte Induction

1. New Zealand White rabbits (~6 mo old, ≥3 kg).
2. 3-cc Syringes (no. of rabbits × 3).
3. 26-Gage needles.
4. Clinical centrifuge with rotor suitable for hematocrit determination (e.g., IEC rotor no. 927).
5. Hematocrit reader, hematocrit tubes, and clay to plug tubes.
6. Acetyl phenyl hydrazine (APH) solution: add 2.5 g of APH to 20 mL of EtOH and then add 50 mL of water. Adjust pH to 7.0 with approx 1 mL of 1 *M* KOH, and bring to 100 mL with water. Filter the solution (0.22 µm), aliquot, and store at –20°C.

#### 2.1.2. Processing of Reticulocyte Lysate

1. Rabbit restrainer.
2. Pentobarbital/Nembutal, 50 mg/mL (4-5 mL per rabbit).
3. Heparin, 1000 U/mL (2 mL per rabbit).
4. 70% Isopropanol.
5. IV tubing (e.g., extension set no. ET-20L (472010) from B/Braun – 21-in long tubing).
6. Reticulocyte wash buffer: 5 m*M* glucose, 0.14 *M* NaCl, 0.05 *M* KCl, 5 m*M* $MgCl_2$ (~500 mL per rabbit).
7. *Staphylococcus aureus* (57) nuclease, 15 U/µL: dissolve in 10 m*M* Tris-HCl, pH 8.0 or RNase-free water. Store in aliquots at –80°C. (Roche, Indianapolis, IN).
8. 0.1 *M* CaCl2.
9. 0.1 *M* EGTA-KOH, pH 7.5.
10. 1 m*M* Hemin stock solution. Combine reagents in the following order: 6.44 mg of hemin (bovine crystalline type I, Sigma, St. Louis, MO), 0.25 mL of 1 *M* KOH, 0.5 mL of 0.2 *M* Tris-HCl, pH 7.0–8.0, 8.9 mL of ethylene glycol, 0.19 mL of 1 *N* HCl, 0.05 mL of $H_20$. Filter (0.22 µm) and store at –20°C.

### 2.2. Preparation of ER Microsomal Membranes

1. Pancreas tissue.
2. 50-mL Potter–Elvejhem tissue homogenizer with Teflon pestles (one loose fitting and one snug).
3. 25-mL Potter–Elvejhem tissue homogenizer (Teflon pestle).
4. Hand-held food grinder.
5. 10X Stock buffer (100 mL): 0.5 *M* KoAc, 60 m*M* $Mg(oAc)_2$, 10 m*M* EDTA, 0.5 *M* triethanol-amine-acetate, pH 7.5.
6. Buffer A: 1:10 dilution of 10X stock buffer containing 0.25 *M* sucrose, 1 m*M* dithiothreitol (DTT), 0.5 m*M* phenylmethylsulfonyl fluoride (PMSF). Add DTT and PMSF immediately before use.
7. Buffer B: 1:10 dilution of 10X stock buffer containing 1.3 *M* sucrose.
8. Buffer C (100 mL): 0.25 *M* sucrose, 1 m*M* DTT, 50 m*M* triethanolamine-acetate, pH 7.5. Add DTT immediately before using.
9. 1% Sodium dodecyl sulfate (SDS), 0.1 *M* Tris-HCl, pH 8.0.

## 2.3. In Vitro Transcription/Translation

### 2.3.1. Transcription

1. 5X Transcription buffer: 30 m*M* $MgCl_2$, 10 m*M* spermidine, 200 m*M* Tris-HCl, pH 7.5.
2. 10X NTPs: 5 m*M* ATP, 5 m*M* CTP, 5 m*M* UTP, 1 m*M* GTP, 20 m*M* Tris-HCl, pH 8.0. (Nucleotides from Roche Molecular Biochemicals, Indianapolis, IN.)
3. Diguanosine triphosphate (Promega, Madison, WI): dissolve 25 U in 300 μL of sterile, $ddH_2O$ (~5 m*M*) and store at –80°C.
4. 0.1 *M* DTT (Roche Molecular Biochemicals, Indianapolis, IN).
5. 10 mg/mL of tRNA (bovine liver type XI, Sigma, St. Louis, MO).
6. 20 U/mL of RNAse inhibitor (Promega, Madison, WI).
7. 10 U/mL of SP6 polymerase (Epicenter, Madison, WI).

### 2.3.2. Translation

1. 20X Translation buffer: 2*M* KoAc, 16 m*M* $Mg(oAc)_2$, 40 m*M* Tris-acetate, pH 7.5.
2. 5X Emix: 5 m*M* ATP, 5 m*M* GTP, 60 m*M* creatine phosphate, 1 m*M* mixture of 19 amino acids except methionine, and 5 μCi/μL of Trans35S-label (ICN, Costa Mesa, CA). Aliquots are typically made in 100- to 200-μL volumes, adjusted to pH 7.5 with Tris base and stored at –80°C.
3. RRL (hemin- and nuclease-treated; *see* **Subheading 3.1., item 7**).
4. 10 mg/mL of tRNA (bovine liver type XI, Sigma, St. Louis, MO).
5. 20 U/mL of RNase inhibitor (Promega, Madison, WI).
6. 4 mg/mL of creatine kinase (CK) (Sigma, St. Louis, MO): dissolve in 50% glycerol, 10 m*M* Tris-acetate, pH 7.5.
7. Microsomal membranes (nuclease treated; *see* **Subheading 3.2., item 6**).
8. SDS loading buffer (SDS-LB): 4% SDS, 2 m*M* EDTA, 10% sucrose, 0.05% bromophenol blue, 1 *M* DTT, 100 m*M* Tris-HCl, pH 8.9.
9. 1–2 m*M* aurin tricarboxylic acid (ATA) (Sigma, St. Louis, MO). Adjust to pH approx 7.0 with Tris base.

## 2.4. In Vitro Degradation and Membrane Release Assays

1. 10X Degradation buffer: 10 m*M* ATP, 120 m*M* creatine phosphate, 30 m*M* DTT, 50 m*M* MgC12, 100 m*M* Tris-HCl, pH 7.5.
2. 0.1*M* Sucrose buffer: 0.1*M* sucrose, 100 m*M* KCl, 5 m*M* $MgCl_2$, 1 m*M* DTT, 50 m*M* *N*-(2-hydroxyethylpiperazine-*N*′-(2-ethanesulfonic acid) (HEPES)–NaOH, pH 7.5.
3. 0.5*M* Sucrose cushion: 0.5 *M* sucrose, 100 m*M* KoAc, 5 m*M* $Mg(oAc)_2$, 1 m*M* DTT, 50 m*M* HEPES–NaOH, pH 7.5.
4. RRL (untreated. *See* **Subheading 3.1., step 6**).
5. Creatine kinase, 4 mg/mL (*see* **Subheading 2.3.2., item 6**).
6. 20% Trichloroacetic acid (TCA).
7. Ready-Safe scintillation fluid (Beckman, Fullerton, CA).
8. 10 m*M* MG132 (in DMSO). (Sigma, St. Louis, MO. listed as "*n*-cbz-leu-leu-leu-al").

## 2.5. Affinity Adsorption of RRL ERAD Components

### 2.5.1. Purification of His-Tagged Recombinant Proteins

1. Ni-NTA agarose (e.g. Qiagen, Inc., Valencia, CA).
2. *E. coli* strain BL21(DE3).

3. Expression vector (e.g. pET, pTrc, pQE).
4. L-broth: 7.5 g of NaCl, 7.5 g of tryptone, 15 g of yeast extract. Bring to a volume of 1.5 L with water. Autoclave.
5. Ampicillin, 8 mg/mL (stock).
6. 20% Glucose.
7. 1 *M* Isopropyl β-D-1-thiogalactopyranoside (IPTG) (Sigma, St. Louis, MO).
8. Buffer D: 300 m*M* NaCl, 5% glycerol, 0.5 m*M* PMSF, 2 m*M* β-mercaptoethanol, 50 m*M* Tris-HCl, pH 7.5. Add PMSF and β-mercaptoethanol prior to using.
9. Buffer E: buffer D with 500 m*M* imidazole.
10. French press (e.g., SLM-Aminco).
11. Econo-column (1 cm × 10 cm; cat. no. 737-1011) and flow adaptor (1 cm × 1–7 cm; cat. no. 738-0014) (e.g., Bio-Rad, Hercules, CA).
12. Fast protein liquid chromatograph (FPLC) (e.g., Pharmacia model 500) with fraction collector.
13. Dialysis tubing (e.g., Spectra/Por 12–14 kDa molecular weight cutoff (MWCO); Spectrum Laboratorie Inc., Rancho Dominguez, CA).
14. Dialysis buffer: 100 m*M* NaCl, 1 m*M* β-mercaptoethanol, 25 m*M* Tris-HCl, pH 7.5.
15. Centriplus concentrators (e.g., Amicon, Millipore Corp., Bedford, MA).

#### *2.5.2. RRL Affinity Depletion*

1. Ni-NTA agarose (Qiagen, Inc., Valencia, CA).
2. His-tagged recombinant protein.
3. RRL (not nucleased, +/– hemin).
4. TNB buffer: 100 m*M* NaCl, 1 m*M* β-mercaptoethanol, 25 m*M* Tris-HCl, pH 7.5.
5. Wash buffer: TNB buffer + 20 m*M* imidazole.
6. High salt elution buffer: 1 *M* NaCl, 1 m*M* β-mercaptoethanol, 25 m*M* Tris-HCl, pH 7.5.
7. Imidazole elution buffer: TNB buffer + 500 m*M* imidazole.

### 2.6. Inactivation and Reconstitution of Microsomal Membranes

#### *2.6.1. Proteolysis of Microsomal Membranes*

1. Microsomal membranes (nucleased) (*see* **Subheading 3.2., item 6**).
2. 4 mg/mL of trypsin (stock) (Sigma, St. Louis, MO).
3. 0.1 *M* PMSF (in DMSO).
4. Buffer C (*see* **Subheading 2.2., item 8**).
5. Buffer F: 1 m*M* DTT, 50 m*M* triethanolamine–acetate, pH 7.5.
6. 4*M* KoAc.
7. PK cushion: 0.5 *M* sucrose, 500 m*M* KoAc, 0.5 m*M* PMSF, 5 m*M* $Mg(oAc)_2$, 1 m*M* DTT, 50 m*M* HEPES–NaOH, pH 7.5. Add PMSF just before using.
8. 0.5 *M* Sucrose cushion (**Subheading 2.4., item 3**).

#### *2.6.2. Alkylation of Microsomal Membranes*

1. Microsomal membranes (nucleased) (*see* **Subheading 3.2, step 6**).
2. 0.5 *M* Sucrose cushion (–DTT): *see* **Subheading 2.4., item 3**. Do not add DTT.
3. Buffer C (–DTT): *see* **Subheading 2.2., item 8**. Do not add DTT.
4. 50 m*M* *N*-Ethylmaleimide (NEM) (Sigma, St. Louis, MO)
5. 1 *M* DTT.

#### 2.6.3. SR α Repopulation of Treated Microsomal Membranes

1. Proteolysed or alkylated microsomal membranes (**Subheadings 3.6.1.** and **3.6.2.**).
2. Recombinant SRP receptor α-subunit (**Subheading 3.5.1.**).
3. Buffer C: *See* **Subheading 2.2., item 8**.
4. 0.5 *M* Sucrose cushion: *See* **Subheading 2.4., item 3**.

## 3. Methods

### 3.1. Preparation of RRL

The following procedure, used routinely in our laboratory, is based on a protocol previously described by Jackson et al. *(22)* and takes roughly 1 wk from start to finish. Typically 5–10 rabbits are processed simultaneously. Preparing RRL *de novo* permits the composition and quality of reaction conditions to be controlled and optimized. Different preparations may vary significantly in terms of activity. The lysate from each animal is processed and stored separately and assayed independently for efficiency in translation and degradation reactions. We have found that the best preparations for degradation are often the least suitable for translation. The reason for this is unknown, but likely reflects biological differences between animals and/or the amount of endogenous free hemin present in RRL (*see* **Note 1**).

Commercial preparations of RRL are available for in vitro translation (Promega TNT systems, L2080; Nuclease treated RRL, L4960; and Untreated RRL, L4151). However, if large numbers of experiments are to be performed it is often more economical to prepare this reagent because one rabbit yields between 20 and 40 mL of RRL. It should be emphasized that with the exception of Untreated RRL (Promega, L4151) these commercial products contain exogenous hemin, which stimulates RRL translation but is also a potent proteasome inhibitor. The concentration of hemin in commercial preparations is not specified, but based on preparation of Untreated RRL (Promega, cat. no. L4151) for use in protein synthesis, it seems that 20 μ*M* hemin is Promega's standard treatment. Because most commercial products are optimized for translation they are generally not suitable to study degradation (*see* **Note 2**).

1. All steps involving the handling and care of animals should be approved by the institutional animal review board.
2. Inject rabbits subcutaneously in scruff of the neck on three consecutive days (d 1, 2, and 3) with 2 mL of APH solution.
3. On d 4–7 monitor animals for health, activity, and food intake. Hematocrits should be measured in selected animals by collecting blood samples in a capillary tube (via ear vein puncture). Centrifuge blood at approx 8000*g* for 5 min and read hematocrit directly. The hematocrit should normally not fall below 20%.
4. On d 8, place rabbits in a restrainer, cannulate an ear vein and inject 2 mL of heparin (2000 U) followed by a lethal dose of pentobarbital (200–250 mg) 1 min later. When rabbit has lost eyelid reflex and respiration has ceased, remove the animal from restrainer, transect left ribs with a scalpel, open the chest cavity, and puncture the left ventricle directly with a 16-gage needle attached to IV extension tubing. Collect blood in a 250-mL centrifuge bottle on ice. Perform all subsequent steps on ice.

5. Centrifuge blood at 2500*g* at 4°C for 10 min and discard supernatant (aspirate buffy coat layer of white blood cells). Wash cells three times with 150 mL of ice-cold reticulocyte wash buffer followed by centrifugation at 2500*g* for 10 min. After the third wash resuspend cells in small volume of wash buffer, transfer to a 50-mL conical tube, centrifuge at 3000*g* for 10 min, and remove the supernatant. Add one volume of ice-cold $H_2O$ and shake very vigorously for 30 s to lyse cells. Repeat shaking every 5 min three times. Centrifuge the lysate at 15,000*g* for 10 min and carefully decant supernatant. Lysate can be frozen in liquid nitrogen at this stage and stored at –80°C for several years with minimal loss of activity. Set aside several small samples (0.5 mL) to test relative activity in translation and/or degradation assays.
6. For degradation reactions the lysate is used without further modification. This is untreated RRL (not to be confused with Promega's product of the same name). (*See* **Note 2.**)
7. For translation, hemin is added to a final concentration of 40 μ*M*, and RRL RNA is digested by nuclease treatment. To perform the latter, a 1-mL aliquot of RRL is quickly thawed in a 25°C water bath. Ten microliters of 0.1 *M* $CaCl_2$ and 10 μL of micrococcal nuclease are added. The sample is incubated for 8–10 min at 25°C, and 20 μL of 0.1 *M* EGTA is added. Samples are aliquoted, frozen in liquid nitrogen, and stored at –80°C for 2–3 mo.

### 3.2. Preparation of ER Microsomes

Canine pancreas microsome preparation is based on the procedure described by Walter and Blobel *(23)*. All procedures are carried out on ice in a 4°C cold room.

1. Quickly remove pancreas from euthanized animal and immediately place into 100 mL of ice-cold buffer A (*see* **Note 3**). Trim away fat, connective tissue and blood vessels, and record the weight. Grind pancreas by hand using a coarse grinder until a pulpy consistency is achieved. Collect fragments together with four volumes (4 mL/g of original tissue) of ice-cold buffer A.
2. Homogenize this mixture in a 50 mL Potter–Elvehem tissue homogenizer attached to a high speed motor with three or four passes of a loose-fitting Teflon pestle, and return homogenate to ice. Homogenize solution a second time with one to three passes of a tight-fitting Teflon pestle. Sample must be kept ice cold during the entire procedure.
3. Divide homogenate into 50-mL polyallomer tubes and centrifuge at 600*g* for 10 min. Decant supernatant into fresh 50-mL polyallomer tubes and centrifuge at 10,000*g* for 10 min. Decant the resulting supernatant into a 150-mL beaker, taking care not to contaminate the supernatant with the loose pellet.
4. Carefully layer approx 15–20 mL of this supernatant fraction over a 5- to 7-mL cushion of buffer B and centrifuge at 150,000*g* for 3 h. Aspirate supernatant and gently resuspend pellets (by hand in a 25-mL homogenizer) in buffer C (1/2 mL per gram of starting material). When the solution is uniform, aliquot and freeze 1-mL aliquots in liquid nitrogen. Microsomes are stable for several years when stored at –80°C.
5. Determine the $OD_{280}$ by dissolving 10 μL of membrane suspension in 990 μL of 1% SDS, 0.1 *M* Tris-HCl, pH 8.0. This provides a useful reference based on crude protein content. $OD_{280}$ concentration of the final membrane prep should be approx 50–100.
6. Prior to use, digest endogenous mRNA with micrococcal nuclease as described for RRL (*see* **Subheading 3.1., item 7**). Typically, 1 mL of microsomes are treated, frozen in liquid nitrogen as 50- to 100-μL aliquots, and stored at –80°C for up to 3 mo.
7. Test microsomes for translocation and glycosylation (*see* **Note 4**).

## 3.3. In Vitro Transcription/Translation

We routinely assemble reactions *de novo*, which allows us to control conditions for both synthesis and degradation. All reagents, including RRL and microsomal membranes, and containers used for transcription and translation must be kept RNase free. Transcription reactions can be made up fresh or stored at –80°C. The following conditions are used for SP6 RNA polymerase, which usually gives better results for translation than T7 RNA polymerase in our laboratory. Our vector (SPG4T) has been optimized for use in RRL by adding a 5' UTR from *Xenopus laevis* globin gene (TCTTGTTCTTTTTGCAGAAGCTCAGAATAAACGCTCAACTTT GGCAGATCTGAACATG) immediately upstream of the ATG start site *(24)*. DNA constructs are put into this vector after removing the endogenous 5' UTR to place the start codon as close as possible to the end of the *X. laevis* 5' UTR.

### 3.3.1. Transcription

1. Assemble a 20-µL transcription reaction, by combining the following reagents on ice: 4 µL of 5X transcription buffer, 2 µL of 10X NTPs, 2 µL of 0.1 *M* DTT, 2 µL of GpppG, 0.4 µL of tRNA, 0.8 µL of RNase inhibitor, 0.8 µL of SP6 polymerase, 4 µg of plasmid DNA, and 4 µL of $mH_2O$. For larger constructs such as CFTR (4400 bp) use 8 µg of plasmid DNA. RNase-free $H_2O$ is substituted for DNA in mock transcription reactions. Incubate for 1 h at 40°C (37°C for T7 polymerase) and transfer to ice.

### 3.3.2. Translation

1. Translation reactions are usually linked to transcription so that no purification or extraction of RNA is necessary. The composition of translation buffers thus takes into account the ionic contributions of the transcript. Assemble 100 µL of translation reaction by combining 5 µL of 20X translation buffer, 20 µL of 5X Emix, 40 µL of RRL (nucleased and treated with hemin), 1 µl RNase inhibitor, 1 µL of CK, and 20 µL of transcription mixture. The volume of microsomes is determined empirically based on translocation and glycosylation activity (*see* **Subheading 3.2., item 7**). Most translations are performed at 24°C for 1 h. For proteins >80 kD, ATA (an inhibitor of translation initiation) is added to increase the ratio of full-length protein to partially synthesized fragments. Add ATA typically after 15 min (0.05–0.1 m*M* final concentration) and continue incubating at 24°C until maximal expression is achieved (up to 2 h). Transfer reaction to ice (*see* **Note 5**).

### 3.3.3. Optimization

1. Certain components may need to be titrated for optimal translation. Typical concentrations for potassium and magnesium vary between 50–150 m*M* and 0.5–2.5 m*M*, respectively. High concentrations of microsomes inhibit translation, whereas low concentrations yield low translocation and glycosylation efficiencies. Optimal concentration of $Mg^{2+}$, $K^+$, and the timing of ATA addition should be determined empirically to maximize translation efficiency (*see* **Note 6**).

## 3.4. Using RRL for Reconstituting ERAD

### 3.4.1. Collection and Resuspension of Microsomal Membranes

Following translation, membrane proteins are isolated by pelleting microsomes through a sucrose cushion to remove unincorporated radiolabel. This isolation step should reduce background counts in mock translations to <20%. Isolated microsomes are then used to reconstitute degradation reactions.

1. Dilute translation mixture with two volumes of 0.1 *M* sucrose buffer and layer mixture onto three to four volumes of 0.5 *M* sucrose cushion. Centrifuge at 180,000*g* for 10 min. Aspirate the supernatant and resuspend the pellet in twice the original translation volume in 0.1 *M* sucrose buffer by physically dislodging the membrane pellet with a 200-μL pipet tip. Vigorously titurate the solution with the end of the pipette tip placed tightly against the bottom of the tube so that the buffer (and pellet) slowly squeezes through the opening. It is critical to resuspend the pellet uniformly. Recool the tube on ice if necessary (*see* **Note 7**).
2. Layer the resuspended pellet on 0.5 *M* sucrose cushion, and centrifuge at 180,000*g* for 10 min. The final resuspension of the membrane pellet is done in 1/2 the original translation volume in 0.1 *M* sucrose buffer.

### 3.4.2. In Vitro Degradation Assay

Substrate degradation can be monitored by two techniques: (1) direct visualization by SDS-polyacrylamide gel electrophoresis (SDS-PAGE) or (b) scintillation counting. The former will indicate the disappearance of full-length protein that occurs either by ubiquitination or cleavage, whereas the latter can be used to determine the rate and extent to which substrate is converted into TCA-soluble fragments (i.e., end products of proteasome-mediated degradation).

1. For a 100-μL reaction, combine on ice: 10 μL of 10X degradation buffer, 72 μL of untreated RRL (*see* **Subheading 3.1., step 6**), 2 μL of creatine kinase, and 18 μL of resuspended membranes. Use one reaction as a control and others to test various experimental conditions (e.g., depleted RRL [*see* **Subheading 3.5.2.**], proteasome inhibitors, and/or treated membranes [*see* **Subheading 3.6.**]). It is critical to set up a parallel degradation reaction containing membranes from a mock translation for each membrane treatment tested.
2. Prior to the start of the reaction ($T_0$), remove 3-μL aliquots from each reaction mixture and place one in 2.5 mL of scintillation fluid (duplicates are usually done for total counts), another in 300 μL of 20% TCA (on ice), and a third in 30 μL of SDS-LB (*see* **Note 8**).
3. Incubate at 37°C. At desired time points, remove 3-μL aliquots and place one in SDS-LB and the other into 300 μL of 20% TCA. Reactions usually proceed for 3–4 h or until maximum degradation is achieved.
4. Precipitate TCA samples on ice (minimum of 30 min) and centrifuge at 16,000*g* for 15 min (4°C). Transfer 250 μL of supernatant to 2.5 mL of scintillation fluid and count in a scintillation counter.
5. Samples collected into SDS-LB are incubated for 30 min at 37°C, centrifuged at 16,000*g* for 2 min, separated on polyacrylamide gels, and processed for autoradiography.

#### 3.4.2.1. Calculating Percentage of Protein Degraded

Determine total $^{35}S$ in the degradation reaction by counting 3 μL of reaction mixture directly in scintillation fluid. Because TCA supernatants contain 250/300 of the sample, counts are corrected by a factor of 1.2 to yield "corrected TCA soluble counts." Calculate the percent of protein degraded at each time point using the following formula:

$$\% \text{ degraded} = \{[(Tn - T_0) - (Mn - M_0)]/[(T - T_0) - (M - M_0)]\}*100$$

where

$Tn$ = corrected TCA soluble counts in samples taken at time point, n.

$T_0$ = corrected TCA soluble counts measured at time 0 (i.e., background TCA soluble counts).

$T$ = total counts in degradation reaction.

$Mn$ = corrected TCA soluble counts from mock samples taken time point, *n*.

$M_0$ = corrected TCA soluble counts from mock degradation at time 0 (i.e., background TCA soluble counts present in mock degradation).

$M$ = total counts in mock degradation reaction. (The factors $(Mn - M_0)$ and $(M - M_0)$ represent nonspecific counts). (*See* **Notes 9** and **10**.)

### 3.4.3. ER Release of Integral Membrane Proteins During ERAD

The membrane release assay is similar to the degradation assay, except that each sample is pelleted through a sucrose cushion, and the supernatant is analyzed for total counts and/or TCA-soluble counts released from the membranes. Protein fragments can be analyzed by SDS-PAGE as well. Without proteasome inhibitors, all released counts are usually TCA soluble, but TCA insoluble fragments are generated with progressive proteasome inhibition.

1. Before starting the degradation reaction ($T_0$), mix a 7-μL aliquot from each reaction with 13 μL of 0.1 *M* sucrose buffer. Layer the sample on 30 μL of 0.5 *M* sucrose cushion. Centrifuge at 180,000*g* for 10 min.
2. Remove the supernatant and aliquot 10 μL into 2.5 mL of scintillation cocktail, 20 μL into 20 μL of SDS-LB (*see* **Note 11**), and 10 μL into 300 μL of 20% TCA (*see* **Note 8**).
3. Keep TCA-precipitated samples on ice a minimum of 30 min and centrifuge at 16,000*g* for 15 min (4°C). Transfer 250 μL of supernatant to 2.5 mL of scintillation fluid and count in a scintillation counter.
4. Solubilize the membrane pellet in 50 μL of 1% SDS, 0.1 *M* Tris-HCl, pH 8, and add 20 μL to 20 μL of SDS-LB. (*See* **Note 11**.)
5. Incubate degradation reactions at 37°C for 3–4 h and process 7-μL samples as in **steps 1–4** at desired time points.

#### 3.4.3.1. Calculating ERAD-Mediated Substrate Release From Membranes

Count 3 μL of degradation mixture directly in scintillation fluid to determine total $^{35}S$ in the reaction. Total $^{35}S$ in the supernatant is determined by counting 10 μL of supernatant after pelleting 7 μL of degradation mixture. Because 10 μL of supernatant contains 1.4 μL of degradation mixture ((10 μL/50 μL)*7 μL), the supernatant counts are multiplied by a factor of 2.14, giving "corrected totaL counts in the supernatant." Calculate the percent substrate released into the supernatant at each time point using the following formula:

$$\% \text{ substrate released} = \{[(TSn - TS_0) - (MSn - MS_0)]/[(T - TS_0) - (M - MS_0)]\} * 100$$

where

$TSn$ = corrected total counts in supernatant at time point, *n*.

$TS_0$ = corrected total counts in supernatant at time 0 (i.e., background).

$T$ = total counts in degradation reaction.

*MSn* = corrected total counts in mock supernatant at time point, *n*. (mock degradation reaction).

$MS_0$ = corrected total counts in mock supernatant at time 0 (i.e., background in mock degradation reaction).

*M* = total counts in mock degradation reaction. [The factors ($MSn - MS_0$) and ($M - MS_0$) represent nonspecific counts.] (*See* **Note 9**.)

#### 3.4.3.2. Calculating TCA-Soluble Counts Released From Membranes

Correct the TCA-soluble counts from supernatant samples by a factor of 2.14 as in **Subheading 3.4.3.1**. An additional correction by a factor of 1.2 is made because only 250/300 of the sample in TCA supernatants is counted. Using the "corrected TCA-soluble counts," calculate the percent TCA-soluble fragments released into the supernatant using the following formula:

$$\%\ \text{TCA soluble fragments released} = \{[(An - A_0) - (Bn - B_0)]/TC\} * 100$$

where

*An* = corrected TCA-soluble counts in supernatant samples taken at time points, *n*.

$A_0$ = corrected TCA-soluble counts in supernatant samples at time 0 (i.e., background).

*Bn* = corrected TCA-soluble counts in mock supernatant samples taken time points, *n* (mock degradation).

$B_0$ = corrected TCA soluble counts in mock supernatant sample at time 0 (i.e., background in mock degradation). The ($Bn - B_0$) represents nonspecific counts.

*TC* = total insoluble counts in the reaction at $T = 0$. Calculated as in **Subheading 3.4.2.1.**: $(T - T_0) - (M - M_0)$. (*See* **Note 9**.)

### 3.5. Affinity Adsorption of RRL ERAD Components

RRL can be manipulated by affinity adsorption using immobilized recombinant proteins. One approach is to deplete components necessary for ERAD using ubiquitin-conjugating proteins as bait. For example, an N-terminal ubiquitin fusion of UbcH5a (Ub-UbcH5a) depleted >95% of Ufd1 from RRL in a single adsorption step **(Fig. 1A)**. His-tagged bait is first adsorbed onto Ni-NTA resin, which is then incubated in RRL to remove endogenous binding partners. Adsorbed proteins can be identified by elution and mass spectrometry. Using this approach, we found that RRL contained numerous Ub-UbcH5a binding proteins including Ufd2 (an E3 ubiquitin ligase), Isopeptidase T (a deubiquitinating enzyme), and p97 (an AAA-ATPase involved in ERAD) **(Fig. 1B)**. (*See* **Note 12**.)

#### 3.5.1. Purification of His-Tagged Recombinant Proteins

Our laboratory uses several different expression vectors that include the pET family (Novagen), pTrc (Invitrogen), and the pQE family (Qiagen) and which allow the inducible expression (with IPTG) of the recombinant protein in *E. coli*. They also contain either an N- or C-terminal His tag. A T7 promoter allows the protein to be expressed in RRL or other cell-free systems.

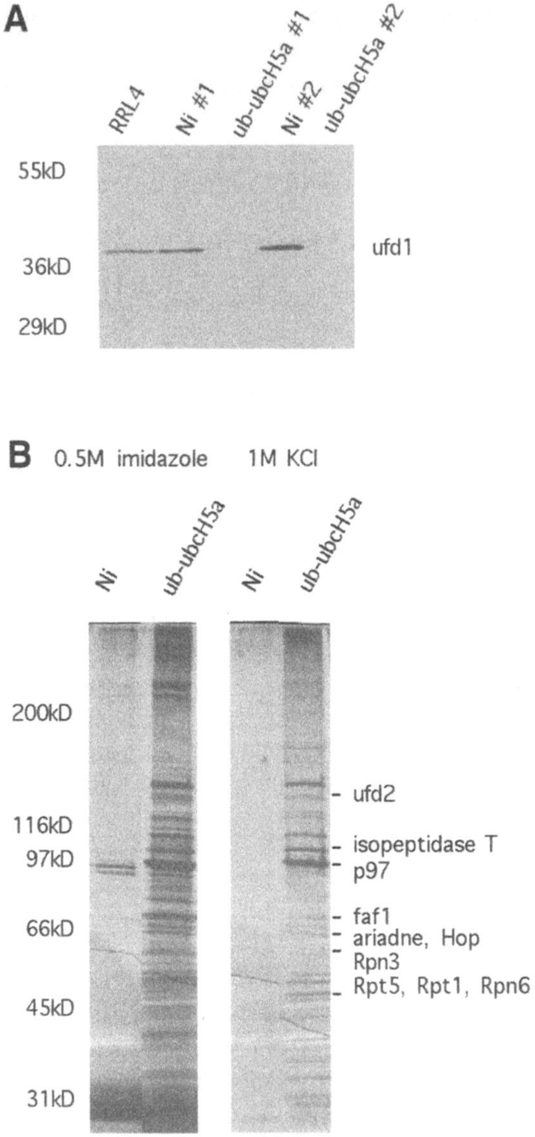


Fig. 1. Affinity depletion of RRL. RRL was depleted with Ub-UbcH5a or Ni-NTA beads (*see* **Subheading 3.5.2.**). **(A)** The depleted RRL was immunoblotted for ufd1. Essentially complete depletion of ufd1 was obtained from a single adsorption step. **(B)** The beads used to deplete RRL were washed and eluted with either 0.5 *M* imidazole or 1 *M* KCl. The proteins were separated by SDS-PAGE on a 7–12% gel and silver stained. Bands were excised, trypsinized, and identified by MS analysis.

1. Transform BL21(DE3) *E. coli* with the plasmid containing your construct by heat shock (42°C for 1 min) and plating on LB/Amp plates (37°C).
2. Pick one colony and inoculate a 40 mL overnight culture: 39 mL LB + ampicillin (200 μg/mL final). Incubate in a 37°C shaker.
3. The next day inoculate a 1-L culture containing approx 950 mL of LB, ampicillin (200 μg/mL), 10 mL of 20% glucose, and 20 mL of overnight culture. Incubate at 37°C with continuous shaking until the $OD_{600}$ is between 0.4 and 0.5. Induce expression of the recombinant protein by adding 800 μL of 1 *M* IPTG to the culture. Incubate at 37°C for 3 h, and immediately place on ice.
4. Pellet bacteria by centrifuging at 2400*g* for 10 min. Resuspend the pellets in approx 80 mL of 1X PBS and then centrifuge at 2000*g* for 30 min. Decant the supernatant and freeze the pellets at –20°C overnight.
5. Resuspend the pellets in lysis buffer (buffer D + 10% buffer E) and lyse using a French press (1250 psi; lyse twice). Pellet the membrane debris from the lysate by centrifuging at 19,000*g* for 40 min. Decant the cleared lysate into a 50-mL conical tube.
6. Load the lysate at a flow rate of 1 mL/min using either a peristaltic pump or FPLC onto a Ni-NTA column (5 mL beads in a 1 × 10 cm column) equilibrated in 10% buffer E. After loading, wash the column with 50 mL of 5% buffer E, followed by a 125-mL elution with a linear imidazole gradient (5% buffer E to 100% buffer E). Collect 25 × 5-mL fractions of the elution. Typically proteins elute between fractions 3 and 8. Analyze 10 μL of each fraction by SDS-PAGE followed by Coomassie blue staining. Pool the fractions containing your protein (typically five or six fractions or 25–30 mL) and dialyze overnight at 4°C in 4 L of dialysis buffer.
7. Concentrate the dialyzed material using Centriplus concentrators with an appropriate MWCO. Store at –80°C in aliquots of 10–500 μL. Quantitate the recombinant protein using a suitable assay (i.e., Bradford or Lowry). Typically, we obtain 10 mg/L of culture, but actual yield varies significantly depending on expression and purification efficiency. (*See* **Note 13.**)

### *3.5.2. RRL Affinity Depletion*

1. Wash 100 μL of Ni-NTA beads with 2 × 1 mL of TNB buffer. Add 10 mg of His-tagged protein and bring volume to 1 mL with TNB buffer. Leave the control beads in 1 mL of TNB buffer. Rotate the beads at 4°C for at least 30 min to bind the His-tagged protein. Then wash the beads with 6 × 1 mL TNB buffer. (*See* **Note 14.**)
2. After final wash remove as much buffer as possible from the beads, and add 500 μL of RRL. Rotate at 4°C at least 4 h. Centrifuge briefly (16,000*g* for ~10 s) and remove approx 400 μL of the affinity depleted RRL (be careful to avoid contaminating the depleted RRL with beads containing the bait protein). Store at –80°C in 100- to 200-μL aliquots. Efficiency of RRL depletion can be tested by immunoblotting and/or silver staining.
3. To identify bound proteins, remove the remaining supernatant and wash the beads 6–8 × 1 mL of wash buffer.
4. Elute the beads with 500 μL of elution buffer. We use two methods of elution. Imidazole elution buffer (**Subheading 2.5.2., item 7**) will elute virtually all bound RRL proteins and His-tagged bait. Because the excess His-tagged bait (and globin from RRL) often poses a problem in subsequent identification, we also elute with graded concentrations of high salt (0.5 *M*, 1 *M*, 1.5 *M*, and 2 *M* KCl or NaCl). 1 *M* salt is sufficient to disrupt most protein–protein interactions and release RRL proteins (**Fig. 1B**).

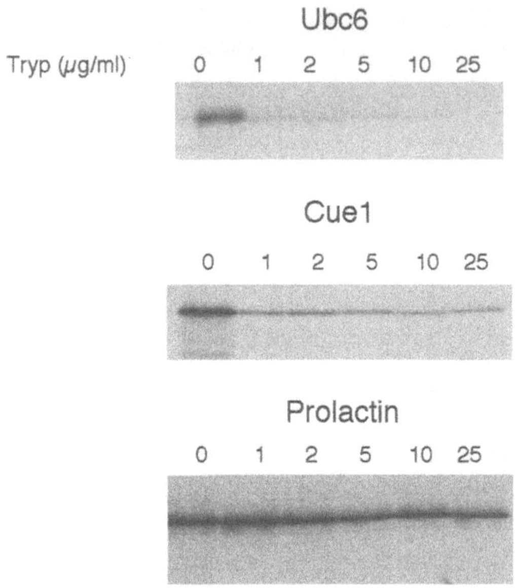


Fig. 2. Trypsinization of microsomal membranes. The indicated proteins were translated in vitro in the presence of ER microsomes (*see* **Subheading 3.3.2.**). The membranes were pelleted through a 0.5 *M* cushion and subjected to limited proteolysis with increasing concentrations of trypsin (*see* **Subheading 3.6.1.**). The membranes were then solubilized in SDS-LB and resolved by SDS-PAGE and autoradiography. Ubc6, an ER-localized E2 enzyme, is highly sensitive to trypsinization, even at 1 mg/mL of trypsin. Cue1 is slightly less sensitive, and the secretory protein prolactin is protected within the lumen of the ER.

5. To prepare eluates for analysis by mass spectrometry, the eluted proteins are concentrated by TCA precipitation (20% TCA, centrifuged at 16,000*g* for 10 min, washed with 50% acetone, and resuspended in SDS-LB, analyzed by SDS-PAGE, and silver stained (*see* **Note 15**). Selected protein bands are manually excised from the gel and trypsinized, and the resulting fragments are analyzed by LC–MS/MS or matrix-assisted laser desorption ionization/time-of-flight mass spectrometry (MALDI/TOF MS) and identified by mass fingerprinting. Specific conditions for sample preparation should be determined in conjunction with a suitable MS facility.

### 3.6. Biochemical Inactivation and Reconstitution of Translocation Activity

One of the strengths of the RRL system is that membrane components can also be manipulated biochemically. Here we describe two techniques, proteolysis and alkylation ***(25)***, that can be used for analyzing membrane components of the ERAD pathway. Proteolysis results in the cleavage of many peripheral and integral membrane proteins, and alkylation modifies accessible cysteines. For example, microsomes are rendered incompetent for translocation when SRP receptor α-subunit (SRα) is cleaved or alkylated ***(26)***. Up to 80% of translocation activity can be restored, however, by adding recombinant SRα to treated membranes. The treated membranes can then be tested for release and removal of other endogenous membrane proteins and used in degradation reactions. As shown in **Fig. 2**, the E2 enzyme, ubc6, is extremely sensitive to trypsinization with >90% removed by incubation in 1 μg/mL of trypsin, and >95% cleaved in 25 μg/mL of trypsin. In vitro expressed Cue1 is slightly less sensitive although the majority is also cleaved from membranes by 1 μg/mL of trypsin. The secretory control protein prolactin is essentially fully protected within the microsomal lumen.

### 3.6.1. Proteolysis of Microsomal Membranes

This procedure is performed on ice and in a cold room. It is necessary to empirically determine the optimal concentration of protease for a given microsomal membrane preparation. We have obtained similar results with both trypsin and proteinase K.

1. Dilute 100 μL of membranes in buffer F followed by the addition of 0, 1, 2, 5, 10, and 25 μL of 100 μg/mL trypsin for a total volume of 1 mL. Mix thoroughly and incubate on ice for 1 h.
2. Add 10 μL of 50 m*M* PMSF (0.5 m*M* final concentration) to each reaction and incubate on ice for 15 min to inactivate the protease.
3. Add 144 μL of 4*M* KoAc (0.5 *M* final concentration) to each reaction and then layer the mixture on 1.2 mL of PK cushion. Centrifuge for 10–15 min at 180,000*g*. Resuspend the membrane pellet in 200 μL of buffer C (*see* **Subheading 3.4.1., item 1** and **Note 7**).
4. Layer the resuspended pellet on 300 μL of 0.5 *M* sucrose cushion and centrifuge at 180,000*g* for 10–15 min.
5. Resuspend the pellets in 100 μL of buffer C. Freeze in liquid nitrogen and store at –80°C in 20- to 25-μL aliquots. SRα repopulation can also be done at this time (*see* **Subheading 3.6.3.**).
6. To determine the amount of protein remaining associated with the membranes, remeasure the $OD_{280}$ as described in **Subheading 3.2., item 5**.

#### 3.6.1.1. Alkylation of Microsomal Membranes

1. Layer 50 μL of membranes on 300 μL of 0.5 *M* sucrose cushion (–DTT) and centrifuge for 10 min at 180,000*g*. Resuspend the pellet in 50 μL of buffer C (–DTT).
2. Split the resuspended membranes into two 25-μL aliquots. To one add 1 μL of 50 m*M* NE*M* (2 m*M* final) and to the other add 1 μL of $ddH_2O$. Incubate at 24°C for 30 min.
3. Add 0.5 μL of 1 *M* DTT (20 m*M* final) to each tube to quench the reaction. Freeze in liquid nitrogen and store at –80°C in 10- to 25-μL aliquots. SRα repopulation can be done also be done at this time (*see* **Subheading 3.6.3.**). The final step of the repopulation protocol is pelleting the membranes through a 0.5 *M* sucrose cushion and resuspending in buffer C, so the high concentration of DTT is not maintained. Also, it has been shown that excess DTT does not affect translocation *(25)*.

#### 3.6.1.2. SR A Repopulation of Treated Microsomal Membranes

Both membrane treatments described disrupt ER targeting due to cleavage or modification of SRα *(27)*. By adding recombinant SRα to the treated membranes, this activity can be restored **(Fig. 3)**. We use a soluble His-tagged fragment missing the N-terminal 152 amino acids that is equivalent to the fragment produced by trypsinization *(26)*. The concentration of SRα needed to repopulate a particular membrane prepa-

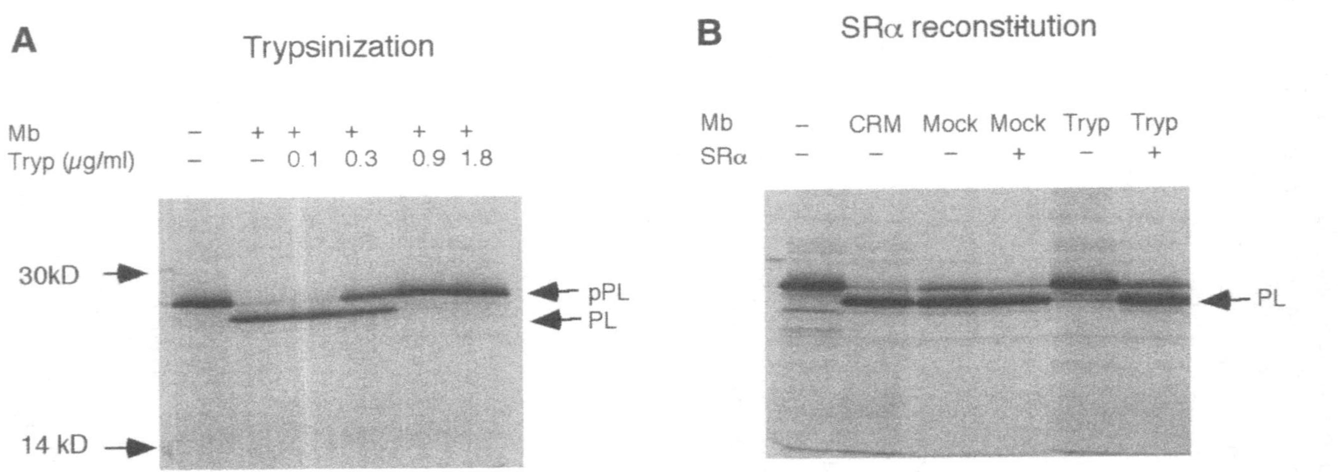


Fig. 3. Reconstitution of SRP dependent translocation in proteolyzed microsomes. **(A)** Microsomes were trypsinized (*see* **Subheading 3.6.1.**) and used for in vitro translations of preprolactin. Signal sequence cleavage no longer occurs at 0.9 μg/mL of trypsin, indicating microsomes are incompetent for translocation. **(B)** Adding recombinant SRα to trypsinized microsomes (*see* **Subheading 3.6.3.**) restores translocation activity as demonstrated by signal sequence cleavage of preprolactin.

ration must be determined empirically as excess SRα will inhibit translation. This can be tested by repopulating mock treated membranes with recombinant SRα.

1. To repopulate 20 μL of membranes add 1 μg of recombinant SRα/μL of membranes. Incubate on ice for 30–60 min.
2. Bring the volume up to 100 μL with buffer C, and layer the mixture on 300 μL of 0.5 *M* sucrose cushion. Centrifuge at 180,000*g* for 10 min to remove unbound SRα.
3. Resuspend the pellet in 20 μL of buffer C. Freeze in liquid nitrogen and store at –80°C.

## 4. Notes

1. Repeated freeze–thaw cycles will decrease RRL activity. Three or four cycles are generally well tolerated as long as samples are thawed quickly, and refrozen in liquid nitrogen. The quality of RRL may vary between rabbits due to slight variations in cytosolic factors such as endogenous hemin, and age and maturity of the reticulocytes. Therefore we process lysate from each rabbit separately. The most active preparations for translation and/or degradation are stored and used appropriately. Lower yields can result from failure to completely exanguinate the animal or inefficient lysis of reticulocytes. The latter problem is usually due to inadequate shaking of the lysate and will be manifest by a very large pellet in the second 15,000*g* centrifugation in the RRL protocol.
2. Most commercial RRL preparations are designed for performing translations and, therefore, contain added hemin (Promega TnT system and nucleased RRL). These preparations are incompatible with proteasome degradation assays because hemin inhibits ATPase activity of the 19S subunit ***(28)***. Promega's "untreated" RRL appears to contain no hemin. However, different lots may need to be tested to find one with optimal activity. In the presence of hemin, degradation is inhibited but ubiquitination is not, so degradation substrates usually accumulate as ubiquitinated intermediates. It should be noted that unsupplemented RRL may also contain some endogenous free hemin. This may account for differences in degradation efficiency observed for different rabbits. The best RRL preps in our hands will convert 60–70+% of in vitro synthesized CFTR into TCA soluble fragments in 4 h at 37°C.
3. Traditionally canine pancreatic microsomes have been used to study in vitro translation of secreted and transmembrane proteins. However, preparations can also be made from pig or sheep pancreas by following essentially the same procedure ***(29)***. Proteolysis or autolysis of the pancreas can be a significant problem so the pancreas should be removed as quickly as possible following euthanasia.
4. Like RRL, canine pancreas membrane preparations vary widely in their efficiency of translocation, and it may take several attempts to achieve a satisfactory preparation. A suitable preparation, however, generates 10–30 mL of microsomes and is sufficient for thousands of typical translation and/or degradation reactions. Optimal preparations yield ≥90% translocation and approx 80% core glycosylation efficiency at a final $OD_{280}$ concentration of 4–10 in the translation reaction.
5. Ubiquitination of membrane proteins generally does not occur at 24°C (temperature used for translation). However, ubiquitin ladders rapidly form cotranslationally when polytopic membrane proteins are translated at a higher temperature or in the absence of microsomes (our unpublished observations). For example, CFTR translated without microsomes is cotranslationally incorporated into a covalent high molecular weight complex, and aquaporin translated in the absence of microsomes forms clear ubiquitin ladders. Cotranslational ubiquitination of CFTR has been reported to occur at 30°C but was observed for only a small fraction of total protein ***(17)***.

6. A common problem in translating large proteins (e.g., CFTR) is the generation of incomplete translation products. This can occur if (a) insufficient ATA is added, (b) ATA is added too late, or (c) transcript concentration is too high. In addition to inhibiting translation initiation, ATA has been shown to inhibit the ATPase activity of the 26S proteasome ***(28)***. Additional articles by Jackson and Hunt (22) and Promega's technical manual part TM232 (can be downloaded from their website) contain further information concerning the optimization of RRL translation.
7. Uniform resuspension of microsomes is necessary to avoid variation in experimental results. This is one of the most difficult steps in the degradation procedure and may require significant practice. When resuspension is complete, the solution should appear faintly opalescent and there should be no particulate membrane clumps visible. If foaming of the solution occurs, transfer the liquid portion to a fresh tube. If the resuspended microsomes have significantly less radiolabeled protein than the original translation (as determined by SDS-PAGE autoradiography), the following should be considered. (a) A significant fraction of the microsomal pellet may have been lost during resuspension. (b) Protein was synthesized but did not insert into the membranes. The latter can occur if microsomes are suboptimal or if the translation reaction was not mixed sufficiently.
8. Because RRL is so concentrated, the protein will rapidly precipitate in 20% TCA. Therefore, expel the sample from the pipette with the tip pressed against the bottom of the tube, break up large pieces of precipitate with the tip, and vortex immediately after addition of sample to TCA to assure adequate dispersion of TCA-soluble peptide fragments. Periodically vortex-mixing the samples during the 30-min incubation on ice is also recommended.
9. Radioactivity released from membranes in the absence of ATP results from nonspecific association of Trans35S-Label to microsome components and a small amount of background translation in RRL. The relative contribution of ATP-independent release is assessed by comparing standard (containing transcript) and "mock" translations. Nonspecific counts also vary depending on the membrane prep (from 15% to 60%), and several preps may be evaluated before finding one with a low background. Nucleasing microsomes for up to 20 min may help reduce the background without affecting activity. Also, EDTA stripping of microsomes ***(23)*** may reduce background. Incubate microsomes in 25 m*M* EDTA for 15 min on ice, pellet through a 0.5 *M* sucrose cushion, and resuspend in buffer C. When analyzed under reducing conditions by SDS-PAGE, microsomes from mock translations appear largely devoid of radiolabeled protein. Release of nonspecific counts into the TCA-soluble fraction is also independent of ATP and RRL, and is likely attributable to reversible association of components in Trans35S-Label with components of microsomal membranes. Release of TCA-soluble radioactivity from mock reactions is thus equivalent to the ATP-independent release of radioactivity from standard translations. In other words, nonspecific release is not related to substrate degradation, and can be subtracted from each corresponding sample.
10. Under optimal conditions, 50–70% of total CFTR is converted to TCA-soluble fragments in 4 h. Factors affecting degradation include (a) the RRL prep (does it contain hemin), (b) the membrane prep, (c) ATP-regenerating system, and (d) translation efficiency. RRL degradation activity can also be easily assessed using commercially available radiolabeled substrates such as [$^{14}$C]lysozyme ***(16)***.
11. Release of proteins from the ER membrane can also be followed by SDS-PAGE. However, unless a proteasome inhibitor such as MG132 is being used the fragments in the supernatant will be too small to visualize.

12. For RRL depletion, the smallest bead volume possible should be used because degradation activity decreases significantly with dilution. We use a 5:1 (RRL/bead volume) ratio as a starting point. This causes less <20% loss of degradation for the substrate CFTR. Control RRL is always "depleted" in parallel using empty Ni-NTA beads. Depleted RRL can then be compared with a mock depletion in standard degradation assays (*see* **Subheading 3.4.2.**) to determine if depleted factors are involved in substrate degradation.
13. We typically purify only soluble proteins because of the likelihood of forming inclusion bodies with membrane proteins. Removing transmembrane segments from single spanning membrane proteins (e.g., ubc6) has also worked. Qiagen's manual, the QIAexpressionist, and references therein provide a good source of information on solubilizing inclusion bodies, various buffer conditions, and so forth.
14. It may be necessary to titrate the amount of bait protein to load on the Ni-NTA beads. Qiagen indicates that the beads have a 10 mg/mL binding capacity for a 20-kDa protein.
15. Not all silver staining protocols are compatible with MS. Typically, glutaraldehyde is used during the sensitization step, and formaldehyde is a component of the silver impregnation buffer. However, if MS analysis is desired the protocol used for staining must be modified by omitting these reagents at the indicated steps. In addition, 0.05% formaldehyde is used during the development step ***(30)***.

## References


1. Brodsky, J. and McCracken, A. (1997) ER-associated and proteasome-mediated protein degradation: how two topologically restricted events came together. *Trends Cell Biol.* **7,** 151–155.
2. Kostova, Z. and Wolf, D. H. (2003) For whom the bell tolls: protein quality control of the endoplasmic reticulum and the ubiquitin-proteasome connection. *EMBO J.* **22,** 2309–2317.
3. Sommer, T. and Wolf, D. (1997) Endoplasmic reticulum: reverse protein flow of no return. *FASEB J.* **11,** 1227–1233.
4. Kopito, R. (1997) ER quality control: the cytoplasmic connection. *Cell* **88,** 427-430.
5. Hampton, R. Y. (2002) ER-associated degradation in protein quality control and cellular regulation. *Curr. Opin. Cell Biol.* **14,** 476–482.
6. Lippincott-Schwartz, J., Bonifacino, J., Yuan, L., and Klausner, R. (1988) Degradation from the endoplasmic reticulum: disposing of newly synthesized proteins. *Cell* **54,** 209–220.
7. McCracken, A. and Brodsky, J. (2003) Evolving questions and paradigm shifts in endoplasmic-reticulum-associated degradation (ERAD) *BioEssays* **25,** 868–877.
8. Wilson, C., Farmery, M., and Bulleid, N. (2000) Pivotal role of calnexin and mannose trimming in regulating the endoplasmic reticulum-associated degradation of the major histocompatibility complex class I heavy chain. *J. Biol. Chem.* **275,** 21224–21232.
9. Taxis, C., Hitt, R., Park, S. H., Deak, P. M., Kostova, Z., and Wolf, D. H. (2003) Use of modular substrates demonstrates mechanistic diversity and reveals differences in chaperone requirement of ERAD. *J. Biol. Chem.* **278,** 35903–35913.
10. Horwich, A., Weber-Ban, E., and Finley, D. (1999) Chaperone rings in protein folding and degradation. *Proc. Natl. Acad. Sci. USA* **96,** 11033–11040.
11. Plemper, R. and Wolf, D. (1999) Retrograde protein translocation: ERADication of secretory proteins in health and disease. *Trends Biol. Sci.* **24,** 266–270.
12. Jarosch, E., Taxis, C., Volkwein, C., et al. (2002) Protein dislocation from the ER requires polyubiquitination and the AAA-ATPase Cdc48. *Nat. Cell Biol.* **4,** 134–139.

13. Hershko, A. and Ciechanover, A. (1998) The ubiquitin system. *Annu. Rev. Biochem.* **67,** 425–479.
14. Gusarova, V., Caplan, A. J., Brodsky, J. L., and Fisher, E. A. (2001) Apoprotein B degradation is promoted by the molecular chaperones hsp90 and hsp70. *J. Biol. Chem.* **276,** 24891–24900.
15. Qu, D., Teckman, J., Omura, S., and Perlmutter, D. (1996) Degradation of a mutant secretory protein, alpha1-antitrypsin Z, in the endoplasmic reticulum requires proteasome activity. *J. Biol. Chem.* **271,** 22971–22975.
16. Xiong, X., Chong, E., and Skach, W. (1999) Evidence that endoplasmic reticulum (ER)-associated degradation of cystic fibrosis transmembrane conductance regulator is linked to retrograde translocation from the ER membrane. *J. Biol. Chem.* **274,** 2616–2624.
17. Sato, S., Ward, C., and Kopito, R. (1998) Cotranslational ubiquitination of cystic fibrosis transmembrane conductance regulator *in vitro. J. Biol. Chem.* **273,** 7189–7192.
18. Wilson, R., Allen, A. J., Oliver, J., Brookman, J. L., High, S., and Bulleid, N. J. (1995) The translocation, folding, assembly, and redox-dependent degradation of secretory and membrane proteins in semi-permeabilized cells. *Biochem. J.* **307,** 679–687.
19. Lu, Y., Turnbull, I., Bragin, A., Carveth, K., Verkman, A., and Skach, W. (2000) Reorientation of Aquaporin-1 topology during maturation in the endoplasmic reticulum. *Mol. Biol. Cell* **11,** 2973–2985.
20. Kobilka, B. (1990) The role of cytosolic and membrane factors in processing of the human beta-2 adrenergic receptor following translocation and glycosylation in a cell free system. *J. Biol. Chem.* **265,** 7610–7618.
21. Brodsky, J. L., Hamamoto, S., Feldheim, D., and Schekman, R. (1993) Reconstitution of protein translocation from solubilized yeast membranes reveals topologically distinct roles for BiP and cytosolic Hsc70. *J. Cell Biol.* **120,** 95–102.
22. Jackson, R. and Hunt, T. (1983) Preparation and use of nuclease-treated rabbit reticulocyte lysates for the translation of eukaryotic messenger RNA. *Methods Enzymol.* **96,** 50–74.
23. Walter, P. and Blobel, G. (1983) Preparation of microsomal membranes for cotranslational protein translocation. *Methods Enzymol.* **96,** 84–93.
24. Melton, D. A., Krieg, P. A., Bebagliati, M. R., Maniatis, T., Zinn, K., and Green, M. R. (1984) Efficient in vitro synthesis of biologically active RNA and RNA hybridization probes from plasmids containing a bacterial SP6 promoter. *Nucl. Acids Res.* **12,** 7035–7056.
25. Gilmore, R., Blobel, G., and Walter, P. (1982) Protein translocation across the endoplasmic reticulum. I Detection in the microsomal membrane of a receptor for the signal recognition particle. *J. Cell Biol.* **95,** 463–469.
26. Andrews, D. W., Lauffer, L., Walter, P., and Lingappa, V. R. (1989) Evidence for a two-step mechanism involved in assembly of functional signal recognition particle receptor. *J. Cell Biol.* 1**08,** 797–810.
27. Walter, P. and Blobel, G. (1981) Translocation of proteins across the endoplasmic reticulum II. Signal recognition protein (SRP) mediates the selective binding to microsomal membranes of *in-vitro* assembled polysomes synthesizing secretory protein. *J. Cell Biol.* **91,** 551–556.
28. Hoffman, L. and Rechsteiner, M. (1996) Nucleotidase activities of the 26 S proteasome and its regulatory complex. *J. Biol. Chem.* **271,** 32538–32545.

29. Kaderbhai, M., Harding, V., Karim, A., Austen, B., and Kaderbhai, N. (1995) Sheep pancreatic microsomes as an alternative to the dog source for studying protein translocation. *Biochem. J.* **15,** 57–61.
30. Yan, J. X., Wait, R., Berkelman, T., et al. (2000) A modified silver staining protocol for visualization of proteins compatible with matrix-assisted laser desorption/ionization and electrospray ionization-mass spectrometry. *Electrophoresis* **21,** 3666–3672.

# 13

# Deubiquitinating Enzyme Purification, Assay Inhibitors, and Characterization

**Nathaniel S. Russell and Keith D. Wilkinson**

**Summary**

Despite the identification of numerous deubiquitinating enzymes (DUBs) in recent years, the large majority of this class of enzymes has not been well characterized. This chapter describes biochemical methods that can be used to characterize the function and substrate specificity of DUBs. Methods described will include: fluorescence assay using ubiquitin–amidomethylcoumarin (AMC); a high-performance liquid chromatography assay using ubiquitin ethyl ester or ubiquitin fusion peptides as model substrates to monitor DUB activity; and the purification of a recombinant human DUB, isopeptidase T, in *E. coli* using low-temperature expression as well as ion-exchange and affinity chromatography.

**Key Words:** Deubiquitinating enzymes; isopeptidase T; purification; ubiquitin; ubiquitin–amidomethylcoumarin.

## 1. Introduction

Deubiquitinating enzymes (DUBs) are a heterogeneous class of cysteine proteases involved in the regulation of ubiquitin attachment to various substrates and regenerating free ubiquitin in the cell *(1)*. The DUBs also consist of enzymes that act on ubiquitin-like proteins such as Nedd8, SUMO, and ISG15 *(2–4)*. There are six classes of DUBs that have been identified to date. They include the ubiquitin C-terminal hydrolases (UCHs) that remove small peptides from the C-terminus of ubiquitin; ubiquitin-specific processing proteases (UBPs) that can act on monoubiquitin and polyubiquitin chains; and the ubiquitin-like proteases (ULPs) that act on SUMO and Nedd8 *(5,6)*; metalloprotease DUBs such as the JAMM isopeptidase present in the COP9/signalosome and lid of the proteasome; OTU (ovarian tumor) domain DUBs that may act on polyubiquitin; and the CYLD tumor suppressor protein responsible for cylindromatosis *(7–11)*. In light of these recent discoveries, it is likely that other classes of DUBs have yet to be elucidated.

From: *Methods in Molecular Biology, vol. 301, Ubiquitin–Proteasome Protocols*
Edited by: C. Patterson and D. M. Cyr © Humana Press Inc., Totowa, NJ

Despite the dozens of DUBs now known, very few DUBs have been carefully analyzed with respect to their specific substrates or the rates at which they cleave these substrates. By definition, DUBs cleave the ubiquitin domain at the C-terminal glycine. With a few exceptions, DUBs cleave their substrates using a catalytic triad consisting of an active site cysteine that attacks the carbon atom of the scissile bond, a histidine residue that increases the nucleophilicity of the cysteine, and an aspartate residue hydrogen bonded to the histidine ***(12)***. Mutation of these residues in a DUB will inactivate the enzyme ***(13)***.

Theoretically, DUB enzymes can act on a number of substrates, including ubiquitin or other ubiquitin-like proteins such as Nedd8 and SUMO. Ubiquitin substrates include monoubiquitin or polyubiquitin conjugated to a protein, as well as free polyubiquitin ***(1)***. These polyubiquitinated substrates can contain ubiquitin linked through different lysine residues (i.e., the well-known K48 linked chains, as well as K11, K29, and K63 linked chains). Different polyubiquitin chains are expected to have different structures so different DUBs would be expected to be necessary to cleave different types of chains. Although a number of ubiquitin binding domains have been identified in DUBs (UIM and UBA motifs), it is still not clear how DUBs can discriminate between different forms of ubiquitin in the cell ***(14)***. For most DUBs, it is not known if they act on free polyubiquitin, mono or polyubiquitin conjugated to protein, or if they cleave a polyUb chain linked through a particular lysine residue other than K48.

The limits of the knowledge of DUB function and specificity are exemplified in the Ubp class of DUBs identified in *Saccharomyces cerevisiae*. Currently, 16 Ubps have been identified in yeast although little is known about the function of these DUBs beyond their identification as such based on domain homology ***(15)***. Our laboratory and others have characterized Ubp6 and Ubp14, but these particular Ubps were investigated only because the yeast strains containing deletions of these Ubps had distinct phenotypes that could be analyzed ***(16–18)***. Overall, genetic analysis in yeast has not been particularly useful, as many single Ubp deletions in yeast have no major discernible defects. Double and triple deletions of Ubps had few additional phenotypic consequences compared to single deletions ***(15)***. The more subtle phenotypes that may be caused by the lack of Ubps and other DUBs can only be analyzed if one knows what type of phenotype to look for. An alternative approach relies on biochemistry to purify and functionally characterize the DUB of interest.

The characterization of isopeptidase T and its yeast homolog, Ubp14, illustrates the use of biochemical analysis to characterize DUBs. Isopeptidase T was originally identified in a biochemical screen for proteins that bound Ub–Sepharose resin ***(19)***. It was then identified as a DUB by its ability to break down ubiquitin conjugates. Purification of human isopeptidase T and further biochemical analysis determined that its preferred substrate was free branched polyubiquitin chains and that cleavage required the presence of the C-terminal diglycine motif in the ubiquitin molecule ***(20)***. This suggested a role for the enzyme in disassembling polyubiquitin chains. Genetic and biochemical analysis of the yeast homolog confirmed that role and demonstrated that the human and yeast versions were functional homologs ***(17)***. It should be noted that the Δubp14 phenotype was missed in the original screen and only after the biochemical analysis was reported did it become apparent what it was necessary to look for.

A number of biochemical techniques have been developed to identify and functionally characterize other DUBs. Because DUBs always cleave at the C-terminus of ubiquitin, the Varshavsky and Chung groups have made use of ubiquitin fusion proteins to identify DUBs. Chung et al. have made use of the Ub–PEST fusion protein to identify and characterize numerous DUB activities ***(21)***. In this system, the Ub–PEST fusion is radiolabeled with $^{125}I$ and incubated with a lysate or extract expected to have DUB activity. The reaction is then precipitated with trichloroacetic acid (TCA) and the supernatant assayed for radioactivity. The cleaved PEST sequence is TCA soluble, so any DUB activity can be measured by the amount of radioactivity in the soluble fraction.

Varshavsky has identified a number of DUBs through the ingenious use of yeast genomic libraries, N-terminal ubiquitin fusion proteins, and *E. coli* ***(22,23)***. Bacterial strains expressing an Ub-β-galactosidase fusion protein are transformed with the genomic library and plated on X-Gal plates. Colonies that have DUB activity cleave the Ub from the β-gal causing the protein to be short lived. Thus, library inserts coding for active DUBs yield a white colony instead of the expected blue. As *E. coli* have no intrinsic DUB activity, only a protein expressed from a transformed plasmid can have this effect. These plasmids can then be cloned and sequenced for identification of the responsible DUB.

Identified DUBs are often further characterized by sodium dodecyl sulfate-polyacrylamide gel electrophoresis (SDS-PAGE) assays monitoring the deubiquitination of various polyubiquitin or ubiquitinated substrates. The utility of these techniques in DUB identification and characterization is well established, but further characterization of DUB activity often requires purification of the DUB. There are three main ways to measure purified DUB activity: gel assays, HPLC assays, and fluorescence assays.

The focus in this chapter is on techniques applicable to our DUB studies: assays using fluorescent substrates to characterize the substrate specificity of DUBs; high-performance liquid chromatography (HPLC) assays to monitor their enzymatic activity; and biochemical purification of DUBs. The initial part of this chapter will describe fluorescence assays that have been developed using Ub–amidomethylcoumarin (AMC) as a substrate for DUB activity. These assays allow testing of the substrate specificity of DUBs, especially in conjunction with inhibitors such as Ub–aldehyde or Ub–vinyl sulfone. Quantifying the rate of release of the fluorescent tag from the substrate using a luminescence spectrometer allows calculation of the amount of DUB enzymatic activity. The use of inhibitors can allow calculations of $K_i$ as well, thereby revealing the binding constants of ubiquitin substrates and inhibitors. The advantages of these assays are the small amounts of DUB or DUB containing lysate required, the strong signal obtained from the AMC fluorophore, and the rapid and specific inhibition by the aldehyde and vinyl sulfone inhibitors. Although not yet commercially available, Nedd8–AMC or SUMO–AMC could be utilized instead to characterize DUBs whose main substrates are not ubiquitin ***(24,25)***. Second, we describe the use of an HPLC assay using ubiquitin ethyl ester or ubiquitin peptide fusions as substrates to monitor DUB activity during purification, characterization of a purified DUB's activity, or measuring DUB activity in crude lysates. The final technique, biochemical purification of DUBs, involves the recombinant expression and purification of DUB proteins from *E. coli* using ion-exchange and affinity chromatography. Our laboratory has purified a number of DUBs from the UCH and UBP DUB families using these techniques ***(13,20)***. As a general example of DUB purification, we will describe the purification of the recombinant human DUB, Isopeptidase T (or USP5), expressed in *E. coli.*

## 2. Materials

### 2.1. Fluorescence Assays

1. Ubiquitin–AMC (available from Boston Biochem) or other appropriate substrate.
2. Luminescence spectrometer.
3. Water circulator.
4. Temperature probe.
5. 200-μL quartz cuvets.
6. 1X Assay buffer: 50 m*M* Tris-HCl, pH 7.5, 1 m*M* dithiothreitol (DTT), 10 μg/mL of ovalbumin, purified DUB as positive control (ex. we use UCH-L3 or isopeptidase T, both of which are commercially available from Boston Biochem).
7. Putative DUB or cell lysate with DUB activity.
8. Irreversible inhibitor such as ubiquitin aldehyde (Boston Biochem) or ubiquitin vinyl sulfone

### 2.2. HPLC Assays

1. HPLC.
2. C8 Alltima 5-μm HPLC column (Alltech, 88076).
3. HPLC buffer A: 25 m*M* sodium perchlorate, 4% (v/v) of 70% perchloric acid.
4. HPLC buffer B: same as A except 75% acetonitrile (v/v).
5. Ubiquitin ethyl ester (1 mg/mL).
6. Master mix for Ub–ethyl ester assay: 250 m*M* Tris-HCl, pH 7.5, 12.5 m*M* $MgCl_2$, 7.5 m*M* DTT.
7. Sample containing DUB activity to be analyzed (lysate or pure protein).

### 2.3. Isopeptidase T Purification

1. pRSIsoT or similar expression vector to express Isopeptidase T from the lac promoter.
2. BL21-DE3 competent cells (Invitrogen).
3. Ampicillin (100 mg/mL stock solution).
4. LB-ampicillin plates (ampicillin concentration of 100 μg/mL).
5. LB media: 10 g of tryptone, 5 g of yeast extract, 7.5 g of NaCl /L of medium.
6. Isopropyl thiogalactopyranoside (IPTG).
7. Incubator/shaker, temperature adjustable.
8. Cell lysis buffer: 50 m*M* Tris-HCl, pH 8.0, 25 m*M* EDTA, 10 m*M* β-mercaptoethanol + protease cocktail with final concentrations 100 μ*M* benzamidine, 0.5 μg/μL of leupeptin, 1 μg/mL of pepstatin A, 1 μg/mL of chymostatin, 2 μg/mL of antipain A, and 2 μg/mL of aprotinin.
9. Lysozyme.
10. Sonicator.
11. Fast Flow Q Resin (Amersham).
12. Fast protein liquid chromatography (FPLC).

13. FPLC equilibration buffer: 20 m*M* Tris-HCl, pH 7.6, 10 m*M* β-mercaptoethanol, 5 m*M* EDTA.
14. FPLC elution buffer: FPLC equilibration buffer + 350 m*M* NaCl.
15. HPLC.
16. C8 reversed-phase HPLC column.
17. HPLC buffer A: 25 m*M* sodium perchlorate, 4% (v/v) of 70% perchloric acid.
18. HPLC buffer B: same as A except 75% acetonitrile (v/v).
19. Ubiquitin ethyl ester.
20. Master Mix for Ub–ethyl ester assay: 250 m*M* Tris-HCl, pH 7.5, 12.5 m*M* $MgCl_2$, 7.5 m*M* DTT.
21. Amicon ultrafiltration cell + YM-30 filter.
22. Ubiquitin–Sepharose (12 mg of Ub/mL resin).
23. Ub–Sepharose equilibration buffer: 0.5*M* KCl, 50 m*M* Tris-HCl, pH 7.6, 2 m*M* DTT.
24. Ub–Sepharose elution buffer: 25 m*M* ethanolamine, 10 m*M* DTT, pH 9.4.
25. Sephacryl S-200 resin (Amersham).

## 3. Methods

### 3.1. Fluorescence Assay Protocol

1. Turn luminescence spectrometer on to warm up and set temperature of water circulator to 37°C.
2. Set spectrometer parameters to a bandpass of 4 n*M* for excitation and emission, an excitation wavelength of 340 nm, and an emission wavelength of 440 nm (*see* **Note 1**).
3. Prepare a reference sample in a 200-μL quartz cuvet by adding 120 μL of 40 n*M* AMC in 1X assay buffer and place in spectrometer. Allow to thermoequilibrate for 2 min in spectrometer. Adjust slits and voltage to obtain full-scale fluorescence (*see* **Note 2**).
4. Now use 40 n*M* Ub–AMC as a reference sample and run a time trace. Set range to seconds and run time trace for 500 s. (This is the baseline for further assays.) (*See* **Note 3.**)
5. Repeat **step 4** in the presence of the positive control DUB, putative DUB, or cell lysate being tested. The cuvet containing only 40 n*M* Ub–AMC is monitored for 60 s or until baseline fluorescence is obtained. The DUB of interest is added to the cuvet, which is mixed six times by either pipetting or shaking before it is returned to the spectrometer at 100 s (*see* **Note 4**).
6. Test inhibition of the DUB activity by repeating the assay as described in **step 5,** except wait until 100 s after addition of DUB to cuvet or until cleavage of Ub–AMC is in the linear phase. Remove the cuvet from spectrometer, add inhibitor (Ub–aldehyde or ubiquitin vinyl sulfone) in an approximately equimolar amount relative to Ub–AMC substrate, mix six times, and return to spectrometer 40 s after removal. (*See* **Fig. 1** for examples of baseline fluorescence, positive control, and DUB inhibition.)

Analyze data using software packaged with spectrometer or any other appropriate software package that can properly analyze kinetic data. Absolute rates of enzymatic activity can be calculated from the percent of substrate cleaved relative to full-scale AMC reference, and concentration of substrate.

### 3.2. HPLC Assay for Isopeptidase T Activity Using Ubiquitin Ethyl Ester

This assay is very useful for analyzing DUB activity from DUBs in purified form or in bacterial lysates. As bacteria do not have DUBs or an ubiquitin–proteasome system, all activity from this assay is from the recombinant DUB of interest. This is advantageous as DUB activity can be followed from postsonication centrifugation to the final

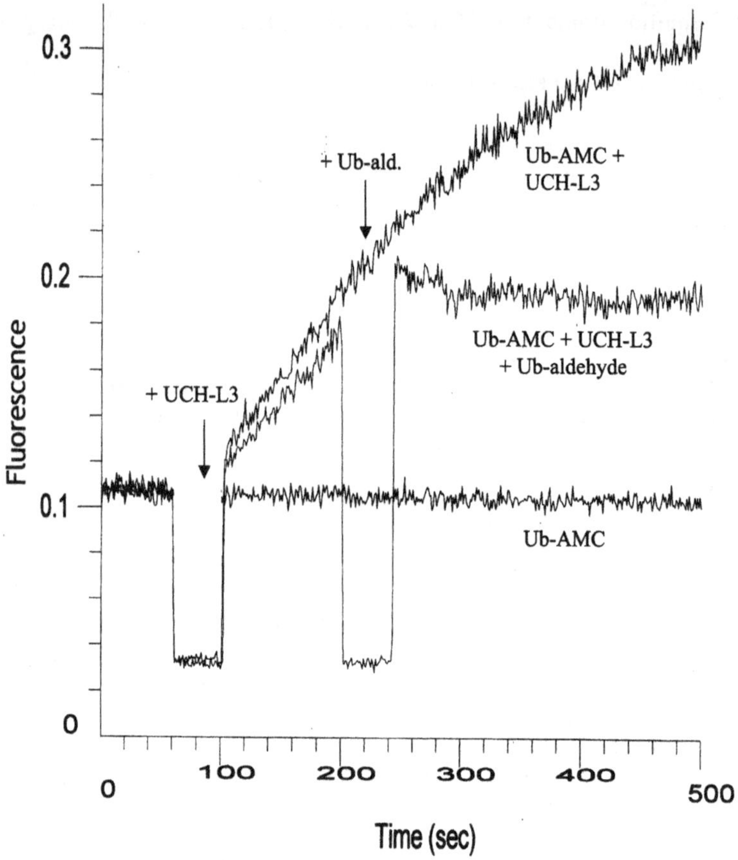


Fig. 1. Fluorescence assays using Ub–AMC as a substrate. Three time trace assays are shown on this plot. Background fluorescence is measured using 40 n*M* Ub–AMC in the absence of enzyme. A $1{:}2 \times 10^4$ dilution of 1 mg/mL of UCH-L3 was used to obtain cleavage of the substrate. Inhibition of substrate cleavage was achieved by addition of Ub–aldehyde to a final concentration equivalent to concentration of substrate.

step in a DUB bacterial expression and purification scheme. As the assay is rapid (20–25 min per sample), it is preferable over a gel based assay especially during protein purification when time is essential. This assay is flexible as it can be applied to any protein or lysate with DUB activity (*see* **Fig. 2**). Various substrates can be substituted for ubiquitin ethyl ester. Substrates our laboratory has used with success include ubiquitin with an extra amino acid at the C-terminus, a short 10-amino-acid peptide, or even Ub-AMC ***(26)***. These have been helpful in characterizing DUBs such as UCH-L1 and UCH-L3 and are especially useful if fluorescent substrates are unavailable.

1. Set up a 10-μL reaction in a 0.5-mL Eppendorf tube using 2 μL of 1 mg/mL of Ub ethyl ester, 4 μL of the Master Mix, and 4 μL of cell lysate or purified DUB. Dilute cell lysate or purified DUB as necessary to obtain the appropriate level of DUB activity.

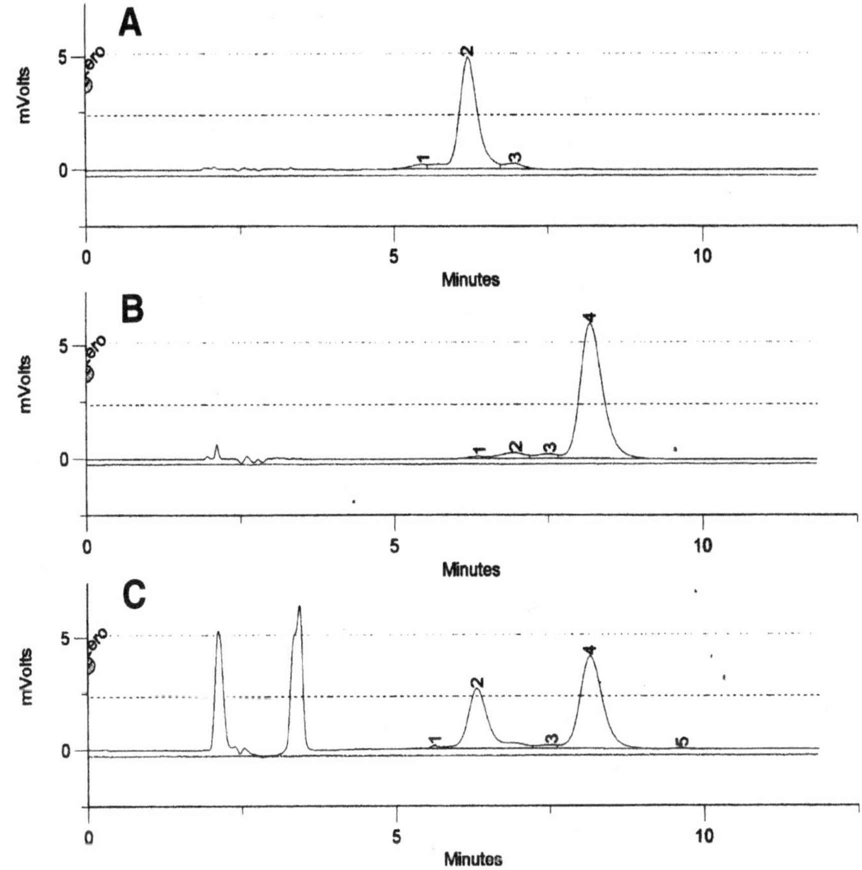


Fig. 2. HPLC assay using ubiquitin ethyl ester as a substrate for DUB activity. (**A**) Purified ubiquitin standard (2 μL of 2 mg/mL of Ub) was analyzed using an isocratic method. Elution time was 6.2 min. (**B**) Purified Ub-ethyl ester substrate (4 μL of 1 mg/mL of ester). Elution time was 8.2 min. (**C**) Analysis of pooled FPLC fractions from DUB purification for DUB activity. Peaks before 5 min result from buffer used in FPLC fractions. DUB activity is present as a portion of Ub–ethyl ester substrate has been converted to ubiquitin seen by the presence of the ubiquitin peak at 6.2 min.

2. Incubate the reaction for 15 min at 37°C.
3. Quench with 40 μL of 0.1 *M* HCl and centrifuge briefly to ensure all of the reaction mix is at the bottom of the tube.
4. Inject onto HPLC with a C8 reversed phase column running on an isocratic gradient of 61% HPLC buffer B at a flow rate of 1 mL/min. Each run should be about 12 min long (*see* **Note 5**).
5. Calculate amount of conversion to ubiquitin by determining percentage of ubiquitin peak area relative to total peak area (ubiquitin peak + Ub– ethyl ester peak). A reference injection of Ub–ethyl ester can be used for calibration. Use the rate of conversion to determine the total amount of DUB activity present (*see* **Note 6**).

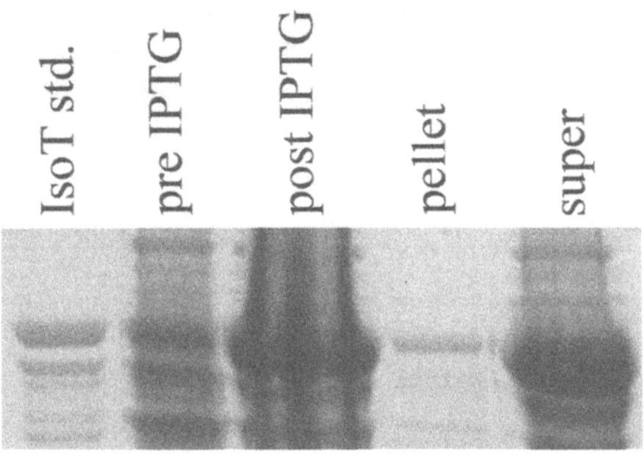


Fig. 3. Isopeptidase T expression and solubilization from *E. coli*. From *left* to *right*: previously purified Isopeptidase T, pre-IPTG induction, post-IPTG induction, pellet from centrifugation after sonication, and clarified cell lysate after sonication *(5)* were resolved on a 10% SDS-PAGE gel stained with Coomassie blue. Pre- and post-IPTG samples were prepared by taking 1 mL of cell culture, pelleting the cells, removing the supernatant, resuspending in 100 µL of 1X SDS sample buffer, and then loading 10 µL onto the gel.

### 3.3. Isopeptidase T Purification

#### 3.3.1. E. coli Growth and Induction

1. Plate BL21-DE3 cells transformed with the IsoT expression vector on LB–ampicillin plates according to standard methods. Inoculate LB–ampicillin (100 µg/mL) media starter cultures with colonies from the plates and grow to mid-log phase ($OD_{600\ nm}$ 0.6–0.8) at 37°C.
2. Inoculate 1.5 L of LB–ampicillin media in a 2.8-L Fernbach flask with 50 mL of starter culture (inoculate eight flasks for a total of 12 L of media) and grow to mid-log phase at 37°C.
3. Pellet cells at 3000*g* for 10 min (*see* **Note 7**).
4. Resuspend cell pellets in 12 L of fresh, 42°C LB–ampicillin media, put back in Fernbach flasks, and incubate at 42°C for 30 min (*see* **Note 8**).
5. Induce cultures with 10 µ*M* IPTG, cool to 15°C, and express 40–48 h at 15°C (*see* **Note 8**).
6. Pellet cells at 3000*g* for 10 min. Cell pellets can be frozen at –80°C at this point.

#### 3.3.2. Cell Pellet Lysis and Sonication

1. Resuspend pellets in 100 mL of cell lysis buffer per liter of cells (1200 mL total).
2. Add lysozyme to 100 µg/mL and incubate at room temperature for 20 min.
3. Sonicate cell lysate in 300-mL aliquots (we use six cycles of 30 s sonication at 48 W followed by a 1-min rest on ice).
4. Centrifuge the sonicated lysate at 23,000*g* for 30 min. (*See* **Fig. 3** for analysis of induction and sonication efficiency.)
5. Test the lysate for isopeptidase T enzymatic activity using ubiquitin ethyl ester or other appropriate substrate such as Ub–AMC.

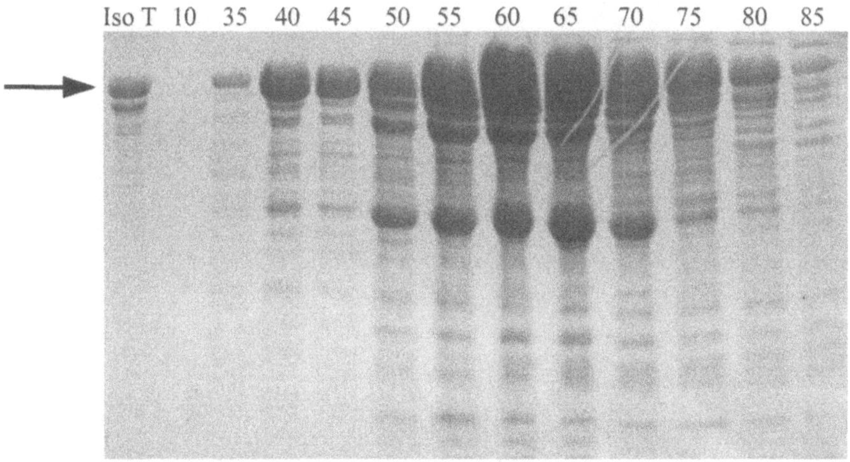


Fig. 4. Ion-exchange chromatography of clarified cell lysate. Ten microliters of each fraction was resolved on a 10% SDS-PAGE gel and stained with Coomassie blue. Fractions to be loaded were determined by results from HPLC analysis of fractions for DUB activity.

### *3.3.3. Initial Purification of Isopeptidase T by Ion-Exchange Chromatography*

1. Perform all chromatography steps at 4°C. Load the supernatant directly onto 100-mL column of Fast Flow Q Sepharose resin equilibrated with FPLC equilibration buffer at 5 mL/min using a peristaltic pump at 4°C (*see* **Note 9**).
2. After loading the lysate, wash the column with 400 mL of FPLC equilibration buffer at 4 mL/min.
3. Elute isopeptidase T from the column with a 500-mL gradient (0–350 m*M* NaCl in FPLC equilibration buffer) at 4 mL/min and collect 7-mL fractions.
4. Monitor fractions for Isopeptidase T activity on HPLC using an appropriate substrate. (*See* **Fig. 4** for SDS-PAGE analysis of FPLC fractions.)
5. Pool FPLC fractions with high activity. This pool starts from the center of the elution peak and extends out on either side to fractions that have equal activity to the lysate loaded onto the FPLC column.

### *3.3.4. Final Purification of Isopeptidase T by Affinity Chromatography and Gel Filtration*

1. Load pooled FPLC fractions onto a 5-mL or larger Ub–Sepharose column equilibrated with Ub–Sepharose equilibration buffer by gravity flow (*see* **Note 10**).
2. After loading, wash resin with five column volumes of Ub–Sepharose equilibration buffer. Elute with Ub–Sepharose elution buffer by gravity flow collecting 4-mL fractions (*see* **Note 11**).
3. Neutralize fractions immediately after elution with 3 *M* Na-acetate to approx pH 7.0.
4. Monitor elution fractions for isopeptidase T activity by HPLC using ubiquitin ethyl ester or other appropriate substrate and then pool fractions with high activity using the same

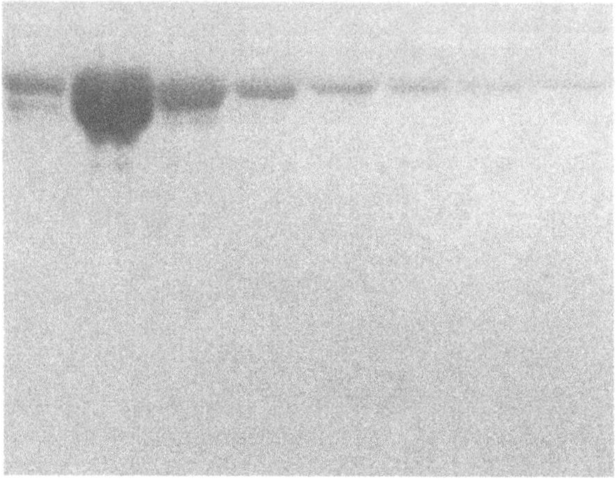


Fig. 5. Affinity chromatography of pooled FPLC fractions using Ub–Sepharose. Ten microliters of each elution fraction was resolved on a 10% SDS-PAGE gel and stained with Coomassie blue.

pooling method used for the FPLC purification step. (*See* **Fig. 5** for SDS-PAGE analysis of elution fractions.)

5. Concentrate pooled fractions with an Amicon ultrafiltration cell using a YM-30 filter (*see* **Note 12**).
6. Dialyze the concentrate into 50 m*M* Tris-HCl, pH 7.5, 10 m*M* β-mercaptoethanol.
7. Perform gel filtration of the purified protein using Sephacryl S-200 resin equilibrated in 50 m*M* Tris-HCl, pH 7.5, 10 m*M* ß-mercaptoethanol buffer; then concentrate, aliquot, and freeze purified protein at –80°C. Specific activity of the protein should be approx 1.25 IU/mg of protein in our esterase assay.

## 4. Notes

1. Settings will vary from machine to machine. The best data are obtained when the combination of the sensitivity of the spectrometer and the dilution of the DUB combine to give a strong signal and an approximately linear rate of substrate cleavage. Adjustments of the voltage or substrate concentration are usually sufficient to achieve these conditions.
2. Ovalbumin is added to the assay buffer as a carrier protein. Because many DUBs have very high enzymatic activities and the Ub–AMC substrate gives a strong signal, very dilute solutions of protein are required to get useful data in this assay. Ovalbumin is added to the assay buffer in order to prevent the DUB of interest and substrate from adhering to the sides of the cuvet, which would prevent obtaining useful data. DTT is also important to prevent oxidation of the active site cysteine present in DUBs, which would reduce the intrinsic DUB activity of the enzyme being analyzed.
3. A negative slope is sometimes observed when analyzing the 40 n*M* Ub–AMC baseline reference sample. This is a result of photobleaching and can be averted by reducing the amount of light to which the Ub–AMC is exposed. Reducing the slit aperture and increasing the high voltage to maintain signal strength should alleviate the problem.

4. We recommend using UCH-L3 as a positive control for Ub–AMC cleavage, as it has a very high specific activity.
5. We generally run our samples at 61% buffer B. This percentage can vary slightly depending on the instrument and batch of buffer being used. Adjust percentage accordingly so that ubiquitin will elute roughly in the middle of the run. Ubiquitin ethyl ester should then elute approximately 2 min after ubiquitin. Any good quality C8 column other than the Alltech Alltima C8 5- µm column that we use will perform well in this assay. Less expensive columns have given mixed results.
6. This assay works best for accurately determining the amount of DUB activity in a sample when roughly 50% of the ethyl ester is converted to ubiquitin. Anything less than 10% or more than 90% conversion to ubiquitin gives unreliable results. Adjust dilutions of sample being analyzed as necessary to fall in the appropriate range.
7. For pelleting cells, we prefer using 1-L plastic bottles (Nalgene). They fit most types of swinging bucket rotors and are quite useful when resuspending cell pellets for the heat shock step, storing pellets at –80°C, and in holding cell lysate for sonication.
8. The 42°C heat shock and low temperature expression steps are essential to obtaining large quantities of soluble protein. Previous attempts at isopeptidase T expression in our laboratory with a standard IPTG induction protocol (0.3–1.0 m*M* IPTG for 3 h at 37°C) resulted in 75–80% of the expressed isopeptidase T being insoluble. The heat shock step may activate chaperones in *E. coli* that assist in protein folding. Once the chaperones have been activated, the low amount of IPTG added induces expression, but cell death or growth inhibition from IPTG toxicity is reduced. This allows for a longer induction period. The low temperature leads to a slower expression of isopeptidase T and when combined with the active chaperones, allows expressed isopeptidase T to be correctly folded by the bacteria. Using this protocol, we obtain 80–95% of recombinantly expressed isopeptidase T in a soluble form. (We would like to thank Eileen Jaffe, Fox Chase Cancer Center, Philadelphia, PA for suggesting this protocol.)
9. The supernatant from the postsonication centrifugation can be very viscous. If necessary, it can be diluted in FPLC equilibration buffer before loading onto the Fast Flow Q resin. Because of the viscosity of the lysate, it is not recommended to use FPLC pumps to load clarified lysate onto the Fast Flow Q column.
10. Five milliliters of 12 mg/mL of Ub–Sepharose will bind approx 60 mg of isopeptidase T. A 12-L prep will yield 120–150 mg of protein so a larger volume of Ub–Sepharose can be used. If necessary, the column can be reequilibrated and the pooled FPLC fractions passed over the column a second time. The column will effectively bind any isopeptidase T still present in the pooled FPLC fractions.
11. Oxidation of the active site thiol can be a problem in this preparation, especially during elution from the Ub–Sepharose column and can cause the loss of significant amounts of enzymatic activity. Loss of activity does not result in a lower yield of protein as isopeptidase T binding to ubiquitin is unaffected by loss of enzymatic activity. Some of the lost isopeptidase T activity can be restored by incubating the enzyme with an excess of DTT. Purified isopeptidase T stored at –80°C will slowly lose enzymatic activity over a period of 3–6 mo.
12. Before loading the pooled elution fractions into the ultrafiltration cell, it is recommended to test the apparatus for leaks with deionized water. This ensures that the cell and filter used are sealed and do not have any holes that would allow protein to escape into the flow-through. It is also recommended to check for isopeptidase T in the flow-through by measuring enzymatic activity to ensure there are no slow leaks that could be missed by the initial test.

## References


1. Wilkinson, K. D. (1997) Regulation of ubiquitin-dependent processes by deubiquitinating enzymes. *FASEB J.* **11,** 1245–1256.
2. Cope, G. A., Suh, G. S., Aravind, L., et al. (2002) Role of predicted metalloprotease motif of Jab1/Csn5 in cleavage of Nedd8 from Cul1. *Science* **298,** 608–611.
3. Li, S. J. and Hochstrasser, M. (1999) A new protease required for cell-cycle progression in yeast. *Nature* **398,** 246–251.
4. Malakhov, M. P., Malakhova, O. A., Kim, K. I., Ritchie, K. J., and Zhang, D. E. (2002) UBP43 (USP18) specifically removes ISG15 from conjugated proteins. *J. Biol. Chem.* **277,** 9976–9981.
5. Li, S. J. and Hochstrasser, M. (2000) The yeast ULP2 (SMT4) gene encodes a novel protease specific for the ubiquitin-like Smt3 protein. *Mol. Cell Biol.* **20,** 2367–2377.
6. Tobias, J. W. and Varshavsky, A. (1991) Cloning and functional analysis of the ubiquitin-specific protease gene UBP1 of *Saccharomyces cerevisiae. J. Biol. Chem.* **266,** 12021–12028.
7. Verma, R., Aravind, L., Oania, R., et al. (2002) Role of Rpn11 metalloprotease in deubiquitination and degradation by the 26S proteasome. *Science* **298,** 611–615.
8. Balakirev, M. Y., Tcherniuk, S. O., Jaquinod, M., and Chroboczek, J. (2003) Otubains: a new family of cysteine proteases in the ubiquitin pathway. *EMBO Rep.* **4,** 517–522.
9. Bignell, G. R., Warren, W., Seal, S., et al. (2000) Identification of the familial cylindromatosis tumour-suppressor gene. *Nat. Genet.* **25,** 160–165.
10. Evans, P. C., Smith, T. S., Lai, M. J., et al. (2003) A novel type of deubiquitinating enzyme. *J. Biol. Chem.* **278,** 23180–23186.
11. Borodovsky, A., Ovaa, H., Kolli, N., et al. (2002) Chemistry-based functional proteomics reveals novel members of the deubiquitinating enzyme family. *Chem. Biol.* 9, 1149–1159.
12. Johnston, S. C., Riddle, S. M., Cohen, R. E., and Hill, C. P. (1999) Structural basis for the specificity of ubiquitin C-terminal hydrolases. *EMBO J.* **18,** 3877–3887.
13. Larsen, C. N., Price, J. S., and Wilkinson, K. D. (1996) Substrate binding and catalysis by ubiquitin C-terminal hydrolases: identification of two active site residues. *Biochemistry* **35,** 6735–6744.
14. Buchberger, A. (2002) From UBA to UBX: new words in the ubiquitin vocabulary. *Trends Cell Biol.* **12,** 216–221.
15. Amerik, A. Y., Li, S. J., and Hochstrasser, M. (2000) Analysis of the deubiquitinating enzymes of the yeast *Saccharomyces cerevisiae. Biol. Chem.* **381,** 981–992.
16. Leggett, D. S., Hanna, J., Borodovsky, A., et al. (2002) Multiple associated proteins regulate proteasome structure and function. *Mol. Cell* **10,** 495–507.
17. Amerik, A., Swaminathan, S., Krantz, B. A., Wilkinson, K. D., and Hochstrasser, M. (1997) In vivo disassembly of free polyubiquitin chains by yeast Ubp14 modulates rates of protein degradation by the proteasome. *EMBO J.* **16,** 4826–4838.
18. Borodovsky, A., Kessler, B. M., Casagrande, R., Overkleeft, H. S., Wilkinson, K. D., and Ploegh, H. L. (2001) A novel active site-directed probe specific for deubiquitylating enzymes reveals proteasome association of USP14. *EMBO* J. **20,** 5187–5196.
19. Hadari, T., Warms, J. V., Rose, I. A., and Hershko, A. (1992) A ubiquitin C-terminal isopeptidase that acts on polyubiquitin chains. Role in protein degradation. *J. Biol. Chem.* **267,** 719–727.
20. Wilkinson, K. D., Tashayev, V. L., O'Connor, L. B., Larsen, C. N., Kasperek, E., and Pickart, C. M. (1995) Metabolism of the polyubiquitin degradation signal: structure, mechanism, and role of isopeptidase T. *Biochemistry* **34,** 14535–14546.

21. Woo, S. K., Lee, J. I., Park, I. K., et al. (1995) Multiple ubiquitin C-terminal hydrolases from chick skeletal muscle. *J. Biol. Chem.* **270,** 18766–18773.
22. Baker, R. T., Tobias, J. W., and Varshavsky, A. (1992) Ubiquitin-specific proteases of Saccharomyces cerevisiae. Cloning of UBP2 and UBP3, and functional analysis of the UBP gene family. *J. Biol. Chem.* **267,** 23364–23375.
23. Varshavsky, A. (2000) Ubiquitin fusion technique and its descendants. *Methods Enzymol.* **327,** 578–593.
24. Gan-Erdene, T., Kolli, N., Yin, L., Wu, K., Pan, Z. Q., and Wilkinson, K. D. (2003) Identification and characterization of DEN1, a deneddylase of the ULP family. *J. Biol. Chem.* **278,** 28,892–28,900.
25. Wu, K., Yamoah, K., Dolios, G., et al. (2003) DEN1 is a dual function protease capable of processing the C-terminus of Nedd8 deconjugating hyper-neddylated CUL1. *J Biol Chem.* **278,** 28,882–28,891.
26. Larsen, C. N., Krantz, B. A., and Wilkinson, K. D. (1998) Substrate specificity of deubiquitinating enzymes: ubiquitin C-terminal hydrolases. *Biochemistry* **37,** 3358–3368.

# II

# Cellular Methods to Study Ubiquitin–Proteasome-Dependent Functions

14

# Measuring Ubiquitin Conjugation in Cells


## Edward G. Mimnaugh and Leonard M. Neckers



### Summary

Protein ubiquitination is crucial to many diverse and critical functions of cells. Although it has been long known that conjugation of ubiquitin to proteins results in their destruction by the proteasome, recently it has become apparent that reversible protein ubiquitination, particularly monoubiquitination, performs regulatory functions in cells, analogous to protein phosphorylation. The most powerful and sensitive technique for measuring specific protein ubiquitination is antiubiquitin immunoblotting of the immunoprecipitated protein after gel electrophoresis. Efficient antibodies recognizing ubiquitinated proteins are now available, making ubiquitin immunoblotting a practical tool for research into the many and varied aspects of this extremely interesting posttranslational protein modification. Here, we describe in detail the steps to follow in order to determine whether a particular protein might become ubiquitinated, or deubiquitinated, and we offer warnings about pitfalls to avoid in antiubiquitin immunoblotting.



**Key Words:** Deubiquitination; gel electrophoresis; immunoblotting; immunoprecipitation; ubiquitin; ubiquitination.


## 1. Introduction

Within recent years, our understanding of the ubiquitin–proteasome pathway has been extended beyond the "trash can" or "degrading machine" descriptions, which have implied that protein ubiquitination and proteasomes perform predominantly housekeeping functions for cells. It is now appreciated that ubiquitination is also a highly versatile and critical regulatory protein modification that functions more like a corporate executive officer than a sanitary engineer. Beyond its important role in clearing misfolded, mutated, or chemically damaged proteins from cells, ubiquitin conjugation participates in such diverse functions as antigen presentation, cell cycle regulation, endocytosis, signal-transduction, apoptosis, DNA damage repair, transcription regulation, fertility, modulation of chromatin structure, and even retroviral infectivity. (For reviews, *see* **refs. *1–6*.**)

The bulk of cellular proteins, both short- and long-lived, are susceptible to ubiquitination at some point. Consequently, this process is precisely orchestrated by a


From: *Methods in Molecular Biology, vol. 301, Ubiquitin–Proteasome Protocols*
Edited by: C. Patterson and D. M. Cyr © Humana Press Inc., Totowa, NJ

coterie of E2 ubiquitin-conjugating enzymes and at least five classes of E3 ubiquitin ligases, each with unique domains, counterbalanced by two major families of deubiquitinating enzymes, the ubiquitin-processing proteases and the ubiquitin C-terminal hydrolases. The detailed enzymology of ubiquitination and deubiquitination can be found in other chapters of this book and in **refs.** ***7–14***. Therefore, the ubiquitin–proteasome pathway should be viewed as an extremely complex, multitask mechanism of organizing vital cellular functions. Clearly, advancement of our understanding of the rich and varied functions of protein ubiquitination will touch on many other aspects of biochemistry and cell biology.

Because ubiquitination is so important to the normal business of cells, when things go wrong, there are serious consequences. Defective protein ubiquitination plays a role in the pathogenesis of human neurodegenerative disorders such as Parkinson's disease, Alzheimer's and Huntington's diseases ***(3)***, in muscle wasting during cachexia ***(15)***, and in cancer ***(16)***, for example, oncogenic HPV-caused cervical carcinoma ***(17)*** and the hereditary von Hippel–Lindau cancer syndrome ***(18)***. Therefore, the study of protein ubiquitination could eventually identify novel therapeutic strategies that are exploitable for the treatment of these and other human diseases.

In this chapter, we provide a detailed methodology for measuring protein ubiquitination by immunoblotting (Western blotting), as originally developed nearly two decades ago by Haas and Bright ***(19)***. Today, immunoblotting remains the technique of choice for measuring protein ubiquitination in cells because of its high specificity and sensitivity, and the relative simplicity with which ubiquitinated proteins may be detected.

## 2. Materials

### 2.1. Sample Preparation

1. TNESV lysis buffer: 50 m*M* Tri-HCl, pH 7.5, 1% v/v Nonidet P-40 detergent, 2 m*M* EDTA, 100 m*M* NaCl, 10 m*M* sodium orthovanadate, supplemented with 1 m*M* phenylmethylsulfonyl fluoride, 10 µg/mL of leupeptin, and 10 µg/mL of aprotinin or preprepared protease inhibitor cocktail tablets (Roche Molecular Biochemicals, Indianapolis, IN). *N*-Acetyl-leucyl-leucyl-norleucinal (ALLnL) (100 m*M*) or other inexpensive proteasome inhibitor can be added to prevent the loss of ubiquitinated proteins by proteasome catabolism, and 10 m*M* iodoacetamide or 10 m*M* *N*-ethylmaleimide can be included to inactivate ubiquitin-cleaving isopeptidases (*see* **Notes 1** and **2**).
2. Phosphate-buffered saline (PBS): dissolve 8.0 g of NaCl, 0.2 g of KCl, 1.44 g of $Na_2HPO_4$, and 0.24 g of $KH_2PO_4$ in approx 800 mL of purified water, adjust the pH to 7.4 with HCl, then dilute to 1.0 L.
3. Reducing gel-loading buffer: 100 m*M* dithiothreitol (DTT), 50 m*M* Tris-HCl, pH 6.8, 10% (w/v) sodium dodecyl sulfate (SDS), 10% (v/v) glycerol, 0.1% (w/v) (a pinch) of bromphenol blue tracking dye ***(20)***. Warm in a 37°C water bath and rock gently to dissolve the SDS.

### 2.2. Immunoprecipitation

1. Antibody selection: immunoprecipitating a specific ubiquitinated protein with an antiubiquitin antibody is impractical because of the abundance of free ubiquitin. In most cases, the protein of interest will be immunoprecipitated, and the subsequent immunoblot after gel electrophoresis will be probed with an antiubiquitin antibody (*see* **Note 3**).

Fortunately, a wide selection of monoclonal and polyclonal antiubiquitin antibodies from a variety of species is now available from commercial sources. Most of the antiubiquitin antibodies we have used in our laboratory perform well.

2. Antibody reconstitution buffer: sterile PBS containing 0.1% w/v sodium azide as a preservative, if the antibody is not already in solution.
3. Protein A-coated Sepharose microbeads: rehydrate Protein A-immobilized Sepharose CL-4B beads (1.5 g) (Pharmacia Biotechnology, Piscataway, NJ, USA) overnight in 50 mL of TNESV buffer containing 1% (w/v) bovine serum albumin at 4°C with rotation to keep the beads in suspension. The beads are centrifuged at 1000*g* for 5 min, and then washed twice in the same buffer by resuspending and centrifuging. Twenty micrograms of rabbit antimouse IgG (RAM) (ICN/Cappel Biomedical, Aurora, OH, USA) is added to a 10-mL aliquot of the washed resuspended beads (equivalent to 0.3 g of dry beads), and they are rotated at 4°C for an additional 2 h. Finally the RAM-coated beads are washed three times in TNESV and stored at 4°C. Antibody-linked Protein A- or Protein G-coated microbeads for immunoprecipitations are also available from commercial suppliers.
4. Vertical rotator for mixing the vial contents during immunoprecipitations.

## 2.3. Gel Electrophoresis

1. Full-sized or minigel electrophoresis chamber and power supply.
2. 10% Resolving polyacrylamide gel: mix 10 mL of 30% (w/v) acrylamide 0.8% (w/v) *N, N'*-methylene–*bis*-acrylamide mixture (ProtoGel, National Diagnostics, Atlanta, GA), 7.5 mL of 1.5 *M* Tris buffer, pH 8.8, 12.5 mL of water, 0.3 mL of 10% w/v SDS, 0.1 mL of 10 % (w/v) ammonium persulfate solution, and 20 μL of *N, N, N', N'*-tetramethylethylenediamine (TEMED) (Bio-Rad, Hercules, CA, USA), in this order. Precast gels are also available from commercial suppliers (*see* **Note 4**).
3. Water-saturated *n*-butanol.
4. Whatman 3MM paper.
5. Polyacrylamide stacking gel: mix 1.5 mL of 30% (w/v) acrylamide 0.8% (w/v) *bis*-acrylamide mixture, 2.5 mL of 0.5 *M* Tris-HCl, pH 6.8, 6.0 mL of deionized water, 0.1 mL of 10% (w/v) SDS, 0.1 mL of 10% (w/v) ammonium persulfate, and 8 μL of TEMED.
6. Laemmli's sample gel-loading buffer: 80 m*M* Tri-HCl, adjusted to pH 6.8, 100 m*M* DTT, 10% (w/v) SDS, 10% (v/v) glycerol, with a small amount (a pinch) of bromphenol blue indicator dye (~0.2% w/v) *(20)*.
7. Tris-glycine-SDS gel-running buffer: 25 m*M* Tris base, 192 m*M* glycine, 0.1% (w/v) SDS. Do not adjust the final pH, which should be approx 8.3.
8. Prestained protein standards: molecular weight standards are available from a variety of commercial suppliers. We use prestained broad range protein standards from Bio-Rad from 250 to 10 kDa apparent molecular mass.

## 2.4. Electrophoretic Transfer of Ubiquitinated Proteins to Membranes

1. Wet-type vertical gel transfer box.
2. Choice of nitrocellulose or PVDF membranes.
3. Towbin wet-transfer buffer: 25 m*M* Tris, 0.15% (w/v) SDS, 0.2 *M* glycine, and 20% (v/v) methanol. Do not adjust the pH. For an alternative transfer buffer that is useful for transferring free ubiquitin, as well as ubiquitinated proteins, use CAPS buffer: 25 m*M* cyclohexylaminopropanesulfonic acid, pH 10.0, and 20% (v/v) methanol.
4. Gel-soaking buffer: 63 m*M* Tris base, pH 6.8, 2.3% (w/v) SDS, and 5% (v/v) 2-mercaptoethanol.

### 2.5. Heat-Activation of Blotted Ubiquitinated Proteins

1. Autoclave or hot plate to boil water.
2. Glass tray.

### 2.6. Blocking the Membrane and Probing Immunoblots for Ubiquitinated Proteins

1. Blocking solution: 5% (w/v) fat-free dry milk in 10 m*M* Tris-HCl, pH 7.5, 50 m*M* NaCl, 2.5 m*M* disodium EDTA with 0.1% (v/v) Tween-20 detergent.
2. Antibody diluting solution: 2.5% (w/v) fat-free dry milk in TNE wash solution.
3. TNE wash solution: 10 m*M* Tris-HCl, pH 7.5, 2.5 m*M* disodium EDTA, 50 m*M* NaCl with 0.05% (v/v) Tween-20 detergent.
4. Horseradish peroxidase (HRP)-conjugated secondary antibodies are available from several commercial manufacturers. Remember that the secondary HRP-conjugated antibody must immunoreact with the primary antibody in a species-specific fashion.
5. Whatman 3MM paper.
6. Luminol-based chemiluminescence immunoblot development solution kit (commercially available from Pierce (Rockford, IL, USA) or Amersham (Piscataway, NY, USA).
7. Quality film, for example, X-OMAT AR film purchased from Eastman Kodak (Rochester, NY, USA).
8. Film cassettes.
9. Dark room with an automatic film developer or a chemiluminescence imaging system.

## 3. Methods

The majority of investigators studying protein ubiquitination in cells rely on antiubiquitin immunoblotting to detect ubiquitin-protein conjugates. The immunoblotting method and variations described here have been adapted, with modifications, from an original description of the immunochemical detection of ubiquitin-protein conjugates by Haas and Bright ***(19)***, with refinements by Wilkinson ***(21)***. Measuring ubiquitination of a particular specific protein substrate is accomplished by first immunoprecipitating the ubiquitinated target protein from cell lysates using an antibody directed against the individual protein under study. An alternative approach is the use of expression vector-generated epitope-tagged ubiquitin, such as FLAG–ubiquitin, hemagglutinin (HA)-ubiquitin or even Myc-ubiquitin, which is immunoprecipitated with an antibody that efficiently recognizes the epitope tagged ubiquitin–substrate complex. Immunoprecipitated samples are subjected to SDS-polyacrylamide gel electrophoresis (SDS-PAGE) (10% or 4–20% gradient gel), and the proteins are electrophoretically transferred to nitrocellulose or polyvinylidene difluoride (PVDF) membranes. The ubiquitinated proteins on the blot are incubated with an antiubiquitin antibody, which immunoreacts with "higher molecular weight forms" of the protein of interest, located above the usual position of the substrate protein on the blot. These represent ubiquitinated forms of the protein. In the situation where epitope-tagged ubiquitin is expressed, an antibody that recognizes the ubiquitin tag is used as the primary probe. Species-matching HRP-coupled secondary antibody is then reacted with the primary antibody, and the ubiquitinated protein bands are visualized by HRP-coupled, luminol-based chemiluminescence.

### *3.1. Sample Preparation*

Polyubiquitination of proteins dramatically shortens their half-lives because they are marked for rapid proteasomal digestion, or alternatively, some monoubiquitinated cell surface receptor proteins are internalized and destroyed by lysosomal proteases. In addition, linear or branched ubiquitin chains attached to proteins are susceptible to trimming by abundant ubiquitin-cleaving isopeptidases *(13)*. The entire process of protein ubiquitination is, in theory, reversible, as is the case for histone H2A and H2B monoubiquitination. Consequently, several precautions should be taken to prevent adventitious cleavage of ubiquitin from proteins during sample preparation. This is easily accomplished by supplementing the cell lysis buffer with a cocktail of several protease inhibitors in combination with a proteasome inhibitor and a thiol-reactive agent to inactivate deubiquitinating enzymes (*see* **Notes 1** and **2**).

1. Typically, cells grown to subconfluency in flasks or plates are gently washed twice with ice-cold PBS, then lysed on ice into 0.5–1.0 mL of TNESV lysis buffer with added inhibitors, for 20 min. Mammalian cells grown in suspension should be collected from growth medium by centrifugation, washed twice in PBS and lysed in the centrifuge tubes on ice for 20 min. Vortex-mix the tubes after aspirating the PBS wash to break up the cell pellets prior to adding the lysis buffer, and vortex-mix the cells vigorously with lysis buffer for approx 10 s. Yeast cells can be lysed directly into neat ethanol containing 50 m*M* *N*-ethylmaleimide and glass beads with vigorous vortex-mixing for several minutes *(22)*.
2. Lysed cells are then scraped from the plastic surface with a spatula (or removed from centrifuge tubes) and transferred into plastic vials to be centrifuged at 14,000*g* for 20 min. Many cytosolic and membrane-bound ubiquitinated proteins are found in the resulting clarified, detergent-soluble supernatant fraction. The remaining pellet following centrifugation is mostly chromatin, remnants of cellular membranes, cytoskeletal proteins, nuclear matrix, and detergent-insoluble protein aggregates. Some heavily ubiquitinated proteins may aggregate to the extent they become insoluble in mild detergent, and these will cosediment in the pellet fraction *(23)*.
3. The detergent-insoluble pellet should be resuspended by sonication or vigorous mechanical homogenization in lysis buffer before assaying for protein and electrophoretic separation of the ubiquitinated proteins. Aggregated ubiquitinated proteins will dissolve when the samples are heated in the Laemmli's reducing gel-loading buffer. If isolated cell membrane fractions are used, these too, should be sonicated (*see* **Note 5**).

### *3.2. Immunoprecipitation*

Before conducting immunoprecipitations, the concentration of the proteins in the cell lysate or other sample should be determined by a preferred protein assay.

1. Immunoprecipitation of protein is conveniently carried out using 1.5-mL disposable microcentrifuge tubes with flip-top or screw top caps that are available from a variety of manufacturers. All steps are carried out on ice, and a refrigerated centrifuge should also be used. Alternatively, the entire procedure can be conducted in a cold room at 4°C.
2. Aliquot the lysates into prenumbered tubes. Because only a small percentage of a particular protein under investigation may be ubiquitinated, a significantly greater amount of lysate protein should be immunoprecipitated than would normally be used if the immunoblot were probed for the targeted protein of interest, rather than for ubiquitin. For example, 50 μg of cell lysate protein is sufficient to immunoprecipitate epidermal

growth factor-receptor from A431 tumor cell lysate prior to epidermal growth factor (EGF) receptor immunoblotting, however, as much as 1.0 mg of cell of lysate should be immunoprecipitated with the intention of visualizing ubiquitinated forms of the EGF receptor by antiubiquitin immunoblotting.

3. Add the primary immunoprecipitating antibody to the samples, mix gently by inversion and incubate on ice for 2 h or more. This step can usually be conveniently conducted overnight. Although the antibody concentration that is optimal will depend on the efficiency of the specific antibody used, usually 4–5 µg of primary antibody per 1.0 mg of sample protein is adequate to maximize the pull-down of ubiquitinated forms of the protein.
4. Add the secondary immunoprecipitating antibody to the samples. Secondary antibodies are usually linked to a support such as Protein A- or Protein G-coated resin microbeads. The antibody should be directed against the same species and subclass isotype as the primary antibody. If the primary antibody has been purified from immunized rabbit serum, the antibody–antigen complex will bind directly to Protein A–Sepharose without the need for a secondary resin-conjugated antibody. Usually 50–100 µL of antibody beads in a 1:1 bead suspension in PBS or lysis buffer per milliliter of cell lysate is sufficient to collect essentially all of the primary antibody–antigen complex. Of course, this depends on the amount of antigen in the sample. Some primary antibodies are available prelinked to support beads, eliminating the need for a secondary antibody for completing the immunoprecipitations.
5. During immunoprecipitation, the samples should be agitated constantly on a mechanical vertical rotator to prevent the antibody-conjugated beads from settling.
6. Centrifuge the tubes at 10,000*g* at 4°C for 1 min to pellet the beads.
7. Aspirate the supernatant liquid above the beads carefully using a syringe and blunt needle, pipet, or similar device. Leave behind a small amount of supernatant rather than risk disturbing the bead pellet.
8. Wash the sedimented beads with 1.0 mL of lysis buffer by vortex-mixing and recentrifuging at 10,000*g* for 1 min. The inclusion of protease or proteasome inhibitors in the wash buffer is optional and is usually not necessary because proteases will have been removed with the aspirated lysate.
9. Repeat the wash procedure three times for a total of four washes.
10. Elute the immunoprecipitated protein-immune complex from the beads into 50–100 µL of Laemmli reducing gel-loading/sample buffer.
11. Vortex-mix the samples vigorously, place them into a boiling water bath or heating block at 95°C for 5 min, then chill them on ice for 10 min.
12. Centrifuge the samples for 1 min at 10,000*g* to sediment the beads from the eluted immunoprecipitated proteins.
13. Aspirate the supernatants for protein resolution by gel electrophoresis using gel-loading pipet tips. The entire sample or only a portion is then transferred to the well of a gel.

### 3.3. Gel Electrophoresis

Ubiquitinated proteins can usually be resolved by standard single-dimensional SDS-PAGE using 8% or 10% polyacrylamide gels and Tris–glycine–SDS gel-running buffer. Gradient gels may be used and provide the opportunity to separate high molecular weight ubiquitinated proteins.

1. Immunoprecipitated proteins must be released from the immunoprecipitated antibody complexes, heat denatured and linearized in Laemmli sample loading buffer prior to SDS-PAGE.

2. Commercial, precast minigels (*see* **Note 4**) or conventional 14 cm × 14 cm × 1.5 mm gels cast in the laboratory may be used to resolve ubiquitinated proteins. For a 10% resolving gel, mix 10 mL of 30% (w/v) acrylamide 0.8% (w/v) *N,N'*-methylene *bis*-acrylamide mixture (ProtoGel, National Diagnostics, Atlanta, GA), 7.5 mL of 1.5 *M* Tris buffer, pH 8.8, 12.5 mL of water, 0.3 mL of 10% (w/v) SDS, 0.1 mL of 10% (w/v) ammonium persulfate solution, and 20 µL of TEMED, in this order. Pour the gel to a height of approx 11.5 cm and gently pipet a layer of water-saturated *n*-butanol (~0.5 mL) onto the top of the resolving gel to break up bubbles and to exclude oxygen, which will slow the polymerization of the gel. After the gel solidifies (~1 h), decant the butanol, rinse the gel with deionized water, and adsorb any remaining water using a portion of Whatman 3MM paper as a blotter. Insert a comb between the glass plates to form the wells and pour a 2.5-cm 4.5% acrylamide stacking gel on top of the polymerized resolving gel. The stacking gel is prepared by mixing 1.5 mL of 30% (w/v) acrylamide 0.8% (w/v) *bis*-acrylamide mixture, 2.5 mL of Tris buffer, pH 6.8, 6.0 mL of deionized water, 0.1 mL of 10% (w/v) SDS, 0.1 mL of 10% (w/v) ammonium persulfate, and 8 µL of TEMED. Allow the stacking gel to polymerize for 1 h or more before removing the comb, and then fill the wells completely with 1X gel-running buffer.
3. Fill the gel box about one third to one half full with gel-running buffer, insert the gel into the box, and fill the upper buffer chamber with gel-running buffer.
4. Carefully add the immunoprecipitated samples to the *bottom* of the wells using "gel loading" pipet tips. Include an aliquot of prestained molecular weight standards in one of the lanes to monitor the progression of electrophoresis and to provide references to estimate the molecular weight of the antiubiquitin signals in the sample lanes on the subsequent immunoblots.
5. For 10% minigels, use a constant voltage of 100–125 V for approx 1–1.5 h; for larger conventional 10% gels, use 40–50 V overnight for approx 12–14 h. Higher voltages will heat the gel and possibly distort the bands. To use higher voltage (faster electrophoresis), chill the gel-running buffer by circulating cold water through the heat-exchange device that is usually supplied with gel boxes.

### 3.4. Electrophoretic Transfer of Ubiquitinated Proteins to Membranes

In our experience, electrophoretic transfer of ubiquitinated proteins from SDS-polyacrylamide gels to nitrocellulose membranes by the "semidry" method usually is unsatisfactory. This is probably because polyubiquitinated proteins can have molecular masses as high as 300–400 kDa, and these do not elute easily from high percentage gels. We strongly recommend transfer of ubiquitinated proteins from gels by the "wet-transfer" method using the Tris–glycine–SDS–methanol buffer, as originally described by Towbin *(24)*.

1. Nitrocellulose or PVDF membranes should be fully hydrated in 1X transfer buffer or deionized water for 1–2 h. This will help to maximize the binding of ubiquitinated proteins to the membrane. Note that PVDF membranes must first be wet in methanol for approx 10 s before immersing them in water.
2. Full-sized gels should be electrotransferred in Towbin wet-transfer buffer for 16–20 h (overnight) at 40–50 V (usually ~300 mAmp) and at 4°C in a refrigerated cold box or cold room. Transfer of ubiquitinated proteins out of minigels onto membranes requires less time, but these, too, may be transferred overnight.

3. The electroelution of ubiquitinated proteins from SDS-polyacrylamide gels may be enhanced by soaking the gels in 2.3% SDS, 5% 2-mercaptoethanol, 63 m*M* Tris-HCl, pH 6.8, for approx 30 min prior to electrotransferring ***(25)***. This step is optional (*see* **Note 6**).

### 3.5. Heat Activation of Ubiquitinated Proteins on Immunoblots

This is an important step when using certain antiubiquitin antibodies. Detection of ubiquitinated proteins on immunoblots can be remarkably enhanced by autoclaving the membranes in deionized water for 30–40 min ***(26)***. Ubiquitinated proteins on membranes can also be activated by heating the blotted membrane at 75°C in an oven ***(19)***, by boiling the blot in deionized water, or by incubating the membrane in guanidine-denaturing buffer (*see* **Note 7**).

### 3.6. Blocking and Probing Immunoblots for Ubiquitinated Proteins

1. After the membrane has been heat-activated by autoclaving, it should be submerged in room temperature deionized water for several minutes to cool.
2. Gently rock the membrane in a tray, at room temperature, for 2 h, with sufficient "dry milk blocking solution" to cover the membrane surface completely. Alternatively, the membrane can be blocked overnight in a cold room.
3. Incubate the membrane at room temperature with the primary antiubiquitin antibody diluted in "antibody diluting solution" for 1–2 h. Exposing the membrane to the primary antibody may also be done overnight in a cold room with constant rocking (*see* **Notes 3, 8,** and **9**).
4. Wash the membrane in 1X TNE washing buffer for 30 min with five changes of the wash solution. It is not a problem if the membranes are washed for longer times.
5. Incubate the immunoblot with an appropriate HRP-conjugated secondary antibody diluted in "antibody diluting solution" for 1–2 h with constant gentle rocking.
6. Repeat the washing step as described in **step 4.**
7. Remove the membrane from the wash buffer and gently blot any excess wash buffer by sandwiching the membrane between two sheets of Whatman 3MM paper for a few seconds.
8. Immerse the immunoblot membrane in a small volume of freshly prepared luminol-based chemiluminescence development solution with gentle rocking for 1 min. Mix equal volumes of reagent A and reagent B according to the manufacturer's instructions (*see* **Note 10**).
9. Remove the membrane with blunt forceps, allow the excess solution to drain for 10–15 s, blot excess reagent with Whatman paper and seal the membrane inside a transparent plastic bag or plastic wrap.
10. Tape the protected membrane to the inside of a cassette and expose the immunoblot to good quality film in the dark room. The first exposure should be for 5 or 10 s; then longer or shorter exposures should be done, depending on the quality of the first developed exposed film.
11. Ubiquitinated forms of proteins will migrate more slowly through the gel than the unmodified target protein, ideally at intervals of approx 8 kDa above the location of the protein under investigation, reflective of the successive addition of ubiquitin molecules. In many instances, however, ubiquitinated proteins, especially if they are ubiquitinated at multiple sites, will visualize as a smear extending upward to the top of the gel (*see* **Note 11** and **Fig. 1**).

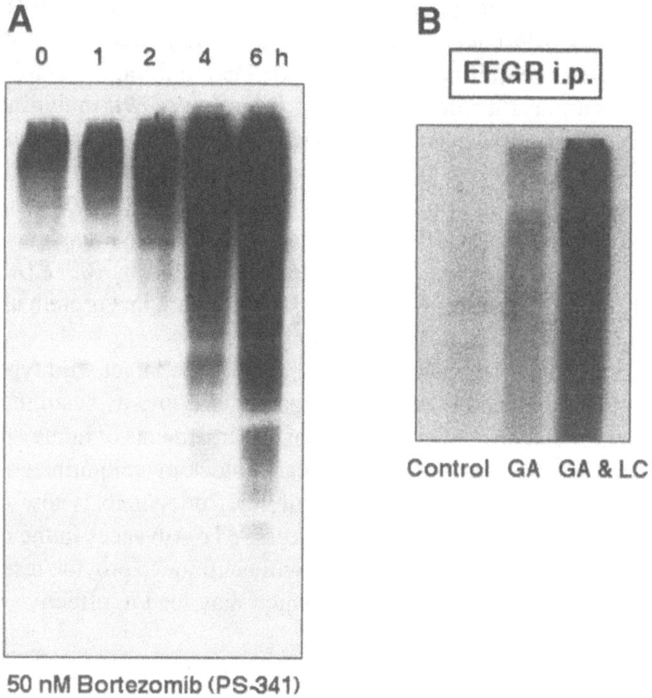


Fig. 1. Antiubiquitin immunoblots. In **(A),** MCF-7 human tumor cells were treated with the proteasome inhibitor, Bortezomib (PS-341), at 50 n*M* for various times. Cells were lysed into TNESV buffer with protease inhibitors, and 20 μg of clarified cell lysates were subjected to SDS-PAGE, electrotransferred to a nitrocellulose membrane and immunoblotted for ubiquitin. The dramatic increase in ubiquitination represents proteins that would normally be cleared by the proteasome. In **(B),** A431 human tumor cells were treated with 5 μ*M* Geldanamycin (GA) or a combination of 5 μ*M* GA plus 10 μ*M* lactacystin (LC) for 24 h (*see* **Note 14**). Cells were lysed and the EGF receptor was immunoprecipitated from 1.0 mg of lysate protein using an anti-EGF receptor antibody, electrophoresed, and transferred to a membrane to be immunoblotted for ubiquitin. In both immunoblots, visualization was by chemiluminescence.

12. If free ubiquitin is to be measured, care should be taken to prevent it from running off the gel as it migrates in SDS-PAGE as if it were a 5- or 6-kDa protein (*see* **Note 12**).
13. If the protein of interest has not previously been shown to be ubiquitinated, it is a good idea to verify positive results by using one or two different antibodies, preferably from different species (*see* **Note 13**).

### *3.7. Conclusion*

The number of publications, both primary research papers and reviews, on protein ubiquitination has grown exponentially within the past several years. For example, significant advances have been made in the pairing of ubiquitin–ligases with their substrates. Yet, while it is known that the VHL-complex ubiquitinates Hif-1α, MDM2 ubiquitinates p53 and Cbl ubiquitinates the EGF receptor, there are literally hundreds of ubiquitin ligases with unknown substrates. Other important exciting discoveries have pushed protein ubiquitination into unexpected and unpredictable new areas: (1) Ubiquitination plays a role in retroviral infectivity ***(70)***, including HIV ***(71)***; (2) ubiquitination can activate, as well as inactivate, certain transcription factors with or without proteolysis ***(72–76)***; (3) adaptor proteins interact with ubiquitin ligases to regulate monoubiquitination of plasma membrane receptors ***(77–79)***; and (4) histone H2B monoubiquitination signals the recruitment of a methyltransferase to nucleosomes to methylate histone H3 and causes telomere silencing ***(80–82)***. Given these discoveries, it should not be surprising to find in the future that protein ubiquitination participates in other new and novel cell functions.

With respect to ubiquitin-related diseases, attempts to target wild type or mutated E2 ubiquitin-conjugating and E3 ubiquitin-ligating enzymes by small molecules may quickly lead to new and effective pharmacological treatments of human malignancies or protein misfolding diseases. On the other hand, blocking ubiquitinated protein degradation through the use of the proteasome inhibitor, Bortezomib, is now a new therapy for multiple myeloma and possibly other cancers ***(83)***. Advances in the measurement of protein ubiquitination by immunoblotting will continue to provide researchers with the results to answer important questions, which may lead to effective therapies for human diseases.

## 4. Notes

1. Blocking enzyme-catalyzed deubiquitination in cell lysates. Inclusion of peptide aldehyde protease inhibitors such as *N*-acetyl-leucyl-leucyl-norleucinal (ALLnL) or Z-leucyl-leucyl-leucyl-CHO (MG132) in the cell lysis buffer at 50 μ*M* will preclude unwanted proteasomal digestion of ubiquitinated proteins during sample preparation. Although clasto-lactacystin is a more potent and an irreversible proteasome inhibitor ***(27)***, it is too expensive to be practical for this use. Cells also contain two major families of ubiquitin-specific thiol proteases, named Ubp (Ubiquitin processing proteases) and Uch (Ubiquitin C-terminal hydrolases) ***(8–10)***. These enzymes are capable of cleaving ubiquitin from proteins in cells and in cell lysates. Including thiol-reactive *N*-ethylmaleimide or iodoacetate at 10 m*M* in the lysis buffer will inactivate the ubiquitin-cleaving enzymes by alkylating their active site cysteine residues ***(21,28,29)***. This class of proteases is also sensitive to manganese ion, and inhibition of their deubiquitinating activity by adding 5 m*M* manganese ions in vitro has been reported ***(30,31)***. Ubiquitin aldehyde, now available from several commercial sources, is another potent inhibitor of ubiquitin-cleaving hydrolases that can be used to prevent inadvertent deubiquitination. Chilling all buffers and cell lysates and keeping the vials on ice also inhibits the activity of deubiquitinating enzymes.
2. Use of proteasome inhibitors to promote cellular accumulation of ubiquitinated proteins. Proteasome inhibition in cells stabilizes ubiquitinated proteins that normally are rapidly degraded via the ubiquitin–proteasome pathway ***(1,32,33)***. Proteasome inhibition may be absolutely necessary to detect ubiquitinated forms of some extremely short-lived proteins. ALLnL at 100 μ*M*, 50 μ*M* MG132, or 10 μ*M* clastolactacystin will effectively cross the plasma membrane of cells in culture and prevent proteasomal degradation of ubiquitinated proteins. Because ALLnL also inhibits calpains in addition to the

proteasome, *N*-acetyl-Leu-Leu-Met-H peptide should be used as a negative control to rule out participation of alternative proteolytic pathways, since it has inhibitory activity toward cathepsins and calpains, but does not inhibit the proteasome. Most recently, boronic acid-containing peptides such as PS-341, now called Bortezomib, have been developed (Millennium Pharmaceuticals, Cambridge, MA), and these are extremely potent inhibitors of the proteasome ***(34)***. Affinity Research Products recently offered several new proteasome inhibitors. The first, designated MG262, is the boronic acid derivative of MG132 and has much greater specificity for the proteasome than for other proteases such as cathepsin B. Epoxomicin is another specific and irreversible inhibitor of the chymotrypsin-like activity of the proteasome, and epoxomicin has essentially no inhibitory activity toward nonproteasomal proteases ***(35)***. A third inhibitor is aclacinomycin A, an anticancer agent that inhibits the chymotrypsin-like activity of the proteasome ***(36)***.

Although proteasome inhibitors are useful in the study of protein ubiquitination and degradation, they should be used with caution because of their inherent cytotoxic properties. ALLnL, clastolactacystin, PS-341 and other proteasome inhibitors, when used at concentrations sufficient to inhibit the proteasome, are lethal to many tissue culture cell lines within 12–24 h. Proteasome inhibitors cause cell death by apoptosis, probably through multiple secondary mechanisms ***(25,37,40)***. In addition, both ALLnL and clastolactacystin indirectly inhibit messenger RNA transcription and DNA replication by as much as 50% within 4 h ***(25)***, raising the likely possibility that the cellular level of some proteins might even be diminished by proteasome inhibitors.

On the other hand, because the signals for promoting protein ubiquitination are independent of proteasomal protease activity ***(41–43)***, blocking the degradation of proteins with proteasome inhibitors will not automatically promote massive protein ubiquitination. The multiple proteasomal proteases are localized inside the lumen or "barrel" of the proteasome, while the S5a polyubiquitin binding subunit is part of the 19S regulatory complex that sits at the entrance to the proteasome interior ***(44)***. Therefore, completely inhibiting proteasomal protease activities does not ensure that ubiquitin will remain attached to proteins when they interact with the proteasome regulatory complex. Ubiquitin may still be susceptible to cleavage from proteins by proteasome-associated ubiquitin hydrolase activity ***(45)*** or by deubiquitinating enzymes that are independent of the proteasome ***(13)***.

3. Antibody selection. Antiubiquitin antibodies now are available from a broad selection of commercial sources from Abcam to Zymed Labs. An extensive and interactive list of sources and characteristics, including specificity, of antiubiquitin antibodies can be found on the World Wide Web, *see www.alzforum.org/res/com/ant/ubiquitin/UBIQUITIN-tableC.html*. We prefer not to make specific recommendations, and rather than endorse a particular source for an antiubiquitin antibody, we recommend that researchers select several antibodies from different suppliers and compare the antibody performance. Most antiubiquitin antibodies react with both free ubiquitin and polyubiquitinated proteins. Unique antiubiquitin antibodies with novel characteristics have been developed, such as one with the ability to specifically recognize monoubiquitinated histone H2A ***(46)***. Antiubiquitin antibodies may be available gratis on request from investigators. A few have been recently commercialized. For example, an antibody originally developed by Hideyoshi Yokosawa of Hokkaido University, Japan ***(47,48)*** that specifically recognizes only polyubiquitinated proteins but not monoubiquitinated proteins or free ubiquitin is now available from Affiniti Research Products, United Kingdom. Antibodies to epitope-tagged ubiquitin coupled to beads for immunoprecipitations are commercially available.

4. Precast gels. A variety of gels at different polyacrylamide percents as well as gradient gels are available from commercial sources. While the use of these gels may be prohibitive in some labs because of their expense, they are time saving. They generally perform well, unless they are used beyond their expiration date. One liability is that they are not universal because they usually require use of a particular gel box, also sold by the gel manufacturer.
5. Immunoprecipitation of membrane-bound proteins. Including 1% (w/v) SDS in the TNESV lysis buffer will help to remove ubiquitinated proteins from cell membranes. The lysates will be very viscous or even gelatinous because 1% (w/v) SDS denatures chromatin. Lysates should be passed several times through a 20-gage needle or sonicated to break up the DNA. Prior to immunoprecipitating the protein under study, the samples must be diluted 10-fold in SDS-free lysis buffer to decrease the concentration of SDS to 0.1% (w/v). This is necessary to prevent the SDS from denaturing the immunoprecipitating antibody.
6. Reducing proteins in the gel to enhance transfer to membranes. Heavily ubiquitinated proteins move very slowly through gels, and consequently, they are difficult to elute onto membranes. Exposing gels to 2.3% SDS, 5% 2-mercaptoethanol, 63 m*M* Tris-HCl, pH 6.8, for 30 min facilitates the efflux of ubiquitinated proteins from high percentage polyacrylamide gels. The 2-mercaptoethanol reduces any protein disulfide bonds that may have reformed in the gel during electrophoresis, and the SDS makes the proteins more soluble. Soaking gels in this solution causes them to expand considerably, and this increase in pore size may also facilitate the release of ubiquitinated proteins from the gel.
7. Heat-inactivation of ubiquitin on membrane blots. Purified ubiquitin is usually SDS-denatured prior to being conjugated to hemocyanin or gamma globulin, and this immunogen is frequently used to generate antiubiquitin antibodies. Consequently, an antibody generated by this technique may fail to recognize its epitope in native, nondenatured ubiquitin, whether it is free or conjugated to proteins. Some time ago, Swerdlow and coworkers ***(26)*** reported the novel observation that boiling or autoclaving nitrocellulose membranes with adherent free ubiquitin or ubiquitinated proteins from HeLa cell lysates greatly improved their immunoreactivity. These investigators concluded that heat denaturation of ubiquitin revealed latent antigenic sites that were hidden within the mostly globular intact ubiquitin molecule. Ubiquitinated proteins bound to the surface of the membrane should therefore be heat activated by autoclaving when using an antibody generated against denatured ubiquitin. Often, this is proprietary information that is not available. Importantly, heat activation should be done prior to blocking the membrane to prevent the heat denaturation of milk proteins, which will interfere with subsequent immunoblotting. Membranes should be autoclaved for 30–40 min, sandwiched between several sheets of Whatman 3MM paper, at the bottom of a glass tray containing deionized water. Autoclaving may wrinkle the membrane somewhat, but do not attempt to autoclave the membrane between two glass plates. Unfortunately, this will severely corrugate the blot, possibly because of trapped steam or uneven heat transfer from the glass to the membrane, making subsequent immunoblotting difficult, if not impossible.

   Autoclaving the membrane may greatly intensify the resulting immunoblotting signal. However, this procedure may not be universally effective for all anti-ubiquitin antibodies. A pilot experiment should be conducted to determine whether autoclaving the membrane to denature ubiquitin increases the immune reactivity of a particular antibody toward ubiquitinated proteins.

If an autoclave is not available, the membrane blots may be heat activated by boiling them in a glass tray for about 30 min. An alternative treatment to promote ubiquitin epitope exposure and increase the sensitivity of antiubiquitin immunoblots is to immerse the membrane in 6 *M* guanidine-HCl, 20 m*M* Tris-HCl, pH 7.5, 1 m*M* PMSF, and 5 m*M* DTT, for 30 min at room temperature, before blocking ***(49)***. Heat activation and guanidine activation both denature ubiquitin and facilitate the recognition of latent ubiquitin epitopes by antiubiquitin antibodies resulting in intensification of the antiubiquitin immunoblot signal 20-fold or greater ***(26)***.

8. Antibody epitope masking by ubiquitin. In theory, the addition of multiple copies of ubiquitin may interfere with the recognition of the protein epitope by the immunoprecipitating antibody. If an initial attempt to determine the ubiquitin status of a particular protein is inconclusive, epitope masking by polyubiquitin chains may have occurred. In this situation, it is worthwhile to attempt to immunoprecipitate the target protein with an alternative antibody that recognizes a different epitope or use a polyclonal antibody. Epitope masking by ubiquitination theoretically can also occur on immunoblots probed for the protein being ubiquitinated.
9. Use of epitope-tagged ubiquitin. In some instances, epitope-tagged ubiquitin can be used to increase the specificity and sensitivity of ubiquitinated protein measurement ***(50)***. Ubiquitin covalently linked to hemagglutinin (HA), c-Myc protein, hexahistidine, biotin, or even glutathione transferase is frequently added to in vitro biological systems, such as reticulocyte lysate, that contains the enzymes necessary to ubiquitinate proteins. Epitope-tagged ubiquitin effectively competes with unmodified ubiquitin for enzymatic conjugation to newly synthesized or exogenously added purified or immunoprecipitated proteins. Epitope-tagged ubiquitin can be introduced into cells by transient transfection with a vector containing an insert coding for fused epitope-tagged ubiquitin, such as HA-ubiquitin. Immunoprecipitation with a highly efficient anti-HA antibody will efficiently isolate HA-tagged ubiquitinated proteins from cell lysates. Following separation of the immunoprecipitated, epitope-tagged ubiquitinated proteins by SDS-PAGE, immunoblotting with an antibody directed toward the particular protein in question can be used to detect ubiquitinated forms, which will be seen as higher molecular weight bands of the particular protein. This technique has been used to demonstrate ubiquitination of the short-lived MATa2 yeast transcription regulator ***(51)***, c-Jun transcription factor ***(52)***, and the p27 cyclin-dependent kinase inhibitor ***(53)***. One can also visualize epitope-tagged ubiquitinated proteins on immunoblots by probing with an antibody that recognizes the epitope tag fused to the ubiquitin molecule.
10. Detection reagents: the most sensitive visualization technique for anti-ubiquitin Western blotting is luminol-generated chemiluminescence. Both Amersham (Piscataway, NY, USA) and Pierce (Rockford, IL, USA) sell easy-to-use chemiluminescence detection kits. This technique requires that the primary antibody be decorated with a species-appropriate HRP-conjugated secondary antibody, which catalyzes light generation via the hydrogen peroxide-dependent, HRP-catalyzed oxidation of luminol. Occasionally, the light signal may be too intense when the reagents in these kits are used full strength. In this situation, the chemiluminescence detection reagents can be diluted 1:1 to 1:5 with deionized water. This will attenuate the light output from the immunoblot. An alternative technique is to place an undeveloped film between the immunoblot and the detection film, and of course, develop only the detection film. This procedure will diminish an overly strong signal approximately fivefold.

11. Smearing on antiubiquitin immunoblots is not a failed experiment. Antiubiquitin immunoblots of samples immunoprecipitated using an antibody against the protein under investigation may appear as discrete bands, at intervals of approx 8 kDa, extending upward from the expected location of the protein. This "ladder" of bands with slower mobility results from the successive addition of monomeric ubiquitin molecules to targeted proteins to form chains containing as many as 20 or more ubiquitin molecules (an addition of 170 kDa molecular mass). More often, however, antiubiquitin immunoblotting visualizes ubiquitinated proteins as an unappealing smear stretching upwards toward the top of the membrane. This phenomenon is especially true for many membrane-anchored proteins that are heavily glycosylated.

    For example, antiubiquitin immunoblots of the glycosylated plasma membrane tyrosine kinase, ErbB2 ***(54)*** or the EGF receptor ***(55)*** will invariable yield smears, no matter how thoroughly the immunoprecipitations might be washed. *N*-linked oligosaccharide chains of various lengths and heterogeneous ubiquitin chains, possibly conjugated to more than one lysine residue within an individual protein molecule, make it difficult to resolve ubiquitinated glycosylated proteins into sharp bands. Incubating cell lysates with endoglycosidase-F to cleave oligosaccharides from glycosylated ubiquitinated proteins may diminish this smearing and improve immunoblot quality ***(56)***. On the other hand, cytosolic non-glycosylated ubiquitinated proteins such as IκB ***(57)***, p53 ***(58)***, and B-catenin ***(59)*** also appear as smears instead of a "ladder of bands" in antiubiquitin immunoblots. Smears have also been obtained when epitope-tagged or biotin-conjugated ubiquitin have been used to probe for the ubiquitination of c-Jun ***(52–60)***, the p27 cyclin-dependent kinase inhibitor ***(53)***, IκBα ***(61)*** and hypoxia-inducible transcription factor 1α ***(62)***.

    Collectively, these observations indicate that smearing is a frequent characteristic of antiubiquitin immunoblots. The smearing is a consequence of different-length ubiquitin chains attached to a substrate protein at a single site, several different length polyubiquitin chains conjugated to multiple lysine residues within the protein, or the combination of both types of ubiquitin modifications. Sometimes bands can be seen within the smears on antiubiquitin immunoblots, and these bands probably represent the more abundant multiples of the ubiquitinated protein.

12. Measurement of unconjugated ubiquitin. In addition to ubiquitin–protein conjugates, it may be necessary to measure unconjugated ubiquitin. Because ubiquitin is heat stable and denaturation resistant, unconjugated ubiquitin can be separated from the majority of ubiquitinated proteins by scraping cells into a small volume (0.5–1.0 mL) of deionized water and boiling the sample in a glass tube for 10 min. Many denatured ubiquitinated and nonubiquitinated proteins rapidly coagulate at this temperature and they can be easily separated from free ubiquitin by low speed centrifugation (1000*g* for 5 min). An additional benefit is that boiling rapidly denatures deubiquitinating enzymes including the proteasome and prevents any artifactual increase in free ubiquitin in the sample. Unconjugated ubiquitin remains in the supernatant fraction, and it can be separated from any remaining ubiquitinated proteins by 16% or 18% polyacrylamide SDS-PAGE. Soaking high percentage (18–20%) SDS-polyacrylamide gels in 2.3% SDS, 5% 2-mercaptoethanol, 63 m*M* Tris-HCl, pH 6.8, buffer for 30 min prior to electrotransferring to the nitrocellulose membrane and using 25 m*M* CAPS, pH 10, with 20% (v/v) methanol transfer buffer will facilitate the elution of free ubiquitin from the gel to the membrane surface. As with ubiquitinated proteins, autoclaving free ubiquitin on membranes before immunoblotting greatly enhances its immune detection. It should be noted that although

monomeric ubiquitin has a molecular mass of 8.6 kDa, its mobility on SDS-PAGE causes it to appear as a 5- or 6-kDa polypeptide.

13. Verification of ubiquitination. When measuring ubiquitination of a protein for the first time, it is a good idea to verify the initial result by probing separate immunoblots with one or two other anti-ubiquitin antibodies, preferably antibodies from different species that recognize different ubiquitin epitopes. Another validation approach is to use cells that contain a temperature-sensitive mutated E1 ubiquitin-activating enzyme, provided these cells express the particular protein under study. When ts-mutant E1 cells, such as ts-20 ***(63)*** or ts-85 ***(64)***, are heated from the permissive temperature of 33°C to the nonpermissive temperature of 39°C, the E1 enzyme becomes unstable and no longer activates ubiquitin, the first step in protein ubiquitination. Ubiquitination of the protein under investigation is verified by observing a dramatic decrease in its ubiquitination on immunoblots after the ts cells have been incubated at the nonpermissive temperature for 4 h.
14. Disruption of protein chaperones with Geldanamycin. Hsp90-containing molecular chaperone complexes stabilize many newly synthesized and certain long-lived mature proteins ***(65)***. Disrupting the client protein–heat shock protein chaperone complexes with the Hsp90-binding ansamycin drug, Geldanamycin (GA), stimulates the ubiquitination of chaperoned proteins ***(65,66)***. Treating cells with GA has been shown to greatly increase ubiquitination of the ErbB2 membrane receptor tyrosine kinase ***(54)***, c-Raf-1 serine-threonine kinase ***(67)***, certain mutated forms of p53 ***(68)*** and the mutated cystic fibrosis transmembrane conductance regulator protein ***(69)***. Thus, GA can be used as an experimental tool to stimulate ubiquitination of those proteins that depend upon Hsp90 chaperone interactions for their stability. Treating cells with 1 μ*M* GA combined with a proteasome inhibitor such as PS-341 at 25 n*M* should further increase the likelihood of accumulating ubiquitinated forms of proteins stabilized by Hsp90 chaperone complexes.

## Acknowledgment

The authors apologize in advance to any researchers whose published work we failed to acknowledge in this chapter. For a fine article on measuring protein ubiquitination in yeast, *see* **ref. *22***.

## References

1. Finley, D. and Chau, V. (1991) Ubiquitination. *Annu. Rev. Cell Biol.* **7,** 25–69.
2. Bonifacino, J. S. and Weissman, A. M. (1998) Ubiquitin and the control of protein fate in the secretory and endocytic pathways. *Annu. Rev. Cell Dev. Biol.* **14,** 19–57.
3. Schwartz, A. L. and Ciechanover, A. (1999) The ubiquitin-proteasome pathway and pathogenesis of human diseases. *Annu. Rev. Med.* **50,** 57–74.
4. Baarends, W. M., Roest, H. P., and Grootegoed, J. A. (1999) The ubiquitin system in gametogenesis. *Mol. Cell Endocrinol.* **151,** 5–16.
5. Jesenberger, V. and Jentsch, S. (2002) Deadly encounter: ubiquitin meets apoptosis. *Nat. Rev. Mol. Cell Biol.* **3,** 112–121.
6. Wojcik, C. (2002) Regulation of apoptosis by the ubiquitin and proteasome pathway. *J. Cell. Mol. Med.* **6,** 25–48.
7. Ciechanover, A. and Schwartz, A. L. (1998) The ubiquitin-proteasome pathway: the complexity and myriad functions of protein death. *Proc. Natl. Acad. Sci. USA* **95,** 2727–2730.

8. D'Andrea, A. and Pellman, D. (1998) Deubiquitinating enzymes: a new class of biological regulators. *Crit. Rev. Biochem. Mol. Biol.* **33,** 337–352.
9. Hershko, A. and Ciechanover, A. (1992) The ubiquitin system for protein degradation. *Annu. Rev. Biochem.* **61,** 761–807.
10. Hershko, A. and Ciechanover, A. (1998) The ubiquitin system. *Annu. Rev. Biochem.* **67,** 425–479.
11. Pickart, C. M. (2001) Mechanisms underlying ubiquitination. *Annu. Rev. Biochem.* **70,** 503–533.
12. Weissman, A. M. (2001) Themes and variations on ubiquitylation. *Nat. Rev. Mol. Cell Biol.* **2,** 169–178.
13. Wilkinson, K. D. (1997) Regulation of ubiquitin-dependent processes by deubiquitinating enzymes. *FASEB J.* **11,** 1245–1256.
14. Yang, Y. and Yu, X. (2003) Regulation of apoptosis: the ubiquitous way. *FASEB J.* **17,** 790–799.
15. Jagoe, R. T. and Goldberg, A. L. (2001) What do we really know about the ubiquitin-proteasome pathway in muscle atrophy? *Curr. Opin. Clin. Nutr. Metab. Care* **4,** 183–190.
16. Spataro, V., Norbury, C., and Harris, A. L. (1998) The ubiquitin-proteasome pathway in cancer. *Br. J. Cancer* **77,** 448–455.
17. Scheffner, M. and Whitaker, N. J. (2003) Human papillomavirus-induced carcinogenesis and the ubiquitin-proteasome system. *Semin. Cancer Biol.* **13,** 59–67.
18. Pugh, C. W. and Ratcliffe, P. J. (2003) The von Hippel-Lindau tumor suppressor, hypoxia-inducible factor-1 (HIF-1) degradation, and cancer pathogenesis. *Semin. Cancer Biol.* **13,** 83–89.
19. Haas, A. L. and Bright, P. M. (1985) The immunochemical detection and quantitation of intracellular ubiquitin-protein conjugates. *J. Biol. Chem.* **260,** 12464–12473.
20. Laemmli, U. K. (1970) Cleavage of structural proteins during the assembly of the head of bacteriophage T4. *Nature (Lond.)* **227,** 680–685.
21. Wilkinson, K. D. (1990) Detection and inhibition of ubiquitin-dependent proteolysis. *Methods Enzymol.* **185,** 387–397.
22. Laney, J. D. and Hochstrasser, M. (2002) Assaying protein ubiquitination in *Saccharomyces cerevisiae. Methods Enzymol.* **351,** 248–257.
23. Johnston, J. A., Ward, C. L., and Kopito, R. R. (1998) Aggresomes: a cellular response to misfolded proteins. *J. Cell Biol.* **143,** 1883–1898.
24. Towbin, H., Staehelin, T., and Gordon, J. (1979) Electrophoretic transfer of proteins form polyacrylamide gels to nitrocellulose sheets: procedure and some applications. *Proc. Natl. Acad. Sci. USA* **76,** 4350–4365.
25. Mimnaugh, E. G., Chen, H. Y., Davie, J. R., Celis, J. E., and Neckers, L. (1997) Rapid deubiquitination of nucleosomal histones in human tumor cells caused by proteasome inhibitors and stress response inducers: effects on replication, transcription, translation and the cellular stress response. *Biochemistry* **36,** 14418–14429.
26. Swerdlow, P. S., Finley, D., and Varshavsky, A. (1986) Enhancement of immunoblot sensitivity by heating of hydrated filters. *Anal. Biochem.* **156,** 147–153.
27. Fenteany, G., Standaert, R. F., Lane, W. S., Choi, S., Corey, E. J., and Schreiber, S. L. (1995) Inhibition of proteasome activities and subunit-specific amino-terminal threonine modification by lactacystin. *Science* **268,** 726–731.
28. Papa, F. R. and Hochstrasser, M. (1993) The yeast DOA4 gene encodes a deubiquitinating enzyme related to a product of the human tre-2 oncogene. *Nature* **366,** 313–319.
29. Falquet, L., Paquet, N., Fruitger, S., Hughes, G. J., Hoarig-Van, K., and Jaton, J.-C. (1995) cDNA cloning of a human 100 kDa de-ubiquitinating enzyme – the 100 kDa human

de-ubiquitinase belongs to the ubiquitin C-terminal hydrolase family-2 (UCH2). *FEBS Lett.* **376,** 233–237.

30. Mayer, A. N. and Wilkinson, K. D. (1989) Detection, resolution, and nomenclature of multiple ubiquitin carboxyl-terminal esterases from bovine calf thymus. *Biochemistry* **28,** 166–172.
31. Baek, S. H., Woo, S. K., Lee, J. I., et al. (1997) New de-ubiquitinating enzyme, ubiquitin C-terminal hydrolase 8, in chick skeletal muscle. *Biochem. J.* **325,** 325–330.
32. Ciechanover, A., Finley, D., and Varshavsky, A. (1984) The ubiquitin-mediated proteolytic pathway and mechanism of energy-dependent intracellular protein degradation. *J. Cell. Biochem.* **24,** 137–163.
33. Fenteany, G. and Schreiber, S. L. (1998) Lactacystin, proteasome function and cell fate. *J. Biol. Chem.* **273,** 8545–8548.
34. Adams, J. (2003) The proteasome: structure, function, and role in the cell. *Cancer Treat. Rev.* **29,** 3–9.
35. Meng, L., Mohan, R., Kwok, B. H., Elofsson, M., Sin, N., and Crews, C. M. (1999) Epoxomicin, a potent and selective proteasome inhibitor, exhibits *in vivo* antiinflammatory activity. *Proc. Natl. Acad. Sci. USA* **96,** 10403–10408.
36. Figueiredo-Pereira, M. E., Chen, W. E., Li, J., and Johdo, O. (1996) The antitumor drug aclacinomycin A, which inhibits the degradation of ubiquitinated proteins, shows selectivity for the chymotrypsin-like activity of the bovine pituitary 20 S proteasome. *J. Biol. Chem.* **271**, 16455–16459.
37. Imajoh-Ohmi, S. I., Kawaguchi, T., Sugiyama, S., Tanaka, K., Omura, S., and Kikuchi, H. (1995) Lactacystin, a specific inhibitor of the proteasome, induces apoptosis in human monoblast U937 cells. *Biochem. Biophys Res. Commun.* **217,** 1070–1077.
38. Drexler, H. C. A. (1997) Activation of the cell death program by inhibition of proteasome function. *Proc. Natl. Acad. Sci. USA* **94,** 855–860.
39. Orlowski, R. Z. (1999) The role of the ubiquitin-proteasome pathway in apoptosis. *Cell Death Differ.* **6,** 303–313.
40. Mimnaugh, E. G., Kayastha, G., McGovern, N. B., et al. (2001) Caspase-dependent deubiquitination of monoubiquitinated nucleosomal histone H2A induced by diverse apoptogenic stimuli. *Cell Death Differ.* **8,** 1182–1196.
41. Ciechanover, A. and Schwartz, A. L. (1994) The ubiquitin-mediated proteolytic pathway: mechanisms of recognition of the proteolytic substrate and involvement in the degradation of native cellular proteins. *FASEB J.* **8,** 182–191.
42. Pickart, C. M. (1997) Targeting of substrates to the 26S proteasome. *FASEB J.* **11,** 1055–1066.
43. Laney, J. D. and Hochstrasser, M. (1999) Substrate targeting in the ubiquitin system. *Cell* **97,** 427–430.
44. Deveraux, Q., Ustrell, V., Pickart, C., and Rechsteiner, M. (1994) A 26S proteasase subunit that binds ubiquitin conjugates. *J. Biol. Chem.* **269,** 7059–7061.
45. Eytan, E., Armon, T., Heller, H., Beck, S., and Hershko, A. (1993) Ubiquitin C-terminal hydrolase activity associated with the 26 S protease complex. *J. Biol. Chem.* **268,** 4668–4674.
46. Vassilev, A. P., Rasmussen, H. H., Christensen, E. I., Nielsen, S., and Celis, J. E. (1995) The levels of ubiquitinated histone H2A are highly upregulated in transformed human cells: Partial colocalization of uH2A clusters and PCNA/cyclin foci in a fraction of cells in S-phase. *J. Cell Sci.* **108,** 1205–1215.
47. Fujimuro, M., Sawada, H., and Yokosawa, H. (1994) Production and characterization of monoclonal antibodies specific to multi-ubiquitin chains of polyubiquitinated proteins. *FEBS Lett.* **349,** 173–180.

48. Fujimuro, M., Sawada, H., and Yokosawa, H. (1997) Dynamics of ubiquitin conjugation during heat-shock response revealed by using a monoclonal antibody specific to multi-ubiquitin chains. *Eur. J. Biochem.* **249,** 427–433.
49. Pagano, M. (1997) Cell cycle regulation by the ubiquitin pathway. *FASEB J.* **11,** 1067–1075.
50. Ellison, M. J. and Hochstrasser, M. (1991) Epitope-tagged ubiquitin: a new probe for analyzing ubiquitin function. *J. Biol. Chem.* **266,** 211150–21157.
51. Hochstrasser, M., Ellison, M. J., Chau, V., and Varshavsky, A. (1991) The short-lived MATα2 transcriptional regulator is ubiquitinated *in vivo. Proc. Natl. Acad. Sci. USA* **88,** 4606–4610.
52. Treier, M., Staszewski, L. M., and Bohmann, D. (1994) Ubiquitin-dependent c-jun degradation in vivo is mediated by the δ domain. *Cell* **78,** 787–798.
53. Pagano, M., Tam, S. W., Theodoras, A. M., et al. (1995) Role of the ubiquitin-proteasome pathway in regulating abundance of the cyclin-dependent kinase inhibitor p27. *Science* **269,** 682–685.
54. Mimnaugh, E. G., Chavany, C., and Neckers, L. (1996) Polyubiquitination and proteasomal degradation of the p185*c-erb*$^{B-2}$ receptor protein-tyrosine kinase induced by geldanamycin. *J. Biol. Chem.* **271,** 22796–22801.
55. Galcheva-Gargova, Z., Theroux, S. J., and Davis, R. J. (1995) The epidermal growth factor receptor is covalently linked to ubiquitin. *Oncogene* **11,** 2649–2655.
56. Hiller, M. M., Finger, A., Schweiger, M., and Wolf, D. H. (1996) ER degradation of a misfolded luminal protein by the cytosolic ubiquitin-proteasome pathway. *Science* **273,** 1725–1728.
57. Chen, Z., Hagler, J., Palombella, V. J., et al. (1995) Signal-induced site-specific phosphorylation targets IκBα to the ubiquitin-proteasome pathway. *Genes Dev.* **9,** 1586–1597.
58. Maki, C. G., Huibregtse, J. M., and Howley, P. M. (1996) In vivo ubiquitination and proteasome-mediated degradation of p53. *Cancer Res.* **56,** 2649–2654.
59. Bonvini, P., Nguyen, P., Trepel, J., and Neckers, L. M. (1998) *In vivo* degradation of N-myc in neuroblastoma cells is mediated by the 26S proteasome. *Oncogene* **16,** 1131–1139.
60. Musti, A. M., Treier, M., and Bohmann, D. (1997) Reduced ubiquitin-dependent degradation of c-Jun after phosphorylation by MAP kinases. *Science* **275,** 400–402.
61. Roff, M., Thompson, J., Rodriquez, M. S., et al. (1996) Role of IκBα ubiquitination in signal-induced activation of NF-κB *in vivo. J. Biol. Chem.* **271,** 7844–7850.
62. Huang, L. E., Gu, J., Schau, M., and Bunn, H. F. (1998) Regulation of hypoxia-inducible factor1α is mediated by an $O_2$-dependent degradation domain via the ubiquitin-proteasome pathway. *Proc. Natl. Acad. Sci. USA* **95,** 7987–7992.
63. Kulka, R. G., Raboy, B., Schuster, R., et al. (1988) A chinese hamster cell cycle mutant arrested at G2 phase has a temperature-sensitive ubiquitin-activating enzyme, E1. *J. Biol. Chem.* **263,** 15726–15731.
64. Deveraux, Q., Wells, R., and Rechsteiner, M. (1990) Ubiquitin metabolism in ts85 cells, a mouse carcinoma line that contains a thermolabile ubiquitin activating enzyme. *J. Biol. Chem.* **265,** 6323–6329.
65. Neckers, L., Mimnaugh, E. G., and Schuhlte, T. W. (1999) The Hsp90 chaperone family. In *Handbook of Experimental Pharmacology,* Vol. 136: *Stress Proteins* (Latchman, D. S., ed.), Springer-Verlag, Berlin, pp. 9–42.
66. Whitesell, L., Mimnaugh, E. G., DeCosta, B., Myers, C. E., and Neckers, L. M. (1994) Inhibition of heat shock protein HSP90-pp60*v-src* heteroprotein complex formation by benzoquinone ansamycins: essential role for stress proteins in oncogenic transformation. *Proc. Natl. Acad. Sci. USA* **91,** 8324–8328.

67. Schulte, T. W., An, W. G., and Neckers, L. M. (1997) Geldanamycin-induced destabilization of Raf-1 involves the proteasome. *Biochem. Biophys. Res. Commun.* **239,** 665–669.
68. Whitesell, L., Sutphin, P., An, W. G., Schulte, T., Blagosklonny, M. V., and Neckers, L. (1997) Geldanamycin-stimulated destabilization of mutated p53 is mediated by the proteasome *in vivo*. *Oncogene* **14,** 2809–2816.
69. Loo, M. A., Jensen, T. J., Cui, L., Hou, Y., Chang, X. B., and Riordan, J. R. (1998) Perturbation of Hsp90 interaction with nascent CFTR prevents its maturation and accelerates its degradation by the proteasome. *EMBO J.* **17,** 6879–6887.
70. Strack, B., Calistri, A., Accola, M. A., Palu, G., and Gottlinger, H. G. (2000) A role for ubiquitin ligase recruitment in retrovirus release. *Proc. Natl. Acad. Sci. USA* **97,** 13063–13068.
71. Ott, D. E., Coren, L. V., Chertova, E. N., Gagliardi, T. D., and Schubert, U. (2000) Ubiquitination of HIV-1 and MuLV Gag. *Virology* **278,** 111–121.
72. Conaway, R. C., Brower, C. S., and Conaway, J. W. (2002) Emerging roles of ubiquitin in transcription regulation. *Science* **296,** 1254–1258.
73. Hoppe, T., Matuschewski, K., Rape, M., Schlenker, S., Ulrich, H. D., and Jentsch, S. (2000) Activation of a membrane-bound transcription factor by regulated ubiquitin/proteasome-dependent processing. *Cell* **102,** 577–586.
74. Kaiser, P., Flick, K., Wittenberg, C., and Reed, S. I. (2000) Regulation of transcription by ubiquitination without proteolysis: Cdc34/SCF(Met30)-mediated inactivation of the transcription factor Met4. *Cell* **102,** 303–314.
75. Muratani, M. and Tansey, W. P. (2003) How the ubiquitin-proteasome system controls transcription. *Nat. Rev. Mol. Cell Biol.* **4,** 192–201.
76. Hitchcock, A. L., Krebber, H., Frietze, S., Lin, A., Latterich, M., and Silver, P. A. (2001) The conserved npl4 protein complex mediates proteasome-dependent membrane-bound transcription factor activation. *Mol. Biol. Cell* **12,** 3226–3241.
77. Hicke, L. (2001) A new ticket for entry into budding vesicles-ubiquitin. *Cell* **106,** 527–530.
78. Hicke, L. (2001) Protein regulation by monoubiquitin. *Nat. Rev. Mol. Cell Biol.* **2,** 195–201.
79. Shih, S. C., Prag, G., Francis, S. A., Sutanto, M. A., Hurley, J. H., and Hicke, L. (2003) A ubiquitin-binding motif required for intramolecular monoubiquitylation, the CUE domain. *EMBO J.* **22,** 1273–1281.
80. Sun, Z. W. and Allis, C. D. (2002) Ubiquitination of histone H2B regulates H3 methylation and gene silencing in yeast. *Nature* **418,** 104–108.
81. Dover, J., Schneider, J., Tawiah-Boateng, M. A., et al. (2002) Methylation of histone H3 by COMPASS requires ubiquitination of histone H2B by Rad6. *J. Biol. Chem.* **277,** 28368–28371.
82. Ng, H. H., Xu, R. M., Zhang, Y., and Struhl, K. (2002) Ubiquitination of histone H2B by Rad6 is required for efficient Dot1-mediated methylation of histone H3 lysine 79. *J. Biol. Chem.* **277,** 34655–34657.
83. Adams, J. (2003) Potential for proteasome inhibition in the treatment of cancer. *Drug Discov. Today* **8,** 307–315.

# 15

# Assays for Proteasome Assembly and Maturation

**R. Jürgen Dohmen, Markus K. London, Christoph Glanemann, and Paula C. Ramos**

**Summary**

The 20S proteasome is a complex multisubunit protease that is present in all phylae of life. Eukaryotic 26S proteasomes, which are composed of 20S proteasomes and 19S activator complexes, mediate the degradation of ubiquitylated proteins. Biogenesis of proteasomes involves a coordinated expression of proteasome genes as well as numerous assembly and maturation steps. Activation of proteolytic sites occurs via autocatalytic processing of the N-terminal propeptides of β subunits. This process is coupled to the dimerization of half-proteasome precursor complexes and, in eukaryotes, requires the presence of the Ump1 maturation factor to occur efficiently. After activation of proteolytic sites the encased Ump1 is degraded rapidly. Here we describe methods that track assembly and maturation of proteasomes in bacteria and eukaryotic cells. Assembly intermediates and mature forms of the proteasome present in cells at steady state are analyzed by gel filtration and immunoblotting after sodium dodecyl sulfate (SDS)- and native polyacrylamide gel electrophoresis (PAGE). The kinetics of proteasome assembly is followed by pulse chase detection of β subunit maturation or of Ump1 degradation.

**Key Words:** Degradation; gel filtration; native polyacrylamide gel electrophoresis; proteasome maturation; pulse chase; ubiquitin, Ump1.

## 1. Introduction

20S proteasomes are complex barrel-shaped threonine proteases that exist in all phylae of life. They are composed of four rings of seven subunits each, two outer rings formed by α subunits and two inner rings formed by β subunits. The active sites including the N-terminal threonine residues reside in the β subunits. 26S proteasomes found in eukaryotes are composed of 20S proteasomes (core particle, CP) to which two 19S activator complexes (also termed regulatory particles, RP) are attached at either end. Attachment of the RP enables the proteasome to degrade ubiquitylated proteins. The assembly and maturation of this complex protease is a complicated multistep process that is only partly understood (reviewed in **ref. *1***). Assembly proceeds via

From: *Methods in Molecular Biology, vol. 301, Ubiquitin–Proteasome Protocols*
Edited by: C. Patterson and D. M. Cyr © Humana Press Inc., Totowa, NJ

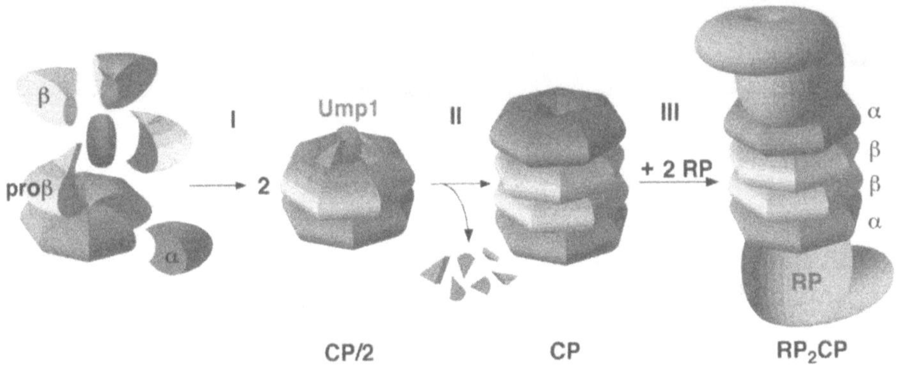


Fig. 1. Model of proteasome assembly and maturation. In step I, α and β subunits assemble into half-proteasome precursor complexes (CP/2) that are characterized by the presence of unprocessed, propeptide-bearing β subunits and maturation factor Ump1. In step II, two half-proteasomes dimerize to form the 20S core particle (CP), which coincides with the processing of β subunits' propeptides and the degradation of Ump1. Attachment of two regulatory particles (RP) to the CP results in the formation of the 26S proteasome ($RP_2CP$).

half-proteasome precursor complexes (CP/2) composed of one α-ring and one ring containing unprocessed, propeptide-bearing precursor forms of the active site β subunits (**Fig. 1**). In eukaryotes, half-proteasome precursor complexes, in addition, are characterized by the presence of the Ump1 protein, a dedicated chaperone that *u*nderpins *m*aturation of the *p*roteasome *(2,3)*. The dimerization of two such precursor complexes triggers the autocatalytic processing of β subunits *(4)*. The presence of Ump1 is required for these maturation steps to occur efficiently. On maturation of the active sites, Ump1 is rapidly degraded characterizing it also as the first substrate of the nascent eukaryotic proteasome *(2)*.

In this chapter we describe several methods that allow to follow assembly and maturation of proteasomes in wild-type and mutant yeast cells. These methods have proven to be suitable for studying the biogenesis of proteasomes also in bacteria or mammalian cells. The distribution of assembly intermediates and mature forms of the proteasome at steady-state is assayed by gel filtration combined with activity determination or immunoblot analyses. The kinetics of proteasome assembly and maturation is followed by pulse chase analysis of β-subunit processing and of Ump1 degradation.

## 2. Materials

### 2.1. Medium

1. Yeast synthetic minimal medium with 2% dextrose (SD): 6.7 g yeast nitrogen base without amino acids, 2% glucose, bring up to 1 L with $H_2O$ and appropriate dropout supplements for yeast strain selection (available from BD Biosciences).

### 2.2. Protein Extraction

1. Extraction buffer: 50 m*M* Tris-HCl, pH 7.5, 2 m*M* ATP, 5 m*M* $MgCl_2$, 1 m*M* dithiothreitol (DTT), 15% (v/v) glycerol.
2. Mortar and pestle.

### 2.3. Gel Filtration

1. Equilibration buffer: identical to extraction buffer.
2. Proteasome substrates:
   a. Succinyl-Leu-Leu-Val-Tyr-7-amido-4-methylcoumarin (Suc-LLVY-AMC).
   b. *t*-Butyloxycarbonyl-Leu-Arg-Arg-7-amido-4-methylcoumarin (Boc-LRR-AMC).
   c. Acetyl-Gly-Pro-Leu-Asp-7-amido-4-methylcoumarin (Ac-GPLD-AMC).

Stock solutions (100X) in dimethyl sulfoxide (DMSO) contain 10 mg/mL of substrate. Store frozen at –20°C.

### 2.4. Native Polyacrylamide Gel Electrophoresis (PAGE)

1. Acrylamide stock solution: 30% acrylamide, 0,8% *bis*-acrylamide.
2. Separating gel buffer (TBE) (4X): 360 m*M* Tris, pH 8.3, 320 m*M* boric acid, 0.4 m*M* EDTA.
3. MAD (10X): 50 m*M* $MgCl_2$, 10 m*M* ATP, 10 m*M* DTT.
4. Stacking gel buffer: 1 *M* Tris-HCl, pH 6.8.
5. Sucrose solution (10% w/v).
6. *N,N,N',N'*-Tetramethylethylenediamine (TEMED).
7. Ammonium persulfate solution (10% w/v).
8. Loading buffer (4X): 200 m*M* Tris-HCl, pH 6.8, 60% (v/v) glycerol, 0.05% (w/v) bromophenol blue.
9. Electrophoresis buffer: 90 m*M* Tris, 80 m*M* boric acid (pH will be ~8.3), 0,1 m*M* EDTA, 5 m*M* $MgCl_2$, 1 m*M* ATP, and 1 m*M* DTT.
10. Blotting buffer: 25 m*M* Tris, 192 m*M* glycine, 20% methanol, 0.1% sodium dodecyl sulfate (SDS).

### 2.5. Immunoblot Analysis

1. Fixation solution: 10% (v/v) acetic acid, 25 % (v/v) isopropanol.
2. Phosphate-buffered saline (PBS): 4.3 m*M* $Na_2HPO_4$, 1.4 m*M* $KH_2PO_4$, 2.7 m*M* KCl, 137 m*M* NaCl.
3. Low-salt washing solution (PBST): PBS with 0.1% (v/v) Tween-20.
4. Blocking solution: PBST with 3% (w/v) nonfat dry milk powder.
5. High-salt washing solution: 1 *M* NaCl, 10 m*M* $Na_2HPO_4$, 0.5% (v/v) Tween-20.

### 2.6. Pulse-Chase Analysis

1. [$^{35}$S]Methionine, or a mixture of [$^{35}$S]methionine and [$^{35}$S]cysteine, for in vivo labeling of proteins (pulse).
2 Pulse solution: 2X SD medium without methionine.
3. Chase solution: 2X SD with 10 m*M* methionine, cycloheximide (0.1 mg/mL)
4. Lysis buffer: 50 m*M* *N*-(2-hydroxyethyl)piperazine-*N'*-(2-ethanesulfonic acid) (HEPES)–NaOH, pH 7.5, 5 m*M* EDTA, 0.15 *M* NaCl, 1% (v/v) Triton X-100.
5. Glass beads: acid washed, diameter 425–600 μm.
6. Filter paper discs for the determination of TCA-precipitable counts. We use round filter papers (diameter 9 mm) commonly referred to as "antibiotic assay test discs." The paper

should be highly absorbent to ensure that the spotted protein solution does not leak through.
7. Trichloroacetic acid (TCA).
8. Protein A–Sepharose, 50% in $H_20$.
9. Sodium acetate.

## 3. Methods

### 3.1. Generation of Protein Extracts From Frozen Cells

Various cell extraction procedures have been employed to generate preparations of active proteasomes. In our hands, a procedure (described below for budding yeast cells) that involves grinding of cells frozen in liquid nitrogen using mortar and pestle yielded the highest content of active proteasome *(2)*. The same procedure can also be applied to bacterial and mammalian cells.

#### 3.1.1. Preparation of the Cells

1. Grow *Saccharomyces cerevisiae* in 200 mL of selective SD medium to an optical density measured at 600 nm ($OD_{600}$) of 1.0 ± 0.2.
2. Harvest the cells by centrifugation at 3000*g* for 5 min at 4°C. Resuspend the cells in 40 mL of cold distilled water and transfer to a 50-mL Falcon tube. Collect cells by centrifugation as previously, and determine the wet weight (200-mL cultures will yield 0.3–0.4 g).
3. Freeze in liquid nitrogen and store at –80°C until further use (*see* **Note 1**).

#### 3.1.2. Protein Extraction

1. Take the Falcon tube from the –80°C, and keep it for around 30 s at room temperature. To release the pellet from the tube walls, gently knock the tube onto the bench. Do not let the cells thaw to avoid repeated freeze–thaw cycles.
2. Drop the loose cell pellet into a mortar containing liquid nitrogen.
3. Grind the cells to a fine powder with a pestle, always in presence of liquid nitrogen.
4. Add extraction buffer (2 mL/g wet wt) to the cell powder (*see* **Note 2**). Because the mortar is very cold, it is natural that your buffer will freeze. Continue to grind until the suspension is unfrozen. This procedure can be done at room temperature since the mortar will be precooled with liquid nitrogen and the porcelain keeps the temperature very low. Extraction in a cold room is not recommended, as it takes much longer.
5. Transfer the suspension to microfuge tubes and centrifuge for 10 min at 4°C at maximum speed to pellet the cell debris.
6. Transfer supernatant to ultracentrifuge tube and centrifuge at 60,000*g* for 30 min at 4°C to sediment membranes and residual cell debris.
7. The supernatant is your crude extract containing soluble proteins. Take an aliquot to measure the protein content.

### 3.2. Analysis of Proteasomal Complexes by Gel Filtration

To analyze proteasomal complexes as they are depicted in **Fig. 1** in cell extracts, we find fractionation by gel filtration in an FPLC system followed by immunoblot analysis the most convenient method *(2)*. It is fast and highly reproducible due to constant flow rates and automatic fraction collection thus allowing to correlate protein complex sizes with fraction numbers.

### *3.2.1. Gel Filtration*

1. Equilibrate a Superose 6 column (*see* **Note 3**) connected to an FPLC system with equilibration buffer at a flow rate of 0.3 mL/min at 4°C. Because the buffer contains 15% glycerol, higher flow rates result in too high pressure, which may lead to damaging the column.
2. Establish a program to separate proteasomal complexes in the fast protein liquid chromatography (FPLC) with isocratic elution using equilibration buffer at a flow rate of 0.3 mL/min and collecting 0.6-mL fractions.
3. Dilute your sample with extraction buffer to 1 mg/200 μL.
4. Pass the extract through a filter (0.45 μm pore size, precooled at 4°C).
5. Load a 200-μL loop with the sample, and start the run.
6. Take the samples from the fraction collector (that should be at 4°C) and place them on ice as soon as the run is completed.

### *3.2.2. Measuring Proteasomal Peptidase Activity*

Because peptidase activity is restricted to 20S and 26S proteasomes, and is absent from the precursor complexes, its detection can be used to identify mature proteasomes in fractions of cell extracts. The 26S proteasome generally has a higher specific activity than the 20S particle due to the opening of the so-called gate in the α-ring, which is induced by the RPs ***(5,6)***. Gate opening in the 20S proteasome, however, can also be induced in vitro by low levels of SDS ***(5,7)***. Proteasomal peptidase cleavage activity is commonly detected using peptide substrates with fluorogenic leaving groups. Depending on the peptide substrate, one can distinguish the three characteristic activities, namely the chymotrypsin-like (cleavage after large hydrophobic residues), the trypsin-like (cleavage after basic residues) and the postacidic activity (cleavage after acidic residues), which are mediated by the active sites residing in the β5, β2, and the β1 subunit, respectively ***(1)***. An example of monitoring the chymotrypsin-like activity in fractions obtained by gel filtration is shown in **Fig. 2A**.

1. For assaying chymotrypsin-like activity, reactions are set up in a total volume of 100 μL as follows: 80 μL of extraction buffer are mixed with 10 μL of a fraction from the gel filtration and 10 μL of a 1:10 dilution in extraction buffer of Suc-LLVY-AMC stock solution. (If activity in crude extracts is to be determined, approx 5 μg of extract protein is diluted with extraction buffer to a volume of 90 μL before the substrate is added.)
2. For assaying trypsin-like activity, reactions are set up with 40 μL of a fraction, 50 μL of extraction buffer, and 10 μL of a 1:10 dilution in extraction buffer:Boc-LRR-AMC stock solution. (If activity in crude extracts is to be determined, approx 20 μg of extract protein is diluted with extraction buffer to a volume of 90 μL before substrate is added.)
3. For assaying postacidic activity, proceed as for trypsin-like activity but using Ac-GPLD-AMC as a substrate.
4. In each case, the release of fluorescent 7-amido-4-methylcoumarin (AMC) is measured at 440 nm using an excitation wave length of 380 nm (*see* **Note 4**). One unit (U) of a proteolytic activity is defined as 1 μmol of AMC produced per minute under the conditions described.

### *3.2.3. Immunoblot Analysis*

1. Take 75-μL aliquots from the fractions obtained by gel filtration, add 25 μL of 4X loading buffer ***(8)*** and boil for 5 min.
2. Prepare a 4%/12% discontinuous 0.75-mm SDS polyacrylamide gel ***(8)*** with 20 slots.

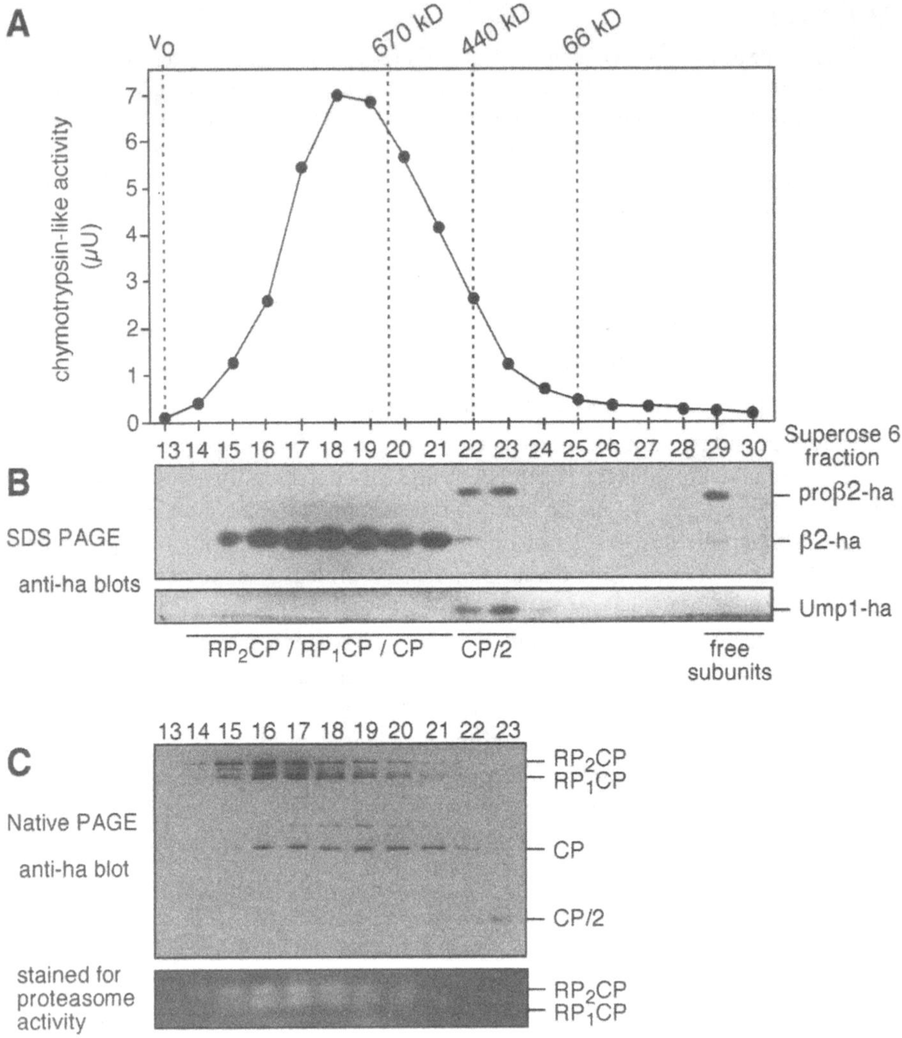


Fig. 2. Analysis of *S. cerevisiae* proteasomal complexes by gel filtration and gel electrophoresis. **(A)** Profile of chymotrypsin-like proteasomal activity in fractions obtained by Superose 6 gel filtration. Void volume ($v_0$) and position of marker proteins (thyroglobulin, ferritin, and bovine serum albumin) are indicated. **(B)** Immunoblot analysis of epitope-tagged proteasomal complexes in Superose 6 fractions after SDS PAGE. In the *upper panel*, the β2 subunit tagged with two ha epitopes (β2-ha) is detected. The *lower panel* shows an otherwise identical experiment, in which the strain carried an ha-tagged version of the maturation factor Ump1 (Ump1-ha). **(C)** Native PAGE analysis of proteasomal complexes in fractions obtained by Superose 6 chromatography. *Upper panel*, immunoblot detecting β2-ha. *Lower panel*, detection of proteasomal activity in a native gel.

3. Load 40 µL of the samples prepared from fractions 12–30. The remainder is frozen and can be used if a repetition of the immunoblot analysis is required.
4. Set up the gel in running buffer (*see* **Note 5**).
5. After electrophoresis, blot proteins to polyvinylidene fluoride (PVDF) membranes, for example, using a semidry blotting system for 30 min at 0.8 mA/cm$^2$ (**ref. *9***; *see* **Note 6**).
6. Incubate the membrane in fixation solution for 5 min.
7. Rinse the membrane with water to remove the fixation solution.
8. Incubate the membrane for 1 h in blocking solution.
9. Incubate the membrane in blocking solution plus primary antibody for 1 h at room temperature or overnight at 4°C.
10. Wash the membrane twice for 5 min with PBST.
11. Wash twice for 5 min with high-salt washing buffer.
12. Incubate the membrane for 15 min in blocking solution.
13. Incubate membrane in blocking solution plus secondary antibody linked to horseradish peroxidase for 1–2 h.
14. Wash membrane twice for 5 min with PBST.
15. Wash for 10 min with high-salt washing buffer.
16. Wash membrane three times for 5 min with PBST.
17. Wash once with PBS for 1 min to remove Tween-20.
18. Incubate blot with chemiluminescence substrate for 1 min and expose to X-ray film.

### 3.3. Analysis of Proteasomal Complexes by Nondenaturing PAGE

To determine the nature of proteasomal complexes in cell crude extracts or fractions thereof, nondenaturing PAGE has proven to be very useful ***(10)***. This technique allows to detect assembly intermediates such as half-proteasome precursor complexes (CP/2) as well as assembled 20S proteasomes with or without attached regulatory particles (CP, $RP_1CP$, or $RP_2CP$). The use of discontinuous polyacrylamide gels with stacking gels improves the resolution as well as the sharpness of the protein bands. The following protocols describe the preparation of the gel, the electrophoresis, and the detection of proteasomes by an in-gel activity assay or by immunoblotting.

#### 3.3.1. Preparation of the Gel

1. Set up the gel casting system.
2. Prepare the separating gel (4.5%) by mixing the following:

| | |
|---|---|
| a. TBE (4X) | 3.75 mL |
| b. Acrylamide stock | 2.25 mL |
| c. Sucrose (10%) | 3.75 mL |
| d. MAD (10X) | 1.50 mL |
| e. $H_20$ | 3.75 mL |
| f. Ammonium persulfate (10%) | 0.05 mL |
| g. TEMED | 10 µL |

3. Mix carefully and fill in the solution to about 4 cm from the top of the glass plates. Carefully overlay it with isopropanol (2–3 mm layer).
4. Let the gel polymerize for at least 30 min.
5. Prepare a stacking gel (2.5%) by mixing the following:

| | |
|---|---|
| a. 1 *M* Tris-HCl, pH 6.8 | 0.50 mL |
| b. Acrylamide stock | 0.83 mL |

| | | |
|---|---|---|
| c. | $H_2O$ | 8.70 mL |
| d. | Ammonium persulfate (10%) | 0.05 mL |
| e. | TEMED | 10 μL |

6. Pour off the isopropanol overlayering the gel, rinse twice with water, and fill in the stacking gel solution up to the top of the glass plates. Insert the well forming comb between the glass plates.
7. Let the stacking gel polymerize for at least 30 min.
8. Remove the comb and fill the slots with electrophoresis buffer (*see* **Note 7**).

### 3.3.2. Gel Electrophoresis

Protein samples are obtained by mechanical disruption of cells using glass beads or by grinding in a mortar in the presence of liquid nitrogen using a pestle (*see* **Subheading 3.1.**).

1. Dilute protein samples with 4X loading buffer. Ten micrograms of total *S. cerevisiae* protein per gel lane is usually sufficient for immunological detections. To analyze samples obtained by gel filtration (*see* **Subheading 3.2.**), we use 40 μL of each fraction.
2. Load the samples on the gel, connect the power supply to the gel apparatus and adjust the current to 15 mA. Continue the electrophoresis until the bromophenol blue reaches the bottom of the gel. This usually takes around 4–5 h. Remove the gel from the glass plates and proceed either with a detection of proteasomal activity in the gel, or by transfer of the proteins to a membrane for immunological detection (*see* **Note 8**).

### 3.3.3. Detection of Proteasomal Activity in the Gel

After gel electrophoresis, active proteasome complexes can be visualized in the gel by using the fluorogenic proteasome substrate Suc-LLVY-AMC ***(11)***.

1. Remove the gel from the glass plates, cut off the stacking gel and incubate the separating gel for 20 min at 37°C in 20 mL of electrophoresis buffer containing 200 μL of Suc-LLVY-AMC stock solution (*see* **Note 8**).
2. Proteasome bands can be visualized by exposure to UV light (360 nm).

### 3.3.4. Immunological Detection

After transfer of the proteins to the membrane, components of the proteasome can be detected immunologically by standard procedures using specific antisera, or if cells expressing epitope-tagged variants of specific proteins were used, by using commercially available monoclonal antibodies. An antibody recognizing Ump1 will detect proteasomal precursor complexes, but not the 20S or 26S complexes (unless mutants with impaired proteasomal activity are used). Antibodies recognizing an α or a β subunit detect half-proteasomes as well as 20S and 26S complexes (**Fig. 2**), whereas antibodies against components of the 19S regulatory particle will only detect the 26S complexes.

1. Remove the stacking gel, and transfer the proteins onto PVDF membranes for 2 h at 0.8 mA/cm$^2$ using a standard blotting procedure (*see* **Notes 6** and **8**).
2. Proceed with the detection following the protocol described in **Subheading 3.2.3.**

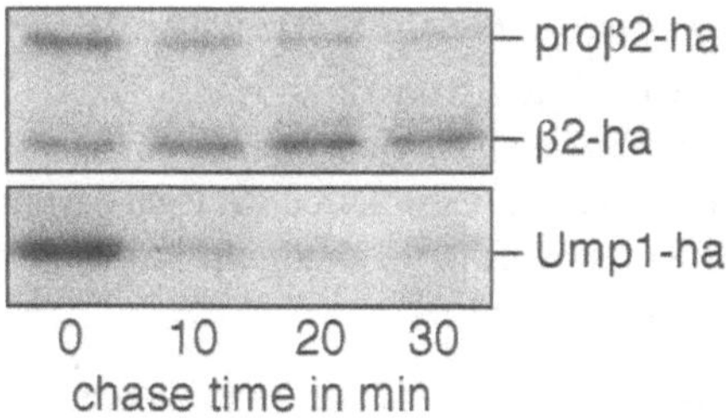


Fig. 3. Pulse-chase analysis of proteasome maturation. **Top:** Detection of N-terminal processing of β2-ha. **Bottom:** Degradation of maturation factor Ump1.

### 3.4. Following Proteasome Maturation by Pulse-Chase Analysis

A key event in proteasome biogenesis is the maturation of its active sites, which occurs by autocatalytic processing of the active β subunits (β1, β2, and β5). In this process, their propeptides are cleaved off yielding N-terminal threonine residues that are essential components of the active sites. The processing events coincide with the dimerization of two half-proteasome precursor complexes. Synchronously or nearly so, maturation factor Ump1 is degraded by the activated proteasomes. The kinetics of these processes can be followed in a pulse-chase experiment *(2)*. We can subdivide the entire experiment into four sections: (1) the actual pulse-chase labeling, (2) protein extraction and determination of the amount of labeled protein, (3) immunoprecipitation of the protein of interest, and (4) separation of these proteins by SDS-PAGE followed by their detection. In the following we present a protocol for each step.

To enable specific immunoprecipitation of the proteasomal proteins of interest after the pulse chase, in the example shown in **Fig. 3** we used strains expressing C-terminally tagged versions of proteasome subunit β2/Pup1 or of Ump1. The epitope tags we used here were double ha tags *(2)*.

#### 3.4.1. In Vivo Pulse Chase

1. Grow yeast culture in minimal medium to a density of approx $1.5 \times 10^7$ cells/mL (midexponential phase, this equals an optical density at 600 nm of about 1). Harvest cells from 25 mL of culture by centrifugation at 3000*g* for 5 min.
2. Wash the cell pellet once with 800 μL of pulse solution.
3. Resuspend the cells in 600 μL of pulse solution, and transfer the sample into a 1.5-mL microcentrifuge tube. Add 250–500 μCi of [$^{35}$S]methionine to pulse label proteins for 5 min with agitation in a thermoshaker at 30°C.
4. Pellet the cells briefly in a microcentrifuge and discard the supernatant.
5. Resuspend the cells in 400 μL of chase solution. Immediately take a 100-μL aliquot as your 0 min time point sample, pipetting it into a 1.5-mL microcentrifuge tube on ice containing 600 μL of lysis buffer with protease inhibitors (*see* **Note 2**) and 460 mg of glass beads.
6. Collect further 100 μL samples after 10, 20, and 30 min in the same manner.
7. Here, you can either proceed or snap-freeze and store the samples at –80°C.

### 3.4.2. Protein Extraction and Determination of TCA-Precipitable Counts

The cells are lysed mechanically with glass beads. These steps are carried out at 4°C. To assess differences between the protein extracts in their concentration of labeled proteins, TCA-precipitable counts are measured for each extract (*see* **Note 9**). Differences are then compensated for by using volumes with equal amounts of TCA-precipitable counts in the immunoprecipitation.

1. Vortex the cell suspension with glass beads for 3 × 1 min with 1 min pause intervals on ice.
2. Centrifuge the samples for 10 min at 12.000*g*, and transfer the supernatants into fresh tubes.
3. Spot 3 µL of the protein extract onto a filter paper disc and air dry it for 10 min. Dip the filter disc once into a 10% TCA solution at room temperature (RT) and let it dry on a filter paper. Repeat this step.
4. Boil the filter disc in a 10% TCA solution at approx 95°C for 5 min (*see* **Note 10**).
5. Dip the filter disc into a 10% TCA solution (RT) and let it dry on a filter paper. Repeat this step.
6. Dip the filter disc twice into 100% EtOH and dry it thoroughly under a heat lamp.
7. Immerse filter disc into scintillation fluid and measure counts per minute (cpm) with a liquid scintillation counter.

### 3.4.3. Immunoprecipitation

The immunoprecipitations shown in **Fig. 3** were carried out with a monoclonal anti-ha antibody (16B12) and Protein A–Sepharose beads (*see* **Note 11**).

1. Continue with a volume corresponding to 5–10 × $10^6$ cpm per sample or more (*see* **Note 12**).
2. To each sample add a suitable amount of antibody and mix gently. Incubate the samples over night on ice.
3. Add 40 µL of a Protein A–Sepharose slurry to each sample and incubate gently rotating for 4 h at 4°C.
4. Centrifuge 2 s to pellet the Protein A–Sepharose beads. Carefully remove supernatant such that any loss of beads is avoided and wash the Protein A–Sepharose-bound immunocomplexes three times with 800 µL of precooled lysis buffer containing 0.1% SDS (see **Note 13**).
5. Add 20 µL of 2X loading buffer *(8)* to the sediment, boil for 3 min, and centrifuge the beads for 30 s.
6. Here, you can either proceed or snap-freeze and store the samples at –80°C.

### 3.4.4. Electrophoresis and Detection by Autoradiography

1. The protein samples are now subjected to standard denaturing gel electrophoresis (SDS-PAGE, *see* **Note 5**). To keep the amounts comparable, it is important to load the protein samples quantitatively.
2. Remove the gel from the glass plates and fix proteins by incubating the gel in 50% methanol, 10% acetic acid, 5% formaldehyde for 30 min. Rinse gel in water, incubate in 5% glycerol for 10 min and dry on a gel dryer.
3. For detection of the radioactively labeled proteins, two detection methods can be employed: phosphoimaging or autoradiography. While phosphoimaging is the faster detection method and provides a broader range of linearity for quantification, direct exposure of the dried gel to X-ray film will give a higher resolution. Commonly, we first expose the film to a phosphoimaging plate for up to 24 h to quantify the signals and to estimate the time suitable for exposure to X-ray film (*see* **Note 14**).

## 4. Notes

1. As an alternative to freezing the cell pellets in Falcon tubes, one can immediately resuspend them in the extraction buffer (2 mL/g wet wt) and produce droplets releasing the suspension in a beaker filled with liquid nitrogen. Collect the frozen tablets and store at –80°C.
2. If you intend to study proteolytic activity of the proteasome, the use of protease inhibitors is not recommended. If the composition and distribution of proteasomal complexes are to be studied, a combination of protease inhibitors such as phenylmethylsulfonyl fluoride (PMSF), antipain, aprotinin, chymostatin, leupeptin and pepstatin, or a protease inhibitor cocktail may be added to inhibit nonspecific proteases in the extract.
3. Superose 6 (Amersham Biosciences) has a range of separation from 5 kDa to 5 MDa enabling the analysis of proteins including free proteasomal subunit (21 kDa and larger) as well as the approx 2 MDa 26S proteasome. This gel filtration column is therefore used routinely to fractionate proteasomal complexes in our laboratory as well as in many others. However, other columns such as Sephacryl S400 (Amersham) can be used as an alternative.
4. Ideally, the release of fluorescent product is monitored over time, and the activity is calculated from the slope of the linear part of the curve. If the available fluorimeter does not allow to monitor kinetics, the reaction can be stopped after 60 min with 900 µL of ethanol followed by an end point measurement of the product released.
5. A 12% gel with a gel length of approximately 16 cm will give a good resolution. Adjust the anode buffer to 0.1 *M* sodium acetate to increase the resolution and the sharpness of low molecular weight protein bands ***(12)***.
6. Semidry blotting is the standard technique in our laboratory but other blotting procedures are suitable as well.
7. Remove the comb very slowly otherwise the slots are often damaged. During slow removal of the comb sometimes air bubbles appear between the slots. Usually, however, the slots are intact, and the bubbles disappear after the comb is removed completely.
8. Because the gel is very fragile, it requires careful handling. One safe method to take the intact gel off the glass plate is as follows. Gently press a sheet of Whatman 3MM paper onto the gel. The gel can than be peeled of easily from the glass plate along with the Whatman paper. Place gel and Whatman paper either in extraction buffer for activity detection, or in blotting buffer for electrotransfer. The gel will float off the paper, which can then be removed.
9. We recommend the determination of TCA-precipitable counts. However, an alternative is to just strictly use the same sample volumes at all steps.
10. Boil the 10% TCA solution with precaution! Use boiling stones to avoid overheating and work under a fume hood.
11. The amount of antibody suitable for immunoprecipitation varies between antibodies. It is therefore adviced to determine the amount of a particular antibody that is sufficient to quantitatively immunoprecipitate the protein of interest. Depending on the antibody subtype used, either Protein A–Sepharose or Protein G–Sepharose should be selected according to the recommendations of the supplier.
12. Depending on the efficiency of the following immunoprecipitation and the autoradiography steps you might be able to use less radioactive material. $5 \times 10^6$ cpm per sample is an average amount to start with that gives us satisfying results.
13. During the wash steps once also transfer the suspensions into fresh tubes. This will help reducing the background as some proteins tend to stick to the centrifuge tube wall and do not come off during the wash steps.

14. If the signal intensity is insufficient, it can be increased by one of the following procedures. Before drying, treat the gel with commercially available enhancer solution for the detection of radioactivity. Alternatively, intensifying screens for low-energy β emitters can be used to enhance autoradiography.

## Acknowledgments

This work was supported by grants from the Deutsche Forschungsgemeinschaft (Do 649) to R.J.D., and from Fundação para a Ciência e Tecnologia (POCTI/32621/BME/2000) to P.C.R. C.G. was supported by a fellowship from the NRW graduate school program.

## References

1. Heinemeyer, W., Ramos, P. C., and Dohmen, R. J. (2004) The ultimate nano-scale mincer: assembly, structure and active sites of the 20S proteasome core. *Cell. Mol. Life Sci.* **61,** 1562–1568.
2. Ramos, P. C., Höckendorff, J., Johnson, E., Varshavsky, A., and Dohmen, R. J. (1998) Ump1p is required for maturation of the 20S proteasome, and becomes its substrate upon completion of the assembly. *Cell* **20,** 489–499.
3. Burri, L., Höckendorff, J., Boehm U., Klamp, T., Dohmen, R. J., and Lévy, F. (2000) Identification and characterization of a mammalian protein interacting with 20S proteasome precursors. *Proc. Natl. Acad. Sci. USA* **97,** 10348–10353.
4. Chen, P. and Hochstrasser, M. (1996) Autocatalytic subunit processing couples active site formation in the 20S proteasome to completion of assembly. *Cell* **86,** 961–972.
5. Groll, M., Bajorek, M., Kohler, A., et al. (2000) A gated channel into the proteasome core particle. *Nat. Struct. Biol.* **7,** 1062–1067.
6. Whitby, F. G., Masters, E. I., Kramer, L., et al. (2000) Structural basis for the activation of 20S proteasomes by 11S regulators. *Nature* **408,** 115–120.
7. Dahlmann, B., Becher, B., Sobek, A., Ehlers, C., Kopp, F., and Kuehn, L. (1993) In vitro activation of the 20S proteasome. *Enzyme Protein* **47,** 274–284.
8. Laemmli, U. K. (1970) Cleavage of structural proteins during the assembly of the head of bacteriophage T4. *Nature* **227,** 680–685.
9. Towbin, H., Staehelin, T., and Gordon, J. (1979) Electrophoretic transfer of proteins from polyacrylamide gels to nitrocellulose sheets: procedure and some applications. *Proc. Natl. Acad. Sci. USA* **76,** 4350–4354.
10. Hough, R., Ratt, G., and Rechsteiner, M. (1987) Purification of two high molecular weight proteases from rabbit reticulocyte lysate. *J. Biol. Chem.* **262,** 8303–8313.
11. Glickman, M. H., Rubin, D. M., Fried, V. A., and Finley, D. (1998) The regulatory particle of the *Saccharomyces cerevisiae* proteasome. *Mol. Cell. Biol.* **18,** 3149–3162
12. Christy, K. G., La Tart, D. B., and Osterhoudt, H. W. (1989) Modifications for SDS PAGE of proteins. *BioTechniques* **7,** 692–693.

# 16

# N-Terminal Ubiquitination

Aaron Ciechanover

**Summary**

An important step in the ubiquitin proteolytic cascade is specific recognition of the substrate by a member of the ubiquitin ligases family of proteins—an E3, that is followed by generation of the polyubiquitin degradation signal. For most substrates, it is believed, though it has been demonstrated experimentally only for a few, that the first ubiquitin moiety is conjugated, via its C-terminal $Gly^{76}$ residue, to an $\varepsilon$-$NH_2$ group of an internal Lys residue. Recent findings indicate that for several proteins, the first ubiquitin moiety is fused linearly to the $\alpha$-$NH_2$ group of the N-terminal residue. Important biological questions relate (1) to the evolutionary requirement for an alternative mode of ubiquitination, (2) to the identity of the set of proteins in the proteome that undergoes N-terminal ubiquitination, and (3) to the relationship between N-terminal ubiquitination and N-terminal acetylation. In this chapter we describe methods that will enable researchers to identify this novel mode of ubiquitination.

**Key Words:** Conjugation; degradation; methylated ubiquitin; N-terminal acetylation; N-terminal ubiquitination; ubiquitin, ubiquitin aldehyde.

## 1. Introduction

Conjugation of ubiquitin to the protein target proceeds via a three-step cascade mechanism. Following activation by the ubiquitin-activating enzyme, E1, the activated ubiquitin moiety is transferred by one of several E2 enzymes (ubiquitin-carrier proteins or ***ub***iquitin-***c***onjugating enzymes, UBCs) to the substrate that is specifically bound to a member of the ubiquitin-protein ligase family of proteins, E3. The E3s belong to several distinct families, based on commonly shared structural motifs and mechanism of action. As the specific substrate-recognizing components of the ubiquitin system, the E3s play a key role in the conjugation process. It is not surprising therefore that with the myriad substrates of the ubiquitin system, several hundred different E3s have been identified in the human genome database.

In most cases, it is believed, though it has been shown in a direct manner for only a few target proteins, that the first ubiquitin moiety is conjugated to an internal Lys residue where an isopeptide bond is generated between the carboxyl group of the

From: *Methods in Molecular Biology, vol. 301, Ubiquitin–Proteasome Protocols*
Edited by: C. Patterson and D. M. Cyr © Humana Press Inc., Totowa, NJ

C-terminal $Gly^{76}$ residue of ubiquitin and an ε-$NH_2$ group of the Lys residue. Successive addition of activated ubiquitin moieties to internal $Lys^{48}$ of the previously conjugated moiety leads to generation of a polyubiquitin chain, that serves, in the case of the $Lys^{48}$-based chain, as the 26S proteasome recognition signal. Other chains such as $Lys^{63}$-based chain serve non-proteolytic purposes such as activation of transcription. Monoubiquitination or conjugation of a single moiety of a ubiquitin-like protein also serve nonproteolytic functions such as routing of proteins to their subcellular destinations or regulating the endocytic pathway. For recent reviews on basic mechanisms of ubiquitination, see **refs. *1–4***.

Thus, it is clear that for ubiquitination, each protein substrate must have a specific E3 recognition motif and one or more ubiquitin-anchoring groups. Surprisingly, only in a few substrates a specific Lys residue has been identified. In IκBα, for example, Lys21 and 22 have been shown to be important ***(5)***. In contrast, in cyclin B ***(6)*** and in the ζ chain of the T cell receptor cytosolic tail ***(7)***, any single Lys residue, even if artificially inserted, can serve as a ubiquitin acceptor. Moreover, in only a handful of cases it has been shown directly, via chromatographic or mass spectrometric analyses, that ubiquitin is indeed anchored to a specific Lys residue (*see,* e.g., **refs. *8, 9***). In most other cases studied, and there are not too many, the assumption that an internal Lys serves as the polyubiquitin chain anchor is based on mutational analyses.

Surprisingly, substitution of all nine internal Lys residues in the myogenic transcriptional switch protein MyoD did not affect significantly its stability either in vivo nor in vitro ***(10)***. Degradation of the lysine-less (LL) protein in cells was inhibited by proteasome inhibitors, and was accompanied by accumulation of ubiquitin conjugates. In vitro, LL MyoD was ubiquitinated, and its degradation required polyubiquitin chain formation. Selective chemical modification of the N-terminal group (by carbamoylation) but not of internal lysines (by guanidination) of bacterially expressed purified MyoD blocked degradation almost completely. Fusion of a 6 X Myc tag to the N-terminal but not to the C-terminal residue of WT MyoD stabilized the protein both in vivo and in vitro (***10*** and unpublished data). Since all the internal lysines are present in this MyoD molecule, this finding ruled out the possibility that substitution of all the internal Lys residues "forced" ubiquitination at the N-terminal residue that would not have occurred otherwise. Taken together, these results strongly suggested that the first event in ubiquitin-mediated degradation of MyoD involves fusion of the first ubiquitin moiety to the N-terminal residue of MyoD **(Fig. 1)**. Additional ubiquitin moieties are then added, most probably, to $Lys^{48}$ of the previously conjugated moiety. This conclusion is based on the finding that methylated ubiquitin, in which all internal Lys residue were modified, blocked both conjugation and degradation of MyoD ***(10)***. N-terminal tagging by Myc stabilizes the protein, most probably, by blocking access of ubiquitin to the specific ubiquitination site at the N-terminus. Using a similar, though not a complete, set of experiments, 12 additional proteins have been identified recently that appear to undergo N-terminal ubiquitination: (1) the ***h***uman ***p***apilloma***v***irus-16 (HPV-16) E7 oncoprotein ***(11)***, (2) the ***l***atent ***m***embrane ***p***rotein 1 (LMP1) ***(12)***, and (3) 2A (LMP2A) ***(13)*** of the ***E***pstein ***B***arr ***v***irus (EBV), involved in viral activation from latency, (4) the ***c***ell cycle-***d***ependent ***k***inase (CDK) inhibitor p21 ***(14,15)***, (5) the extracellular signal-regulated kinase 3, ERK3 ***(15)***, the ***i***nhibitors of

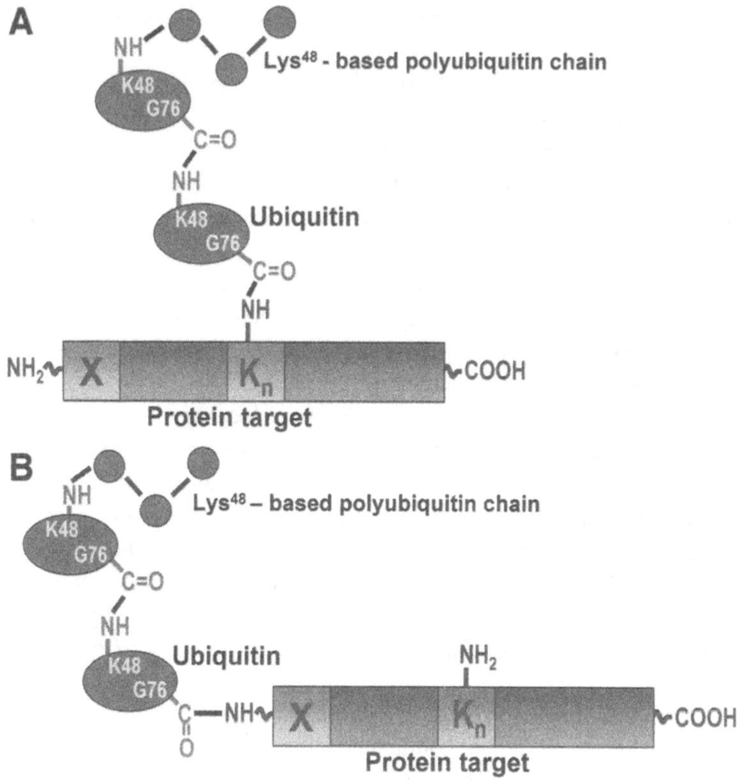


Fig. 1. Ubiquitination on an internal lysine and on the N-terminal residue of the target substrate. (A) The first ubiquitin moiety is conjugated, via its C-terminal $Gly^{76}$ residue, to the $\varepsilon$-$NH_2$ group of an internal lysine residue of the target substrate ($K_n$). (B) The first ubiquitin moiety is conjugated to a free $\alpha$-$NH_2$ group of the N-terminal residue, X. In both cases, successive addition of activated ubiquitin moieties to internal $Lys^{48}$ on the previously conjugated ubiquitin moiety leads to the synthesis of a polyubiquitin chain which serves as the degradation signal for the 26S proteasome.

differentiation (6) Id2 ***(16)*** and (7) Id1 ***(17)***, two proproliferative Helix-Loop-Helix proteins, (8) hydroxymethylglutaryl-Coenzyme A reductase (HMG-CoA reductase) ***(18)***, the first and key regulatory enzyme in the cholesterol biosynthetic pathway, (9) $p19^{ARF}$, the mouse Mdm2 inhibitor and (10) $p14^{ARF}$, its human homologue ***(19)***, (11) the HPV-58 E7 oncoprotein, and (12) the cell cycle regulator $p16^{INK4a}$ ***(20)***. Importantly, $p14^{ARF}$, HPV-58 E7 and $p16^{INK4a}$ are naturally occurring lysine-less proteins (NOLLPs) that must undergo N-terminal ubiquitination for their ubiquitin- and proteasome-mediated degradation to occur, as the $\alpha NH_2$ is their only potential ubiquitination site. Searching the database reveals the existence of several hundreds of NOLLPs. It should also be noted that in the case of p21, it is the N-terminally HA-tagged protein that was shown to be N-terminally ubiquitinated. Recent evidence casts a doubt whether the native cellular protein undergoes such a modification and argues that the degradation of the protein is proteasome-dependent, yet ubiquitin-independent ***(21,22)***.

It should be noted that while the internal lysine residues are not essential for degradation of these substrates, for several of them, including MyoD ***(10)*** and HPV-16 E7 ***(11)***, we have shown that they nevertheless play a modulating role. Thus, while the LL MyoD is still ubiquitinated, the modification is less efficient then that observed for its WT counterpart, and its degradation is slowed down by an approx two- to threefold compared to the WT protein ***(10)***.

For E7-16 ***(11)***, LMP1 ***(12)***, Id2 ***(16)***, and MyoD (unpublished), it has been shown that truncation of a short N-terminal segment stabilized the proteins, suggesting that the entire domain, beyond the single N-terminal residue, plays a role in governing the stability of the proteins. It may do so by allowing the mobility/flexibility necessary for the N-terminal residue to serve as a ubiquitin acceptor. It can also serve as a recognition domain for the E3. There is no homology between the N-terminal domains of these three proteins, suggesting that if the three N-terminal domains serve as recognition motifs, they are being recognized by different components of the ubiquitin system. Interestingly, the LMP2A E3 was identified as a member of the NEDD4 family of HECT domain ligases, AIP4 and/or WWP2 ***(13)***. The PY motif of LMP2A recognized by the E3 resides in the N-terminal domain of the molecule, supporting the hypothesis that in these proteins, the E3-binding domain may reside in close proximity to the ubiquitination site.

Is there a direct evidence for N-terminal ubiquitination? Recently, direct evidence has been provided that ubiquitin is conjugated indeed to the N-terminal residue of the target substrate. Using mass spectrometric analysis of tryptic peptides of the HPV-58 E7 ubiquitin adduct, we have isolated a fusion peptide that spans the C-terminal domain of ubiquitin and the N-terminal domain of E7 ***(20)***. Similar, though less direct, data have been reported for HA-tagged p21 ***(14,15)*** and the ARF proteins ***(19)***, showing that the ubiquitin moiety is fused to the N-terminal tag. As noted, the attachment of ubiquitin to the HA tag of the artificially constructed p21 appears to be confined to the tag, and the conclusion that the native cellular protein is also targeted via N-terminal ubiquitination ***(14,15)*** does not appear to hold at present ***(21,22)***.

N-terminal ubiquitination is a novel pathway, distinct from the N-end rule pathway ***(23)***. In the latter, the N-terminal residue serves as a recognition and binding motif to the ubiquitin ligase, E3α, however, ubiquitination occurs on an internal lysine(s). In contrast, in the N-terminal ubiquitination pathway, modification occurs on the N-terminal residue, whereas recognition probably involves a downstream motif. It should be mentioned that in yeast, using the model fusion protein ubiquitin-Pro-β-galactosidase, a new proteolytic pathway was described, designated the UFD (***u***biquitin ***f***usion ***d***egradation) pathway ***(24)***. The stably fused ubiquitin moiety (that in this exceptional case where Pro is the linking residue, is not removed by ubiquitin C-terminal hydrolases), functions as a degradation signal, where its $Lys^{48}$ serves as an anchor for the synthesized polyubiquitin chain. This pathway involves several enzymes, UFD 1–5, some of which appear to be unique and are not part of the "canonical" UPS for soluble proteins. It is possible that N-terminal ubiquitination is the most upperstream event in the UFD pathway that was discovered using an artificial chimeric ubiquitin-protein model substrate: N-terminal ubiquitination, in contrast, is the native mode of ubiquitination that may provide substrates to the UFD pathway.

The physiological significance of N-terminal ubiquitination is still obscure. Naturally occurring lysine-less proteins (*see* earlier) that are degraded by the ubiquitin system must traverse this pathway. Many such proteins, mostly viral, can be found in the database. For lysine-containing proteins, they must contain a free, unmodified N-terminal residue. Such proteins constitute approx 25% of all cellular proteins, while the remaining approx 75% are N-α-acetylated. The number of N-termini free proteins may be even higher than 25% and may reach up to 70% according to some estimates. Whether the N-terminal residue will be acetylated is dependent on the structure of the N-terminal domain of the mature protein. This is determined by the combined activity of ***m***ethionine ***a***mino***p***eptidases (MAPs) and ***N***-terminal ***a***cetyl***t***ransferases (NATs) that is dependent on the specific sequence of up to the first ten N-terminal residues (reviewed in **ref. *25***). Thus, it is possible to predict which proteins will be potential substrates of the N-terminal ubiquitination pathway. Internal C-terminal fragments of N-α-acetylated proteins can also be modified by ubiquitin at their N-terminal residue following limited processing. Many proteins, such as the NF-κB precursors p105 and p100 or caspase substrates, are processed initially in a limited manner, generating a C-terminal fragment with a newly exposed N-terminal residue. In all these lysine-containing proteins, the naturally occurring free N-termini as well as the processing products, the assumption is that their internal lysines are not accessible, from whatever reason, for ubiquitination, and it is only the N-terminal residue that can be modified.

In summary, to demonstrate that a protein is targeted via ubiquitination of its N-terminal residue, one or more of the following criteria should be fulfilled:

1. A lysine-less (naturally occurring or generated via site directed mutagenesis of all of its internal lysine residues) protein is conjugated and degraded in a cell-free reconstituted system in vitro in a ubiquitin-dependent manner.
2. The same protein is degraded in cells in vivo following its ubiquitination, and inhibition of the proteasome leads not only to its stabilization, but also to accumulation of specific high molecular mass ubiquitin adducts.
3. Site-directed replacement of selective residues in the N-terminal domain, so that the protein will be now be acetylated by NAT(s), should stabilize it both in a cell-free assay, in vitro, and following expression in cells, in vivo. In parallel, it should not undergo ubiquitination or ubiquitination should decrease significantly. Similarly, carbamoylation of the N-terminal residue should also render the protein resistant to ubiquitination and degradation in a cell free assay.
4. Fusion of a long tag such as 6 x Myc (we have experience with this tag; HA undergoes N-terminal ubiquitination; *see* **refs. *14,15,21,22***) to the N-terminal residue but not to the C-terminal residue of the protein, should stabilize it, both in vitro and in vivo.
5. Truncation of the N-terminal segment of the test protein (first 15–30 residues) will stabilize it, both in vitro and in vivo. Experimental establishment of some of these criteria is described in the following subheadings.

## 2. Materials

### 2.1. Conjugation and Degradation of Protein Substrates in an In Vitro, Cell-Free Reconstituted Ubiquitin System

1. New Zealand white rabbits (preferably females) of approx 2 kg body wt (2–3 mo old) for preparation of reticulocyte lysate.
2. Phenylhydrazine, 2-deoxyglucose, and 2,4-dinitrophenol (Sigma).
3. Cultured cells (HeLa, 293, Cos) for preparation of nucleated cell extract
4. $MgCl_2$, $NaN_3$, and NaF (Sigma).
5. Nitrogen Cavitation Bomb (Parr Instrument Company, 211 53rd Street, Moline, IL 61265-9984. Tel: 1-800-872-7720 or 309-762-7716; Fax: 309-762-9453)
6. Diethylaminoethyl cellulose (DEAE cellulose: DE-52, Whatman).
7. Ammonium sulfate, enzyme grade.
8. Dialysis tubing.
9. For iodination, purified protein dissolved (at ~1–10 mg/mL) in water or a buffer. Make sure that the buffer does not contain free amino or hydroxyl groups, Tris-HCl, for example, as this may result in iodination of these groups. Because the buffer is in a large molar excess over the protein, the protein will not be labeled.
10. Radiolabeled Na[$^{125}$I] at a specific activity of 100–350 mCi/mL (Amersham-Pharmacia Biotech) and unlabeled NaI (10 m*M* stock solution).
11. Chloramine-T and Na-metabisulfite stock solutions of 10 and 20 mg/mL, respectively (both from Sigma; freshly dissolved in 0.05 *M* NaPi, pH 7.5).
12. Desalting column (low molecular mass exclusion gel filtration matrix; Pharmacia-Amersham Biotech).
13. cDNA template coding for the test protein and driven by the RNA polymerase promoters SP6, T7, or T3. The cDNA can code for a lysine-less (naturally occurring or generated via site directed mutagenesis of all of its internal lysine residues) protein, an N-terminal segment-truncated protein, an N- or C-terminally tagged protein, or any protein in which the N-terminal domain was manipulated so that a conclusion can be drawn as for the possibility that it is targeted via N-terminal ubiquitination (see earlier under Introduction).
14. [$^{35}$S]Methionine (1000 Ci/mmol at 10 mCi/mL; Amersham-Pharmacia Biotech).
15. In vitro translation–transcription kit (reticulocyte lysate or wheat germ extract (TNT®; Promega).
16. RNasin®, ribonuclease inhibitor (40 U/µL; Promega).
17. Nuclease-free water.
18. Dithiothreitol (DTT).
19. Krebs– Ringer phosphate solution (Gibco).
20. *N*-(2-hydroxyethyl)piperazine-*N*'-(2-ethanesulfonic acid) (HEPES), pH 7.5.
21. ATP regenerating system: ATP (0.1 *M* stock solution), phosphocreatine (PC; 0.5 *M* stock solution), and creatine phosphokinase (CPK; 10 mg/mL stock solution).
22. ATPγS (Sigma).
23. Ubiquitin and ubiquitin aldehyde (UbAl) (BIOMOL International, previously Affiniti-Research Products, Ltd.).
24. Hexokinase (Roche; 1500 U/mL; specific activity of approx 450 U/mg), ammonium sulfate slurry. Centrifuge an aliquot the slurry and resuspend to the same volume of the original aliquot with 20 m*M* Tris-HCl buffer, pH 7.6. Dilute in the same buffer. Stock solution in the buffer can be stored at 4°C for at least 4 weeks.

### 2.2. Conjugation and Degradation of Proteins in Cells In Vivo

1. Cultured cells in a monolayer or in suspension.
2. [$^{35}$S]Methionine (1000 Ci/mmol at 10 mCi/mL; Amersham-Pharmacia Biotech, ICN, or NEN).

3. Methionine-free medium (the same medium in which the cells are grown but lacking methionine. Dulbecco's modified Eagle medium [DMEM], minimum essential medium [MEM], or Richter's Improved MEM Insulin [RPMI] medium are the media that are most frequently used for most cells and can be substituted for one another for the short labeling period provided that they lack methionine).
4. Dialyzed serum (the same serum in which cells are grown).
5. Lactacystin (or its lactone homolog, clastolactacystin β-lactone) or MG132 (Z-Leu-Leu-Leu-H), or epoxomicin, or Z-Leu-Leu-Leu-vinyl sulfone, or Z-Leu-Leu-Leu-B$(OH)_2$ dissolved in DMSO. Stock solutions are of approx 10 m*M* and final concentration in cultured cells is approx 10–20 µ*M*. Please consult the catalogs of the suppliers (such as BIOMOL International, previously Affinit-Research Products Ltd).
6. Cycloheximide (Sigma).
7. CHO-E36 and CHO-ts20 E1 ts mutant cells.
8. Potassium phosphate and potassium chloride (Sigma).
9. Buffer A: 5 m*M* potassium phosphate, pH 7.0, 1 m*M* DTT.
10. Buffer B: 20 m*M* Tris-HCl, pH 7.2, 1 m*M* DTT.

## 3. Methods

### 3.1. Preparation of Cell Extracts for Monitoring Conjugation and Degradation

To conjugate or degrade a protein substrate in vitro, one has to utilize the appropriate cell extract. Rabbit reticulocyte lysate contains all the enzymes required for conjugation and degradation of most proteins and can be therefore used in most cases. Reticulocytes contain a relatively small number of proteins and do not have lysosomes from which proteases can leak during preparation of the extract. Unlike the case in cultured cells lysate, one can obtain reticulocyte lysate in a relatively large amount. Also, the lack of requirement for tissue culture media and sera make this lysate significantly less expensive than its nucleated cultured cells counterpart. All these attributes make this lysate an ideal extract in which one can test conjugation and proteolysis of the studied protein. For monitoring conjugation and degradation of labeled proteins in the crude extract, it is not necessary to deplete ATP from the cells prior to the preparation of the extract. This will be necessary, however, in order to reconstitute the cell-free proteolytic system and to monitor dependence of the proteolytic process on the addition of exogenous ubiquitin. It will also be necessary in order to monitor conjugation of labeled or tagged ubiquitin to different substrates. Depletion of ATP from cells leads to deubiquitination of most proteins. Once such an ATP-depleted lysate is fractionated over the anion-exchange resin diethylaminoethyl (DEAE)-cellulose, ubiquitin is eluted in fraction I, the unadsorbed, flow-through material that contains also certain E2 enzymes. Fraction II, the high-salt eluate, contains E1, the remaining E2s, all the E3s, and the 26S proteasome.

#### 3.1.1. Preparation of Reticulocyte Lysate

1. Inject rabbits subcutaneously with 10 mg/kg of phenylhydrazine (freshly dissolved in phosphate buffered saline [PBS]) on d 1, 2, 4, and 6.
2. Bleed the rabbits from the ear artery or vein or from the heart (following anaesthesia) on d 8. Induction of reticulocytosis is dramatic and >90% of the circulating red blood cells are reticulocytes as determined by methylene blue or brilliant cresyl blue staining.

3. Wash the cells three times with ice-cold PBS, and using a Pasteur pipet, aspirate carefully the thin layer of white blood cells ("buffy coat") that overlays the pelleted red blood cells.
4. Lyse the cells in 1.6 volumes (of pelleted cells volume) of ice-cold $H_2O \times 2$ (double distilled water) containing 1 m*M* DTT.
5. Centrifuge at 80,000*g* for 1 h at 4°C to remove particulate material.
6. Collect the supernatant and freeze in aliquots at –70°C.
7. To deplete ATP, cells are washed twice in PBS and resuspended in one volume of Krebs–Ringer phosphate buffer containing 20 m*M* 2-deoxyglucose and 0.2 m*M* 2,4-dinitrophenol. Following incubation accompanied by gentle shaking for 90 min at 37°C, the cells are washed twice in PBS, and lysed and centrifuged as described in **steps 4** and **5**. Supernatant is collected and frozen as described in **step 6**.

#### 3.1.2. Preparation of Extract From Cultured Cells

All procedures are carried out at 4°C.

1. Wash cells three times in 20 m*M* HEPES, pH 7.5-saline buffer (150 mL NaCl), and resuspend to a concentration of $10^7$–$10^8$/mL in 20 m*M* HEPES, pH 7.5, that contains also 1 m*M* DTT.
2. Cavitate cells in a high-pressure nitrogen chamber. For HeLa cells, the best conditions are 1000 psi for 30 min. However, these conditions may vary among different cell species. Make sure that most of the cells are disrupted by visualizing the suspension in a light microscope before and after cavitation. Following disruption, one should observe intact nuclei and cell debris.
3. Centrifuge the homogenate successively at 3000*g* and 10,000*g* for 15 min, and then at 80,000*g* for 60 min. The supernatant is collected and frozen at –70°C.
4. To deplete ATP, cells are washed twice in HEPES–saline buffer and resuspended in Krebs– Ringer phosphate buffer (to a density of $10^7$ cells/mL) in the presence of 2-deoxyglucose, 2,4-dinitrophenol (as described earlier), 20 m*M* NaF, and 10 m*M* of $NaN_3$. Following incubation for 60 min at 37°C, cells are washed twice in HEPES–saline, resuspended in HEPES–DTT (1 m*M*), and lysed and centrifuged as described in **steps 2** and **3**.

### 3.2. Fractionation of Cell Extract to Fraction I and Fraction II

As described earlier, fractionation of the lysate into fraction I and fraction II separates ubiquitin from many of the other components of the system, thus enabling one to examine the dependence of conjugation and degradation on the addition of exogenous ubiquitin and certain E2 enzymes. To fractionate the lysate, ATP-depleted lysate is resolved on a DEAE-cellulose column. In the ATP-depleted lysate, all the ubiquitin is free. It was released from conjugates by isopeptidases during the incubation in the presence of the glycolysis and respiration inhibitors. In the absence of ATP, reconjugation is inhibited. Under these conditions, ubiquitin is resolved in fraction I, and fraction II is dependent for its conjugating and proteolytic activities on the addition of exogenous ubiquitin. In cell extracts from which ATP was not depleted, the ubiquitin that is still conjugated to endogenous protein substrates will adsorb to the anion-exchange resin DEAE (via the protein substrate moiety) and will elute in fraction II. During incubation, this bound ubiquitin will be released by the activity of isopeptidases and will be available for conjugation to other proteins, including the test substrate we examine. Therefore, it will be difficult to demonstrate ubiquitin-dependent conjugation and degradation in fraction II that is prepared from an extract from which ATP was not depleted. In addition, the bound ubiquitin, when released, will dilute any added labeled or tagged ubiquitin, and thus decrease the detectable signal in the specific biosynthesized ubiquitin adducts.

#### 3.2.1. Fractionation of Cell Extracts Into Fraction I and Fraction II

All procedures are carried out at 4°C.

1. Swell the resin in 0.3 *M* potassium phosphate, pH 7.0, for several hours. Use enough resin to adsorb all the proteins in the extract that can be bound. As a rule, use 0.6 resin volume per volume of reticulocyte lysate or 1 mL resin/approx 5 mg of protein of nucleated cell extract (in principle, one can use also a chromatographic system such as fast protein liquid chromatography [FPLC, Pharmacia-Amersham Biotech] with a MonoQ column, although, for resolution of large quantities, the DEAE resin procedure is advantageous).
2. Load the resin onto a column and wash with 10 column volumes of buffer A.
3. Load the extract. Once all the material is loaded, elute fraction I with buffer A. When resolving reticulocyte lysate, collect only the dark red fraction. When resolving cell extract, collect only the fractions with the highest absorption at 280 nm. Freeze fraction I in aliquots at –70°C.
4. Wash the column extensively with buffer A containing 20 m*M* KCl. When resolving reticulocyte lysate, make sure all the hemoglobin is eluted. When resolving nucleated cell extract, wash until the absorbency at 280 returns to baseline.
5. Elute fraction II with 2.5 column volumes of buffer B containing 1 *M* KCl.
6. Add ammonium sulfate to saturation (~70 g/L of solution) and swirl on ice for 30 min.
7. Centrifuge at 15,000 rpm for 15 min.
8. Resuspend pellet in 0.2–0.3 the volume of the original extract. At times, it will be impossible to dissolve all the proteins. This is not essential. They will be dissolved during dialysis.
9. Dialyze against two changes of buffer B. Dialysis should be carried out on ice. Remove particulate material by centrifugation at 15,000 rpm for 15 min. Freeze in aliquots at –70°C.

### 3.3. Labeling of Proteolytic Substrates

In most cases, monitoring the conjugation and/or degradation of a specific protein substrate requires its labeling. The fate of the protein can also be followed via Western blot analysis using specific antibodies directed against the test protein (Western blot analysis is not described here. Yet, the conjugation and degradation assays for labeled proteins [*see* **Subheadings 2.3.4.** and **3.5.**] can be applied in an almost identical manner for unlabeled proteins, followed via immune techniques). Two methods of labeling have proved to be useful, iodination and biosynthetic incorporation of labeled amino acid such as [$^{35}$S]methionine. Iodination is utilized mostly when a purified recombinant or a pure commercial protein are available. The main advantage of the method is the high specific radioactivity that can be attained. The disadvantage of the method is that one needs a pure protein. Also, during iodination, unless it is carried out using the Bolton–Hunter reagent, the protein can be damaged from the oxidizing agent (chloramine-T) used to activate the iodide. In addition, during storage, the labeled substrate may be subjected to radiochemical damage from the isotope. A different method of labeling utilizes incorporation of $^{35}$S-labeled methionine to a protein that is synthesized in a cell-free system from its corresponding mRNA. The generated protein is native; however, the specific activity obtained is relatively low. Also, the labeled protein is contained in the crude extract in which it is synthesized and it is not pure. This extract contains, among other proteins, enzymes of the ubiquitin system that may interfere in the reconstitution of a cell-free system from purified components.

### 3.3.1. Radioiodination of Proteins

1. Add the following reagents in the following order to 1.5-mL microcentrifuge (Eppendorf) tube. The volume of the reaction mixture can vary from 20 to 100 µL.
   a. NaPi, pH 7.5, final concentration of 100 m*M*.
   b. Protein substrate, 10–500 µg.
   c. 50 nmol of unlabeled NaI.
   d. 0.1–2.0 mCi of radiolabeled $Na^{125}I$.
   e. 10–50 µg of chloramine-T solution.
2. Vortex-mix once and incubate for 1–2 min at room temperature.
3. Add 20–100 µg of Na-metabisulfite solution (twofold the amount of chloramine-T added) and mix.
4. To remove unreacted radioactive iodine, resolve the mixture over a desalting column equilibrated with 10 m*M* Tris-HCl, pH 7.6, and 150 m*M* NaCl. Collect fractions (in a fraction collector or manually) of approx 10% of column volume each. The radioactive protein is typically eluted in fraction 4 (void volume of the column which is ~35% of the column's total volume). Identify, via counting, the fractions that contain the labeled protein.
5. Store in aliquots at –18°C.

### 3.3.2. Biosynthetic Labeling of Proteins

This is the most frequently used procedure to label substrate proteins and follow their fate in vitro. To label proteins biosynthetically, one can first synthesize the specific mRNA on its cognate cDNA template, using the appropriate RNA polymerase. Following digestion of the cDNA, the extracted mRNA can be translated in vitro in reticulocyte or wheat germ extracts. Alternatively, one can use a coupled transcription–translation cell-free extract that synthesizes the mRNA and translates it simultaneously. Such systems are available commercially (TNT®; Promega). Biosynthesis is carried out basically according to the manufacturer's instructions. In principle, it is preferred to use a wheat germ extract. This extract lacks many, although not all, of the mammalian E3 enzymes. Therefore, in most cases, a protein synthesized in this extract can be used in experiments in which a cell-free system is reconstituted from purified enzymes, and in particular, when the role of a specific E3 is tested. A protein synthesized in reticulocyte lysate may be "contaminated" in many cases with its cognate endogenous E2 and/or E3 enzyme(s). These enzymes, which are being carried to the reconstituted system, may interfere with the examination of the role of an exogenously added E2 or E3 in the conjugation of the translated protein. Yet, at times, one must use the reticulocyte lysate, as the translation efficiency in the wheat germ extract may be extremely low. In that case, the "contaminating" E2 or E3 in the reticulocyte lysate can be inactivated by *N*-ethylmaleimide (NEM; 10-min incubation at room temperature in a final concentration of 10 m*M* of freshly prepared solution). Because E1, all known E2s, and some of the E3s (HECT domain-containing) have an essential –SH group, the alkylating agent inactivates them. The NEM is then neutralized by the addition of DTT (final concentration of 7.5 m*M*). It should be noted that this procedure can, at times, denature/inactivate the substrate. Also, the more prevalent RING finger-containing E3s are not inactivated. In most cases, the NEM-treated substrate can still be utilized and reproduces faithfully the behavior of the native substrate.

When monitoring conjugation without an attempt to identify the E2 or E3, the in vitro translated substrate can be used without further processing. This is also true in many cases when the degradation of the labeled substrate is followed by monitoring its disappearance in PhosphorImager-analyzed gels after sodium dodecyl sulfate-polyacrylamide gel electrophoresis (SDS-PAGE). However, as the degradation of certain proteins is not always efficient, it may be difficult to follow with accuracy the disappearance of 10–30% of a labeled protein band in a gel. In this case, it will be necessary to monitor the release of radioactive material into Trichloroacetic acid (TCA)-soluble fraction. Here, to avoid high background the excess of unincorporated labeled methionine in the preparation of the translated protein must be removed. This can be achieved via chromatography over DEAE exactly as described above for fractionation of lysate into resin-unadsorbed (fraction I) and adsorbed fractions (fraction II). The vast majority of the labeled proteins will resolve in fraction II, while the labeled amino acid will be eluted in fraction I, which must be washed extensively with buffer B. If the labeled protein is eluted in fraction I, changing the pH may lead to its adsorption. Alternatively, extensive dialysis of the labeled protein (in the crude extract in which it was synthesized) against a solution of 20 m*M* Tris-HCl, pH 7.6, and 150 m*M* NaCl that contains also 1 m*M* of unlabeled methionine will also remove efficiently the labeling amino acid.

### 3.4. Conjugation of Proteolytic Substrates In Vitro

To demonstrate that the degradation of a certain protein proceeds in a ubiquitin-dependent manner, it is essential to demonstrate the intermediates in the process, ubiquitin–protein adducts. Typically, incubation of the labeled protein in a complete cell extract in the presence of ATP will lead to the formation of high molecular mass adducts that can be detected following resolution of the mixture in SDS-PAGE. To increase the amount of the adducts generated, one can use two approaches, independently or simultaneously. The nonhydrolyzable ATP analog, adenosine-5'-*O*-(3-thiotriphosphate) (ATPγS) can be used instead of ATP *(26)*. The ubiquitin-activating enzyme, E1, can catalyze activation of ubiquitin in the presence of the analog, as it utilizes the α-β high-energy bond of the nucleotide that is cleavable also in the ATPγS analog. In contrast, assembly and activity of the 26S proteasome complex requires the β-γ bond that cannot be cleaved in the analog. Caution should be exercised, however, when utilizing the ATP analog. Often, phosphorylation of the target protein is required in order for the ubiquitin ligase to recognize it and tag it with ubiquitin. In these cases, the analog cannot substitute for the hydrolyzable native nucleotide, ATP. An additional approach to increase the amount of generated conjugates in a cell-free system is to use ubiquitin aldehyde (UbAl), a specific inhibitor of certain ubiquitin C-terminal hydrolases, isopeptidases *(27)*. This derivative is available from BIOMOL International (previously Affiniti-Research Products).

### 3.4.1. In Vitro Conjugation Assay

1. Add the following reagents to 0.5-mL microcentrifuge (Eppendorf) tubes. The volume of the reaction mixture can vary from 10.0 to 50.0 µL. Addition of all the reagents should be carried out on ice:
   a. 50 m*M* Tris-HCl, pH 7.6.
   b. 5 m*M* $MgCl_2$.
   c. 2 m*M* DTT
   d. 5.0–30 µL of reticulocyte lysate or 50–200 µg of complete cell extract protein.
   e. 2.5–10 µg of ubiquitin.
   f. 0.5–2.0 µg UbAl.
   g. ATP-regenerating (0.5 m*M* ATP, 10 m*M* phosphocreatine and 10 µg creatine phosphokinase) system or 2–5 m*M* ATPγS.
   h. For depletion of endogenous ATP, the system should contain, instead of the ATP-regenerating system, 10 m*M* 2-deoxyglucose and 0.1–0.5 U of hexokinase.
   i. Substrate. Use either a labeled protein (25,000–100,000 cpm) or an unlabeled substrate in an amount that is more than sufficient to detect by Western blot analysis (100–2000 ng). This is important, as the conjugates will be much less abundant than the substrate.
2. Incubate the mixture for 30–60 min at 37°C and resolve via SDS-PAGE (7.5–12.5% acrylamide depending on the mol wt of the target substrate).
3. Detect high molecular mass conjugates by PhosphorImager analysis (labeled proteins) or via enhanced chemiluminescence (ECL) following Western blot (for unlabeled substrates) using a specific primary antibody against the test protein and a secondary tagged antibody.

There are several ways to demonstrate that the high molecular mass adducts generated are indeed ubiquitin conjugates of the test protein.

1. It is expected that the adducts will not be generated in an ATP-depleted system.
2. Generation of the conjugates of the specific substrate should be inhibited in a cell-free system by the addition of increasing amount of methylated ubiquitin (MeUb; *[27]*; available from BIOMOL International, previously Affiniti-Research Products Ltd.). This reductively methylated derivative of ubiquitin lacks free amino groups and therefore cannot generate polyubiquitin chains. It serves therefore as a chain terminator in the polyubiquitination reaction, and consequently as an inhibitor in this reaction ***(10,28)***.
3. Adducts can be precipitated from the reaction mixture with an antibody directed against the test protein, and following SDS-PAGE, can be detected with an anti-ubiquitin antibody (available from BIOMOL International, Chemicon, Zymed, Sigma and several other suppliers). Alternatively, the reaction can be carried out in the presence of HA-, Myc-, or His-tagged ubiquitin, and the immunoprecipitate can be detected following SDS-PAGE, with an antibody against the appropriate tag.
4. A cell-free system can be reconstituted from purified or isolated components of the ubiquitin system, and the formation of conjugates can be followed, dependent on the addition of these components. Rather then adding a complete cell extract, it is possible to add fraction II (50–200 µg; derived from ATP-depleted cells) and free or tagged ubiquitin (2.5–10 µg: same amount as added to supplement the complete extract; *see* above). Because fraction II is devoid of ubiquitin, formation of conjugates that is dependent on the addition of exogenous ubiquitin will strongly suggest that the high molecular mass derivatives generated are indeed ubiquitin adducts of the test substrate. Since not all E2 enzymes are present in fraction II, it may be necessary, at times, to add to the reconsti-

tuted system purified UbcH5a, b, or c, UbcH7, or UbcH8 (available from BIOMOL International, previously Affiniti-Research Products, Ltd.). In most cases, one of the UbcH5 enzymes will be able to reconstitute activity.

### 3.5. Degradation of Proteolytic Substrates In Vitro

With several exceptions, cell-free systems for monitoring degradation of proteolytic substrates are similar to those used for monitoring their conjugation. In proteolytic assays, however, unlike in conjugation assays, ATP (and not ATPγS) must be used, as activity of the 26S proteasome complex is dependent on cleavage of the high energy β-γ bond (*see* above). ATP is added along with ATP-regenerating system as described above. Also, UbAl is not added. Following incubation for 2–3 h at 37°C, the reaction mixture is resolved via SDS-PAGE, and disappearance of the substrate can be monitored either via PhosphorImager analysis (in case the protein substrate is radioactively labeled), or via Western blot analysis (in case of unlabeled substrate). Control reactions are complete mixtures that have been incubated on ice, and mixtures that were incubated at 37°C in the absence of ATP. At times, degradation efficiency is low, and it is difficult to follow the reduction in the amount of a protein band in gel analysis. In these cases, it is necessary to monitor the appearance of radioactivity in trichloroacetic acid (TCA)-soluble fraction (*see* previously). Here, only radioactive substrate can be used. Radio-iodinated proteins can be used directly. In vitro translated proteins must undergo DEAE fractionation or extensive dialysis in order to remove the excess of unincorporated labeled methionine (*see* above). At the end of the incubation, a carrier protein (10–25 μL of 100 mg/mL solution of bovine serum albumin [BSA]) is added, followed by the addition of 0.5 mL of ice-cold TCA (20%). Following mixing, the reaction is incubated on ice for 10 min and centrifuged (5 min at 15,000*g*). The supernatant is collected and the radioactivity is determined in either β-scintillation counter (for methionine) or a γ-counter (for iodine-labeled substrates). Control reactions are again complete mixtures that have been incubated on ice, and mixtures that were incubated at 37°C in the absence of ATP.

### 3.6. Involvement of the Ubiquitin System in the Degradation of Proteins In Vivo: Effect of Specific Proteasomal Inhibitors and Inactivation of E1 on the Stability of Proteins in Intact Cells

All the known proteolytic substrates of the ubiquitin system are degraded, following generation of the covalently conjugated polyubiquitin chain, by the 26S proteasome complex. The opposite notion, that all substrates of the 26S proteasome must be ubiquitinated prior to their recognition by the enzyme is true in all but one established case, that of ornithine decarboxylase, ODC ***(29)***. This enzyme is degraded by the 26S complex without prior ubiquitination. A noncovalent association with another protein, antizyme, renders ODC susceptible to degradation by the proteasome. It is possible that p21 is also degraded by the proteasome in a proteasome-dependent, yet ubiquitin-independent manner ***(21,22)***, yet this has to be established more firmly. The core catalytic subunit of the 26S enzyme is the 20S proteasome complex, and inhibition of this complex inhibits all proteolytic activities of the 26S proteasome. To test whether a certain protein substrate is degraded by the 20S proteasome, it is possible to inhibit the enzyme, both in vitro and in vivo. Inhibition of the 26S proteasome in a cell free system requires higher concentrations of the inhibitor (two- to fivefold) compared to the concentrations required to inhibit the enzyme in cultured cells. Also, as noted earlier, for accumulation of ubiquitin adducts in cell free systems, it is possible to inhibit the activity of the proteasome by the utilization of the nonhydrolyzable ATP analog adenosine-5'-*O*-(3-thiotriphosphate) (ATPγS; *see* in **Subheading 3.4.**). Stabilization of a protein under such conditions is a strong indication that the protein is indeed degraded by the 26S proteasome. Furthermore, inhibition of the 20S proteasome may lead to accumulation of ubiquitin adducts of the test protein that cannot be detected under conditions of rapid degradation when the proteasome is active. Detection of such intermediates serves as a strong evidence that the protein is degraded by the 26S proteasome complex following tagging by ubiquitin.

## 3.7. Determination of the Stability (Half-Life) of a Protein in Cells: Effect of Proteasome Inhibitors

### 3.7.1. Pulse-Chase Labeling and Immunoprecipitation

1. Wash cells twice in a methionine-free medium at 4°C.
2. Add methionine-free medium that contains dialyzed serum (serum is added in the concentration used for growing the cells).
3. Incubate for 1 h (to remove endogenous methionine) and remove the medium (by aspiration for adherent cells and following centrifugation at 800*g* for 10 min for cells in suspension) and add fresh methionine-free medium with serum. To save on labeled methionine, for adherent cells add medium to barely cover the cells (1–1.5.mL for a 60-mm dish. For cells in suspension, resuspend cells to $2 \times 10^6$/mL).
4. Add labeled methionine (50–250 μCi/mL) and continue the incubation for 0.5–1 h (pulse).
5. Add the inhibitor to the experimental dishes. Lactacystin and its lactone homolog should be added to a final concentration of 5–20 μ*M*, while MG132 to a final concentration of 50–100 μ*M*. The inhibitor should be added for 0.5 h (the last 0.5 h of the labeling period, pulse; always make sure that the cells are labeled for at least 15 min before addition of the inhibitor).
6. Remove the labeling medium (containing also the inhibitor in some of the samples).
7. Add ice-cold complete medium that contains, in addition to the inhibitor, also 2 m*M* of unlabeled methionine and wash the cells twice.
8. Add prewarmed complete medium (that contains the inhibitor and 2 m*M* of unlabeled methionine) and continue the incubation for the desired time periods (chase).
9. Withdraw samples at various time points and monitor degradation/stabilization of the target protein by immunoprecipitation followed by PhosphorImaging analysis. High molecular mass conjugates of the labeled protein should be precipitated by the specific antibody directed against the target protein under study. To avoid proteolysis of the conjugates by ubiquitin C-terminal hydrolases, it is recommended to dissolve the cells in a detergent-containing lysis buffer at 100°C. Also, the buffer should contain 10 m*M* NEM to inhibit the ubiquitin recycling enzymes. The NEM should be neutralized following cooling by addition of 7.5 m*M* DTT or 15 m*M* of β-mercaptoethanol.

Instead of using pulse-chase labeling and immunoprecipitation, one can use cycloheximide (20–100 μg/mL diluted from 20–100 mg/mL freshly dissolved solution) to stop general protein synthesis and follow specific protein degradation via Western lot analysis. The advantage of the approach is that it does not necessitate the use of radioactive material and immunoprecipitation, and one can load whole cell extract onto the gel. The utilization of the proteasome inhibitors is similar to that described (*see* **Sub-**

**heading 3.7.1.**) for the pulse-chase experiment. The disadvantage is that the potential interference of a drug in the proteolytic process. Thus, if the drug inhibits the synthesis of a short-lived ubiquitin ligase, E3, involved in targeting the test protein, the protein can be stabilized or further destabilized, dependent on the role of the ligase (whether direct or indirect).

A complementary approach to the utilization of proteasome inhibitors, which results in stabilization of ubiquitin system substrates and accumulation of ubiquitin adducts, is the use of cells that harbor a temperature sensitive mutation in the ubiquitin-activating enzyme E1, the first enzyme in the ubiquitin proteolytic cascade. At the nonpermissive temperature the cells fail to conjugate the target proteins which are consequently stabilized. Such cells can be, for example, the CHO-E36 (WT) and CHO-ts20 (E1 ts mutant) *(12)*. The experimental approach used can be either pulse-chase labeling and immunoprecipitation or cycloheximide chase (*see* **Subheading 3.7.1.** and above).

## Acknowledgments

Research in the laboratory of A.C. is supported by grants from Prostate Cancer Foundation (PCF) Israel-Centers of Excellence Program, the Israel Science Foundation–Centers of Excellence Program, a Professorship funded by the Israel Cancer Research Fund, ICRF (USA), and the Foundation for Promotion of Research in the Technion. Infrastructural equipment has been purchased with the support of the Wolfson Charitable Fund, Center of Excellence for studies on *Turnover of Cellular Proteins and its Implications to Human Diseases*.

## References

1. Pickart, C. M. (2001) Mechanisms of ubiquitination. *Annu. Rev. Biochem.* **70,** 503–533.
2. Weissman, A. M. (2001) Themes and variations on ubiquitylation. *Nat. Rev. Cell Mol. Biol.* **2,** 169–179.
3. Schwartz, D. C. and Hochstrasser M. (2003) A superfamily of protein tags: ubiquitin, SUMO and related modifiers. *Trends Biochem. Sci.* **28,** 321–328.
4. Huang, D. T., Walden, H., Duda, D., and Schulman, B. A. (2004) Ubiquitin-like protein activation. *Oncogene* **23,** 1958–1971.
5. Scherer, D. C., Brockman, J. A., Chen, Z., Maniatis, T., and Ballard, D. W. (1995) Signal-induced degradation of IκBα requires site-specific ubiquitination. *Proc. Natl. Acad. Sci. USA* **92,** 11259–11263.
6. King, R. W., Glotzer, M., and Kirschner, M. W. (1996) Mutagenic analysis of the destruction signal of mitotic cyclins and structural characterization of ubiquitinated intermediates. *Mol. Biol. Cell* **7,** 1343–1357.
7. Hou, D., Cenciarelli, C., Jensen, J. P., Nguygen, H. B., and Weissman, A. M. (1994) Activation-dependent ubiquitination of a T cell antigen receptor subunit on multiple intracellular lysines. *J. Biol. Chem.* **269,** 14244–14247.
8. Goldknopf, I. L. and Busch, H. (1977) Isopeptide linkage between nonhistone and histone 2A polypeptides of chromosomal conjugate-protein A24. *Proc. Natl. Acad. USA* **74,** 864–868.
9. Gronroos, E., Hellman, U., Heldin, C. H., and Ericsson, J. (2002) Control of Smad7 stability by competition between acetylation and ubiquitination. *Mol. Cell* **10,** 483–493.
10. Breitschopf, K., Bengal, E., Ziv, T., Admon, A., and Ciechanover, A. (1998) A novel site for ubiquitination: the N-terminal residue and not internal lysines of MyoD is essential for conjugation and degradation of the protein. *EMBO J.* **17,** 5964–5973.
11. Reinstein, E., Scheffner, M., Oren, M., Schwartz, A. L., and Ciechanover, A. (2000) Degradation of the E7 human papillomavirus oncoprotein by the ubiquitin-proteasome system: targeting via ubiquitination of the N-terminal residue. *Oncogene* **19,** 5944–5950.

12. Aviel, S., Winberg, G., Massucci, M., and Ciechanover, A. (2000) Degradation of the Epstein-Barr virus latent membrane protein 1 (LMP1) by the ubiquitin-proteasome pathway: targeting via ubiquitination of the N-terminal residue. *J. Biol Chem.* **275,** 23491–23499.
13. Ikeda, M., Ikeda, A., and Longnecker, R. (2002) Lysine-independent ubiquitination of the Epstein-Barr virus LMP2A. *Virology* **300,** 153–159.
14. Bloom, J., Amador, V., Bartolini, F., DeMartino, G., and Pagano, M. (2003) Proteasome-mediated degradation of p21 via N-terminal ubiquitinylation *Cell* **115,** 1–20.
15. Coulombe, P., Rodier, G., Bonneil, E., Thibault, P., and Meloche, S. (2004) N-terminal ubiquitination of extracellular signal-regulated kinase 3 and p21 directs their degradation by the proteasome. *Mol. Cell. Biol.* **24,** 6140–6150.
16. Fajerman, I., Schwartz, A. L., and Ciechanover, A. (2004) Degradation of the Id2 developmental regulator: targeting via N-terminal ubiquitination. *Biochem. Biophys. Res. Commun.* **314,** 505–512.
17. Trausch-Azar, J. S., Lingbeck, J., Ciechanover, A., and Schwartz, A. L. (2004) Ubiquitin-proteasome-mediated degradation of Id1 is modulated by MyoD. *J. Biol. Chem.* **279,** 32,614–32,619.
18. Doolman, R., Leichner, G. S., Avner, R., and Roitelman, J. (2004) Ubiquitin is conjugated by membrane ubiquitin ligase to three sites, including the N terminus, in transmembrane region of mammalian 3-hydroxy-3-methylglutaryl coenzyme A reductase: implications for sterol-regulated enzyme degradation. *J. Biol. Chem.* **279,** 38184–38193.
19. Kuo, M. L., den Besten, W., Bertwistle, D., Roussel, M. F., and Sherr, C. J. (2004) N-terminal polyubiquitination and degradation of the Arf tumor suppressor. *Genes and Dev.* **18,** 1862–1874.
20. Ben-Saadon, R., Fajerman, I., Ziv, T., Hellman, U., Schwartz, A. L., and Ciechanover, A. (2004) The tumor suppressor protein $p16^{INK4a}$ and the human papillomavirus oncoprotein E7-58 are naturally occurring lysine-less proteins that are degraded by the ubiquitin system: direct evidence for ubiquitination at the N-terminal residue. *J. Biol. Chem.* **279,** 41,414–41,421.
21. Sheaff, R. J., Singer, J. D., Swanger, J., Smitherman, M., Roberts, J. M., and Clurman, B. E. (2000) Proteasomal turnover of p21Cip1 does not require p21Cip1 ubiquitination. *Mol. Cell* **5,** 403–410.
22. Chen, X., Chi, Y., Bloecher, A., Aebersold, R., Clurman, B. E., and Roberts, J. M. (2004) N-acetylation and ubiquitin-independent proteasomal degradation of p21 (Cip1). *Mol. Cell* **16,** 839–847.
23. Varshavsky, A. (1996) The N-end rule: Functions, mysteries, uses. *Proc. Natl. Acad. Sci. USA* **93**, 12142–12149.
24. Johnson, E. S., Ma, P. C., Ota, I. M., and Varshavsky, A. (1995) A proteolytic pathway that recognizes ubiquitin as a degradation signal. *J. Biol. Chem.* **270,** 17442–17456.
25. Polevoda, B. and Sherman, F. (2003) N-terminal acetyltransferases and sequence requirements for N-terminal acetylation of eukaryotic proteins. *J. Mol. Biol.* **325,** 595–622.
26. Johnston, N. L. and Cohen, R. E. (1991) Uncoupling ubiquitin-protein conjugation from ubiquitin-dependent proteolysis by use of β, γ-nonhydrolyzable ATP analogues. *Biochemistry* **30,** 7514–7522.
27. Hershko, A. and Rose, I. A. (1987) Ubiquitin-aldehyde: a general inhibitor of ubiquitin-recycling processes. *Proc. Natl. Acad. Sci. USA* **84,** 1829–1833.
28. Hershko, A. and Heller, H. (1985) Occurrence of a polyubiquitin structure in ubiquitin-protein conjugates. *Biochem. Biophys. Res. Commun.* **128,** 1079–1086.
29. Murakami, Y., Matsufuji, S., Kameji, T., et al. (1992) Ornithine decarboxylase is degraded by the 26S proteasome without ubiquitination. *Nature* **380,** 597–599.

# 17

# Quantitating Defective Ribosome Products

**Shu-Bing Qian, Jack R. Bennink, and Jonathan W. Yewdell**

**Summary**

The fidelity of the protein production process is monitored by quality control machinery, which ensures that aberrant proteins are not deployed throughout the cell. A significant fraction (upwards of 30%) of proteins are degraded by proteasomes shortly after their synthesis, and we have termed such proteins defective ribosomal proteins (DRiPs). It is of interest and importance to characterize qualitatively and quantitatively this cohort of rapidly degraded nascent proteins. Quantitating DRiPs entails employing a standard pulse-chase protocol using radiolabeled amino acids. Protein degradation kinetics can be determined by either acid precipitation or SDS-PAGE. The introduction of proteasome inhibitors enables quantitation of proteasome-mediated protein degradation in vivo.

**Key Words:** Defective ribosome products; proteasome; proteasome inhibitor; protein degradation; pulse-chase; quantitative analysis.

## 1. Introduction

Polypeptides emerging from the ribosome must fold into a reasonably stable three-dimensional structure and maintain this structure throughout their functional lifetime. Misfolded proteins spell trouble to cells in their potential to malfunction and/or inappropriately associate with other cellular constituents. The fidelity of protein biosynthesis is monitored by cellular quality control machinery that identifies defective proteins *(1–3)*, which are predominantly destroyed by the proteasome, an abundant multicatalytic protease capable of degrading virtually any protein substrate into oligopeptides *(4–6)*. Most oligopeptides are then further degraded into free amino acids by the concerted actions of other endopeptidases and aminopeptidases (cells generally lack carboxypeptidase activities). A small fraction of proteasome-generated peptides are preserved, however, after finding their way to the endoplasmic reticulum where they bind to class I molecules of the major histocompatibility complex. Peptide class I complexes are whisked away to the cell surface for perusal by $CD8^+$ T cells, which monitor gene expression in this manner for the presence of viruses and other intracellular pathogens *(7–9)* and the aberrant expression of cellular proteins.

From: *Methods in Molecular Biology, vol. 301, Ubiquitin–Proteasome Protocols*
Edited by: C. Patterson and D. M. Cyr © Humana Press Inc., Totowa, NJ

Protein biosynthesis plays a critical role in cellular metabolism. In many cell types, protein synthesis represents the single largest consumer of cellular energy. To understand the cellular economy it is therefore crucial to determine the efficiency of protein biosynthesis. Extending the previous findings of Wheatley and colleagues ***(10)***, we provided evidence that a significant fraction (upwards of 30%) of proteins are degraded by proteasomes shortly after their synthesis ***(11,12)***. We presume many of these proteins are destroyed because of their inability to attain a stable conformation, and with this in mind, we have termed such proteins defective ribosomal proteins (DRiPs) ***(13,14)***.

It is of interest and importance to characterize qualitatively and quantitatively this cohort of rapidly degraded nascent proteins. To what extent are the degraded proteins truly defective? What is the nature of the defects? How does biosynthetic efficiency vary among different proteins? How does the abundance of DRiPs vary in different cell lines and under different conditions? Surely the most important question is the relevance of experiments with cultured cells to the workings of cells in intact animals. These questions will take years to answer. Part of the answer will come from determining the overall DRiP rate—a method we detail in this chapter.

Quantitating DRiPs entails employing a standard pulse-chase protocol using radiolabeled amino acids. Protein degradation is quantitated by collecting acid precipitable radiolabeled material on filters or by resolving radiolabeled proteins by sodium dodecyl sulfate-polyacrylamide gel electrophoresis (SDS-PAGE). A complementary method is to measure soluble radioactivity released by cells. These techniques have changed little in the past 25–50 yr, with the notable exceptions of the introduction of:

1. Multisample filter pads that greatly reduce the labor associated with filter-based scintillation counting assays.
2. Phosphor imaging techniques to accurately quantitate radioactivity present in gels.
3. Membrane-permeable compounds capable of specifically inhibiting the proteasome in viable cells ***(15)***. This enables quantitation of proteasome-mediated protein degradation of newly synthesized proteins and represents the single most important innovation in studying protein degradation in decades.

## 2. Materials

### 2.1. Proteasome Inhibitors

There are many commercially available proteasome inhibitors that can be used to quantitate DRiPs. Important considerations are potency, cost, and specificity, which vary widely among the inhibitors. A general caveat for using proteasome inhibitors is that chemically distinct inhibitors should be used in parallel experiments to establish as best as possible that the phenomenon studied is a direct result of blocking proteasomes and is not due to interference with other cellular processes. The most commonly used proteasome inhibitors are carbobenzoxy-leucyl-leucyl-leucinal (zLLL, also known as MG132) ***(16)*** and lactacystin (LC) ***(17)*** (*see* **Note 1**), whose strengths, respectively, are low cost and high specificity. Although less widely used, epoxomicin is also highly specific, and though relatively costly by weight, is far more potent than LC, and consequently offers the best BFB (bang for buck) ***(18)***. Stock solutions of zLLL and LC are made in dimethyl sulfoxide (DMSO) and stored at –20°C. The working concentration of both inhibitors is approx 20 μ*M*, though optimal concentration should be determined for each cell line used. Equal concentrations of solvent should be included as control in all experiments.

### 2.2. Cell Types

The effectiveness of proteasome inhibitors varies among cells for uncertain reasons. An effort should be made to determine the optimal inhibitor concentration, defined as the minimal concentration that completely (or nearly so) blocks cellular degradation of a proteasome substrate, such as noncleavable ubiquitin fusion protein *(19)*. It is necessary to establish a dose– response curve for each cell line–proteasome inhibitor combination (note to yeastophiles: for *Saccharomyces cerevisiae,* it is necessary to use a strain [i.e., JN284] with increased permeability to small molecules).

### 2.3. Radiolabeling and Chasing

1. Complete medium: Dulbecco's modified Eagle's Medium (DMEM, Invitrogen, Carlsbad, CA) containing 10% fetal bovine serum.
2. Phosphate-buffered saline (PBS): pH 7.4, 140 m*M* NaCl, 8 m*M* $Na_2HPO_4$, 1.5 m*M* $KH_2PO_4$.
3. Trypsin solution (Invitrogen, Carlsbad, CA).
4. Labeling medium: L-methionine-free DMEM (Invitrogen, Carlsbad, CA) containing 1X glutamine and 20 m*M* *N*-(2-hydroxyethyl)piperizine-*N*-(2-ethanesulfonic acid) (HEPES).
5. Redivue L-[$^{35}$S]-methionine (10 mCi/mL, Amersham Pharmacia, Sweden).
6. Stopping buffer: PBS containing 1 mg/mL L-methionine; keep ice-cold.
7. Chasing medium: complete DMEM medium containing 20 m*M* HEPES and 1 mg/mL of L-methionine.

### 2.4. TX-100 Fractionation

1. Extraction buffer: 50 mM Tris-HCl, pH7.4, 150 mM NaCl, 1 mM EDTA, and 1% TX-100 (Pierce, Rockford, IL).
2. Proteinase inhibitor cocktail (Boehringer Mannheim, Indianapolis, IN).
3. DNAse (Boehringer Mannheim, Indianapolis, IN). Add to extraction buffer freshly.
4. DC protein assay (Bio-Rad, Hercules, CA).

### 2.5. Trichloroacetic Acid (TCA) Precipitation and Scintillation Counting

1. 10% TCA (w/v): to a bottle containing 500 g of TCA, add 227 mL of $H_2O$. The resulting solution will contain 100% (w/v) TCA. A 10X dilution will be 10% (w/v).
2. 70% Ethanol/$H_2O$.
3. DEAE glass fiber filtermat and plastic sample bag (Wallac, Turku, Finland).
4. Scintillation liquid (Wallac, Turku, Finland).
5. Microbeta counter (Wallac, Turku, Finland).

### 2.6. SDS-PAGE

1. SDS-PAGE is performed using the Laemlli buffer system.
2. Sample buffer is prepared as 2X or 4X solution.
3. ProSieve 50 gel solution (BioWhittaker Molecular Applications, Rockland, ME).
4. 1.5 *M* Tris-HCl, pH 8.8: for preparing resolving gels.
5. 1.0 *M* Tris-HCl, pH 6.8: for preparing stacking gels.
6. Gel fixation solution: glacial acetic acid methanol–water (10:20:70).

7. Gel drier apparatus (Bio-Rad, Hercules, CA).
8. PhosphorImager (Molecular Dynamics, Sunnyvale, CA).
9. Mylar protective film PMS 1.2 (Perseasions, Elburn IL).
10. Biomax MR film (Kodak, Rochester, NY).

## 3. Methods

The methods described in this subheading outline (1) pulse-chase radiolabeling, (2) TX-100 fractionation, (3) TCA precipitation and scintillation counting, (4) SDS-PAGE, and (5) quantitative analysis.

### *3.1. Radiolabeling of Cells*

The isotopes most commonly used to radiolabel amino acids are $^{35}S$, $^{3}H$, and $^{14}C$. Owing to its extremely low specific activity, $^{14}C$ (5730-year half-life), should be used only in special circumstances. Although $^{3}H$ has a 50-fold lower specific activity than $^{35}S$ because of its longer half-life (12 yr vs 87 d), this is mitigated by a number of factors (described in next paragraph) that make it a perfectly useful tracer for scintillation-based detection methods. The low energy of $^{3}H$ β-particles makes their detection in gels difficult, and although special phosphor imager screens are available, in our experience they are insufficiently sensitive to be of much use for detecting radiolabeled proteins in dried gels. Thus, gel-based phosphor imager-based quantitation of proteins requires the use of $^{35}S$ labeling (note, however, that $^{3}H$ can be visualized in gels relatively efficiently by fluorography using preflashed film exposed at –80°C).

Although various amino acids are commercially available and may be used, [$^{3}H$]leucine and [$^{35}S$]methionine are the workhorses of pulse-radiolabeling. Both are essential amino acids, making it easier to control the labeling conditions. Commonly used tissue culture media deficient in either of these amino acids are available from a number of commercial sources. [$^{3}H$]Leucine offers three other advantages over other tritiated amino acids, including (1) higher specific activity (because of the presence of four labeled H atoms), (2) being most abundant amino acid in the proteome (9.6%; by comparison Met at 2.3% is among the least abundant residues), and (3) reaching linear incorporation rates into proteins within seconds of its addition to cells *(20)*. These factors combine to narrow the theoretical specific activity (i.e., disintegrations per protein) with Met to no more than 2.5-fold. On top of this, for reasons we do not understand fully, labeling of proteins with [$^{35}S$]Met results in 1/10 of the expected specific activity based on the rate of protein synthesis determined using [$^{3}H$]Leu (and confirmed by non-radioactive-based methods).

Despite [$^{3}H$]Leu's numerous qualities, owing to its ease of detection and quantitation via the phosphor imager, [$^{35}S$]methionine will be of wider applicability to most readers, and we therefore use it as an example in the following description (for labeling with [$^{3}H$]Leu, simply substitute Met for Leu in what follows).

1. Aspirate the medium from logarithmically growing cell cultures. Cells should be washed with PBS prior to resuspension in growth media for counting (*see* **Note 2**).
2. Based on cell counting, cell concentration should be adjusted to approx $10^7$ cells/mL. We generally use approx $10^5$ cells for each time point (*see* **Note 3**).

3. Split cells into two aliquots: one for inhibitor treatment and the other for a solvent control (*see* **Note 4**).
4. Prepare pulse medium with or without proteasome inhibitor. Typically, 250 μCi of L-[$^{35}$S]methionine is added to 100 μL of Met-free DMEM (*see* **Note 5**). Cells can be most easily manipulated at this point in Eppendorf tubes.
5. Quickly pellet the cells by centrifuging for 30 s at 3400*g* and aspirate the supernatant. Add an appropriate volume of prewarmed pulsing medium containing L-[$^{35}$S]methionine. For a 5-min pulse, $10^6$ cells can be suspended in 100 μL of media. Rapidly resuspend the cell pellet and incubate the cells at 37°C using a water bath or heat block (*see* **Note 6**).
6. Terminate radiolabeling by adding 10 vol of ice-cold PBS containing 1 mg/mL of cold L-methionine. Pellet the cells in the cold and wash twice using ice-cold PBS containing 1 mg/ml of L-methionine.
7. Resuspend cells in normal growth medium with and without proteasome inhibitor. Cells can be further incubated en masse or split into aliquots for different time points. The latter is less convenient but more accurate in equalizing cell numbers per time point. A 96-well plate can be used for collecting many samples.
8. At various times, pellet cells and keep the aspirated supernatant to measure protein secretion and release of TCA-soluble radioactivity. After wishing twice with ice-cold PBS, transfer the cell pellets to dry ice.

### 3.2. Cell Fractionation

Following physical disruption of cells by multiple freeze thawing, DRiPs selectively distributes into material that pellets following centrifugation for 2 h at 150,000*g* *(11)*. Once pelleted, this material resists solubilization in mild detergents, requiring SDS or other denaturing detergents. We recently found that approx two thirds of the rapidly degraded proteins are soluble in Triton X-100 (TX-100), which enables immunological characterization of this material in solution. TX100 fractionation also reveals biochemical properties of DRiPs that can be manipulated by altering cellular physiology using inhibitors or genetic manipulation of cellular gene expression.

1. Suspend the cell pellet in ice-cold TX-100 extraction buffer containing protease inhibitor cocktail. Typically, add 100 μL of extraction buffer per $1 \times 10^5$ cells.
2. Incubate the mixture on ice for 30 min and vortex briefly every 10 min.
3. Centrifuge (microfuge) at 15,000 rpm for 20 min at 4°C.
4. Transfer the supernatant to a fresh tube (TX-100 soluble fraction).
5. Homogenize the TX-100 insoluble pellet with the same volume of extraction buffer with 2% SDS and 10 U DNase by triturating repeatedly to lyse nuclei and shear the released DNA. Incubate the mixture at 95°C for 5 min. For a 96-well plate, seal the plate with Titer-Tops (Diversified Biotech, Boston, MA), and incubate in an 80°C oven for 20 min.
6. Centrifuge (microfuge) at 16,000*g* for 1 min. This is to remove the condensate from the Titer-Top and to ensure by visual inspection that the pellets are completely dissolved. Detectable pellets should be suspended by tituration until a smooth pipeting solution is obtained.
7. Quantitate the amount of extracted proteins using Bio-Rad DC protein assay to normalize for cell loses.

### 3.3. TCA Precipitation and Scintillation Counting

Radiolabeled proteins are most simply separated from free amino acids (or oligopeptides) by TCA precipitation. The following method describes TCA precipitation of proteins on a glass fiber filtermat (**Fig. 1**).

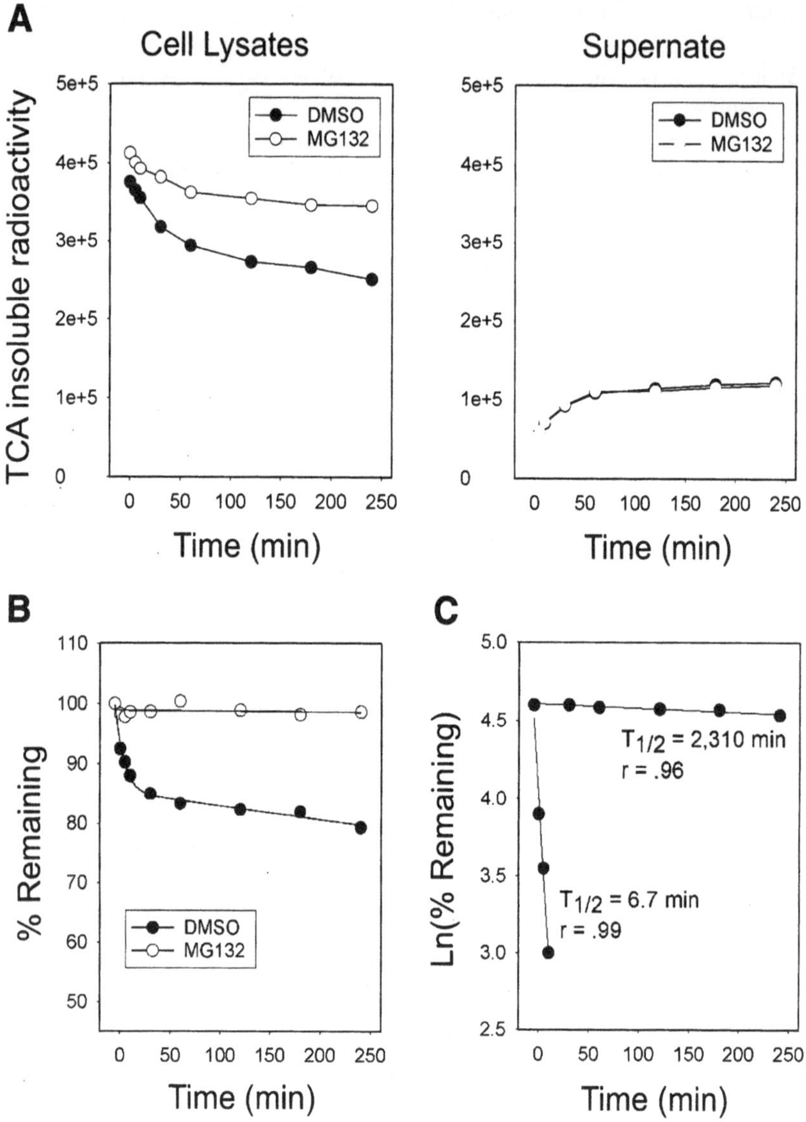


Fig. 1. Detection of DRiPS by TCA precipitation and scintillation counting. **(A)** In the absence (DMSO) or presence of 20 μ*M* MG132, HeLa cells were labeled with L-[$^{35}$S]Met for 5 min, and chased for various times as indicated. Whole cell lysates and supernatants containing secreted material were precipitated by 10% TCA onto filtermats and the insoluble radioactivity was counted. The $t_{1/2}$ for appearance of secreted proteins calculated from linear regression analysis of the appearance of TCA-insoluble counts in the supernatant in the first six time points is 27.1 min ($r$ = .97). **(B)** The counts of both cell lysates and supernatants were plotted as the percentage of initial incorporation using the value obtained with MG132 to estimate the true rate of protein synthesis. **(C)** Half-lives were calculated for the short- and long lived fractions by linear regression analysis of logarithmic decay curves derived from data shown in (B), assigning 15% of total counts to the short-lived population. Note that MG132 was added during the pulse labeling, resulting in an underestimate of the short-lived fraction (which is closer to 30%) caused by a lag in blocking proteasome activity.

1. Pipet 5 μL of lysate or supernatant on to a designated square of the filtermat. Perform in triplicate (*see* **Note 7**).
2. Dry filtermat at 80°C for 30 min (*see* **Note 8**).
3. Place filtermat in 50 mL of 10% TCA for 10 min on an orbital shaker.
4. Rinse twice with 50 mL of 70% ethanol for 10 min each time.
5. Dry filtermat at 80°C for 30 min (*see* **Note 8**).
6. Place filtermat in a plastic sample bag.
7. Seal the sample bag with the heat sealer directly adjacent to the filtermat, and cut short end off sample bag.
8. Pipet 6 mL of scintillation fluid into the sample bag.
9. Ensure that filtermat completely wets with the scintillation fluid then remove excess fluid from the bag.
10. Seal the sample bag with heat sealer directly adjacent to filtermat, and trim excess sample bag.
11. Tape sample bag into Microbeta 96-well counter tray.
12. Count.

### 3.4. SDS-PAGE

SDS-PAGE represents an alternative way to quantitate DRiPs that enables quantitation of proteins of defined size and allows for detection of size-bias in DRiPs. It is especially helpful when combined with TX-100 fractionation **(Fig. 2)**. It is important to maintain the stacking gel, as large/aggregated DRiPs may selectively run in or at the top of the stacking gel. Although it is possible to quantitate radioactivity in gels by X-ray film based methods, this is plagued by nonlinearities, and a phosphor imager (Bio-Rad, Fuji, or Amersham-Molecular Dynamics) offers huge advantages in accuracy and ease.

1. A standard protocol is used for SDS-PAGE. For higher resolution, mid-size 1.5-mm gels made of ProSieve 50 are recommended.
2. Assemble the gel apparatus, and fill the chambers with Tris–glycine running buffer. Remove any bubbles trapped along the bottom of the gels in between the glass plates.
3. Remove the gel from the electrophoresis apparatus and keep the stacking gel with the resolution gel. Fix the gel at room temperature in 5–10 vol of glacial acetic acid–methanol–water (10:20:70).
4. Dry the gel under vacuum with a Bio-Rad Model 583 or equivalent gel dryer for 2 h at 80°C.
5. Cover the gel with PMS1.2 protective Mylar to prevent contaminating the screen (*see* **Note 9**).
6. Expose the dried gel to phosphor imager screen for 4 h or overnight depending on radioactivity (*see* **Note 10**).
7. Read the screen with a scanner, and analyze it using manufacturer provided software.
8. X-ray film still rules for resolution of bands in gels. To produce autoradiographs, place the gel in an X-ray cassette, and expose it to Kodak Biomax film for a time period that gives the desired band intensity.

### 3.5. Quantitative Analysis

DRiPs are detected by the kinetics of protein degradation. In the absence of proteasome inhibitors, there is a rapid loss of a sizable cohort of labeled proteins detected either by acid precipitable radioactivity **(Fig. 1)** or the radioactivity of proteins in SDS-PAGE **(Fig. 2)**. The loss of radioactivity should be blocked by the addi-

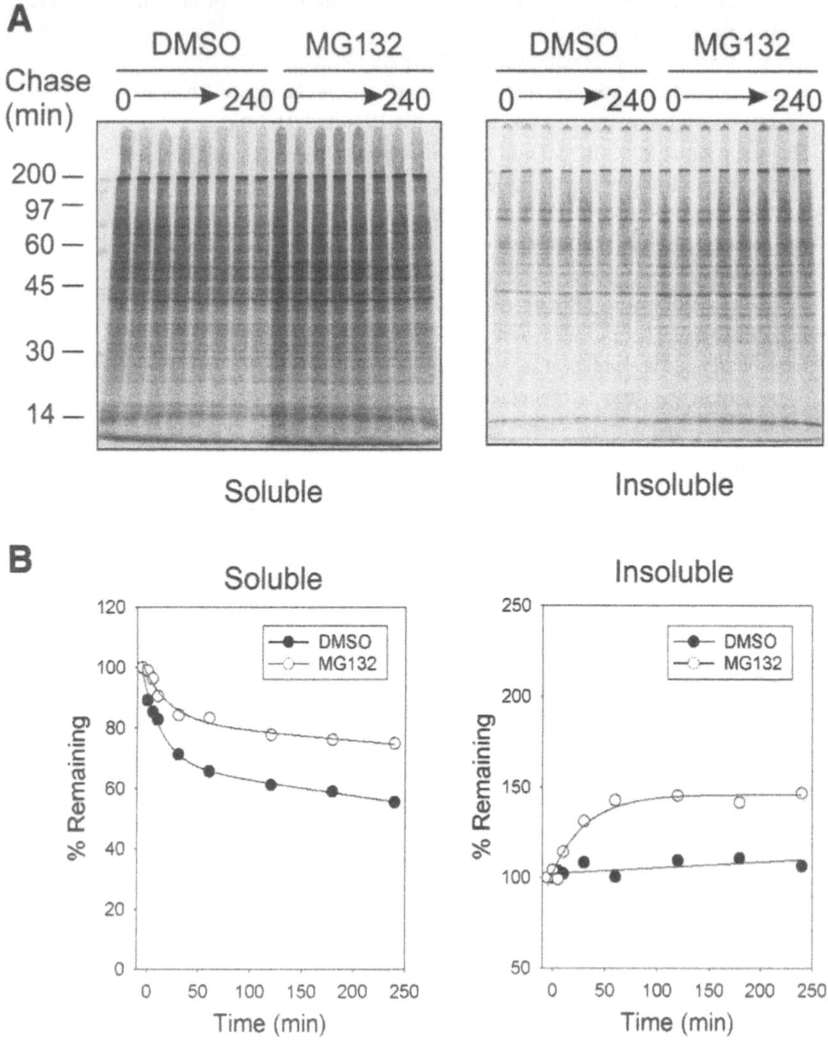


Fig. 2. Detection of radiolabeled proteins on different fractions by SDS-PAGE. (**A**) HeLa cells were labeled with L-[$^{35}$S]Met for 5 min, and chased for various time points as indicated. After fractionation by TX-100, proteins from both soluble and insoluble fractions were separated on a 12% SDS-PAGE. (**B**) The radioactivity from both soluble and insoluble fractions was quantitated using a phosphor imager.

tion of proteasome inhibitors to cells. Plotting the natural log (ln) of the remaining fraction vs. time reveals that the data can be fitted by two lines corresponding to the loss of short-lived proteins and long-lived proteins. From the slope *(m)* of the curves the half-lives can be calculated from the standard equation for random decay:

$$t_{1/2} = .693/m$$

The fraction of short- and long-lived proteins degraded should be normalized by the relative size of each pool. In the data shown in **Fig. 1C**, the short-lived pool is taken as 15% of the total, with the rest constituting the long-lived pool. Because the length of labeling represents a substantial fraction of the $t_{1/2}$ of short-lived proteins (5 min vs 6.7 min), the zero time point underrepresents the amount of biosynthesized material. This can be estimated by the amount of radioactivity recovered in the presence of proteasome inhibitors, which is plotted in the short-lived pool as –5 min.

## 4. Notes

1. The action of MG-132 is rapidly reversed on its removal. It is also a potent inhibitor of thiol proteases such as cathepsin B and calpains. LC and epoxomicin are both irreversible proteasome inhibitors.
2. If possible, it is better to use cells that grow in suspension cultures, as this avoids artifacts associated with cell removal from the plastic substrate. That being said, many adherent cell lines do not seem to mind spending a few hours in suspension, although it is important to establish that reasonable cellular viability (>90%) is maintained throughout the conditions employed. Adherent cells can be removed either by physical (scraping) or chemical means (EDTA with or without trypsin). We generally avoid scraping cells because it usually kills more cells than chemical removal.
3. The exact cell number required will vary depending on the sensitivity of detection, the rate of protein synthesis for different cell lines and the time used for pulse labeling. Note that 1-mm thick SDS-PAGE gels can accommodate approx $10^6$ cell equivalents before overloading of more abundant protein bands occurs.
4. Cells may be pretreated with proteasome inhibitors prior to radiolabeling. Although this will increase the effectiveness of the inhibitors (which require various incubation periods to reach maximal effect) it will also increase the down stream effects of the inhibitors, including interfering with protein synthesis itself. As a rule of thumb, if very brief pulse labeling is used (1 min), cells should be pretreated with inhibitors for a few min before the radiolabel is added. It is better to treat for a shorter time and accept partial proteasome inhibition, while recognizing that the DRiP fraction is probably underestimated.
5. Using Met-free DMEM as pulse medium will increase the specific activity of L-[$^{35}$S]methionine in nascent protein. However, methionine starvation before radiolabeling is to be avoided in quantitating DRiPs, because methionine depletion can affect the fidelity of protein synthesis (where Met has a particularly important role as the initiating amino acid). Artifacts arising from Met starvation can be completely avoiding by adjusting the concentration of total Met with unlabeled Met to match the concentration present in growth medium. This is the method of choice for scintillation counting because of its high sensitivity (the same applies for labeling with [$^3$H]Leu). Detection of radiolabel in gels, however, is considerably less sensitive, and often it is necessary to label in the absence of unlabeled Met.
6. For pulse-chase experiments of up to 4 h, it is not necessary to incubate cells in a $CO_2$ incubator. Suspending cells in small volumes for radiolabeling results in the use of lower amounts of L-[$^{35}$S]-Met and will save money and decrease the amount of radioactive material to be disposed of. Lower volumes for labeling and chasing also enable the use of microfuge tubes, which enables the handling of more samples.
7. Lysates are preferred to whole cells because of radioactivity self-absorption artifacts resulting from the high macromolecule concentrations in individual cells adsorbed to the filter mats.

8. Filtermats can also be quickly dried by microwaving for 5 min on high.
9. Saranwrap and similar products absorb up to 40% of $^{35}S$ β emissions, while absorption with PMS1.2 is <2%.
10. For accurate quantitation between gels it is important to expose gels to the same screen since sensitivity can vary markedly between screens.

## References


1. Wickner, S., Maurizi, M. R., and Gottesman, S. (1999) Posttranslational quality control: folding, refolding, and degrading proteins. *Science* **286,** 1888–1893.
2. Hartl, F. U. and Hayer-Hartl, M. (2002) Molecular chaperones in the cytosol: from nascent chain to folded protein. *Science* **295,** 1852–1858.
3. Frydman, J. (2001) Folding of newly translated proteins in vivo: the role of molecular chaperones. *Annu. Rev. Biochem.* **70,** 603–647.
4. Coux, O., Tanaka, K., and Goldberg, A. L. (1996) Structure and functions of the 20S and 26S proteasomes. *Annu. Rev. Biochem.* **65,** 801–847.
5. Voges, D., Zwickl, P., and Baumeister, W. (1999) The 26S proteasome: a molecular machine designed for controlled proteolysis. *Annu. Rev. Biochem.* **68,** 1015–1068.
6. DeMartino, G. N. and Slaughter, C. A. (1999) The proteasome, a novel protease regulated by multiple mechanisms. *J. Biol. Chem.* **274,** 22123–22126.
7. Rock, K. L. and Goldberg, A. L. (1999) Degradation of cell proteins and the generation of MHC class I-presented peptides. *Annu. Rev. Immunol.* **17,**739–779.
8. Yewdell, J. W. and Bennink, J. R. (2001) Cut and trim: generating MHC class I peptide ligands. *Curr. Opin. Immunol.* **13,** 13–18.
9. Goldberg, A. L., Cascio, P., Saric, T., and Rock, K. L. (2002) The importance of the proteasome and subsequent proteolytic steps in the generation of antigenic peptides. *Mol. Immunol.* **39,** 147–164.
10. Wheatley, D. N., Giddings, M. R., and Inglis, M. S. (1980) Kinetics of degradation of "short-" and "long-lived" proteins in cultured mammalian cells. *Cell Biol. Int. Rep.* **4,**1081–1090.
11. Schubert, U., Anton, L. C., Gibbs, J., Norbury, C. C., Yewdell, J. W., and Bennink, J. R. (2000) Rapid degradation of a large fraction of newly synthesized proteins by proteasomes. *Nature* **404,** 770–774.
12. Princiotta, M. F., Finzi, D., Qian, S. B., et al. (2003) Quantitating protein synthesis, degradation, and endogenous antigen processing. *Immunity* **18,** 343–354.
13. Yewdell, J. W., Anton, L. C., and Bennink, J. R. (1996) Defective ribosomal products (DRiPs): a major source of antigenic peptides for MHC class I molecules? *J. Immunol.* **157,** 1823–1826.
14. Yewdell, J. W. (2001) Not such a dismal science: the economics of protein synthesis, folding, degradation and antigen processing. *Trends Cell Biol.* **11,** 294–297.
15. Kisselev, A. F. and Goldberg, A. L. (2001) Proteasome inhibitors: from research tools to drug candidates. *Chem. Biol.* **8,** 739–758.
16. Rock, K. L., Gramm, C., Rothstein, L., et al. (1994) Inhibitors of the proteasome block the degradation of most cell proteins and the generation of peptides presented on MHC class I molecules. *Cell* **78,** 761–771.
17. Craiu, A., Gaczynska, M., Akopian, T., et al. (1997) Lactacystin and clasto-lactacystin beta-lactone modify multiple proteasome beta-subunits and inhibit intracellular protein degradation and major histocompatibility complex class I antigen presentation. *J. Biol. Chem.* **272,** 13437–13445.

18. Meng, L., Mohan, R., Kwok, B. H., Elofsson, M., Sin, N., and Crews, C. M. (1999) Epoxomicin, a potent and selective proteasome inhibitor, exhibits in vivo antiinflammatory activity. *Proc. Natl. Acad. Sci. USA* **96,** 10403–10408.
19. Qian, S. B., Ott, D. E., Schubert, U., Bennink, J. R., and Yewdell, J. W. (2002) Fusion proteins with COOH-terminal ubiquitin are stable and maintain dual functionality in vivo. *J. Biol. Chem.* **277,** 38818–38826.
20. Wheatley, D. N. and Inglis, M. S. (1985) Turnover of nascent proteins in HeLa-S3 cells and the quasi-linear incorporation kinetics of amino acids. *Cell Biol. Int. Rep.* **9,** 463–470.

# 18

# Endoplasmic Reticulum-Associated Protein Quality Control and Degradation

## *Screen for ERAD Mutants After Ethylmethane Sulfonate Mutagenesis*

**Antje Schäfer and Dieter H. Wolf**

### Summary

Proteins destined for secretion in eukaryotic cells enter the endoplasmic reticulum (ER) in an unfolded state and are properly folded in this organelle and sent to their final destination. Misfolded or orphan proteins are retained in the ER by a quality control system, retrotranslocated into the cytosol and degraded. Soluble and membrane proteins were found to require a basic machinery for elimination. It is composed of (1) the E1 (ubiquitin activating), E2 (ubiquitin conjugating), and E3 (ubiquitin ligase) enzymes, which polyubiquitinate the substrate proteins during retrotranslocation; (2) the trimeric AAA-ATPase complex Cdc48-Ufd1-Npl4p, which liberates the polyubiquitinated proteins from the ER; and (3) the 26S proteasome, finally degrading the misfolded proteins. Additional components for degradation of soluble or membrane proteins may vary depending on the nature of malfolded proteins *(1)*. It is therefore of utmost importance to gain insight into the different components of the ER protein quality control and degradation system required for the elimination of the substrate variety. Protein quality control of the ER and subsequent degradation are evolutionarily highly conserved from yeast to human. The yeast *Saccharomyces cerevisiae* is therefore an elegant model organism for a search of new components of the ER quality control and degradation machinery, because it is easily amenable to genetic and molecular biological experimentation.

In this chapter, a genetic approach is presented, which leads to the isolation of mutants and to the identification of proteins involved in protein quality control and ER-associated degradation (ERAD). The method resides in ethylmethane sulfonate (EMS) mutagenesis of a yeast strain followed by screening for stabilization of soluble ERAD substrates, two mutated and consequently malfolded vacuolar enzymes, carboxypeptidase yscY (CPY*) and proteinase yscA (PrA*). Both malfolded proteins are retained in the ER lumen and become substrates of the ERAD machinery *(2,3)*.

**Key Words:** Endoplasmic reticulum; endoplasmic reticulum-associated degradation; mutants; mutant screen; protein quality control; yeast genetics.

From: *Methods in Molecular Biology, vol. 301, Ubiquitin–Proteasome Protocols*
Edited by: C. Patterson and D. M. Cyr 

## 1. Introduction

This chapter describes the screening of yeast mutants defective in ERAD, which led to the identification of the ERAD components Der1p *(2)*, Der2p (Ubc7p, ubiquitin-conjugating enzyme E2), and Der3p (ubiquitin–protein ligase E3). For this purpose the wild-type strain YAF6 (*see* **Note 1**) is mutated using ethylmethane sulfonate (EMS) *(4)*, which results in transition of G-C base pairs (*see* **Subheading 3.1**). The mutants are screened in colony-immunoblot tests for stabilization of the two soluble ERAD substrates CPY* and PrA* as described in **Subheading 3.2.** In **Subheading 3.3.** the identification and cloning of the *DER1* gene is presented using a genomic DNA library.

## 2. Materials

1. LB: 1% peptone, 0.5% yeast extract, 0.5% NaCl, pH 7.0, 100 μg/μL of ampicillin.
2. YPD: 1% yeast extract, 2% peptone, 2% D-glucose. SPO: 0.1% yeast nitrogen base without amino acids, 0.05% D-glucose, 1.0% potassium acetate, supplemented with amino acids as required *(5)*.
3. SC: synthetic complete medium, 0.67% yeast nitrogen base, without amino acids, 2% D-glucose, as liquid media or as plates including 2% agar.
4. *Saccharomyces cerevisiae* strain YAF6 (*prc1-1; pra1Δss; leu2-3,112*) expressing the ERAD substrates CPY* and PrA*.
5. *Escherichia coli* strain DH5α.
6. Cycloheximide.
7. 30 m*M* Sodium azide.
8. Ethylmethane sulfonate (EMS), mutating agent.
9. 0.1*M* Potassium phosphate buffer, pH 8.0.
10. Nitrocellulose membrane (diameter 82 mm).
11. Petri dishes.
12. 3MM Whatman paper.
13. Lysis solution: 0.1% sodium dodecyl sulfate (SDS), 0.2 *M* NaOH, 0.5% β-mercaptoethanol.
14. TBS-T buffer: 150 m*M* NaCl, 0.05% Tween-20, 50 m*M* Tris-HCl, pH 7.4.
15. Reaction buffer: 60 mg of 4-chloro-1-naphthol in 20 mL of methanol p.a.; add 80 mL of TBS, 30 μL of 30% $H_2O_2$, freshly prepared.
16. Anti-CPY and anti-PrA antibodies (*see* **Note 2**).
17. Sodium dodecyl sulfate-polyacrylamide gel electrophoresis (SDS-PAGE) equipment.
18. Western-blot equipment and buffers, peroxidase coupled antibodies, ECL-System.
19. Yeast DNA library based on the yeast strain K2007 *(6)*.
20. Replica plating block with velvet.
21. Restriction enzymes, T4-DNA ligase and buffers according to the manuals of the preferred company.
22. DNA gel equipment and DNA gel extraction system from agarose (Qiaex-II extraction kit, Qiagen, Germany).

## 3. Methods

### 3.1. Mutagenesis

The mutagenesis of the parental strain YAF6 followed the protocol of Lawrence *(4)*.

1. Grow cells overnight at 30°C in 5 mL of YPD up to the stationary phase.
2. After collecting cells by centrifugation wash twice with sterile potassium phosphate buffer, pH 8.0, and resuspend the cells in 5 mL of the same buffer.
3. Add 150 µL of EMS and shake vigorously for 30 s (*see* **Notes 3** and **4**).
4. Incubate the cells for 1 h at room temperature and shake cells at 20-min. intervals.
5. Wash cells twice with sterile potassium phosphate buffer, pH 8.0.
6. Dilute with 5 mL of potassium phosphate buffer, pH 8.0, containing 10% YPD.
7. Count cells using a *Neubauer* hemocytometer (0.1 mm).
8. Test the cell survival rate by growing a calculated number of 200, 500, and 1000 cells per YPD plate within 2 d at 23°C. Keep the mutagenized cell stock in a refrigerator at 4°C.
9. Count the surviving cells (*see* **Note 5**).
10. Plate approx 8000 to approx 10,000 cells from the cell stock on YPD plates (about 100 surviving cells per plate) and let them grow 3 d at 23°C (*see* **Note 6**).
11. Generate replica plates, first for mutant screening (*see* **Subheading 3.2.**) and second for isolation of cells identified as mutant clones.

### 3.2. Detection of Mutated Cell Clones ("Colony Immunoblot")

The cells will be grown on plates and lysed directly on nitrocellulose membranes *(7)*. Quality control and ERAD-defective mutant colonies will be detected on the basis that CPY* and PrA* antigenic material remains undegraded. Specific binding of antibodies against CPY* and PrA* is visualized using secondary antibodies and the catalytic activity of coupled peroxidase.

1. Generate replica plates using fresh YPD plates covered with nitrocellulose membranes and grow cells for 2 d at 23°C (*see* **Note 7**).
2. Remove the membranes with grown cell clones and place overnight on SPO plates to induce CPY* and PrA* protein content (*see* **Note 8**).
3. Place membranes for additional 10 h on fresh SPO plates containing 4 µg/mL of cycloheximide to block protein synthesis.
4. Place membranes on round filter papers (Whatman 3MM) soaked with 2.8 mL of lysis solution in Petri dishes (82 mm in diameter) and incubate for 1 h at room temperature.
5. Remove the cells by washing the membranes with a sharp water jet (*see* **Note 9**).
6. Incubate the membranes two times for 30 min in TBS-T buffer while shaking.
7. Incubate the membranes with anti-CPY or anti-PrA antibodies in TBS-T buffer overnight at 4°C.
8. Wash membranes three times with TBS-T buffer and incubate with secondary, peroxidase-coupled antibodies (1:1500) for 4 h at room temperature.
9. Wash membranes three times with TBS-T buffer and once with TBS without detergent.
10. For detection soak one round Whatman filter paper per membrane with freshly prepared reaction buffer and place membrane on the top of it.
11. The enzymatic color reaction starts after approx 10 min and the expected mutants are indicated by blue precipitates (**Fig. 1A**). The mutant colonies may show up by releasing a faint blue (**Fig. 1A**, mutants 2–7) to dark blue color (**Fig. 1A**, mutant 1).
12. Stop the reaction after approx 60 min by washing the membranes in distilled water followed by air-drying.

To finally verify ER quality control or ERAD-deficiency of the detected mutants an additional test is indicated. Cell clones of interest will be streaked or spotted on YPD plates and treated as before from **step 1** to **12** of **Subheading 3.2.** using antibod-

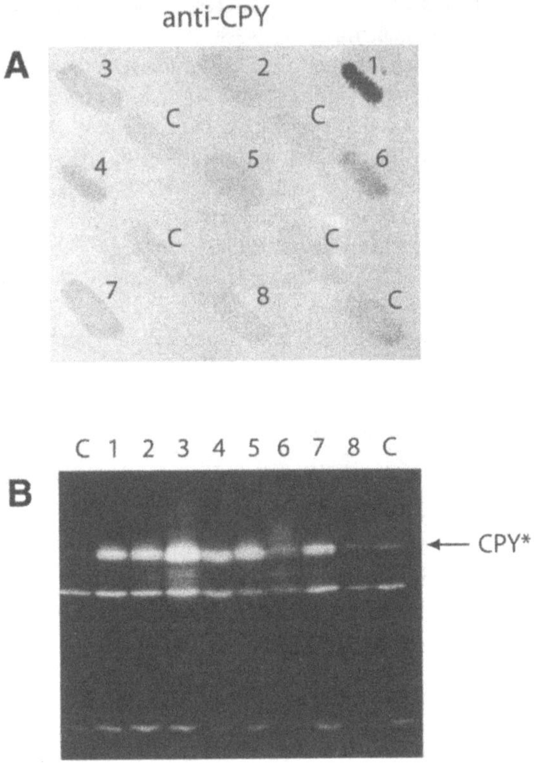


Fig. 1. Colony-immunoblot **(A)** and Western-blot analysis **(B)** performed from eight mutant strains that show stabilization of CPY*. For control, the nonmutated strain YAF6 **(C)** is presented. Mutants indicated by blue precipitates in immunoblot analysis (B) pointed to ERAD deficiency of the corresponding mutants (no. 1–7).

ies raised against an ERAD substrate; do not forget inclusion of a wild-type control on each plate **(Fig. 1A)**. As an additional control equal amounts of crude extracts of mutant and wild-type cells incubated on SPO medium containing cycloheximide were subjected to SDS-PAGE. Subsequent immunoblotting shows CPY* **(Fig. 1B)** and/or PrA* levels in cells. Positive, staining clones should be subjected to biochemical tests measuring the degradation kinetics of CPY* or PrA* that is, performing cycloheximide decay experiments (*see* **Subheading 3.3.**) or in a more quantitative fashion, by pulse-chase analysis (a detailed protocol is published in **ref. *8***).

### 3.3. Cycloheximide Decay Experiment

1. Grow cells to logarithmic phase in SC medium supplemented with amino acids as required.
2. After collecting 8–10 $A_{600}$ of cells by centrifugation, resuspend the cells in 2–2.5 mL of fresh SC medium.
3. Add cycloheximide to the cell suspension to yield a concentration of 100 μg/mL.

4. Take 2–2.5 $A_{600}$ of cells at the time points 0 min, 30 min, 60 min, and 120 min and mix with same volume of 30 m*M* sodium azide (*see* **Note 10**). Keep cells on ice.
5. After final sampling, centrifuge the cells.
6. Prepare cell extracts using alkaline lysis *(3)* and subject to SDS-PAGE followed by immunoblot analysis.

### 3.4. Cloning of the Wild-Type Genes Necessary for ERAD

The corresponding wild-type genes can be cloned by complementation of the recessive mutant phenotype using a library of yeast genomic DNA. For the isolation of the *DER1* gene a genomic DNA library was used, based on a centromere, *LEU2* containing plasmid kindly donated by F. Cvrcková *(6)*. This library contains genomic DNA fragments of 3–9 kb in length of the yeast strain K2007.

After transformation of the plasmids containing the genomic library according to the standard methods *(9)* cells were grown on SC medium lacking leucine as described under **Subheading 3.2.**, resulting in about 40,000 colonies on a total of 320 plates. Colonies that can no longer be stained with anti-CPY and anti-PrA antibodies on nitrocellulose membranes were retested by Western blot analysis. The corresponding plasmids were isolated from yeast and rescued in *E. coli* according to standard methods *(10)*. All isolated plasmids were mapped using the enzymes *Hin*dIII and *Eco*RI and an identical fragment pattern indicated the same genomic insert. The *DER1* gene was located by testing the ability of fragments of a plasmid insert, subcloned into plasmid pRS315 *(10)*, to complement the recessive mutant phenotype.

## 4. Notes

1. Choose a strain that can also be used for screening of a gene library (e.g., pay attention to markers).
2. Antibodies should have no or little cross-reactivity. Prove this by Western-blot analysis.
3. EMS is hardly soluble in potassium phosphate buffer. Therefore shake the cell suspension heavily for 30 s after addition of the mutagen.
4. All equipments, which are contaminated with the strong mutagenic agent EMS, including solutions, have to be incubated with saturated $Na_2S_2O_3$ for 2 h in a flow hood for neutralization of EMS.
5. The survival rate of the mutated cells should be approx 30%.
6. For search of proteins involved in ERAD but vital for growth, membranes that have been placed overnight on SPO plates (**Subheading 3.2., step 2**) should be shifted to 37°C for 2 h and then be replaced on cycloheximide-containing SPO plates (**Subheading 3.2., step 3**) at 37°C.
7. For control the nonmutated strain has to be included and treated under the same conditions.
8. The expression of CPY* and PrA* is increased on starvation for glucose or nitrogen and results in higher *steady-state* levels of both proteins *(2)*. Therefore, to screen for stabilization of CPY* and PrA* due to the deletion of an ERAD component, it is necessary to induce first their expression levels by growth on SPO media (*see* **Subheading 3.2., step 2**) and second inhibit further protein expression by application of cycloheximide (*see* **Subheading 3.2., step 3**)
9. After washing the lysed cells using a water jet, the proteins will remain bound on the membranes.

10. The best suited time points for taking cell samples after adding cycloheximide can vary when using different strains and assaying different proteins. It is proposed to start sampling at 0 min, 60 min, 120 min, and, according to the stabilization of the proteins of interest after immunodetection adapt the sampling times as appropriate.

## Acknowledgments

The authors thank Fatima Cvrcková and Kim Nasmyth for the CEN-LEU2 library. Andreas Finger for developing the screen and his excellent protocols. We also want to thank Sonja Dieter for helpful comments and Wolfgang Heinemeyer for critical reading. The work was supported by grants from the Deutsche Forschungsgemeinschaft (Bonn, Germany) and the Fonds der Chemischen Industrie (Frankfurt, Germany).

## References

1. Kostova, Z. and Wolf, D. H. (2003) For whom the bell tolls: protein quality control of the endoplasmic reticulum and the ubiquitin-proteasome connection. *EMBO J.* **15,** 2309–2317.
2. Finger, A., Knop, M., and Wolf, D. H. (1993) Analysis of two mutated vacuolar proteins reveals a degradation pathway in the endoplasmatic reticulum or a related compartment of yeast. *Eur. J. Biochem.* **218,** 565–574.
3. Knop, M., Finger, A., Braun, T., Hellmuth, K., and Wolf, D. H. (1996) Der1, a novel protein specifically required for the endoplasmatic reticulum degradation in yeast. *EMBO J.* **15,** 753–763.
4. Lawrence, C. W. (2002) Classical mutagenesis techniques. *Methods Enzymol.* **350,** 189–199.
5. Sherman, F. (2002) Getting started with yeast. *Methods Enzymol.* **350,** 3–41.
6. Cvrcková, F. and Nasmyth, K. (1993) Yeast G1 cyclins CLN1 and CLN2 and a GAP-like protein have a role in bud formation. *EMBO J.* **12,** 5277–5286.
7. Lyons, S. and Nelson, N. (1984) An immunological method for detecting gene expression in yeast colonies. *Proc. Natl. Acad. Sci. USA* **81,** 7426–7430.
8. Hilt, W. and Wolf, D. H. (1999) Protein degradation and proteinases in yeast. In *Post-translational Processing. A Practical Approach* (Higgins, S. J. and Hames, B. D., eds.), Oxford University Press, Oxford and New York, pp. 265–317.
9. Gietz, R. D. and Woods, R. A. (2002) Transformation of yeast by lithium acetate/single-stranded carrier DNA/polyethylene glycol method. *Methods Enzymol.* **350,** 87–96.
10. Sikorski, R. S. and Hieter, P. (1989) A system of shuttle vectors and yeast host strains designed for efficient manipulation of DNA in *Saccharomyces cerevisiae. Genetics* **122,** 19–27.

19

# Endoplasmic Reticulum-Associated Protein Quality Control and Degradation

## *Genome-Wide Screen for ERAD Components*

**Antje Schäfer and Dieter H. Wolf**

**Summary**

In this chapter, a genetic approach is presented that leads to the isolation of mutants and to the identification of proteins involved in protein quality control and endoplasmic reticulum-associated degradation (ERAD). The method makes use of a genomic screen of a yeast deletion library (EUROSCARF). Transformation of each of the approx 5000 strains deleted in one nonvital gene each with a CPY* chimera (CTL*, Fig. 1) containing CPY* C-terminally fused to a transmembrane domain and the cytosolic Leu2 protein (3-isopropylmalate dehydrogenase) constitutes the basic screening procedure. Because of a Leu2p deficiency in all deletion strains, cells can grow only when the CTL* chimera is present. As the CPY* module of CTL* will be recognized in ERAD-proficient cells, CTL* will be degraded and the strain is unable to grow. Therefore the absence of genes necessary for ER quality control and ERAD will allow cell growth and indicate the necessity of the respective gene for these processes *(1)*.

**Key Words:** Endoplasmic reticulum; endoplasmic reticulum-associated degradation; genome-wide screen; mutants; mutant screen; protein quality control; yeast genetics.

## 1. Introduction

For the identification of yeast deletion mutants defective in endoplasmic reticulum (ER) quality control or degradation a screening test based on cell growth was elaborated *(1)*. It is based on the degradation of an ER-associated degradation (ERAD) substrate protein, which is necessary for cell growth under selective conditions. For this purpose, a chimeric protein was constructed *(1,2)*, which consists of the ER lumenal CPY* protein fused to a transmembrane domain and cytoplasmic 3-isopropylmalate dehydrogenase, the Leu2 protein (CTL*, **Fig. 1**). To allow sensitive screening the gene chimera was placed under the *GAL4* promoter allowing only low protein expression when cells were grown on glucose (*see* **Note 1**). Cells carrying a

From: *Methods in Molecular Biology, vol. 301, Ubiquitin–Proteasome Protocols*
Edited by: C. Patterson and D. M. Cyr © Humana Press Inc., Totowa, NJ

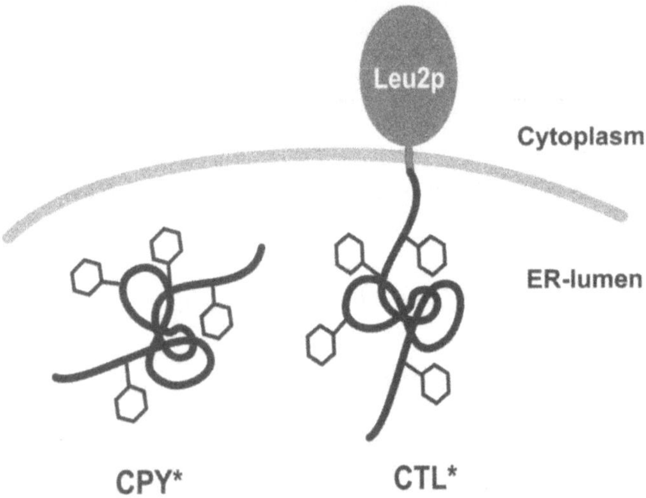


Fig. 1. Schematic representation of the ERAD substrates CPY* and CTL*. The mutated CPY* remains malfolded in the ER lumen. The membrane protein CTL* is a chimeric derivative of CPY*, fused to a transmembrane domain and to Leu2p.

*LEU2* deletion can grow only when the chimeric protein is present: thus only mutant cells defective in a protein quality control or ERAD component can grow. A genome-wide screen can be performed by transforming the CTL*-coding DNA into the approx 5000 individual deletion mutants of the EUROSCARF yeast library (see **Subheading 3.**); (*see* **Note 2**). Examples for components required for ERAD found by this method are the ubiquitin domain proteins Dsk2p and Rad23p *(1)*.

## 2. Materials

1. Media: YPD: 1% yeast extract, 2% peptone, 2% D-glucose; SPO: 0.1% yeast nitrogen base, without amino acids, 0.05% D-glucose, 1.0% potassium acetate supplemented with amino acids as required *(3)*; SC: 0.67% yeast nitrogen base, without amino acids, 2% D-glucose, supplemented with amino acids as required, as liquid media or as plates including 2% agar as required.
2. The EUROSCARF yeast library as a set of diploid deletion strains based on the strain BY4743 (*MATα/MATα*; *his3Δ1/ his3Δ1*; *leu2Δ0/leu2Δ0*; *met15Δ0/MET*15; *LYS2/lys2Δ0*; *ura3Δ0/ura3Δ0*), www.uni-frankfurt.de/fb15/mikro/euroscarf/index/html.
3. 24-Well plates with lids used for tissue cultures (TC-Plate, sterile, No. 662160/ Greiner Bio-one, Germany).
4. 96-Well microplates (PS-Plate, U-shape/ Greiner Bio-one, Germany)
5. Lithium acetate–TE buffer: 10 m*M* Tris-HCl, pH 7.5, 1 m*M* EDTA, 100 m*M* lithium acetate with and without 40–50% PEG3350.
6. Plasmid containing the CTL*-encoding DNA sequence *(1,2)*.
7. Carrier DNA (herring sperm, Roche, Germany).

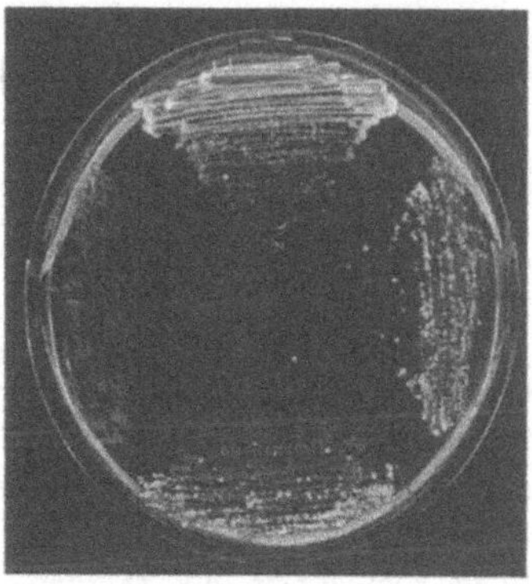


Fig. 2. The BY4743 wild-type strain expressing CTL* fails to grow on SC medium lacking leucine, whereas a strain deleted for the known ERAD component Der3p permits growth (positive control). The deletion strains Δ*dsk2* and Δ*rad23* expressing CTL*, complement the leucine auxotrophy. This indicates the stabilization of CTL* in these deletion strains.

## 3. Methods

### 3.1. Yeast Multitransformation

1. Preculture each deletion strain (*see* **Note 3**) in 0.6 mL of YPD liquid media in 24-well plates with lids for 14–16 h at 30°C while shaking at 150 rpm up to the stationary phase (*see* **Note 4**).
2. Inoculate 60 μL of the preculture in 0.6 mL of fresh YPD in 24-well plates and grow for 2 h at 30°C (shaking) to the logarithmic phase ($A_{600}$ of 0.8–1.5/mL).
3. Centrifuge the cells in 24-well plates for 5–10 min and wash cell sediments with deionized water (sterile).
4. Add 80 μL of lithium acetate–TE buffer, 120 μg of carrier DNA (stock solution, 10 μg/μL) and 2 μg of the *CTL** gene containing plasmid into each well (*see* **Notes 5** and **6**).
5. Incubate for 30–45 min at 30°C while shaking.
6. Add 300 μL of lithium acetate–TE including 50% PEG3350.
7. Incubate for 45 min at 30°C while shaking.
8. Add 50 μL of dimethyl sulfoxide (DMSO) and shake immediately (*see* **Note 7**).
9. Heat shock for 15 min. at 42°C in a water bath (*see* **Note 8**).
10. Centrifuge the cells in the 24-well plates and discard the supernatant.
11. Add 600 μL of synthetic dropout media without uracil (transformation selection media) into each well and let cells grow while shaking at 30°C for 14–16 h.
12. Plate the transformants on synthetic dropout media without uracil plates and grow for 2–3 d at 30°C.
13. Plate the transformants on synthetic dropout media plates without uracil *and* without leucine and incubate cells for 2–3 d at 30°C to induce ERAD.
14. Retest positive yeast strains by plating again on synthetic dropout media plates without uracil and without leucine and incubate at 30°C (**Fig. 2**).

For identification of the corresponding mutated genes check the EUROSCARF strain list. Get further information on the gene product required for ERAD using yeast genome databases, for example, SGD (*Saccharomyces* genome databases, *www.yeastgenome.org*). To verify the involvement of these mutated proteins in ERAD perform additional analyses such as pulse-chase or cycloheximide decay experiments using mutants of additional known ERAD substrate proteins as controls.

## 4. Notes

1. Low expression of CTL* is particularly required as yeast cells can sustain growth even in the presence of minimal amounts of leucine.
2. The set of gene deletion strains was generated in the EUROFAN II project (*www.rz.uni-frankfurt.de/FB/fb16/mikro/euroscarf*) in the BY4743 strain background (*see* **Subheading 2.**), which is derived from the standard strain S288C and has been used for sequencing of the *S. cerevisiae* genome.
3. The EUROSCARF set of gene deletion strains is offered on 96-multiwell microplates on YPD medium. The complete set of deletion strains consists of approx 75 plates. The strains can be used directly for transformation procedures, but it is useful for further experiments to prepare a –80°C stock in similar multiwell microplates containing YPD and 15% glycerol.
4. Use multichannel pipets with sterile tips and tooth picks to handle the procedure more efficiently.
5. Mix lithium acetate–TE buffer, carrier DNA, and plasmid DNA prior to use and add all together in one step using a multichannel pipet.
6. In some instances less than the indicated amounts of carrier DNA and plasmid DNA can be used for sufficient and effective transformations. Start first with higher amounts to practice the experimental procedures.
7. This method can also be performed without addition of DMSO.
8. To protect the cells seal the plastic lids and well plates with Parafilm®.

## Acknowledgments

The authors thank Balasubrahmanyam Medicherla and Zlatka Kostova for the development and utilization of the *CTL** screen, Sonja Dieter for helpful comments, and Wolfgang Heinemeyer for critical reading. The work was supported by grants from the Deutsche Forschungsgemeinschaft (Bonn, Germany) and the Fonds der Chemischen Industrie (Frankfurt, Germany).

## References

1. Medicherla, B., Kostova, Z., Schaefer, A., and Wolf, D. H. (2004) A genomic screen identifies Dsk2p and Rad23p as essential components of ER-associated degradation. *EMBO Rep.* **5,** 692–697.
2. Taxis, C., Hitt, R., Park, S-H., Deak, P. M., Kostova, Z., and Wolf, D. H. (2003) Use of modular substrates demonstrates mechanistic diversity and reveals differences in chaperone requirement of ERAD. *J. Biol. Chem.* **278,** 35903–35913.
3. Sherman, F. (2002) Getting started with yeast. *Methods Enzymol.* **350,** 3–41.

# 20

# Cystic Fibrosis Transmembrane Conductance Regulator as a Model Substrate to Study Endoplasmic Reticulum Protein Quality Control in Mammalian Cells

**J. Michael Younger, Chun-Yang Fan, Liling Chen, Meredith F. N. Rosser, Cam Patterson, and Douglas M. Cyr**

**Summary**

Components of the ubiquitin–proteasome system function on the surface of the endoplasmic reticulum (ER) to select misfolded proteins for degradation. Herein we describe methods that allow for the study of the pathway for proteasomal degradation of the cystic fibrosis transmembrane conductance regulator (CFTR). The experimental system described employs transiently transfected HEK-293 cells and is utilized to monitor the biogenesis of CFTR by Western blot and pulse-chase analysis.

**Key Words:** Cell culture; chaperones; cystic fibrosis transmembrane conductance regulator; endoplasmic reticulum protein quality control.

## 1. Introduction

Endoplasmic reticulum-associated degradation (ERAD) is a process that involves the recognition and degradation of misfolded lumenal and transmembrane proteins via the ubiquitin–proteasome system *(1,2)*. Model substrates that are utilized to study basic principles of the ERAD process include the T-cell receptor α subunit (TCRα) *(3)*, 3-hydroxy 3-methylglutaryl coenzyme A reductase (HMG-CoA reductase, HMGR) *(4,5)*, a mutant form of carboxypeptidase Y *(6,7)*, and cystic fibrosis transmembrane conductance regulator (CFTR) *(8)*.

Misfolded forms of CFTR and ΔF508 CFTR appear to be recognized by the CHIP/Hsc70 E3 ubiquitin ligase complex *(9,10)* and then degraded via a pathway that is blocked by proteasome inhibitors *(11,12)*. CFTR is a 1480-residue glycomembrane protein that contains two membrane spanning domains and cytoplasmic subdomains that include nucleotide binding domain 1, nucleotide binding domain 2, and a regulatory domain. Folding of CFTR is slow and 60–70% of it is selected for ERAD prior to

From: *Methods in Molecular Biology, vol. 301, Ubiquitin–Proteasome Protocols*
Edited by: C. Patterson and D. M. Cyr © Humana Press Inc., Totowa, NJ

reaching the native state and escaping the ER. ΔF508 CFTR contains a deletion of F508 in nucleotide binding domain 1 that causes 100% of it to be degraded ***(13)***. Because CFTR contains transmembrane and cytoplasmic subdomains, the study of its ERAD has the potential to uncover how elements of the cytosolic protein quality control system cooperate with ER-localized proteins to degrade polytopic proteins.

Because ΔF508 CFTR misfolds and is completely degraded via ERAD it serves as an excellent model substrate for basic studies. However, ΔF508 CFTR expression is very low in cultured epithelial cells. Thus, the study of its degradation requires the use of cell lines that stably express ΔF508 CFTR from a strong promoter ***(14)*** or transient cell expression systems that utilize Cos-7 or HEK-293 cell lines ***(10,13)***.

Herein we describe an experimental system to study ERAD of ΔF508 CFTR in a cell expression system that employs HEK-293 cells that are transfected with pcDNA3.1-ΔF508 CFTR. Methods to determine the steady-state level and detergent solubility of ΔF508 CFTR by Western blot are described. In addition, we include a protocol to study the rate of ΔF508 CFTR degradation via pulse-chase analysis.

## 2. Materials

### 2.1. Cell Culture

1. HEK-293 cells are grown in Dulbecco's modified Eagle medium (Gibco, cat. no. 11995-065) supplemented with 10% fetal bovine serum (Sigma, cat. no. F-2442) and 1% penicillin–streptomycin (Gibco, cat. no. 15140-122) (DMEM), at 37°C and 5% $CO_2$ to 85–90% confluency in 100-mm culture dishes. Each well contains approx $7.5 \times 10^6$ cells.
2. Citric saline: 135 m*M* KCl, 150 m*M* Na-citrate, pH 7.4. Filter sterilize and store at 4°C.
3. Minimum essential medium without L-methionine (Sigma, cat. no. M3911) (MEM-Met): prewarm media to 37°C prior to use. Cells are incubated in this media to deplete intracellular methionine prior to incubation with Trans $^{35}$S-Label.
4. Trans $^{35}$S-Label (1200 Ci/mmol; ICN Biomedicals, cat. no. 51006). Trans $^{35}$S-Label is a mixture of [$^{35}$S]methionine and [$^{35}$S]cysteine utilized for radiolabeling of cellular proteins. Supplement MEM-Met with Trans $^{35}$S-Label to a final concentration of 200 μCi/mL.
5. Cycloheximide: Make a 25 mg/mL stock in 100% ethanol and store at 4°C.
6. Phosphate-buffered saline, pH 7.4 (PBS): 135 m*M* NaCl, 2.7 m*M* KCl, 10 m*M* $Na_2HPO_4$, 1.8 m*M* $KH_2PO_4$.
7. ALLN (*N*-acetyl-Leu-Leu-NorLeu-CHO; Calbiochem cat. no. 208719), a proteasome inhibitor that is utilized to block degradation of CFTR. Prepare a 100 m*M* stock of ALLN in ethanol and add it to cell culture media at a final concentration of 200 μ*M*.
8. Effectene Plasmid DNA Cell Transfection Reagent (Qiagen, cat. no. 301427).
9. Highly pure plasmid DNA: purified DNA should be free of endotoxins and contain low levels of genomic DNA and RNA (*see* **Note 1**).

### 2.2. Sample Preparation

1. 10% Bovine serum albumin (BSA) in PBS.
2. Phenylmethylsulfonyl fluoride (PMSF): make a fresh 100 m*M* stock of PMSF in 100% molecular grade ethanol. Maintain at room temperature.
3. Protease inhibitor cocktail (PI): Complete™ Protease inhibitor cocktail (Roche, cat. no. 1697498).

4. Pansorbin cells (Calbiochem, cat. no. 507861): cells used to preclear cell extracts of radiolabeled proteins that precipitate nonspecifically in immunoprecipitations. Wash Pansorbin cells three times with PBS and resuspend the cells in PBS as a 50% slurry. Store at 4°C.
5. CFTR antibody: clone MM13-4 (Upstate Biotechnology, cat. no. 05-581).
6. Protein G–agarose (PG beads) (Roche, cat. no. 1 243 233): resuspend PG beads in PBS supplemented with 1% BSA. Incubate PG beads at 4°C on a rotator for 24 h to block nonspecific binding sites on the beads. Pellet the beads with a microcentrifuge and resuspend them as 50% volume/volume slurry in PBS supplemented with 0.2% BSA. Store at 4°C.
7. Radioimmunoprecipitation analysis (RIPA) buffer: 150 m*M* NaCl, 1% Nonidet P-40 (NP-40) (IGEPAL™), 0.5% deoxycholic acid, 0.2% sodium dodecyl sulfate (SDS), 50 m*M* *N*-(2-hydroxyethyl)piperazine-*N'*-(2-ethanesulfonic acid) (HEPES), pH 7.4. RIPA can be stored at room temperature, but must be ice cold when used to lyse cells. RIPA should be supplemented with 1 m*M* PMSF, PI, and with 0.2% BSA just prior to use in cell lysis.
8. Detergent-soluble fraction (DSF) buffer: 10 m*M* Tris-HCl, 1% Triton X-100, 5 m*M* EDTA, pH 7.5. Store stock solution at room temperature, but use ice cold. Supplement with 1 m*M* PMSF and PI just prior to use.
9. Detergent-insoluble fraction (DIF) buffer: 10 m*M* Tris-HCl, 1% SDS, pH 7.5. Store stock solution at room temperature. Use DIF buffer at room temperature to prevent precipitation of the SDS. Supplement with 1 m*M* PMSF and PI just prior to use.

### 2.3. SDS-PAGE

1. Sodium dodecyl sulfate-polyacrylamide gel electrophoresis (SDS-PAGE) 4X sample buffer: 250 m*M* Tris-HCl, pH 6.8, 8% SDS, 8 m*M* ethylenediaminetetraacetic acid (EDTA), pH 8.0, 40% glycerol, 40 μg/mL of Coomassie blue, and distilled water. This solution can be stored at room temperature for several months. Prior to use, add 80 μL of 98% β-mercaptoethanol, 1 m*M* PMSF, and PI to 1 mL of SDS-PAGE 4X sample buffer. Mix well before use, and discard any material not used within several hours.
2. SDS-PAGE electrode buffer, pH 8.3: 10 m*M* Tris-HCl, 75 m*M* glycine, and 0.1% SDS.
3. 4X SDS-PAGE resolving gel buffer, pH 8.8: 1.5 *M* Tris-HCl, 8 m*M* EDTA, and 0.4% SDS. Adjust to pH 8.8. Store at room temperature.
4. 4X SDS-PAGE stacking gel buffer, pH 6.8: 0.5 *M* Tris-HCl, 8 m*M* EDTA, and 0.4% SDS. Adjust to pH 6.8. Store at room temperature.
5. Acrylamide-*bis*-acrylamide solution: 30% acrylamide, 0.8% *N,N*-methylene *bis*-acrylamide. Store in a dark or aluminum foil covered glass bottle at 4°C.
6. 10% Ammonium persulfate (10% APS).
7. *N,N,N',N'*-Tetramethylethylenediamine (TEMED) (Fisher Scientific, cat. no. BP150-100).
8. 7% SDS-PAGE gels are made by combining gel components with the following volumes in this order: 2.35 mL of acrylamide–*bis*-acrylamide solution, 5.15 mL of water, 2.5 mL of 4X SDS-PAGE resolving gel buffer, pH 8.8, 100 μL of 10% APS, 7.5 μL of TEMED. This 7% gel mixture is cooled on ice prior to adding the TEMED, then mixed well without forming bubbles, and poured between two glass plates of a minigel apparatus (Bio-Rad). Gently add a layer of water to the top surface of the gel to obtain a flat smooth surface. Complete polymerization should occur within 20 min. When the resolving gel is polymerized, pour off the top water layer and add the stacking gel mixture. The stacking gel mixture is composed of the following: 0.6 mL of acrylamide–*bis*-acrylamide solution, 2.35 mL of water, 1 mL of 4X SDS-PAGE stacking gel buffer, pH 6.8, 75 μL of 10% APS,

and 5 µL of TEMED. This stacking gel mixture should be cooled on ice prior to adding TEMED, mixed well, then quickly added to the top of the 7% resolving gel. Well combs should be inserted immediately and stacking gel polymerization should occur within 10 min.

9. SDS-PAGE gel stain (stain): 25% methanol, 10% glacial acetic acid, 2.5 g/L of Coomassie blue. Use distilled $H_2O$ to prepare this solution and thereby minimize the formation of precipitates.
10. SDS-PAGE gel destain: 10% methanol, 10% acetic acid, distilled water.
11. Photo enhancer: 0.5 *M* sodium salicylate, pH 7.4.

### 2.4. Western Blot

1. Western blot transfer buffer: 20% methanol, 0.02% SDS, 20 m*M* Tris-base, 150 m*M* glycine-HCl. Store at room temperature.
2. Ponceau S protein stain (Ponceau S): Dissolve 1 g of Ponceau S in 2 mL of glacial acetic acid and 198 mL of water. Store at room temperature.
3. PBSTrX-100: PBS supplemented with 0.1% Triton X-100.
4. Western blot blocking solution: PBSTrX-100, 4% non-fat dry milk, 0.8% BSA.
5. Antibody solution: antibody diluted into Western blot blocking solution.
6. Nitropure, cast, pure, 0.45-µm nitrocellulose membrane (Osmonics).
7. Whatman type filter paper cut to the approximate size of the transfer apparatus spacer sponges.

## 3. Methods

### 3.1. Preparation of HEK- 293 Cells for Transfection

1. Grow HEK-293 cells to 90% confluency and then detach them by rinsing with 3 mL and incubating for 5 min with 1 mL room temperature citric saline.
2. After cells have detached, dilute the citric saline with 9 mL of DMEM. Pipet several times to break up large clumps of cells.
3. A near-confluent 100-mm culture dish will provide sufficient cells for 10–12, 35-mm wells or culture dishes. To ensure even distribution of cells in the wells, dilute the cells suspended in DMEM so that 3 mL can be added to each well. Allow these cells to adhere 12–14 h, and replace the DMEM with 2 mL of fresh DMEM.
4. Allow the cells to grow to 70% confluency (~$1 \times 10^6$ cells/well). Replace DMEM with 1 mL of fresh DMEM about 2 h prior to transfection.

### 3.2. Transfection of HEK-293 Cells

1. We have found that Qiagen's Effectene transfection reagent yields a transfection efficiency of ≥70%. Transfection efficiency is best if HEK-293 cells are maintained by splitting cells every 3 d (*see* **Notes 2** and **3**).
2. The mammalian expression plasmid pcDNA 3.1-CFTR or -ΔF508 CFTR is introduced into cells with the Effectene Transfection Reagent according to the manufacturer's instructions. In brief, for each 35-mm well of cells add 0–3.0 µg of pcDNA 3.1-CFTR to 200 µL of EC buffer and mix well; add 4 µL of Enhancer reagent/µg of pcDNA3.1 DNA. Mix the cocktail and incubate for 10 min at room temperature. Add 5 µL of Effectene/µg pcDNA 3.1-CFTR, mix, and incubate for 10 min. Dilute the transfection mixture with 800 µL of DMEM.
3. Aspirate the DMEM from the HEK-293 cells and gently add the transfection cocktail. Incubate the cells for 3–4 h at 37°C and 5% $CO_2$. Replace the transfection mixture with fresh DMEM. Incubate for about 20 h.

4. CFTR and ΔF508 CFTR expression levels in HEK-293 cells are a function of post-transfection incubation time and the amount of DNA used to transfect the cells. We typically harvest the cells 20 h after the transfection and utilize 1 μg of pcDNA 3.1-CFTR or -ΔF508 CFTR.
5. Optimal experimental conditions for CFTR and ΔF508 CFTR expression are those that allow the cell to efficiently insert nascent protein into the ER membrane and generate the immaturely glycosylated species that migrates on SDS-PAGE gels with an apparent molecular mass of 140 kDa termed the B-form. Upon folding and escaping the ER, CFTR is modified further to a maturely glycosylated species that migrates with an apparent molecular mass of 160 kDa, and is termed the C-form. Care must be taken to ensure that expression levels of CFTR and ΔF508 CFTR do not exceed the cells capacity to insert them into the ER.
6. The quality of the expression plasmid utilized in transfections is a source of variation in the level of CFTR and ΔF508 CFTR. Therefore, it is wise to determine the level of CFTR and ΔF508 CFTR expression derived from new preparations of DNA, and adjust quantities utilized in transfections accordingly.
7. The age of the cultured HEK-293 cells utilized in transfections has an influence on the efficiency of CFTR and ΔF508 CFTR expression. HEK-293 cells with a passage number in excess of 50 passes yield low levels of CFTR and ΔF508 CFTR expression.

### 3.3. Harvest of Transfected HEK-293 Cells

Harvesting HEK-293 cells is convenient because detachment from the growth surface only requires that cells be bathed in citric saline. This permits cell lysis to be conducted in a microfuge tube and minimizes variability in the protein content of extracts made from different cell preparations.

1. To harvest HEK-293 cells remove the DMEM and incubate cells in 1 mL of ice-cold citric saline for 5–7 min.
2. Detach the cells by repeatedly pipetting the citric saline over the cells. Once the cells are detached, transfer them from each 35-mm culture well to a microcentrifuge tube.
3. Pellet the cells by centrifugation at 800*g* for 3 min at 4°C. Cell pellets should be kept on ice prior to lysis, or frozen in liquid nitrogen for storage and later lysis.

### 3.4. Western Blot Analysis of CFTR and ΔF508 CFTR Expression

Steady-state levels of CFTR and ΔF508 CFTR expression are determined by Western blots of cell extracts produced from transfected HEK-293 cells (**Fig. 1**).

1. Lyse cell pellets by the addition of 100 μL of 2X SB.
2. Incubate the samples at 37°C for 15 min (*see* **Note 4**).
3. Centrifuge the samples at 16,000*g* for 20 min.
4. To remove the viscous pellet, carefully place the pipet tip near the bottom of the tube and pull a portion of the pellet into the tip. Drag the pellet out of each tube, leaving behind the supernatant (*see* **Note 5**).
5. The volumes of individual supernatants will vary, so it is important to denote the volume of the supernatant for each sample. Load an equal volume from each supernatant to a previously prepared 7% SDS-PAGE gel. Electrophorese samples by applying a constant voltage of 100 V to gels for 115 min.
6. To prepare for the wet transfer of proteins in the SDS-PAGE gel to nitrocellulose with a Bio-Rad minigel apparatus:

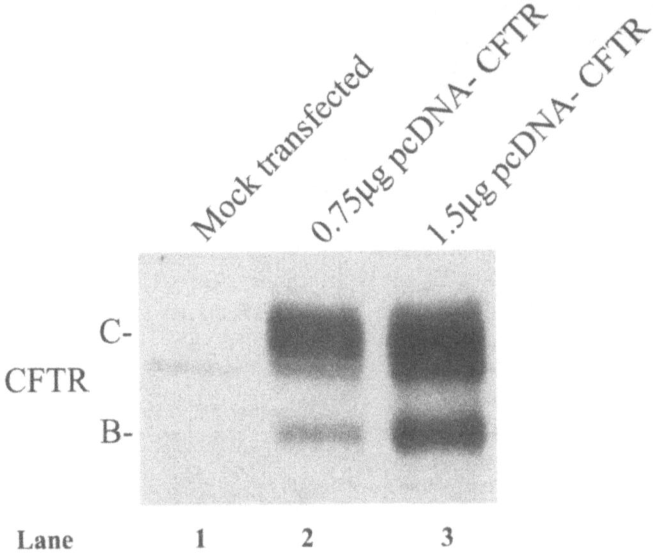


Fig. 1. Western blot analysis of CFTR expression in transfected HEK-293 cells. HEK-293 cells were transiently transfected with the indicated amounts of pcDNA 3.1 or pcDNA 3.1-CFTR. Western blots were then performed as described in **Subheading 3.4.** The B-form denotes the immaturely glycosylated ER localized form of CFTR. The C-form denotes the maturely glycosylated plasma membrane form of CFTR.

   a. Soak a piece of nitrocellulose membrane that is cut to the size of the minigel in Western blot transfer buffer for 10 min.
   b. Soak the transfer apparatus sponges in Western blot transfer buffer.
   c. Cut 3M Whatman filter paper to the same size as the apparatus sponges.
7. On completion of the electrophoresis, soak the SDS-PAGE gel in Western blot transfer buffer for 5 min. Then assemble the gel into the transfer apparatus with a nitrocellulose membrane.
8. Transfer the proteins in the gel to the nitrocellulose by applying a constant voltage of 80 V for 3 h to the apparatus at 4°C.
9. Once the transfer is complete, remove the nitrocellulose membrane from the transfer apparatus and briefly rinse it with distilled water. Stain the proteins on the membrane with Ponceau S for 5 min and destain with water. This step is important for comparison of the protein loads in each lane.
10. Block sites on the nitrocellulose membrane that bind antibodies nonspecifically via incubation of the membrane on a shaker with Western blot blocking solution at room temperature for 3 h, or overnight at 4°C.
11. Incubate the membrane with αCFTR MM13-4 primary antibody at a 1:500 dilution in Western blot blocking solution.
12. The nitrocellulose membrane and the primary antibody solution should be placed in a heat-sealed plastic bag, and rocked for 2–3 h at room temperature or overnight at 4°C.
13. Wash the nitrocellulose with PBSTrX-100 3 × 10 min.

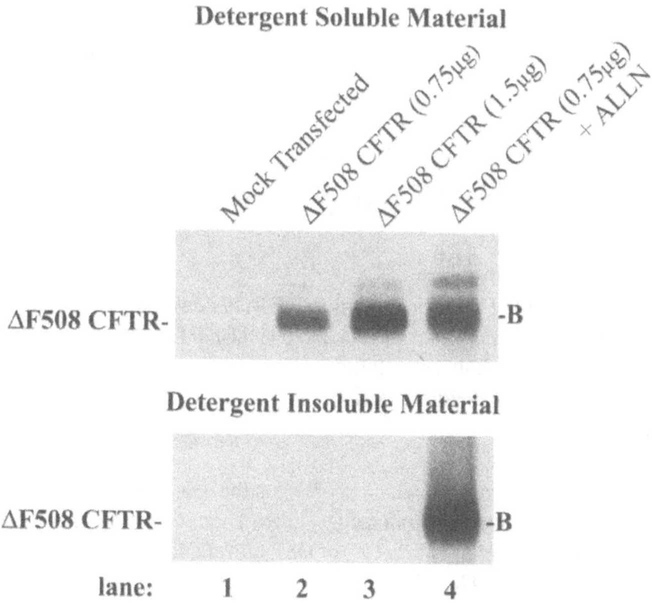


Fig. 2. Western blot analysis of the solubility of ΔF508 CFTR. HEK-293 cells were transfected with pcDNA3.1 or pcDNA3.1-ΔF508 CFTR. Where indicated, 200 μ*M* ALLN was added to cell growth media 12 h prior to harvest. Cells were lysed and prepared for Western blot analysis as described in **Subheading 3.5.** The band labeled ΔF508 CFTR corresponds to the ER localized B-form of ΔF508 CFTR.

14. Block the nitrocellulose again with Western blot blocking solution for 15 min.
15. Incubate the membrane with goat anti mouse sera conjugated to horseradish peroxidase (Bio-Rad) that is diluted 1:3000 with Western blot blocking solution for 40–60 min at room temperature.
16. Wash the nitrocellulose membrane 5 × 5 min with PBSTrX-100 on a shaker.
17. Incubate the nitrocellulose membrane with Enhanced Chemiluminescent Reagent from Amersham Biosciences. Wrap the membrane in plastic wrap, and utilize it to expose X-ray film for 30 s to 2 min. Develop the film and visualize the position of CFTR and ΔF508 CFTR.

### 3.5. Analysis of the Solubility of CFTR and ΔF508 CFTR

Misfolded CFTR and ΔF508 CFTR, whose degradation is blocked, often accumulates in cells as an insoluble aggregate *(11)*. The formation of insoluble CFTR and ΔF508 CFTR aggregates can be detected via the protocol listed below **(Fig. 2)**.

#### 3.5.1. Preparation of Detergent-Soluble and Detergent-Insoluble Fractions of Cell Extracts

1. HEK-293 cell pellets are lysed via incubation in 150 μL of DSF buffer on a rotator for 1 h at 4°C.
2. Centrifuge the samples at 16,000*g* for 15 min at 4°C.

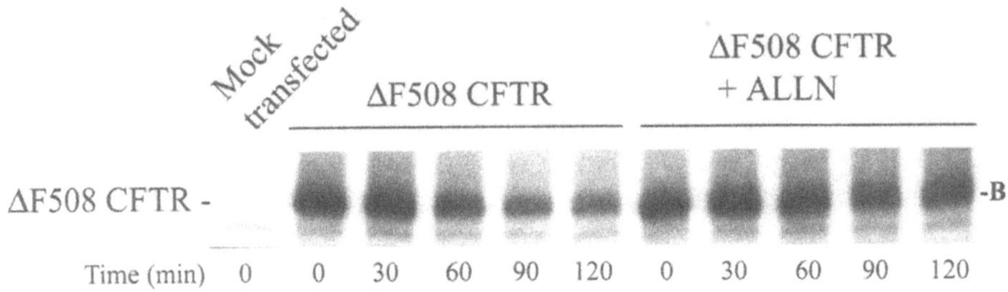


Fig. 3. Kinetics of ΔF508 CFTR degradation. HEK-293 cells were transfected with ΔF508 CFTR and, where indicated, were treated with 200 μ*M* ALLN for 1 hour prior to the labeling of cells with $^{35}$S-translabel. Immunoprecipitation and autoradiography of $^{35}$S-ΔF508 CFTR was carried out as described in **Subheading 3.6.**

3. Transfer 40 μL of supernatant, which represents the 1% Triton X-100 detergent soluble fraction (DSF), to a precooled microfuge tube. Then, carefully remove the remaining supernatant. Rinse the pellet with 50 μL of DSF buffer and spin the sample in a centrifuge at 13,000*g* for 5 min at 4°C. This pellet represents the Triton X-100 insoluble fraction (DIF).

### *3.5.2. Solubilization of the Detergent-Insoluble Material*

1. To solubilize the DIF pellet add 30 μL of room temperature DIF buffer and incubate for 15 min. Light agitation helps expose the pellet(s) to the buffer.
2. Add 120 μL of DSF buffer, do not mix the sample, and return the tubes to ice for 2 min.
3. Sonicate the samples (intensity 6) for 7 s × 2, allowing samples to cool between each. Importantly, samples must be maintained on ice while sonicating to prevent samples from overheating.
4. Pellet any remaining insoluble material by centrifuging at 16,000*g* at 4°C for 5 min.
5. Transfer 40 μL of supernatant to a new precooled tube and store on ice.
6. Add 15 μL of SDS-PAGE 4X sample buffer to each tube containing 40 μL of either DSF or DIF. Mix gently, and incubate at 37°C for 15 min.
7. Load samples on an SDS-PAGE gel and follow **steps 5–17** in **Subheading 3.4.** for Western blot analysis.

## *3.6. Kinetic Analysis of ΔF508-CFTR Degradation in HEK-293 Cells*

To study the rate of ΔF508 CFTR degradation, cellular proteins are labeled with $^{35}$S and the kinetics of the destruction of $^{35}$S-ΔF508 CFTR are determined **(Fig. 3)**.

1. Transfect HEK-293 cells with pcDNA3.1-ΔF508 CFTR as described in **Subheadings 3.1.** and **3.2.**
2. After a 24-h posttransfection period remove the media and wash the cells with 1 mL of prewarmed PBS.
3. Replace the PBS with 2 mL of MEM minus methionine and methionine-starve the cells for 30–40 min in a 5% $CO_2$ atmosphere at 37°C.
4. Remove the MEM minus methionine and replace it with 500 μL MEM minus methionine supplemented with 100 μCi Trans $^{35}$S-Label. Incubate the cells for 20 min at 37°C and 5% $CO_2$.

5. Remove labeling media and wash the cells with 500 µL of DMEM. To conduct the chase, add 2 mL of DMEM supplemented with 25 µg/mL of cycloheximide. Incubate the radiolabeled cells for 0–3 h at 37°C and 5% $CO_2$.
6. To harvest cells, place the 35-mm culture wells on ice, remove the media and add 1 mL of ice- cold citric saline.
7. Collect the detached cells and transfer them into precooled 1.5-mL microfuge tubes. Pellet the cells and remove the supernatant. Cell pellets that are harvested at different time points can be stored on ice for the duration of the chase.
8. When cells incubated for different time periods have all been harvested, they can be lysed via the addition 500 µL of ice-cold RIPA buffer that is supplemented with 0.2% BSA, 1 m*M* PMSF and PI. Cells are resuspended in the RIPA buffer by mixing with a pipet. If the samples are too viscous, additional RIPA buffer can be added to each tube. Incubate samples for 1 h at 4°C on a rocker.
9. Clear the sample material that might stick to Protein A–Sepharose nonspecifically by adding 20 µL of a 50% Pansorbin slurry to each 500 µL of extract. Incubate samples for 15 min at 4°C on a rotator. Centrifuge samples at >16,000*g* for 15 min at 4°C. Transfer 80% of the precleared supernatant to a fresh precooled tube and discard the pellet.
10. Add 3 µL of CFTR MM13-4 antibody to the precleared supernatant and incubate for 1 h on a rotator at 4°C.
11. Add 20 µL of a 50% Protein G–agarose slurry to each sample and incubate for 1 h on a rotator at 4°C.
12. Pellet the Protein G–agarose beads by centrifugation for 1 min at 800*g* and 4°C.
13. Remove the supernatant, then resuspend and wash the pelleted beads three times with ice-cold RIPA buffer.
14. To each pellet add 15 µL of 2X SDS-PAGE sample buffer that had been prewarmed to 37°C and immediately incubate the tubes at 37°C for 10–15 min. Do not boil the samples because this leads to CFTR aggregation.
15. Pellet the agarose beads by centrifugation at 13,000*g* at room temperature for 1 min. Remove 12 µL of the supernatant and load it onto a 7% SDS-PAGE gel.
16. Electrophorese the samples via the application of 100 V to the gel for 115 min.

### 3.7. Detection of Immunoprecipitated and Radiolabeled ΔF508 CFTR by Autoradiography

1. Remove the SDS-PAGE gel from the glass plates and rinse with $dH_2O$.
2. Fix and stain the protein in the gel by incubation in stain for 10 min.
3. Soak the gel in destain until the protein bands on the gel are clearly visible.
4. Rinse the gel with $dH_2O$ for at least 5 min to remove destain from the gel.
5. Soak the gel in 0.5 *M* sodium salicylate for 10 min (*see* **Note 6**).
6. Briefly rinse the gel with $dH_2O$ and place it on Whatman paper to be dried on a slab type gel dryer for 1 h at 80°C.
7. Expose X-ray film to the dried SDS-PAGE gel for 12–18 h at –80°C.
8. Develop the film with a processor and observe the location of the $^{35}S$-ΔF508 CFTR.

## 4. Notes

1. The quality of plasmid DNA is important for the transfection efficiency. A plasmid purification system that yields low levels of contaminating genomic DNA and RNA is recommended. To determine the purity of a plasmid DNA preparation, use a spectrophotometer to observe the optical density of the sample at light wavelength 260 nm ($OD_{260}$) and at

280 nm ($OD_{280}$). The ratio ($OD_{260}/OD_{280}$) should approach a value of 1.8 for the best transfection results.

2. We have found that exceeding 50 passes results in reduced and variable transfection efficiency.
3. Cells should not be allowed to overgrow. Exceeding 95% confluency regularly will result in reduced transfection efficiencies.
4. When cell extracts or cell pellets are treated with sample buffer, they should be heated to 37°C. Boiling the samples for any length of time will cause CFTR to aggregate in the sample buffer.
5. When preparing a cell extract using sample buffer to lyse the cells, the sample will become very viscous. Centrifugation of the sample will separate the viscous material from the supernatant. If this viscous material is not removed from the sample tube, it will interfere with your ability to remove an aliquot of supernatant that can be loaded to the gel.
6. After the SDS-PAGE gel has been destained, it is necessary to rinse the gel with water to remove the acetic acid. If the acetic acid is not sufficiently removed, a precipitate will form on the gel when it is soaked in 0.5 *M* sodium salicylate. This precipitate is caused by the low pH of the gel. The precipitated material can be dissolved by adding 1 *M* Tris, pH 8.0. After dissolving the precipitate, replace the Tris-treated salicylate with fresh 0.5 *M* sodium salicylate.

## References


1. McCracken, A. A. and Brodsky, J. L. (2003) Evolving questions and paradigm shifts in endoplasmic-reticulum-associated degradation (ERAD). *Bioessays* **25,** 868–877.
2. Kostova, Z. and Wolf, D. H. (2003) For whom the bell tolls: protein quality control of the endoplasmic reticulum and the ubiquitin-proteasome connection. *Embo J.* **22,** 2309–2317.
3. Lenk, U., et al. (2002) A role for mammalian Ubc6 homologues in ER-associated protein degradation. *J. Cell Sci.* **115,** 3007–3014.
4. Hampton, R. Y. and Rine, J. (1994) Regulated degradation of HMG-CoA reductase, an integral membrane protein of the endoplasmic reticulum, in yeast. *J. Cell Biol.* **125,** 299–312.
5. Hampton, R. Y. and Bhakta, H. (1997) Ubiquitin-mediated regulation of 3-hydroxy-3-methylglutaryl-CoA reductase. *Proc. Natl. Acad. Sci. USA* **94,** 12944–12948.
6. Hiller, M. M., et al. (1996) ER degradation of a misfolded luminal protein by the cytosolic ubiquitin-proteasome pathway. *Science* **273,** 1725–1728.
7. Plemper, R. K., et al. (1999) Re-entering the translocon from the lumenal side of the endoplasmic reticulum. Studies on mutated carboxypeptidase yscY species. *FEBS Lett.* **443,** 241–245.
8. Cheng, S. H., et al. (1990) Defective intracellular transport and processing of CFTR is the molecular basis of most cystic fibrosis. *Cell* **63,** 827–834.
9. Meacham, G. C., et al. (1999) The Hdj-2/Hsc70 chaperone pair facilitates early steps in CFTR biogenesis. *Embo J.* **18,** 1492–1505.
10. Meacham, G. C., et al. (2001) The Hsc70 co-chaperone CHIP targets immature CFTR for proteasomal degradation. *Nat. Cell Biol.* **3,** 100–105.
11. Ward, C. L., Omura, S., and Kopito, R. R. (1995) Degradation of CFTR by the ubiquitin-proteasome pathway. *Cell* **83,** 121–127.
12. Jensen, T. J., et al. (1995) Multiple proteolytic systems, including the proteasome, contribute to CFTR processing. *Cell* **83,** 129–135.

13. Ward, C. L. and Kopito, R. R. (1994) Intracellular turnover of cystic fibrosis transmembrane conductance regulator. Inefficient processing and rapid degradation of wild-type and mutant proteins. *J. Biol. Chem.* **269,** 25710–25718.
14. Lukacs, G. L., et al. (1994) Conformational maturation of CFTR but not its mutant counterpart (delta F508) occurs in the endoplasmic reticulum and requires ATP. *EMBO J.* **13,** 6076–6086.

21

# Aggresome Formation

Michael J. Corboy, Philip J. Thomas, and W. Christian Wigley

**Summary**

Bulk protein degradation in the cell is catalyzed by the ubiquitin–proteasome system (UPS). At the heart of the UPS is the proteasome, a large multisubunit tightly-regulated protease. The UPS performs key functions in protein quality control by monitoring and eliminating potentially toxic misfolded or damaged proteins. When the capacity of this protease system is exceeded, misfolded protein substrates aggregate and are assembled through an active and regulated process to form an aggresome. Aggresomes are dynamic structures, formed as a general response to an overload of improperly folded proteins. Assembly of aggresomes occurs at the centrosome, a perinuclear structure that also serves as a site for the recruitment and concentration of components of the UPS, including the proteasome, its regulators, and other proteins typically involved in protein quality control. Thus, in addition to other cellular activities, the centrosome may play a central role in protein quality control, sitting at the crossroads of protein folding, degradation, and aggregation.

**Key Words:** Protein misfolding; proteasome; centrosome; protein aggregation; quality control.

## 1. Introduction

The ubiquitin–proteasome system is the primary mechanism for the continual turnover of cellular proteins in eukaryotes. For example, a critical role of the proteasome is the removal of potentially toxic misfolded and damaged proteins. When the degradation capacity of the proteasome is overwhelmed, the misfolded substrates accumulate in the cell as distinct centrosomal inclusions of aggregated protein *(1,2)* that have been termed aggresomes *(3)*. Aggresome formation is not an arbitrary, indiscriminate event, but is instead part of a highly organized and regulated process designed to deliver both the inclusions and the degradation machinery to a single locale. Aggresome formation was first described for the cystic fibrosis transmembrane conductance regulator (CFTR), mutations in which cause cystic fibrosis. Aggresomes have since been shown to form from misfolding and aggregation of an ever-widening spectrum of cytoplasmic, transmembrane and secretory proteins, suggesting a general

From: *Methods in Molecular Biology, vol. 301, Ubiquitin–Proteasome Protocols*
Edited by: C. Patterson and D. M. Cyr © Humana Press Inc., Totowa, NJ

feature of the cell's attempts to deal with these potentially toxic species. In fact, emerging evidence suggests that the formation of aggresomes may actually serve a protective role as a way to confine toxic conformers when degradation lags. This chapter gives an overview of the proteasome's role in the degradation of misfolded proteins, the formation of aggresomes when degradation fails, and details of methods used in our laboratory to study these processes.

### 1.1. The Ubiquitin–Proteasome System

In eukaryotes, the majority of cellular proteins (80–90%) are degraded by the ubiquitin–proteasome pathway. This system generates antigenic peptides for surveillance by the immune system, controls the levels of various short-lived regulatory proteins and transcription factors, and prevents the accumulation of misfolded mutant and damaged proteins ***(4–7)***. Degradation by the 26S particle, an accepted physiological form of the proteasome ***(8)***, is generally an ATP-dependent process requiring the covalent conjugation of ubiquitin to the target protein ***(5,6)***. Ubiquitination is catalyzed by the sequential activity of three classes of enzymes termed ubiquitin-activating (E1), -conjugating (E2), and -ligating (E3) enzymes ***(9)***. The net result of this cascade is an isopeptide bond between the C-terminal glycine of ubiquitin and the ε-amino group of one or more lysine residues on the target protein. Additional ubiquitin moieties are similarly added by sequential processive conjugation to ubiquitin itself, tagging the target protein with a polyubiquitin chain that serves as a marker for 26S proteasomal degradation. The polyubiquitinated substrates are then recognized by the 26S proteasome, deubiquitinated, unfolded and degraded ***(5,6)***.

The 26S proteasome is a 2.5-mDa molecular machine composed of two major subcomplexes, the 20S proteasome and PA700 (or 19S). The barrel-shaped 20S proteasome, which constitutes the catalytic core of the protease, is a paradigm of self-compartmentalization ***(10,11)***. The 20S particle is a 700-kDa cylinder composed of four stacked heptameric rings, two outer α rings and two inner β rings, totaling 28 subunits ***(12,13)***. The eukaryotic proteasome consist of two copies each of seven different α subunits and seven different β subunits that form heterooligomeric rings [($\alpha_1$- $\alpha_7$) ($\beta_1$- $\beta_7$)( $b_1$- $\beta_7$)( $\alpha_1$- $\alpha_7$)]. Three of the β subunits (β1, β2, and β5) contribute the catalytic activities which have been termed chymotrypsin-like, trypsin-like, and postglutamyl peptide hydrolytic activities based on their substrate side chain specificities ***(14,15)***. The overall architecture serves to sequester the catalytic sites within the hollow central cavity of the cylinder. Access to the interior is limited by narrow openings at the ends of the cylinder that limit substrates to short peptides and unfolded proteins ***(12,13,16)***. Therefore, the proteasome is coupled to other accessory complexes that mediate recognition, binding, and unfolding of target proteins. In mammalian cells, these roles are fulfilled by PA700 ***(5,6)*** and another regulatory complex, PA28 ***(17)***.

PA700 is a 700-kDa complex composed of 18–20 distinct polypeptide subunits that binds to one or both of the outer α-rings of the 20S core particle to form the 26S complex ***(10,18,19)***. Binding of PA700 activates the proteasome ***(8,19)***, probably by widening the entry channel, thereby increasing substrate access to the catalytic chamber. PA700 also imparts specificity for ubiquitinated substrates through its polyubiquitin binding ***(20,21)*** and cleavage ***(22,23)*** activities. In addition, PA700 has multiple ATPase activities ***(24)***, and probably enhances degradation of ubiquitinated proteins by unfolding substrates and facilitating their translocation into the central cavity of the protease ***(10,25,26)***. Isolated PA700 also functions as a molecular chaperone in vitro ***(26–28)***, likely a reflection of this ability to recognize and act on misfolded proteins as part of the 26S particle in vivo. Unlike the 20S particle, for which crystallographic structures have been solved from several organisms ***(12,13,29)***, little is known about the detailed three-dimensional structure of PA700. However, the relative arrangement of subunits and some of their functions have been identified ***(5)***. For example, the base of PA700 is composed of six separate AAA ATPase subunits that are the proposed "unfoldases" for assisting the translocation of substrates into the proteasome lumen ***(5,26)***.

A second regulatory complex, PA28, also associates with the 20S proteasome in vivo ***(17,30)***. PA28 (also called the 11S cap) is a 180-kDa multisubunit heterocomplex containing two distinct polypeptides, PA28α and PA28β, arranged in a single hexameric/heptameric ring ***(31–34)***. PA28 binds to and activates the proteasome by opening the access pore ***(33)***, similar to PA700. However, PA28 does not increase processing of ubiquitinated substrates ***(17,30)***, but instead preferentially functions to increase proteasomal processing of small peptides ***(35,36)***. In addition, the expression levels of PA28α, PA28β, and subunits of the transporter for antigen presentation (TAP) are all stimulated by the immunostimulatory cytokine interferon-γ (IFN-γ) ***(4,37,38)***. These observations suggest a role for PA28 in stimulating the processing of substrates for antigen presentation.

### 1.2. Functions of the Proteasome

Proteasome-mediated protein degradation is essential for a number of cellular processes, including immune surveillance, regulation, and quality control ***(5,6)***. The continual turnover of proteins by the proteasome generates peptide antigens for presentation on major histocompatibility complex (MHC) class I molecules. Cleavage by the proteasome results in the production of 8- to 10-residue oligopeptides that are then translocated into the endoplasmic reticulum (ER) by TAP, loaded onto newly synthesized MHC-1 molecules, and delivered to the cell surface ***(4)***. In normal cells, all the antigens presented by MHC-I molecules are derived from cellular proteins and are tolerated as self by the immune system. In virally infected cells, however, the display of antigens derived from the degradation of foreign viral proteins marks the cell for destruction by cytotoxic T cells ***(4)***. As mentioned earlier, IFN-γ stimulation enhances antigen presentation by increasing expression of PA28 and TAP ***(4,5,37,38)***. IFN-γ also promotes formation of so-called "immunoproteasomes" through increased expression of alternative catalytic subunits that are preferentially incorporated into newly assembled 20S particles ***(4,5)***. While the overall degradation activity is unchanged, the immunoproteasomes show altered cleavage site preferences that lead to a different collection of oligopeptides, apparently ones that are more favorable for antigen presentation.

The proteasome also controls the levels of short-lived cell-cycle regulators. Progression through the cell cycle in eukaryotes is controlled by oscillations in the activity of cyclin-dependent kinases (Cdks) ***(39)***. Because Cdks themselves are quite stable, their periodic fluctuations in activity are instead achieved by cyclical changes in the levels of positive regulatory factors, termed cyclins, and negative regulatory factors, termed Cdk inhibitors. Specific cyclins accumulate at different times to activate Cdks, which then mediate passage through various cell-cycle checkpoints, for example, from $G_1$ to S phase. Following checkpoint passage, the cyclins are selectively ubiquitinated and rapidly degraded by the proteasome, and the Cdks inactivate ***(39)***. Cdk inhibitors, in contrast, sequester the kinase in an inactive form even in the presence of the cognizant cyclin ***(39)***. In this case, targeted ubiquitination and degradation by the proteasome relieves inhibition and allows cyclin-mediated activation and cell-cycle progression.

In addition, the ubiquitin–proteasome system regulates the levels of numerous transcription factors, tumor suppressors, and oncoproteins. Perhaps the most well characterized example is nuclear factor κB (NFκB). NFκB is actually a family of inducible transcription factors involved in immunity, inflammation, response to stress, and development ***(40)***. The best-known form is NFκB-1, a heterodimer of p50 and p65 (RelA). In resting cells, the p50/p65 dimer is held in an inactive complex in the cytosol through interaction with the inhibitory protein IκBα. On stimulation by a variety of intra- and extracellular signals, IκBα is phosphorylated, ubiquitinated, and then rapidly degraded by the proteasome ***(40)***. IκBα degradation liberates the p50/p65 dimer, which then translocates to the nucleus and activates expression of target genes. The p50 subunit itself is also produced by the proteasome, via limited cotranslational proteolysis of the precursor p105, releasing the N-terminal p50 moiety while the C-terminal remainder is degraded ***(41–43)***. p50 generation from p105 is the first instance where a protein was shown to be discretely processed by the ubiquitin–proteasome pathway without being completely destroyed, and is believed to be mediated by a previously unknown endoproteolytic activity of the proteasome ***(44)***. Interestingly, then, the proteasome regulates NFκB-1 activity at two levels, both by controlling the amounts of the p50 subunit and the levels of the IκBα inhibitor. The yeast NFκB relatives SPT23 and MGA2 are similarly processed by the proteasome via limited proteolysis ***(45,46)***. The proteasome also regulates (through more traditional degradation) the levels of several transcription factors (e.g., the E2F family) ***(47)***, cellular (but not viral) forms of oncoproteins (Myc, Fos, and Jun) ***(48–51)***, and tumor suppressors (p53 and PTEN) ***(52,53)***.

Finally, the proteasome is the primary protease involved in quality control, responsible for the degradation of misfolded and damaged proteins ***(10)***. The ubiquitin–proteasome system's role in the selective removal of damaged cytosolic and nuclear proteins has long been appreciated. In recent years, though, it has become apparent that the cytosolic proteasome is responsible for the majority of quality control degradation in the secretory pathway, as well ***(54,55)***. Given that the proteasome's role in the surveillance of protein folding and misfolding is at the heart of aggresome formation, this function is discussed in greater detail in **Subheading 1.3.** In addition, although the aggresome formation process is essentially the same regardless of the substrate, both the initial characterization and the experimental protocols in this chapter involve a polytopic transmembrane protein (CFTR) ***(2)***. Therefore, the remaining discussion focuses on the fate of misfolded membrane proteins, with particular emphasis on CFTR *(8)*.

### *1.3. Quality Control of Membrane Protein Folding*

The cystic fibrosis transmembrane conductance regulator (CFTR) is one of the best-characterized systems for studying membrane protein quality control. CFTR, a member of the ABC transporter supergene family, is a polytopic membrane protein composed of two cytosolic nucleotide-binding domains, two transmembrane domains, and a PKA-sensitive regulatory domain ***(56,56)***. In addition to its role in chloride conductance ***(57–59)***, CFTR regulates the activity of several other critical transport systems in the apical membrane of epithelial cells ***(60–62)***. Of particular significance, CFTR regulates bicarbonate secretion mediated by members of the SLC26 family of anion exchangers ***(63)***. Disease-causing mutations in CFTR lead to loss of one or more of these activities due to a variety of molecular mechanisms, the most common of which is defective folding ***(64)***. Several mutations, including the common ΔF508 mutation, result in misfolded CFTR proteins that never reach their proper location in the apical membrane. Even for wild-type CFTR, folding and trafficking are inefficient, with >30% of the wild-type molecules ever reaching mature form in cultured cells ***(65–67)***. In comparison, most of the immature wild-type (~70%) and nearly all of the immature ΔF508 CFTR never reach the membrane but are instead detained by the ER quality control system and degraded.

The folding and maturation of membrane proteins like CFTR begins in the ER. A critical role of ER quality control is to ensure that these nascent proteins achieve their native conformation before being allowed to traffic through the secretory pathway to their ultimate subcellular destination. In general, correctly folded, processed, and assembled proteins are transported, whereas misfolded proteins and persistently unassembled subunits are retained and subsequently degraded by the proteasome ***(55,68)***. Cytosolic proteasomes gain access to and degrade misfolded membrane proteins through a process termed ER-associated degradation (ERAD) ***(54,69,70)***. ERAD is a multistep process beginning with recognition and retention of misfolded proteins in the ER, ubiquitination, dislocation from the membrane, deglycosylation, and finally deubiquitination and degradation by the 26S proteasome. CFTR has been used as a model system for the investigation of ERAD ***(71–74)***, although the molecular mechanisms of several individual steps are still not well understood. For example, the current model for dislocation of misfolded proteins from the ER involves retrograde transport of ubiquitinated substrates through the Sec61 translocon channel ***(55,69,70)***, and there is evidence for such a mechanism in CFTR degradation ***(73–75)***. However, the extent to which extraction and degradation are linked is unclear. In yeast, lumenal ERAD substrates such as carboxypeptidase Y appear to be completely extracted to a cytosolic intermediate before degradation ***(76)***. For transmembrane ERAD substrates like CFTR, on the other hand, studies suggest that dislocation and proteolysis, while not explicitly coupled, are more tightly coordinated ***(77)***. Synchronized extraction and degradation would help prevent cytosolic accumulation of aggregation-prone hydrophobic transmembrane regions.

Such undesired exposure of normally buried hydrophobic regions is typically the initial manifestation of a misfolded protein, whether soluble or transmembrane *(78,79)*. Left unattended, these exposed hydrophobic patches can associate and drive aggregation of the nonnative proteins into insoluble inclusions. Since these aberrantly folded structures are potentially toxic *(80)*, each cellular compartment contains quality control mechanisms to prevent, repair, or eliminate misfolded proteins. Both molecular chaperones (e.g., Hsp70, BiP) and regulatory components of the proteasome (e.g., PA700) rely on binding to the same or similar determinants on the misfolded protein, namely the surface-exposed hydrophobic regions *(78,81)*. Thus, the decision between repair and elimination has been proposed to be a kinetic competition between binding by molecular chaperones, leading to refolding, and binding by regulatory components of the proteasome, leading to degradation. The distinction between the two systems is not quite so clear, however. For example, in addition to its role in preventing aggregation and promoting folding, the chaperone Hsp70 is also required for the ubiquitination and degradation of several proteins *(82)*. Conversely, as mentioned above, PA700 exhibits chaperone-like activity in vitro *(27)*. In any case, when neither the salvage nor degradation systems are capable of dealing with the misfolded protein, the usual result is aggregation.

Several biochemical studies have shown that misfolded CFTR is prone to aggregation, both in vivo and in vitro. Inhibition of the proteasome in cultured cells expressing ΔF508 CFTR leads to the formation of high molecular weight, detergent-insoluble complexes of ubiquitinated CFTR *(3,71)*. In the absence of degradation, CFTR forms similar complexes when expressed in a cell-free system designed to reconstitute ERAD *(75)*. As visualized in vivo by immunocytochemistry, the misfolded CFTR that cannot be degraded accumulates as a large perinuclear inclusion **(Fig. 1)** located at the centrosome and accompanied by proteasome components and molecular chaperones *(2,3)*. The formation of these inclusions is consistent with an ER dislocation process in which extraction from the membrane and degradation are not explicitly coupled *(75,77)*. Proteasome inhibitors such as lactacystin would not necessarily impair dislocation of the ubiquitinated protein. When extraction of CFTR is not quickly followed by or linked to proteolysis, exposure of the large hydrophobic transmembrane domains apparently drives the protein into a nonnative conformation that is recognized by packaging machinery and/or rapidly forms an insoluble aggregate. These distinctive centrosomal inclusions of misfolded protein, replete with quality control machinery, have been termed aggresomes *(3)*.

### *1.4. The Aggresome*

Perhaps not surprisingly, the two studies describing CFTR aggresomes *(2,3)* were not the first reported indications of the structure. Vidair et al. had earlier noted an increase in the aggregation of insoluble, misfolded protein at or near the centrosome in cultured cells subjected to heat shock *(83)*. Similarly, Wojcik and coworkers found that in HeLa cells treated with proteasome inhibitors, massive perinuclear aggregates formed that were rich in ubiquitin and proteasomes *(1,84)*. Furthermore, treatment of cells with the microtubule-disrupting drug nocodazole blocked the perinuclear local-

ber of medically relevant cases disease-causing mutations increase aggresome formation. For example, the membrane protein presenilin 1 (PS1) also forms aggresomes in cultured cells in the absence of proteasome activity, and formation is amplified by the early-onset familial Alzheimer's disease (FAD) mutation A246E ***(3)***. Similar results were seen with the P23H mutant form of rhodopsin, which is associated with autosomal dominant retinitis pigmentosa (ADRP) ***(87,88)***, with peripheral myelin protein 22 (PMP22), an inefficiently folded Schwann cell glycoprotein associated with a number of heritable peripheral neuropathies ***(89)***, and with a polyglutamine expansion in the androgen receptor (AR) associated with spinobulbar muscular atrophy (SBMA) ***(90)***. These structures have also been shown to form from the aggregation of misfolded soluble proteins, as well as from extracted membrane proteins. In addition to the "synthetic" GFP-250 substrate discussed earlier ***(86)*** aggresome formation has also been shown for secreted (surfactant protein C) ***(91)***, and cytosolic (huntingtin exon I, Htn; superoxide dismutase 1, SOD-1) ***(92,93)*** proteins. Interestingly, Htn proteins with pathological polyglutamine repeats (Htn-51Q or Htn-83Q) form aggresomes, but the nonpathological repeat (Htn-25Q) does not ***(92)***. Similarly, mutations in SOD-1 associated with familial amyotrophic lateral sclerosis (FALS) have been reported to give rise to aggresomes, while wild-type SOD-1 does not ***(93)***.

### *1.5. Functional Significance of Aggresomes*

Centrosomal association of quality control components such as the ubiquitin–proteasome system and chaperones is not strictly a function of aggresome formation, but also a general feature of resting cells ***(2,94–97)***. 20S proteasomes, PA700, PA28, ubiquitin, Hsp70, and Hsp90 have all been shown by immunocytochemistry to colocalize with centrosomes in several different cultured cell lines under basal conditions (**Fig. 3**). In addition, purified centrosome-associated proteasomes are active in degrading ubiquitinated proteins and proteasome-specific peptide substrates, and demonstrate the same ATP-dependence and inhibitor profile as soluble proteasomes ***(94)***. When combined with overexpression of mutant CFTR, pharmacological inhibition of the proteasome results in striking recruitment of cytosolic proteasome components to the centrosomal inclusions ***(2)***. This recruitment is accompanied by appreciable expansion of the centrosome and redistribution of proteasomes, γ-tubulin, and chaperones to an insoluble fraction (**Fig. 4**). Therefore, it appears the centrosomal proteolysis centers normally function to concentrate and recruit components of the ubiquitin–proteasome and chaperone systems to balance protein folding and degradation ***(1,2,84,94)*** (**Fig. 5**). Thus, aggresomes likely represent the end point of a normal cellular function gone awry when the burden of misfolded protein cannot be adequately handled.

Such centrosomal proteolysis centers also play a crucial role in antigen processing and, ironically, productive viral assembly. The MTOC is one of the major sites of proteasomal degradation of viral proteins for antigen presentation, along with nuclear promyelocytic leukemia oncogenic domains (PML bodies or PODs) ***(98–100)***. Consistent with this, PML bodies are also enriched with components of the ubiquitin proteasome system, especially PA28 and immunoproteasomes ***(98,101)***. Increased

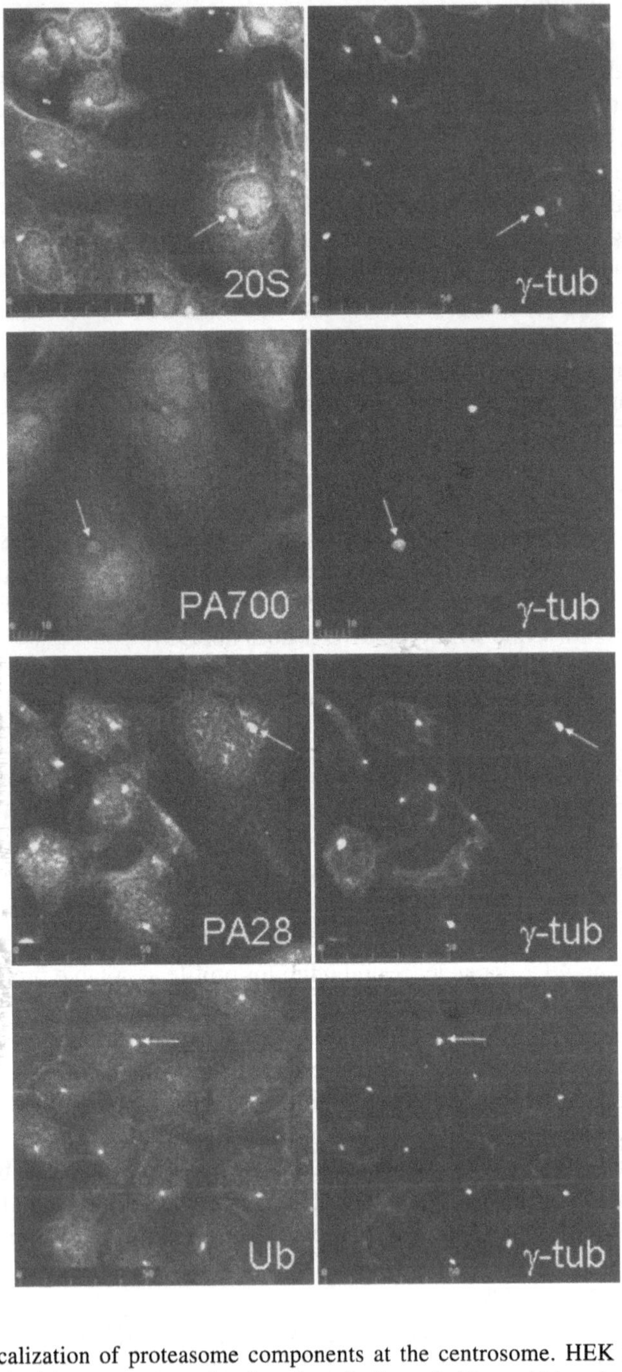


Fig. 3. Localization of proteasome components at the centrosome. HEK 293 cells were immunostained with mouse monoclonal anti-γ-tubulin and either rabbit polyclonal anti-20S proteasome, chicken polyclonal anti-PA700, rabbit polyclonal anti-PA28, or mouse monoclonal anti-ubiquitin antibodies, as indicated in each panel. *Arrows* indicate locations of perinuclear proteasome-enriched centrosomes. Scale bars indicate size in μm. (From **ref. 2**, by copyright permission of The Rockefeller University Press.)

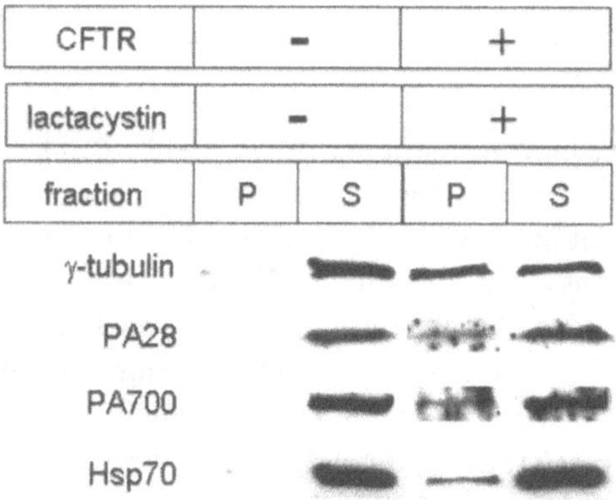


Fig. 4. Coredistribution of the centrosome, Hsp70, and the proteasome in response to inclusion formation. HeLa cells expressing P205S CFTR and treated with 10 μ*M* lactacystin for 12 h and untreated mock-transfected control cells were lysed and separated into supernatant (S) and pellet (P) fractions as described in **Subheading 3.** Fractions were separated by SDS-PAGE and analyzed by Western blotting using the indicated antibodies. (From **ref.** *2*, by copyright permission of The Rockefeller University Press.)

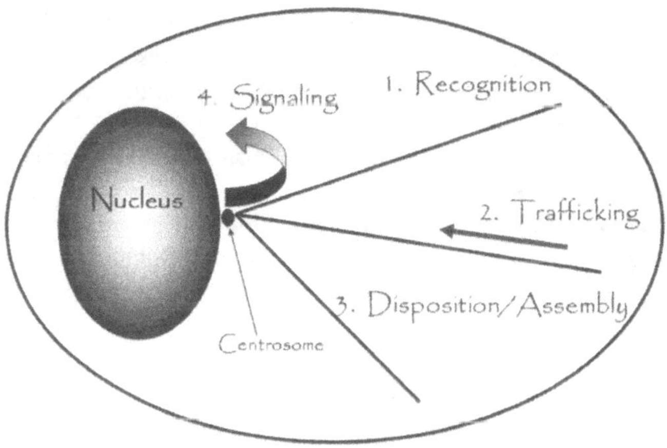


Fig. 5. Schematic of a eukaryotic cell. The nucleus *(gray)* and centrosome *(black)* are as indicated. *Black lines* represent microtubules radiating from the centrosome to the cell periphery. The schematic is drawn to highlight several important questions that remain regarding the centrosomal inclusions described herein and elsewhere: 1: How is the substrate recognized and what proteins are involved? 2: Does trafficking to the centrosome serve to simply deposit aggregated substrates, or also to concentrate them, thereby increasing their exposure/susceptibility to the centrosome-associated proteasomes? 3: Is assembly into vimentin-surrounded perinuclear centrosomal inclusions an active, protein-mediated process, or simply a result of aggregation? 4: What signals arise from the processes occurring at this location and what are their targets?

delivery of viral proteins to the MTOC enhances their processing into MHC-I linked antigens, as seen with a human papilloma virus 16 (HPV-16) E7-γ-tubulin fusion protein ***(102)***. In addition, inflammatory stimulation increases transient aggregation of ubiquitinated proteins at the MTOC ***(100)***, suggesting a regulated delivery system to concentrate viral proteins at the site of their processing. However, this delivery system appears to have been corrupted in some cases for production of viruses rather than degradation. So-called viral factories are high concentrations of viral proteins centered at the MTOC to facilitate viral assembly, which require minus-end directed micotubular trafficking ***(103–105)***. The similarity between these perinuclear assembly centers and aggresomes suggests viral factories represent the subversion of a normal cellular process designed to deliver viral antigens to the MTOC for antigen processing.

Deposition of misfolded protein into aggresome-like cytoplasmic inclusions is a common cytopathological feature in a number of neurodegenerative diseases. The majority of cases are idiopathic, but a number of familial cases can be linked to toxic gain of function mutations ***(106–108)***, which have also been shown to increase aggresome formation in cultured cells, for example, Htn and SOD-1 ***(92,93)***. Protein aggregation in general, and aggresome formation in particular, impairs function of the ubiquitin–proteasome system, as has been demonstrated with ΔF508 and a pathological-repeat length Htn variant ***(109)***. This inhibition suggests a possible mechanism directly linking aggregation to cell death due to the central role of the ubiquitin–proteasome pathway in cell regulation. However, current research has not reached a consensus on whether these inclusions are a cause or a consequence of the pathology. In fact, recent evidence suggests that aggresome formation may even be a protective response, as blocking aggresome formation exacerbates the toxicity of aggregated AR ***(90)***. Clearly, more work is needed to more clearly answer these significant questions. This chapter therefore focuses on techniques used in our laboratory to visualize and characterize aggresome formation using a combination of immunocytochemical and biochemical methods.

## 2. Materials

### 2.1. Centrosome-Associated CFTR Inclusions

#### 2.1.1. Formation: Cell Lines, Plasmids, and Transfection

1. Human embryonic kidney (HEK) 293 cells, HeLa cells (American Type Culture Collection).
2. Dulbecco's modified Eagle's medium (DMEM) (Gibco).
3. Fetal calf serum (Gibco).
4. Tissue culture antibiotics (streptomycin, penicillin) (Gibco).
5. Standard tissue culture plastic ware, glass cover slips.
6. pCMVNot6.2 containing full length human CFTR cDNA (generous gift of Dr. Johanna Rommens [The Hospital For Sick Children, Toronto]).
7. Quick Change Mutagenesis System (Stratagene).
8. Fugene 6 Mammalian Transfection Reagent (Boehringer Mannheim).
9. 10 m*M* Lactacystin and 50 m*M* MG132 (Calbiochem) as stocks in dimethyl sulfoxide (DMSO), stored at –20°C.

### 2.1.2. Analysis

#### 2.1.2.1. Immunocytochemistry

1. Phosphate-buffered saline (PBS).
2. 100% Methanol stored at –20°C.
3. 50 m*M* Glycine in PBS.
4. Blocking medium: 5% goat serum, 1% bovine serum albumin (BSA), 0.1% gelatin in PBS.
5. Primary antibodies: anti-CFTR mouse monoclonal antibody (MAb) 24-1 (R&D Systems); anti-CFTR rabbit pAb R3194 (110); anti-GRP78 (BiP) pAb SPA-826 (StressGen); anti-γ-tubulin mouse MAb (Sigma).
6. Appropriate fluorescently labeled secondary antibodies (Jackson Immunoresearch).
7. Fluorescent microscope (we routinely use a Bio-Rad MRC 1024 confocal microscope).

#### 2.1.2.2. Characterization of Soluble and Insoluble Fractions

1. Trypsin in Hank's balanced salt solution (HBSS) (Gibco).
2. PBS.
3. PBS supplemented with CompleteTM Protease Inhibitor Cocktail (Roche).
4. Syringes fitted with 27-gage needles.
5. Standard protein gel electrophoresis setup, transfer apparatus, and nitrocellulose.
6. Transfer buffer: 25 m*M* Tris, 192 m*M* glycine, pH 8.3, with 20% methanol.
7. TTBS: 20 m*M* Tris-HCl, pH 7.6, 137 m*M* NaCl, 0.05% Tween-20.
8. Blocking buffer: 10% nonfat milk in TTBS.
9. Primary antibodies: anti-γ-tubulin mouse mAb (Sigma); anti-Hsp70 mouse MAb MA3-006 (Affinity Bioreagents); anti-PA700 chicken pAb ***(2)***, and anti-PA28 rabbit pAb ***(111)*** prepared as described.
10. Appropriate horseradish peroxidase (HRP)-labeled secondary antibodies (Jackson Immunoresearch), ECL detection reagents, and Hyperfilm (Amersham).

## 2.2. Centrosome-Associated Proteins

### 2.2.1. Immunocytochemistry

1. Solutions as in **Subheading 2.1.2.1.**
2. Primary antibodies: anti-γ-tubulin mouse MAb and anti-ubiquitin mouse MAb (Sigma); anti-GRP78 (BiP) pAb SPA-826 (StressGen); anti-Hsp70 mouse MAb MA3-006 (Affinity Bioreagents); anti-20S proteasome rabbit pAb ***(17)***, anti-PA700 chicken pAb ***(2)***, and anti-PA28 rabbit pAb ***(111)*** prepared as described.
3. Labeled secondary antibodies and fluorescent microscope as described in **Subheading 2.1.2.1.**

### 2.2.2. Immunoblot Analysis of Purified Centrosomes

1. Cytochalasin D (10 mg/mL), Nocodazole (1 m*M*) (Sigma) as stocks in DMSO, stored at –20°C
2. Trypsin in HBSS (Gibco).
3. TBS.
4. 8% Sucrose in 0.1X TBS.
5. Lysis buffer: 1 m*M* *N*-(2-hydroxyethyl)piperazine-*N*′-(2-ethanesulfonic acid)(HEPES), pH 7.2, 0.5% NP-40, 0.5 m*M* $MgCl_2$, 0.1% β-mercaptoethanol, 1 μg/mL of leupeptin, 1 μg/mL of pepstatin, 1 μg/mL of aprotinin, 1 m*M* PMSF (add Nonidet P-40 and protease

inhibitors fresh just before use). Alternatively, use Complete™ Inhibitor Cocktail tablets, Roche.

6. 1*M* HEPES, pH 7.4.
7. 10 mg/mL of DNase I (Roche) in TBS (store aliquots at –20°C.
8. 40, 50, 60, and 70% sucrose solutions (w/w) in 10 m*M* PIPES, pH 7.2, 0.1% Triton X-100 0.1% β-mercaptoethanol (add β-mercaptoethanol fresh).
9. Electrophoresis and immunoblotting reagents and supplies as described in **Subheading 2.1.2.2.**
10. Primary antibodies: anti-γ-tubulin mouse monoclonal antibody (MAb) (Sigma); anti-Hsp70 mouse MAb MA3-006 and anti-Hsp90 mouse MAb MA3-011 (Affinity Bioreagents); anti-GRP78 (BiP) pAb SPA-826 (StressGen); anti-20S proteasome rabbit pAb ***(17)***, anti-p31 ***(2)***, and anti-PA28 rabbit pAb ***(111)*** prepared as described [to establish purity of the centrosome preparations, we also immunoblot for cytosolic (aldolase), nuclear (lamin B1), and Golgi markers (β-cop)]

## 3. Methods

### *3.1. Centrosome-Associated CFTR Inclusions*

#### *3.1.1. Formation: Cell Lines, Plasmids, and Transfection*

HEK 293 and HeLa cells are maintained in DMEM supplemented with 10% FCS, 50 μg/mL of streptomycin, and 50 U/mL of penicillin at 37°C in humidified 95% air–5% $CO_2$. Plasmids pCMVNot6.2 and pCMVNot6.2-ΔF containing expressible human CFTR cDNAs are purified using standard plasmid purification kits supplied by either Qiagen or Clontech, following protocols supplied by the manufacturer. Additional CFTR mutations (e.g., P205S) are introduced using a suitable mutagenesis method (we prefer the Quick Change Site-Directed Mutagenesis System from Stratagene). The following protocol is for transfecting cells in 35-mm dishes or six-well plates. Reagent volumes and DNA amounts are per well. Scale up or down accordingly for different sized dishes. (*See* **Note 1.**)

1. Plate Cells: plate cells 1 d prior to transfection to achieve approx 50% confluence. For example, split a confluent 60-mm dish 1:12 into six-well cluster plates, using 2 mL of complete DMEM per well.
2. Form DNA–Fugene complexes: for each well, add 12 μL of Fugene directly to 88 μL of serum-free DMEM in an Eppendorf tube and mix by tapping. Do not allow undiluted Fugene to touch the walls of the tube. In a separate tube, place 1 μg of CFTR vector. Add Fugene–DMEM mixture to DNA and mix by tapping. Incubate at room temperature (RT) for 15 min. After incubation, add 900 μL of complete DMEM per tube. We frequently perform a transfection control using, for example, 1 μg of GFP vector.
3. Aspirate medium from cells and replace with 1 mL of complete DMEM. Add DNA–Fugene–DMEM complex mixture (now 1-mL volume, giving 2 mL total per well) directly to wells and swirl gently to mix. Return to the incubator for the appropriate time.
4. Check transfection efficiency, for example, by examining cells for GFP fluorescence.
5. For proteasome inhibition, add lactacystin to final concentration of 10 μ*M* and continue incubation for 2–12 h.

### 3.1.2. Analysis

#### 3.1.2.1. Immunocytochemistry and Inclusion Morphometry

At approx 24 h posttransfection:

1. Release cells from dishes by trypsinization and replate onto glass cover slips. (Alternatively, transfection and analysis can all be performed on cover slips.)

At approx 72 h posttransfection:

2. Rinse cells three times with PBS.
3. Fix and permeabilize cells with 1 mL of ice-cold methanol for 10 min at –20°C (*see* **Note 2**).
4. Rinse three times with PBS.
5. Incubate in 1 mL of 50 m*M* glycine in PBS for 10 min (this and all subsequent steps at RT).
6. Block nonspecific sites with 0.1 mL of blocking medium (*see* **Subheading 2.**) for 1 h.
7. Remove blocking medium and add primary antibody for either CFTR, centrosomes (γ-tubulin), proteasomes (20S, PA700, PA28), or ER (BiP) as a 1:100 dilution in 0.1 mL of blocking medium; incubate 1 h.
8. Remove primary antibody and wash cells three times with blocking medium.
9. Add fluorescently labeled secondary antibody as 1:100 dilution in 0.1 mL of blocking medium; incubate 1 h.
10. Remove secondary antibody and wash cells three times with blocking medium.
11. Repeat **steps 7–10** with appropriate antibodies for counterstaining.
12. Visualize cells and obtain images using Bio-Rad MRC1024 confocal microscope.

#### 3.1.2.2. Characterization of Soluble and Insoluble Fractions

At approx 48 h posttransfection:

1. Add lactacystin to one plate of each set to final concentration of 10 μ*M*; return to incubator and continue growth for 12 h.
2. Wash cells two times with PBS, harvest by trypsinization, and pellet by centrifugation at 4°C.
3. Wash cell pellet two times with PBS and resuspend in 100 μL of PBS with Complete protease inhibitor cocktail (supplemented PBS).
4. Lyse by multiple (10–20 times) passage through a 27-gage needle.
5. Centrifuge the lysate (14,000g, 4°C, 1 h); remove and save supernatant (can be stored at –70°C until analysis).
6. Wash pellets one time with supplemented PBS.
7. Recentrifuge (14,000*g*, 4°C, 30 min); remove and discard the supernatant.
8. Resuspend pellets in 100 μL supplemented PBS (store at –70°C until analysis).
9. Separate pellet and supernatant fractions by electrophoresis on 4–20% sodium dodecyl sulfate-polyacrylamide gel electrophoresis (SDS-PAGE) gels.
10. Transfer proteins to nitrocellulose using transfer buffer (*see* **Subheading 2.**) with 20% methanol, and block with blocking buffer (*see* **Subheading 2.**).
11. Perform immunoblotting using γ-tubulin, PA28, PA700, and Hsp70 antibodies.

## ***3.2. Centrosome-Associated Proteins***

### 3.2.1. Immunocytochemistry

1. Plate cells onto glass cover slips.

At approx 48 h after plating:

2. Fix, permeabilize, and block cells as described in **Subheading 3.1.2.1.**
3. Remove blocking medium and add primary antibody for either centrosomes (γ-tubulin), proteasomes (20S, PA700, PA28), chaperones (Hsp70, Hsp90), ubiquitin, and control subcellular markers (BiP) as a 1:100 dilution in 0.1 mL of blocking medium; incubate 1 h.
4. Remove primary antibody and wash cells three times with blocking medium.
5. Add fluorescently labeled secondary antibody as a 1:100 dilution in 0.1 mL of blocking medium; incubate 1 h.
6. Remove secondary antibody and wash cells three times with blocking medium.
7. Repeat **steps 7–10** with appropriate antibodies for counterstaining.
8. Visualize cells and obtain images using Bio-Rad MRC1024 confocal microscope.

### 3.2.2. Immunoblot Analysis of Purified Centrosomes

1. Treat exponentially growing 293 or HeLa cells for 1 h with 1 μg/mL of cytochalasin D and 0.2 μ*M* nocodazole to disrupt the cytoskeleton.
2. Release cells by trypsinization and collect by centrifugation.
3. Wash the cell pellet one time in TBS and one time in 0.1X TBS/8% sucrose, collecting cells by centrifugation between washes. Resuspend cells in 2 mL of 0.1X TBS–8% sucrose.
4. Add 8 mL of lysis buffer (*see* **Subheading 2.**). Gently shake the suspension and pass five times through a narrow-mouth 10 ml serological pipet to lyse the cells.
5. Centrifuge the lysate (2500*g*, 10 min, RT) to remove swollen nuclei, chromatin aggregates, and unlysed cells.
6. Add HEPES to 10 m*M* and DNase I to 1 μg/mL from concentrated stocks, and incubate for 30 min on ice.
7. Gently underlay solution with 1 mL of 60% sucrose solution (*see* **Subheading 2.**) and centrifuge (10,000*g*, 30 min, RT) to sediment centrosomes onto the cushion.
8. Remove the upper 8 mL of the supernatant and discard. Load the remainder, including the cushion (~3 mL, ~30% sucrose, containing the concentrated centrosomes), onto a discontinuous sucrose gradient consisting of 70, 50, and 40% sucrose solutions (1 mL ea) from the bottom, respectively. Centrifuge the gradient (120,000*g*, 1 h, 4°C). Poke a small hole in the bottom of the tube and collect 300-μL fractions; proceed to **step 9** or store at –70°C.
9. Dilute gradient fractions into 1 mL of 10 m*M* piperazine-1,4-*bis*(2-ethanesulfonic acid) (PIPES), pH 7.2, and pellet centrosomes in a microfuge (14,000 rpm, 15 min, 4°C). Resuspend the centrosome pellets in SDS-PAGE sample buffer, boil for 5 min, and separate samples by electrophoresis on 10% SDS-PAGE gels.
10. Transfer to nitrocellulose as described in **Subheading 3.1.2.3.** and perform immunoblot using antibodies against centrosome (γ-tubulin) and proteins of interest, e.g., proteasome (20S, PA700, PA28), chaperones (Hsp70, Hsp90), and control subcellular markers (e.g., BiP, aldolase, Lamin-B1, β-cop).

## 4. Notes

1. Cell lines selected for transfection analysis of CFTR expression and localization should be at a passage number of between approx 35 and 55 for best results. The reasons for this are twofold. First, at higher passage numbers, transformed cell lines tend to grow much more rapidly than at lower passages, and this can cause cells to form highly confluent cell layers in the time frame of the experiment that may detach, en masse, from the cover slips

during the methanol permeabilization and fixation procedure. This is especially problematic when analyzing HEK 293 cells. Second, cell crowding (not to be confused with cell–cell contacts important for the growth of some lines) can distort cell morphologies and impair interpretation of subcellular staining patterns. This is particularly important for colocalization experiments. For example, the *cis*-Golgi is in close proximity to the centrosome and may appear to colocalize with perinuclear inclusions such as those described here and elsewhere.

2. Methanol fixation, a technique long used by researchers investigating the cytoskeleton, is useful for the analysis of some cytoskeletal staining patterns such as the centrosome and microtubular and actin filament networks. This fixation procedure extracts a significant fraction of the cytoplasm, revealing subtle staining patterns, which would otherwise be masked by large signals originating in the cytoplasmic compartment. For example, staining for 20S proteasomes in detergent permeabilized and paraformaldehyde fixed cells is inadequate to visualize the fraction of this complex that associates with the ER membrane, cytoskeleton, or centrosome.

## Acknowledgments

The authors wish to thank George DeMartino for helpful comments. This work was supported by research grants from the American Heart Association and the NIH-NIDDK to P. J. T., the Cystic Fibrosis Foundation to M. J. C. and P. J. T., and the Haberecht Foundation to W. C. W.

## References

1. Wójcik, C., Schroeter, D., Wilk, S., Lamprecht, J., and Paweleta, N. (1996) Ubiquitin-mediated proteolysis centers in HeLa cells: indication from studies of an inhibitor of the chymotrypsin-like activity of the proteasome. *Eur. J. Cell Biol.* **71,** 311–318.
2. Wigley, W. C., Fabunmi, R. P., Lee, M. G., et al. (1999) Dynamic association of proteasomal machinery with the centrosome. *J. Cell Biol.* **145,** 481–490.
3. Johnston, J. A., Ward, C. L., and Kopito, R. R. (1998) Aggresomes: a cellular response to misfolded proteins. *J. Cell Biol.* **143,** 1883–1898.
4. Rock, K. L. and Goldberg, A. L. (1999) Degradation of cell proteins and the generation of MHC class I- presented peptides. *Annu. Rev. Immunol.* **17,** 739–779.
5. Voges, D., Zwickl, P., and Baumeister, W. (1999) The 26S proteasome: a molecular machine designed for controlled proteolysis. *Annu. Rev. Biochem.* **68,** 1015–1068.
6. Hershko, A. and Ciechanover, A. (1998) The ubiquitin system. *Annu. Rev. Biochem.* **67,** 425–479.
7. Schwartz, A. L. and Ciechanover, A. (1999) The ubiquitin-proteasome pathway and pathogenesis of human diseases. *Annu. Rev. Med.* **50,** 57–74.
8. Ma, C.-P., Vu, J. H., Proske, R. J., Slaughter, C. A., and DeMartino, G. N. (1994) Identification, purification, and characterization of a high molecular weight ATP-dependent activator (PA700) of the 20S proteasome. *J. Biol. Chem.* **269,** 3539–3547.
9. Weissman, A. M. (2001) Themes and variations on ubiquitylation. *Nat. Rev. Mol. Cell Biol.* **2,** 169–178.
10. Coux, O., Tanaka, K., and Goldberg, A. L. (1996) Structure and functions of the 20S and 26S proteasomes. *Annu. Rev. Biochem.* **65,** 801–847.
11. Baumeister, W., Walz, J., Zuhl, F., and Seemuller, E. (1998) The proteasome: paradigm of a self-compartmentalizing protease. *Cell* **92,** 367–380.

12. Lowe, J., Stock, D., Jap, B., Zwickl, P., Baumeister, W., and Huber, R. (1995) Crystal structure of the 20S proteasome from the archaeon *T. acidophilum* at 3.4 A resolution [see comments]. *Science* **268,** 533–539.
13. Groll, M., Ditzel, L., Lowe, J., et al. (1997) Structure of 20S proteasome from yeast at 2.4 A resolution. *Nature* **386,** 463–471.
14. Heinemeyer, W., Fischer, M., Krimmer, T., Stachon, U., and Wolf, D. H. (1997) The active sites of the eukaryotic 20 S proteasome and their involvement in subunit precursor processing. *J. Biol. Chem.* **272,** 25200–25209.
15. Dick, T. P., Nussbaum, A. K., Deeg, M., et al. (1998) Contribution of proteasomal beta-subunits to the cleavage of peptide substrates analyzed with yeast mutants. *J. Biol. Chem.* **273,** 25637–25646.
16. Wenzel, T. and Baumeister, W. (1995) Conformational constraints in protein degradation by the 20S proteasome. *Nat. Struct. Biol.* **2,** 199–204.
17. Ma, C.-P., Slaughter, C. A., and DeMartino, G. N. (1992) Identification, purification, and characterization of a protein activator (PA28) of the 20 S proteasome (macropain) *J. Biol. Chem.* **267,** 10515–10523.
18. Yoshimura, T., Kameyama, K., Takagi, T., et al. (1993) Molecular characterization of the "26S" proteasome complex from rat liver. *J. Struct. Biol.* **111,** 200–211.
19. Adams, G. M., Falke, S., Goldberg, A. L., Slaughter, C. A., DeMartino, G. N., and Gogol, E. P. (1997) Structural and functional effects of PA700 and modulator protein on proteasomes. *J. Mol. Biol.* **273,** 646–657.
20. Deveraux, Q., Ustrell, V., Pickart, C., and Rechsteiner, M. (1994) A 26 S protease subunit that binds ubiquitin conjugates. *J. Biol. Chem.* **269,** 7059–7061.
21. Lam, Y. A., Lawson, T. G., Velayutham, M., Zweier, J. L., and Pickart, C. M. (2002) A proteasomal ATPase subunit recognizes the polyubiquitin degradation signal. *Nature* **416,** 763–767.
22. Yao, T. and Cohen, R. E. (2002) A cryptic protease couples deubiquitination and degradation by the proteasome. *Nature* **419,** 403–407.
23. Verma, R., Aravind, L., Oania, R., et al. (2002) Role of Rpn11 metalloprotease in deubiquitination and degradation by the 26S proteasome. *Science* **298,** 611–615.
24. DeMartino, G. N., Moomaw, C. R., Zagnitko, O. P., et al. (1994) PA700, an ATP-dependent activator of the 20 S proteasome, is an ATPase containing multiple members of a nucleotide-binding protein family. *J. Biol. Chem.* **269,** 20878–20884.
25. Strickland, E., Hakala, K., Thomas, P. J., and DeMartino, G. N. (2000) Recognition of misfolding proteins by PA700, the regulatory subcomplex of the 26 S proteasome. *J. Biol. Chem.* **275,** 5565–5572.
26. Liu, C. W., Millen, L., Roman, T. B., et al. (2002) Conformational remodeling of proteasomal substrates by PA700, the 19 S regulatory complex of the 26 S proteasome. *J. Biol. Chem.* **277,** 26815–26820.
27. Strickland, E., Hakala, K., Thomas, P. J., and DeMartino, G. N. (2000) Recognition of misfolding proteins by PA700, the regulatory subcomplex of the 26 S proteasome. *J. Biol. Chem.* **275,** 5565–5572.
28. Braun, B. C., Glickman, M., Kraft, R., et al. (1999) The base of the proteasome regulatory particle exhibits chaperone-like activity. *Nat. Cell Biol.* **1,** 221–226.
29. Unno, M., Mizushima, T., Morimoto, Y., et al. (2002) The structure of the mammalian 20S proteasome at 2.75 A resolution. *Structure* (Camb.) **10,** 609–618.
30. Dubiel, W., Pratt, G., Ferrell, K., and Rechsteiner, M. (1992) Purification of an 11S regulator of the multicatalytic protease. *J. Biol. Chem.* **267,** 22369–22377.

31. Song, X., Mott, J. D., von Kampen, J., et al. (1996) A model for the quaternary structure of the proteasome activator PA28. *J. Biol. Chem.* **271,** 26410–26417.
32. Ahn, K., Erlander, M., Leturcq, D., Peterson, P. A., Früh, K., and Yang, Y. (1996) In vivo characterization of the proteasome regulator PA28. *J. Biol. Chem.* **271,** 18237–18242.
33. Whitby, F. G., Masters, E. I., Kramer, L., et al. (2000) Structural basis for the activation of 20S proteasomes by 11S regulators. *Nature* **408,** 115–120.
34. Rechsteiner, M., Realini, C., and Ustrell, V. (2000) The proteasome activator 11 S REG (PA28) and class I antigen presentation. *Biochem. J.* 345 Pt 1, 1–15.
35. Dick, T., Ruppert, T., Groettrup, M., et al. (1996) Coordinated dual cleavage induced by the proteasome regulator PA28 lead to dominant MHC ligands. *Cell* **86,** 262.
36. Groettrup, M., Soza, A., Eggers, M., et al. (1996) A role for the proteasome regulator PA28a in antigen presentation. *Nature* **381,** 166–168.
37. Realini, C., Dubiel, W., Pratt, G., Ferrell, K., and Rechsteiner, M. (1994) Molecular cloning and expression of a g-interferon-inducible activator of the multicatalytic protease. *J. Biol. Chem.* **269,** 20727–20732.
38. Ahn, J. Y., Tanahashi, N., Akiyama, K., et al. (1995) Primary structures of two homologous subunits of PA**28,** a γ-interferon-inducible protein activator of the 20S proteasome. *FEBS Lett.* **366,** 37–42.
39. Johnson, D. G. and Walker, C. L. (1999) Cyclins and cell cycle checkpoints. *Annu. Rev. Pharmacol. Toxicol.* **39,** 295–312.
40. Baldwin, A. S., Jr. (1996) The NF-kappa B and I kappa B proteins: new discoveries and insights. *Annu. Rev. Immunol.* **14,** 649–683.
41. Palombella, V. J., Rando, O. J., Goldberg, A. L., and Maniatis, T. (1994) The ubiquitin-proteasome pathway is required for processing the NF-kappa B1 precursor protein and the activation of NF-kappa B. *Cell* **78,** 773–785.
42. Orian, A., Whiteside, S., Israel, A., Stancovski, I., Schwartz, A. L., and Ciechanover, A. (1995) Ubiquitin-mediated processing of NF-kappa B transcriptional activator precursor p105. Reconstitution of a cell-free system and identification of the ubiquitin-carrier protein, E2, and a novel ubiquitin-protein ligase, E3, involved in conjugation. *J. Biol. Chem.* **270,** 21707–21714.
43. Lin, L., DeMartino, G. N., and Greene, W. C. (1998) Cotranslational biogenesis of NF-kappaB p50 by the 26S proteasome. *Cell* **92,** 819–828.
44. Liu, C. W., Corboy, M. J., DeMartino, G. N., and Thomas, P. J. (2003) Endoproteolytic activity of the proteasome. *Science* **299,** 408–411.
45. Hoppe, T., Matuschewski, K., Rape, M., Schlenker, S., Ulrich, H. D., and Jentsch, S. (2000) Activation of a membrane-bound transcription factor by regulated ubiquitin/proteasome-dependent processing. *Cell* **102,** 577–586.
46. Rape, M., Hoppe, T., Gorr, I., Kalocay, M., Richly, H., and Jentsch, S. (2001) Mobilization of processed, membrane-tethered SPT23 transcription factor by CDC48(UFD1/NPL4), a ubiquitin-selective chaperone. *Cell* **107,** 667–677.
47. Campanero, M. R. and Flemington, E. K. (1997) Regulation of E2F through ubiquitin-proteasome-dependent degradation: stabilization by the pRB tumor suppressor protein. *Proc. Natl. Acad. Sci. USA* **94,** 2221–2226.
48. Bonvini, P., Nguyen, P., Trepel, J., and Neckers, L. M. (1998) In vivo degradation of N-myc in neuroblastoma cells is mediated by the 26S proteasome. *Oncogene* **16,** 1131–1139.
49. Tsurumi, C., Ishida, N., Tamura, T., et al. (1995) Degradation of c-Fos by the 26S proteasome is accelerated by c-Jun and multiple protein kinases. *Mol. Cell Biol.* **15,** 5682–5687.

50. Stancovski, I., Gonen, H., Orian, A., Schwartz, A. L., and Ciechanover, A. (1995) Degradation of the proto-oncogene product c-Fos by the ubiquitin proteolytic system in vivo and in vitro: identification and characterization of the conjugating enzymes. *Mol. Cell Biol.* **15,** 7106–7116.
51. Jariel-Encontre, I., Pariat, M., Martin, F., Carillo, S., Salvat, C., and Piechaczyk, M. (1995) Ubiquitinylation is not an absolute requirement for degradation of c-Jun protein by the 26 S proteasome. *J. Biol. Chem.* **270,** 11623–11627.
52. Maki, C. G., Huibregtse, J. M., and Howley, P. M. (1996) In vivo ubiquitination and proteasome-mediated degradation of p53(1) *Cancer Res.* **56,** 2649–2654.
53. Torres, J. and Pulido, R. (2001) The tumor suppressor PTEN is phosphorylated by the protein kinase CK2 at its C terminus. Implications for PTEN stability to proteasome-mediated degradation. *J. Biol. Chem.* **276,** 993–998.
54. Werner, E. D., Brodsky, J. L., and McCracken, A. A. (1996) Proteasome-dependent endoplasmic reticulum-associated protein degradation: an unconventional route to a familiar fate. *Proc. Natl. Acad. Sci. USA* 13797–137801.
55. Plemper, R. K. and Wolf, D. H. (1999) Endoplasmic reticulum degradation. Reverse protein transport and its end in the proteasome. *Mol. Biol. Rep.* **26,** 125–130.
56. Riordan, J. R., Rommens, J. M., Kerem, B.-S., et al. (1989) Identification of the cystic fibrosis gene: cloning and characterization of complementary DNA. *Science* **245,** 1066–1073.
57. Anderson, M. P., Gregory, R. J., Thompson, S., et al. (1991) Demonstration that CFTR is a chloride channel by alteration of its anion selectivity. *Science* **253,** 202–205.
58. Bear, C. E., Duguay, F., Naismith, A. L., Kartner, N., Hanrahan, J. W., and Riordan, J. R. (1991) Cl- channel activity in Xenopus oocytes expressing the cystic fibrosis gene. *J. Biol. Chem.* **266,** 19142–19145.
59. Sheppard, D. N. and Welsh, M. J. (1999) Structure and function of the CFTR chloride channel. *Physiol. Rev.* **79,** S23–S45.
60. Lee, M. G., Wigley, W. C., Zeng, W., et al. (1999) Regulation of Cl-/HCO3- exchange by cystic fibrosis transmembrane conductance regulator expressed in NIH 3T3 and HEK 293 cells. *J. Biol. Chem.* **274,** 3414–3421.
61. Schwiebert, E. M., Benos, D. J., Egan, M. E., Stutts, M. J., and Guggino, W. B. (1999) CFTR Is a Conductance Regulator as well as a Chloride Channel. *Physiol. Rev.* **79,** S145–S166.
62. Choi, J. Y., Muallem, D., Kiselyov, K., Lee, M. G., Thomas, P. J., and Muallem, S. (2001) Aberrant CFTR-dependent HCO3- transport in mutations associated with cystic fibrosis. *Nature* **410,** 94–97.
63. Ko, S. B., Shcheynikov, N., Choi, J. Y., et al. (2002) A molecular mechanism for aberrant CFTR-dependent HCO(3)(-) transport in cystic fibrosis. *EMBO J.* **21,** 5662–5672.
64. Thomas, P. J., Ko, Y. H., and Pedersen, P. L. (1992) Altered protein folding may be the molecular basis of most cases of cystic fibrosis. *FEBS Lett.* **312,** 7–9.
65. Cheng, S. H., Gregory, R. J., Marshall, J., et al. (1990) Defective intracellular transport and processing of CFTR is the molecular basis of most cystic fibrosis. *Cell* **63,** 827–834.
66. Lukacs, G. L., Mohamed, A., Kartner, N., Chang, X.-B., Riordan, J. R., and Grinstein, S. (1994) Conformational maturation of CFTR but not its mutant counterpart (ΔF508) occurs in the endoplasmic reticulum and requires ATP. *EMBO J.* **13,** 6076–6086.
67. Ward, C. L. and Kopito, R. R. (1994) Intracellular turnover of cystic fibrosis transmembrane conductance regulator. *J. Biol. Chem.* **269,** 25710–25718.

68. Ellgaard, L., Molinari, M., and Helenius, A. (1999) Setting the standards: quality control in the secretory pathway. *Science* **286,** 1882–1888.
69. Plemper, R. K. and Wolf, D. H. (1999) Retrograde protein translocation: ERADication of secretory proteins in health and disease. *Trends Biochem. Sci.* **24,** 266–270.
70. Sommer, T. and Wolf, D. H. (1997) Endoplasmic reticulum degradation: reverse protein flow of no return. *FASEB J.* **11,** 1227–1233.
71. Ward, C. L., Omura, S., and Kopito, R. R. (1995) Degradation of CFTR by the ubiquitin-proteasome pathway. *Cell* **83,** 121–127.
72. Jensen, T. J., Loo, M. A., Pind, S., Williams, D. B., Goldberg, A. L., and Riordan, J. R. (1995) Multiple proteolytic systems, including the proteasome, contribute to CFTR processing. *Cell* **83,** 129–135.
73. Bebok, Z., Mazzochi, C., King, S. A., Hong, J. S., and Sorscher, E. J. (1998) The mechanism underlying cystic fibrosis transmembrane conductance regulator transport from the endoplasmic reticulum to the proteasome includes Sec61beta and a cytosolic deglycosylated intermediary. *J. Biol. Chem.* **273,** 29873–29878.
74. Xiong, X., Bragin, A., Widdicombe, J. H., Cohn, J., and Skach, W. R. (1997) Structural cues involved in endoplasmic reticulum degradation of G85E and G91R mutant cystic fibrosis transmembrane conductance regulator. *J. Clin. Invest.* **100,** 1079–1088.
75. Xiong, X., Chong, E., and Skach, W. R. (1999) Evidence that endoplasmic reticulum (ER)-associated degradation of cystic fibrosis transmembrane conductance regulator is linked to retrograde translocation from the ER membrane. *J. Biol. Chem.* **274,** 2626–2624.
76. Plemper, R. K., Deak, P. M., Otto, R. T., and Wolf, D. H. (1999) Re-entering the translocon from the lumenal side of the endoplasmic reticulum. Studies on mutated carboxypeptidase yscY species. *FEBS Lett.* **443,** 241–245.
77. Plemper, R. K., Egner, R., Kuchler, K., and Wolf, D. H. (1998) Endoplasmic reticulum degradation of a mutated ATP-binding cassette transporter Pdr5 proceeds in a concerted action of Sec61 and the proteasome. *J. Biol. Chem.* **273,** 32848–32856.
78. Wickner, S., Maurizi, M. R., and Gottesman, S. (1999) Posttranslational quality control: folding, refolding, and degrading proteins. *Science* **286,** 1888–1893.
79. Agashe, V. R. and Hartl, F. U. (2000) Roles of molecular chaperones in cytoplasmic protein folding. *Semin. Cell Dev. Biol.* **11,** 15–25.
80. Thomas, P. J., Qu, B.-H., and Pedersen, P. L. (1995) Defective protein folding as a basis of human disease. *TIBS* **20,** 456–459.
81. Gottesman, S., Wickner, S., and Maurizi, M. R. (1997) Protein quality control: triage by chaperones and proteases. *Genes Dev.* **11,** 815–823.
82. Bercovich, B., Stancovski, I., Mayer, A., et al. (1997) Ubiquitin-dependent degradation of certain protein substrates in vitro requires the molecular chaperone Hsc70. *J. Biol. Chem.* **272,** 9002–9010.
83. Vidair, C. A., Huang, R. N., and Doxsey, S. J. (1996) Heat shock causes protein aggregation and reduced protein solubility at the centrosome and other cytoplasmic locations. *Int. J. Hyperther.* **12,** 681–695.
84. Wojcik, C. (1997) On the spatial organization of ubiquitin-dependent proteolysis in HeLa cells. *Folia Histochem. Cytobiol.* **35,** 117–118.
85. Wojcik, C. (1997) An inhibitor of the chymotrypsin-like activity of the proteasome (PSI) induces similar morphological changes in various cell lines. *Folia Histochem. Cytobiol.* **35,** 211–214.
86. Garcia-Mata, R., Bebök, Z., Sorcher, E. J., and Sztul, E. S. (1999) Characterization and dynamics of aggresome formation by a cytosolic GFP-chimera. *J. Cell Biol.* **146,** 1239–1254.

87. Rajan, R. S., Illing, M. E., Bence, N. F., and Kopito, R. R. (2001) Specificity in intracellular protein aggregation and inclusion body formation. *Proc. Natl. Acad. Sci. USA* **98,** 13060–13065.
88. Illing, M. E., Rajan, R. S., Bence, N. F., and Kopito, R. R. (2002) A rhodopsin mutant linked to autosomal dominant retinitis pigmentosa is prone to aggregate and interacts with the ubiquitin proteasome system. *J. Biol. Chem.* **277,** 34150–34160.
89. Notterpek, L., Ryan, M. C., Tobler, A. R., and Shooter, E. M. (1999) PMP22 accumulation in aggresomes: implications for CMT1A pathology. *Neurobiol. Dis.* **6,** 450–460.
90. Taylor, J. P., Tanaka, F., Robitschek, J., et al. (2003) Aggresomes protect cells by enhancing the degradation of toxic polyglutamine-containing protein. *Hum. Mol. Genet.* **12,** 749–757.
91. Kabore, A. F., Wang, W. J., Russo, S. J., and Beers, M. F. (2001) Biosynthesis of surfactant protein C: characterization of aggresome formation by EGFP chimeras containing propeptide mutants lacking conserved cysteine residues. *J. Cell Sci.* **114,** 293–302.
92. Waelter, S., Boeddrich, A., Lurz, R., et al. (2001) Accumulation of mutant huntingtin fragments in aggresome-like inclusion bodies as a result of insufficient protein degradation. *Mol. Biol. Cell* **12,** 1393–1407.
93. Johnston, J. A., Dalton, M. J., Gurney, M. E., and Kopito, R. R. (2000) Formation of high molecular weight complexes of mutant Cu, Zn- superoxide dismutase in a mouse model for familial amyotrophic lateral sclerosis. *Proc. Natl. Acad. Sci. USA* **97,** 12571–12576.
94. Fabunmi, R. P., Wigley, W. C., Thomas, P. J., and DeMartino, G. N. (2000) Activity and regulation of the centrosome-associated proteasome. *J. Biol. Chem.* **275,** 409–413.
95. Brown, C. R., Doxsey, S. J., Hong-Brown, L. Q., Martin, R. L., and Welch, W. J. (1996) Molecular chaperones and the centrosome. A role for TCP-1 in microtubule nucleation. *J. Biol. Chem.* **271,** 824–832.
96. Brown, C. R., Hong-Brown, L. Q., Doxsey, S. J., and Welch, W. J. (1996) Molecular chaperones and the centrosome. A role for HSP 73 in centrosomal repair following heat shock treatment. *J. Biol. Chem.* **271,** 833–840.
97. Lange, B. M., Bachi, A., Wilm, M., and Gonzalez, C. (2000) Hsp90 is a core centrosomal component and is required at different stages of the centrosome cycle in Drosophila and vertebrates. *EMBO J.* **19,** 1252–1262.
98. Anton, L. C., Schubert, U., Bacik, I., et al. (1999) Intracellular localization of proteasomal degradation of a viral antigen. *J. Cell. Biol.* **146,** 113–124.
99. Lacaille, V. G. and Androlewicz, M. J. (2000) Targeting of HIV-1 Nef to the centrosome: implications for antigen processing. *Traffic* **1,** 884–891.
100. Lelouard, H., Gatti, E., Cappello, F., Gresser, O., Camosseto, V., and Pierre, P. (2002) Transient aggregation of ubiquitinated proteins during dendritic cell maturation. *Nature* **417,** 177–182.
101. Fabunmi, R. P., Wigley, W. C., Thomas, P. J., and DeMartino, G. N. (2001) Interferon gamma regulates accumulation of the proteasome activator PA28 and immunoproteasomes at nuclear PML bodies. *J. Cell Sci.* **114,** 29–36.
102. Hung, C. F., Cheng, W. F., He, L., et al. (2003) Enhancing major histocompatibility complex class I antigen presentation by targeting antigen to centrosomes. *Cancer Res.* **63,** 2393–2398.
103. Sanchez, V., Greis, K. D., Sztul, E., and Britt, W. J. (2000) Accumulation of virion tegument and envelope proteins in a stable cytoplasmic compartment during human cytomegalovirus replication: characterization of a potential site of virus assembly. *J. Virol.* **74,** 975–986.

104. Sanchez, V., Sztul, E., and Britt, W. J. (2000) Human cytomegalovirus pp28 (UL99) localizes to a cytoplasmic compartment which overlaps the endoplasmic reticulum-golgi-intermediate compartment. *J. Virol.* **74,** 3842–3851.
105. Heath, C. M., Windsor, M., and Wileman, T. (2001) Aggresomes resemble sites specialized for virus assembly. *J. Cell Biol.* **153,** 449–455.
106. Skipper, L. and Farrer, M. (2002) Parkinson's genetics: molecular insights for the new millennium. *Neurotoxicology* **23,** 503–514.
107. Valentine, J. S. and Hart, P. J. (2003) Misfolded CuZnSOD and amyotrophic lateral sclerosis. *Proc. Natl. Acad. Sci. USA* **100,** 3617–3622.
108. Bates, G. (2003) Huntingtin aggregation and toxicity in Huntington's disease. *Lancet* **361,** 1642–1644.
109. Bence, N. F., Sampat, R. M., and Kopito, R. R. (2001) Impairment of the ubiquitin-proteasome system by protein aggregation. *Science* **292,** 1552–1555.
110. Zeng, W., Lee, M. G., Yan, M., et al. (1997) Immuno and functional characterization of CFTR in submandibular and pancreatic acinar and duct cells. *Am. J. Physiol.* **273,** C442–C455.
111. Ma, C.-P., Willy, P. J., Slaughter, C. A., and DeMartino, G. N. (1993) PA28, an activator of the 20S proteasome, is inactivated by proteolytic modification of its carboxyl terminus. *J. Biol. Chem.* **268,** 22514–22519.

# 22

# Detection of Sumoylated Proteins

Roland S. Hilgarth and Kevin D. Sarge

### Summary

Small ubiquitin-related modifier (SUMO) is an ubiquitin-like protein that is covalently attached to a variety of target proteins. Unlike ubiquitination, sumoylation does not target proteins for proteolytic breakdown, but is involved in regulation of protein function, nuclear targeting, and the formation of subcellular structures. Because SUMO is involved in such a plethora of functions and modifies numerous proteins it is important to identify proteins that are sumoylated in order to increase our understanding of how this modification affects protein function and localization. This overview describes techniques utilized for the detection of sumoylated proteins. The techniques covered include immunoprecipitation, an in vitro sumoylation assay, and gel shift mobility assays that have been used to identify SUMO-modified proteins.

**Key Words:** HSF1; HSF2; immunoprecipitation; in vitro modification; sumoylation.

## 1. Introduction

Small ubiquitin-related modifier (SUMO) was discovered as a modifier of mammalian proteins in 1997 *(1,2)*, and SUMO has since been demonstrated to be a modifier of many other cellular proteins *(3,4)*. The machinery involved in the sumoylation pathway is analogous to ubiquitination and requires a specific E1 activating enzyme (SAE1/SAE2), a SUMO-specific E2-conjugating enzyme (Ubc9), and a E3 ligating enzyme *(3)*. The sumoylation target is a lysine that occurs in the consensus motif ΨKXE where Ψ is a hydrophobic amino acid and X is any residue. Because SUMO modification is involved in nuclear targeting, formation of subcellular structures, regulation of transcription factors, and controls protein stability it has become a protein modifier of much interest *(4)*. Here we present three methods commonly used in detecting SUMO modification of proteins. Sumoylated proteins can be detected using immunoprecipitation, in vitro sumoylation assay, and gel mobility shift assay. The proteins we will use to illustrate these techniques are heat shock factor 2 (HSF2) and heat shock factor 1 (HSF1).

From: *Methods in Molecular Biology, vol. 301, Ubiquitin–Proteasome Protocols*
Edited by: C. Patterson and D. M. Cyr © Humana Press Inc., Totowa, NJ

## 2. Materials

1. HeLa cells.
2. Phosphate-buffered saline (PBS).
3. Lysis solution: 0.15 *M* Tris-HCl, pH 6.7, 5% sodium dodecyl sulfate (SDS), and 30% glycerol.
4. Complete protease inhibitor (Roche; Indianapolis, IN).
5. Protein G–Sepharose.
6. HSF2 polyclonal antibody, rabbit IgG, and SUMO-1 antibody.
7. SDS-PAGE load buffer.
8. Polyacrylamide gel electrophoresis (PAGE) equipment and SDS-PAGE solutions.
9. pGEX2T- SUMO-1, pQE30-SUMO-1, and pGEX2T-Ubc9.
10. Ampicillin and LB media.
11. Isopropyl-β-D-thio-galactopyranoside (IPTG).
12. French press.
13. Phenylmethylsulfonyl fluoride (PMSF).
14. Glutathione–agarose and Ni–agarose.
15. Transport buffer: 100 m*M* *N*-(2-hydroxyethyl)piperazine-*N*-(2-ethanesulfonic acid) (HEPES), pH 7.3, 0.1 *M* potassium acetate, 2 m*M* magnesium acetate, 1X protease inhibitor, 50 ng/μL of digitonin, and 1 m*M* dithiothreitol (DTT).
16. 25-Gage needle and syringe.
17. TNT T7 Quick for PCR DNA kit (Promega; Madison, WI).
18. SUMO master mix: 50 m*M* Tris-HCl, pH 7.6, 5 m*M* $MgCl_2$, 1 m*M* ATP, 80 U/μL of creatine phosphokinase, 10 m*M* creatine phosphate, 0.6 U/mL of inorganic pyrophosphatase, 200 μg/mL of pQE30-SUMO-1, 50 μg/mL of Ubc9.
19. Buffer C: 20 m*M* HEPES, pH 7.9, 25% (v/v) glycerol, 0.42 *M* NaCl, 1.5 m*M* $MgCl_2$, 0.2 m*M* EDTA, 0.5 m*M* PMSF, and 0.5 m*M* DTT.
20. Gel shift binding buffer: 10 m*M* Tris-HCl, pH 7.4, 50 m*M* NaCl, 1 m*M* EDTA, 0.5 m*M* DTT, and 5% glycerol.
21. Native electrophoresis solutions.

## 3. Methods

The methods described in this subheading outline the detection of sumoylated proteins using (1) immunoprecipitation techniques and (2) in vitro sumoylation assays. To illustrate the methods utilized by our laboratory we will use HSF2 as our sumoylated protein target for immunoprecipitation and in vitro sumoylation. HSF1 will be used to illustrate the use of the gel mobility shift assay to identify sumoylated proteins.

### *3.1. Immunoprecipitation*

1. Grow HeLa cells to 80–90% confluency on 100-mm × 20-mm tissue culture plates in Dulbecco's modification of Eagle's medium (DMEM) ) supplemented with 10% fetal calf serum (FCS) and 50 μg/mL of gentamicin at 37°C with 5% $CO_2$.
2. For harvesting place cells on ice, remove media by aspiration and add 5 mL of ice-cold PBS.
3. Scrape cells free of plates with a cell scraper and place the 5-mL cell mixture in a 15-mL centrifuge tube. Centrifuge (600*g*; 5 min; 4°C) and remove the supernatant by aspiration.
4. Add 1 mL of ice-cold PBS and transfer cells to a 1.5-mL centrifuge tube and pellet cells (380*g*; 5 min; 4°C) and remove the supernatant by aspiration.

Fig. 1. Detection of sumoylation by immunoprecipitation/SUMO-1 Western blot. HSF2 protein was immunoprecipitated from extracts of HeLa cells followed by western blot using anti-SUMO-1 antibodies. The positions of molecular weight standards are indicated on the *left side* of the panel.

5. Lyse cells in 150 µL of Lysis solution, then dilute 1:10 in PBS–0.5% Nonidet P-40 (NP-40) plus complete protease inhibitor (Roche) and centrifuged (16,000*g*; 10 min; 4°C) to remove cellular debris.
6. While the cell lysate is being centrifuged prepare 30 µL of 50% Protein G–Sepharose as per manufacturer's instructions. After Protein G–Sepharose has been prepared add the above cell lysate to the Protein G–Sepharose and rotate at 4°C for a half hour to preclear the lysate.
7. After preclearing the lysate centrifuge out the Protein G–Sepharose (16,000*g*; 10 min; 4°C ) and transfer the supernatant to a new tube. At this point take 40 µL of cell lysate, place in a separate tube with 12 µL of 4X SDS load buffer, and label input.
8. Divide the remainder of each treatment lysate into two equal amounts in separate 1.5-mL centrifuge tubes. To one of the tubes add 4 µL of HSF2 polyclonal antibody *(5)* and to the other tube add 7 µL of rabbit IgG. Place samples on rotator at 4°C for 1 h after which add 20 µL of PBS washed Protein G–Sepharose and rotate at 4°C for 3 h.
9. Centrifuge the beads (16,000*g*; 10 s; 4°C) and discard supernatant.
10. Wash the beads four times with PBS–0.5% NP-40 plus complete protease inhibitor, collecting the beads by centrifugation after each wash (16,000*g*; 10 s; 4°C). Add 30 µL of SDS load buffer to beads after removing supernatant from final wash.
11. Analyze immunoprecipitated protein by SDS-PAGE, loading 20 µL of sample per lane.
12. Transfer proteins to nitrocellulose by standard electrophoretic techniques and Western blot probing with anti-SUMO-1 antibodies *(6)* (*see* **Fig. 1**). This method can be done utilizing transfected cells (see **Note 2**).

### 3.2. In Vitro Sumoylation Assay

For the in vitro sumoylation assay recombinant proteins and HeLa cell extracts need to be prepared prior to use. Once prepared these reagents can be stored at –80°C for extended periods of time. The expression and purification of recombinant proteins is described in **Subheading 3.2.1.** HeLa cytosolic extract preparation is described in **Subheading 3.2.2.** and the protocol for performing the in vitro sumoylation assay is described in **Subheading 3.2.3.**

### 3.2.1. Expression of Recombinant SUMO-1 and Ubc-9 (SUMO E2 Protein)

The protein expression constructs for Ubc9 and active SUMO-1 were those used by Desterro et al. *(7)*. We utilize two GST-fusion constructs (pGEX2T-SUMO1 and pGEX2T-Ubc9), and one 6-his-fusion construct (pQE30-SUMO1) (*see* **Note 3**). Below is the general protocol for expressing the recombinant proteins needed for the in vitro sumoylation assay. Each recombinant protein to be used in this assay needs to be expressed separately.

1. Transform expression construct into DH10B *E. coli* cells using standard molecular biology methods *(8)*.
2. Plate cells on LB plates containing ampicillin and incubate overnight at 37°C.
3. Select single colonies and grow overnight at 37°C in LB media containing ampicillin.
4. Inoculate individual liters of LB containing ampicillin with aliquots (5 mL) of overnight culture and grow cells to an $O.D._{600nm}$0.6–0.8.
5. Induce the cells with IPTG (1 m*M*) for 3 h.
6. After 3 h of induction harvest cells by centrifugation (4000*g*; 5 min; 4°C) and resuspend pellet in PBS (4°C), then repellet cells. At this point the cell pellet can be extracted or stored at –80°C.
7. To extract protein from the cells they are resuspended in PBS (4°C) with PMSF to a final concentration of 1 m*M*.
8. Pass cells through a French press at 10,000–14,000 psi, and repeat to ensure complete lysis.
9. Centrifuge the cell lysate (30,000*g*; 1 h; 4°C) and retain the supernatant.
10. Purify proteins using standard nondenaturing affinity chromatography techniques suitable for their fusion tags *(8)* (glutathione [GST] agarose affinity chromatography for GST-fusion proteins and Ni-agarose affinity chromatography).
11. Check purified proteins by Coomassie blue staining of an SDS-PAGE gel (GST–SUMO1 = 38 kDa, GST–Ubc9 = 44 kDa, and 6His–SUMO1 = 14 kDa).
12. For the in vitro sumoylation assay the GST–Ubc9 needs to be thrombin cleaved to remove the GST–tag. This can be done following standard protocols *(9)*. Check thrombin cleavage of GST–Ubc9 by Coomassie staining of a SDS-PAGE gel (Ubc9 = 18 kDa).

### 3.2.2. Preparation of HeLa Cytosolic Extracts

Described in the following is the protocol for preparing the HeLa cytosolic extract that is required for the invitro sumoylation assay. This cytosolic extract is the source of the SUMO E1 and E3 machinery which is necessary to get efficient sumoylation in this assay. Recently, progress has been made in purifying recombinant E1 heterodimer enzymes for use in this assay *(10)*.

1. Grow HeLa cells as described in **Subheading 3.1.**
2. Scrape cells off dish with a cell scraper and place in a centrifuge tube. Pellet cells (800*g*; 5 min; 4°C), wash with PBS (4°C), repellet, and resuspend in 10 mL of PBS (4°C).
3. Count cells using a hemocytometer, record number of cells and repellet cells as described in **step 2.**
4. Resuspend pelleted cells (15 μL per $1 \times 10^6$ cells) in 1X transport buffer with digitonin (50 ng/μL) and DTT (1 m*M*). Let cells incubate on ice (5 min).
5. Shear cells by passing them through a 25-gage needle.

6. Pellet cell debris (16,000*g*; 10 min; 4°C). Transfer supernatant to a fresh tube and keep on ice.
7. Resuspend cell debris pellet (5 µL per 1 × $10^6$ cells) with transport buffer (described in **step 4**). Centrifuge (16,000*g*; 10 min; 4°C) and pool with supernatant from **step 6**. Discard the cell pellet.
8. Aliquot HeLa cytosol extract and flash freeze. Store cytosol extract at –80°C until ready to use.

### *3.2.3. In Vitro Sumoylation of Rabbit Reticulocyte System-Translated Proteins*

The in vitro sumoylation assay uses [$^{35}$S]methionine-radiolabeled protein as target template. We generate radiolabeled target protein using the TNT T7 Quick for PCR DNA kit (Promega, Madison WI) following the manufacturers instructions. Described below is the sumo modification procedure to be utilized with radiolabeled translated proteins.

1. Translate target protein fresh before each sumoylation assay following the manufacturer's instructions. Place freshly translated protein on ice until **step 3**.

Make a fresh SUMO master mix.

2. Make a control master mix lacking SUMO-1 (same as the SUMO master mix, but with no SUMO) and make a GST–SUMO-1 master mix (same as the SUMO master mix, but with 200 µg/mL of pGEX2T–SUMO-1) instead of the pQE30–SUMO-1.
3. Set up individual reactions (control, SUMO, and GST–SUMO) using 4 µL of in vitro translated protein, 1 µL of master mix, and 4 µL of HeLa cytosolic extract and incubate at 37°C for 1 h.
4. Terminate the reaction by adding 11 µL of SDS-PAGE load buffer. Store at –20°C until gel electrophoresis.
5. To determine if radiolabeled protein was sumoylated run a one-fourth volume of the sumoylation reactions on SDS-PAGE electrophoresis. Dry the gel on Whatman paper and place on X-ray film.
6. One important experiment to do to give confidence in the in vitro sumoylation assay is to do a reconstitution test where each of the components required for sumoylation (HeLa cytosol, ubc9, SUMO-1) are individually left out of the reaction and compared to a reaction where all components are present (**Fig. 2A**). As shown in this figure, such an experiment can also reveal the relative efficiency of sumoylation of your target protein by different SUMO proteins (e.g., SUMO-1 vs. SUMO-2). Another experiment which increases confidence that your protein is indeed being sumoylated vs. being targeted by some other modification is to compare the effect of using of 6xHis–SUMO vs. GST–SUMO as the donor SUMO, because this gives a predictable size shift between the sumoylated reactions (**Fig. 2B**).
7. To further show the sumoylation site on the protein of interest site-directed mutagenesis should be done on the lysine (changed to arginine) in the sumoylation consensus sequence (ΨKXE) utilizing the Quickchange Site Directed Mutagenesis Kit (Stratagene) (**Fig. 3**).

## *3.3. Gel Mobility Shift Assay*

If the sumoylated protein is a transcription factor a gel shift-supershift approach using SUMO antibodies can be utilized to test whether the DNA-binding form of the

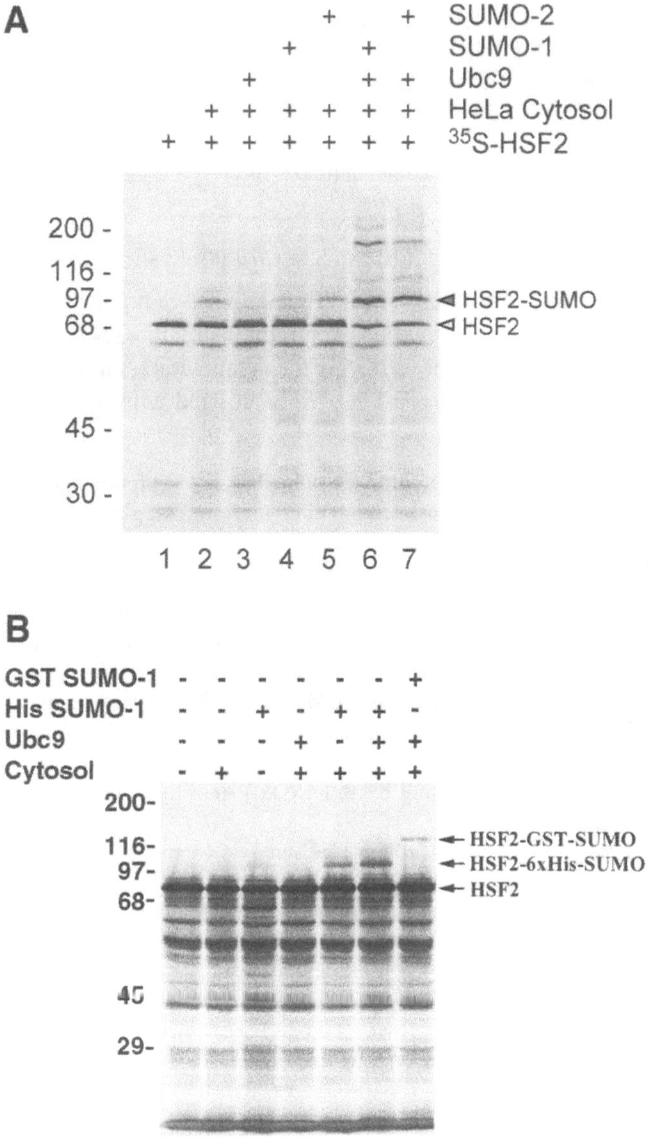


Fig. 2. Analysis of SUMO-1 modification by reconstituted in vitro sumoylation reaction. (**A**) In vitro translated $^{35}$S-labeled HSF2 protein was incubated with HeLa cytosol, Ubc9, SUMO-1, SUMO-2, or with various combinations of each of these, and then subjected to SDS-PAGE followed by autoradiography. The positions of unmodified and SUMO-modified HSF2 are indicated to the right of the panel. (**B**) In vitro translated $^{35}$S-labeled HSF2 protein was subjected to the in vitro SUMO-1 modification assay using either 6xHis–SUMO-1 or GST–SUMO-1 as the SUMO-1 substrate for the reaction.

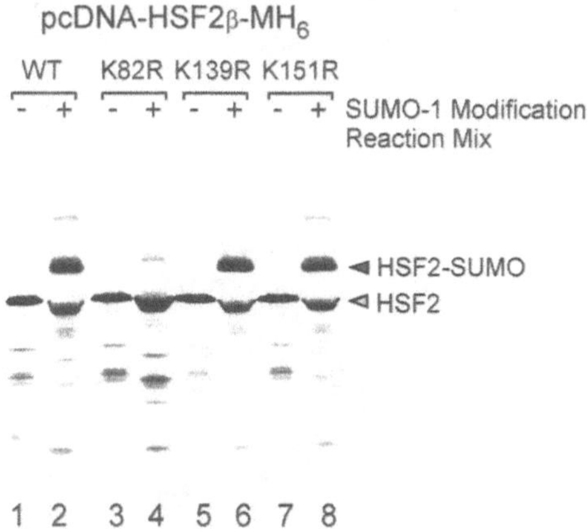


Fig. 3. In vitro translated $^{35}$S-labeled wild-type HSF2 protein and the HSF2 SUMO-1 consensus site mutants K82R, K139R, and K151R were used as substrates in in vitro SUMO-1 modification reactions. The positions of unmodified and SUMO-modified HSF2 proteins are indicated to the *right* of the panel.

protein is sumoylated. We have utilized this technique to study heat shock factor 1 (HSF1) which gains DNA binding ability after sumoylation during heat shock *(11)*.

1. Grow HeLa cells as described in **Subheading 3.1., step 1**.
2. To induce sumoylation of HSF1 heat shock cells (42°C; 1 h).
3. Harvest cells as described in **Subheading 3.1., steps 2–4**.
4. Lyse cells in buffer C.
5. Remove cellular debris by centrifugation (16,000*g*; 10 min; 4°C).
6. Subject the lysate to gel mobility shift assay in the absence or presence of anti-SUMO-1 monoclonal antibodies (21C7) *(6)*.
7. Do the binding reaction in 20 µL of binding buffer. 0.1 ng of $^{32}$P-end-labeled DNA probe, 0.5 µg of poly(dI-dC)-poly(dI-dC), and 10 µg of BSA (*see* **Note 4**).
8. Incubate the reaction at 20°C for 10 min.
9. After incubation analyze the binding reactions by electrophoresis on native 4% polyacrylamide gels in 0.5X Tris-borate–EDTA.
10. After electrophoresis transfer the native gel to Whatmann paper, dry, and visualize by autoradiography (**Fig. 4**).

## 4. Notes

1. SUMO-modified proteins are highly susceptible to SUMO proteases, which has rendered the use of standard immunoprecipitation lysis buffers impractical. This has been previously shown with IκBα *(12)*. The SDS in the lysis buffer described in this protocol inactivates the SUMO proteases allowing for easier detection of sumoylated proteins. SUMO proteases can also be inhibited by the addition of the isopeptidase inhibitors iodoacetamide (200 µ*M*) and *N*-ethylmaleimide (100 µg/mL) to standard lysis buffers such as NP-40 lysis buffer (50 m*M* Tris-HCl, pH 8.0, 150 m*M* NaCl, 1% NP-40) if another lysis buffer besides the SDS-lysis is more desirable. A common complication with the SDS-

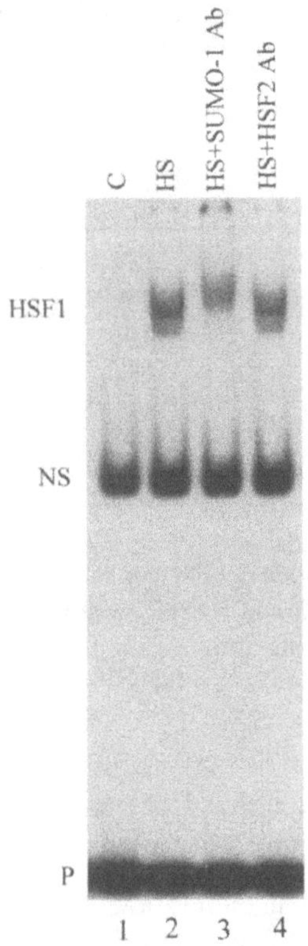


Fig. 4. Gel shift-supershift analysis of HSF1 sumoylation after stress treatment. Extracts of heat-treated HeLa cells (42°C, 1 h) were incubated in the absence or presence of anti-SUMO-1 antibody 21C7 or anti-HSF2 antibody (as a negative control) and then subjected to gel mobility shift assay using a $^{32}$P-labeled heat shock element-containing oligonucleotide probe. NS and P indicate nonspecific DNA-binding activity and free probe, respectively.

lysis described in this protocol is that the cell lysates tend to be very viscous and sticky due to genomic DNA in the lysate. This problem is remedied by brief sonication which shears the DNA and makes the samples easier to manipulate.

2. Investigating sumoylation may also be done using cells transfected with fusion-tagged plasmid constructs (pEGFP, pCDNA, etc.) of the protein thought to be sumoylated with immunoprecipitation utilizing fusion tag antibodies which are readily available from commercial sources (i.e., GFP, Myc, and 6His).
3. SUMO is synthesized as a precursor and processed into an active form ending with a double-glycine motif. The expression constructs utilized here are of the active SUMO-1 (containing amino acids 1–97). If the full-length SUMO-1 is bacterially expressed and used in the in vitro sumoylation assay the assay will not work. The assay should be done using recombinant active forms of SUMO.
4. For detection of HSF1 DNA binding we utilize an oligonucleotide probe that contains four inverted repeats of the heat shock element consensus sequence 5'-nGAAn-3'.

## Acknowledgments

The authors are very grateful to Mike Matunis (Johns Hopkins) for many reagents and advice on the in vitro sumoylation assay, to Chris Lima (Weill Medical College of Cornell University) for his generous gift of constructs encoding recombinant E1 heterodimers, and to the *Journal of Biological Chemistry* for allowing us to reprint figures from two of our papers *(11,13)*. This work was supported by NIH grants GM61053 and GM64606 to K.D.S.


## References


1. Mahajan, R., Delphin, C., Guan, T., Gerace, L., and Melchior, F. (1997) A small ubiquitin-related polypeptide involved in targeting RanGAP1 to nuclear pore complex protein RanBP2. *Cell* **88,** 97–107.
2. Matunis, M. J., Wu, J., and Blobel, G. (1998) SUMO-1 modification and its role in targeting the Ran GTPase-activating protein, RanGAP1, to the nuclear pore complex. *J. Cell Biol.* **140,** 499–509.
3. Verger, A., Perdomo, J., and Crossley, M. (2003) Modification with SUMO. *EMBO Rep.* **4,** 137–42.
4. Kim, K. I., Baek, S. H., and Chung, C. H. (2002) Versatile protein tag, SUMO: its enzymology and biological function. *J. Cell Physiol.* **191,** 257–268.
5. Sarge, K. D., Murphy, S. P., and Morimoto, R. I. (1993) Activation of heat shock gene transcription by heat shock factor 1 involves oligomerization, acquisition of DNA-binding activity, and nuclear localization and can occur in the absence of stress. *Mol. Cell Biol.* **13,** 1392#- #1407.
6. Matunis, M. J., Coutavas, E., and Blobel, G. (1996) A novel ubiquitin-like modification modulates the partitioning of the Ran-GTPase-activating protein RanGAP1 between the cytosol and the nuclear pore complex. *J. Cell Biol.* **135,** 1457–1470.
7. Desterro, J. M., Thomson, J., and Hay, R. T. (1997) Ubch9 conjugates SUMO but not ubiquitin. *FEBS Lett.* **417,** 297–300.
8. Sambrook, J. and Russell, D. (2001) *Molecular Cloning, A Laboratory Manual,* 3rd edit. 3 vols, Cold Spring Harbor Laboratory Press, Cold Spring Harbor.
9. Jaffray, E., Wood, K. M., and Hay, R. T. (1995) Domain organization of I kappa B alpha and sites of interaction with NF-kappa B p65. *Mol. Cell Biol.* **15,** 2166–2172.
10. Tatham, M. H., Jaffray, E., Vaughan, O. A., et al. (2001) Polymeric chains of SUMO-2 and SUMO-3 are conjugated to protein substrates by SAE1/SAE2 and Ubc9. *J. Biol. Chem.* **276,** 35368–35374.
11. Hong, Y., Rogers, R., Matunis, M. J., et al. (2001) Regulation of heat shock transcription factor 1 by stress-induced SUMO-1 modification. *J. Biol. Chem.* **276,** 40263–40267.
12. Desterro, J. M., Rodriguez, M. S., and Hay, R. T. (1998) SUMO-1 modification of IkappaBalpha inhibits NF-kappaB activation. *Mol. Cell* **2,** 233–239.
13. Goodson, M. L., Hong, Y., Rogers, R., Matunis, M. J., Park-Sarge, O. K., and Sarge, K. D. (2001) SUMO-1 modification regulates the DNA-binding activity of heat shock transcription factor 2 (HSF2), a promyelocytic leukemia nuclear body associated transcription factor. *J. Biol. Chem.* **276,** 18,513–18,518.

# 23

# Proteasome Inhibitors in Cancer Therapy

**Robert Z. Orlowski**

**Summary**

Inhibitors of the proteasome have long been used in studies of protein turnover, but in a notable example of successful translational research they have made the leap from the laboratory into the clinical arena. The proteasome inhibitor bortezomib (VELCADE®, formerly known as PS-341), has recently been approved in the United States for treatment of patients with multiple myeloma who have received at least two prior therapies, and have demonstrated disease progression on their last therapy. Furthermore, studies of this agent in other hematologic malignancies and solid tumors are underway, and other proteasome inhibitors for clinical use are under development as well. This chapter provides the reader with guidelines for the optimal clinical administration of VELCADE for its currently approved indication, as well as some suggestions for subsequent management of treatment-related events in these patients.

**Key Words:** Bortezomib; inhibitor; multiple myeloma; proteasome; VELCADE.

## 1. Introduction

The discovery of the proteasome, and its characterization as a multicatalytic proteinase complex (reviewed in **ref.** ***1***), led to the synthesis of inhibitors that were then, and are to this day, used as valuable tools in probing the function of this agent of proteolysis (reviewed in **ref.** ***2***). Exposure of cells to these inhibitors, however, was also noted to induce programmed cell death, or apoptosis, in a variety of human-derived tumor lines (reviewed in **refs.** ***3,4***). In a number of model systems tumor cells appeared to undergo apoptosis preferentially when compared with control, non-transformed cells ***(5–7)***. Such observations, as well as the role of the proteasome in several pathways critical to growth and survival of many tumor cells, including the nuclear factor κB (reviewed in **ref.** ***8***) and p44/42 mitogen activated protein kinase ***(9)***, led to interest in the use of these compounds as drugs for cancer therapy. Demonstration that a proteasome inhibitor could have antitumor activity in an in vivo model of human lymphoma ***(5)***, as well as synthesis of more potent, selective, and specific inhibitors such as PS-341 **(Fig. 1)** ***(10)***, recently renamed bortezomib, or VELCADE,

From: *Methods in Molecular Biology, vol. 301, Ubiquitin–Proteasome Protocols*
Edited by: C. Patterson and D. M. Cyr © Humana Press Inc., Totowa, NJ

spurred development in this field further. Proteasome inhibitors have since been found to have efficacy in a variety of xenograft model systems, and also to enhance the antitumor efficacy of several standard chemotherapeutics (reviewed in **ref. *11***). Phase I clinical trials with bortezomib then demonstrated that this drug could be administered to patients with a manageable toxicity profile ***(12,13)***, and significant antitumor activity was described in multiple myeloma **(Fig. 2)** ***(13)***. These encouraging observations, as well as pioneering preclinical work ***(7,14)***, led to a Phase II trial that confirmed the activity of bortezomib in patients who have received at least two prior therapies, and were progressing on the last of these ***(15)***. This has culminated in the recent approval by the Food and Drug Administration (FDA) of bortezomib for this indication. Moreover, a variety of studies are ongoing in patients with multiple myeloma, other hematologic malignancies (**Fig. 3**), and also with solid tumors, using either bortezomib as a single agent, or in combination with other drugs. Thus, the future of inhibitors of the proteasome as cancer chemotherapeutics seems very promising.

Although a variety of drugs impact on proteasome function as part of their mechanisms of action (reviewed in **ref. *16***), and several agents are in development as specific proteasome inhibitors, at this time the only such drug in clinical use is VELCADE. In keeping with the theme of the *Methods in Molecular Biology* series, this chapter provides an overview of current guidelines and procedures for the clinical use of this proteasome inhibitor in its approved indication, based on experience with this agent at our and other institutions. Those intending to administer bortezomib to patients are also encouraged to consult the package insert for additional details that are beyond the scope of this overview.

## 2. Materials

### *2.1. Supplies*

1. VELCADE (bortezomib) for injection. This is manufactured by Millennium Pharmaceuticals, Inc., and is supplied as single dose vials containing 3.5 mg of bortezomib as a sterile, white or off-white lyophilized cake or powder, along with 35 mg of mannitol, USP.
2. Normal (0.9%) saline, Sodium Chloride Injection, USP, for drug reconstitution.
3. Intravenous access catheters and infusion tubing.

### *2.2. Optional Supplies*

1. Antiemetics, such as ondansetron (Zofran®), prochlorperazine (Compazine®), or dexamethasone (Decadron®), and adjunctive agents such as lorazepam (Ativan®).
2. Normal (0.9%) saline, sodium chloride injection, USP, for intravenous infusions.

## 3. Methods

### *3.1. Patient Selection*

VELCADE was approved by the FDA for treatment of patients with multiple myeloma who have received at least two prior treatment regimens, and who demonstrated disease progression on the last of these modalities. Although studies into the use of this agent as initial therapy for patients with multiple myeloma are ongoing,

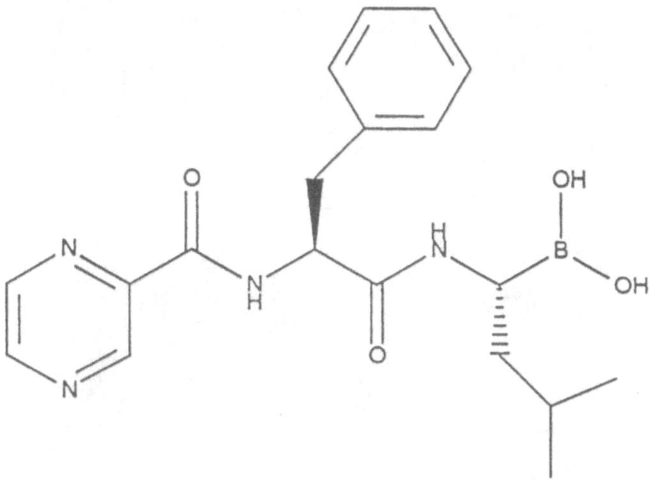


**PS-341**

Fig. 1. The chemical structure of PS-341, since renamed bortezomib (VELCADE).

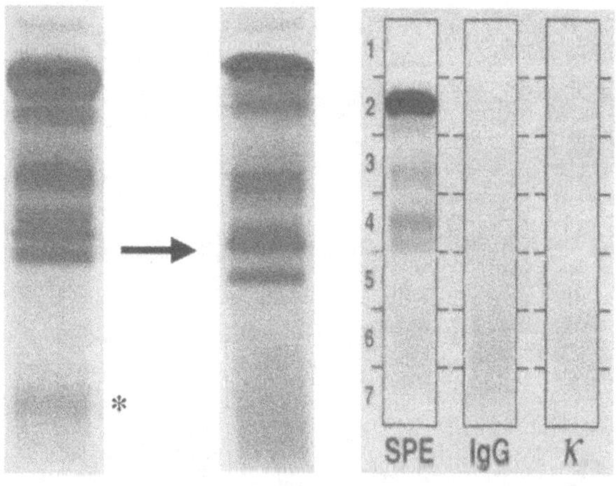


Fig. 2. Response of one patient with multiple myeloma to bortezomib-based therapy. Serum protein electrophoresis from one patient with relapsed, refractory multiple myeloma, showing a large band of restricted mobility (**left panel**, indicated by an *asterisk*) which represented the monoclonal protein made by the patient's neoplastic plasma cells. After three cycles of therapy with bortezomib and liposomal doxorubicin *(22)*, this band disappeared (**middle panel**), and immunofixation revealed a polyclonal immunoglobulin pattern without a monoclonal protein (**right panel**). Please note that this regimen is undergoing further testing in clinical trials, and should be considered investigational at this time.

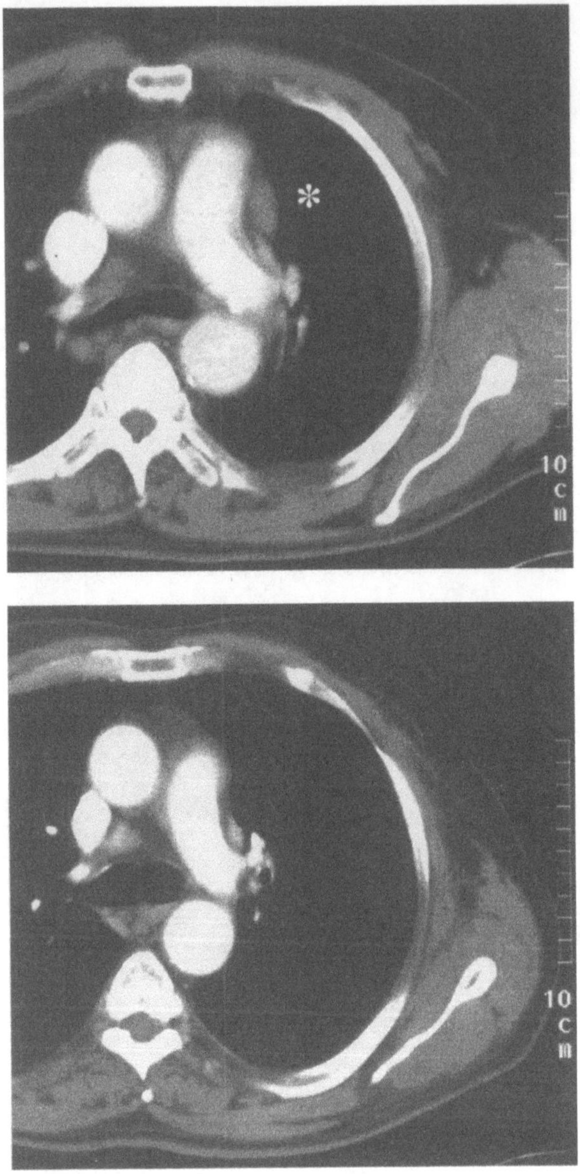


Fig. 3. Response of one patient with non-Hodgkin's lymphoma to bortezomib therapy. Computed tomographic scanning of the chest is shown in one patient with follicular non-Hodgkin's lymphoma documenting paraaortic adenopathy at baseline (**upper panel**, indicated by *asterisk*), which measured 2.5 × 1.2 cm. After 6 wk of therapy with single-agent VELCADE® on a Phase I study at a dose of 1.38 mg/m$^2$ ***(13)***, repeat tomography showed almost complete resolution of this area of adenopathy (**lower panel**). Please note that while VELCADE® is undergoing clinical trials in non-Hodgkin's lymphoma at this time, it has not yet been approved by the FDA for this indication.

as are studies in a variety of other hematologic malignancies and solid tumors, such applications are still considered investigational. Of note from the Phase II studies of this agent is the finding that patients with a wide range of different numbers of prior therapies had an approximately equivalent likelihood of responding to VELCADE *(15)*. Though this study was not powered to answer this specific question, these results nonetheless demonstrate that even patients with advanced, heavily pretreated multiple myeloma can benefit from VELCADE. Therefore, all patients who meet the above noted criteria should be considered candidates for therapy.

With respect to the impact of other factors on the candidacy of patients for VELCADE therapy, placental transfer studies have not been performed, nor have studies with pregnant women. It is therefore recommended that women of childbearing potential use effective contraception to avoid becoming pregnant while on therapy. In regards to renal impairment, studies are ongoing in patients with creatinine clearance values less than 13 mL/min, and also in patients on hemodialysis, but final results of these are not yet available, and so if treated such patients should be monitored especially carefully. Similarly, in clinical trials to date most patients with known human immunodeficiency virus infection, as well as active hepatitis A, B, or C, have been excluded, and therefore there is little information available to guide therapy of such patients, though studies in this area are also ongoing. Because VELCADE is metabolized in part by the liver, patients with any hepatic dysfunction that are treated should also be monitored carefully. Finally, patients with a known allergy to VELCADE, boron-containing compounds, or mannitol, should not be treated with this agent.

### 3.2. Drug Storage and Preparation

Vials of VELCADE are stable until the date indicated on their package, and should be stored in this packaging protected from light at a controlled room temperature of 25°C (77°F), although brief exposure to temperatures between 15 and 30°C (59–86°F) is possible. Once solubilized as described below, VELCADE should be stored either in its original vial, or a syringe under the same conditions. Such material should be used within 8 h of preparation if kept in the original vial, or within 3 h if stored in a syringe.

Single-use vials containing bortezomib should be reconstituted with 3.5 mL of normal (0.9%) saline, sodium chloride injection, USP, using standard aseptic handling techniques, including gloves and protective clothing that are applied to other chemotherapeutics. The resulting solution should be clear and colorless, and if any discoloration or particulate matter is seen on visual inspection the material should not be used, but instead discarded using accepted disposal procedures for chemotherapy drugs.

### 3.3. Drug Dosage and Administration

VELCADE should be dosed at 1.30 mg/m$^2$ per dose, and in the Phase II *(15)* and ongoing Phase III studies has been administered on d 1, 4, 8, and 11, followed by a 10-d rest every 21 d, with this 3-wk period constituting one treatment cycle. We have allowed patients some flexibility in this schedule for their personal needs by delaying dosing up to 24 h, provided that there is an at least 72-h interval maintained between consecutive doses which, based on preclinical studies, is necessary to minimize toxicity ***(10)***. Intravenous access can be obtained for patients through an existing central venous catheter if this is available, but because VELCADE is not a known vesicant placement of a peripheral venous catheter is sufficient. Once this is established, VELCADE should be administered as a rapid infusion over 3–5 s, which we have generally done while normal saline is infusing through this line at 100 mL/h. During clinical trials with VELCADE we have monitored patients with vital sign measurements every 15 min for one h after each injection during cycle 1. For those patients who tolerated their therapy well and showed no infusion-related events in subsequent cycles, this monitoring period was pared down to 15 min.

### 3.4. Management of Infusion-Related Events

#### 3.4.1. Extravasation

Patients receiving VELCADE may occasionally develop local skin irritation at the site of injection, usually manifested by erythema, which was reported in 5% of patients in Phase II clinical trials ***(15)***. Local care consisting of, for example, warm compresses, is generally sufficient to resolve these signs and any associated symptoms. In the event that local tissue infiltration occurs similar measures can be pursued. There is no known antidote for VELCADE, but extravasation events have not to date been associated with tissue necrosis or damage.

#### 3.4.2. Nausea and Emesis

Most patients treated with VELCADE do not have significant infusion-related events comparable to those seen with many other antineoplastics, and we have not therefore routinely used any premedications, nor are any such recommended by the package insert. Some patients, however, will develop nausea and, rarely, emesis, which in some cases can be due to anticipatory nausea. For those patients experiencing such symptoms with their first dosing, routine antiemetic therapy with standard agents including prochlorperazine (Compazine®), serotonin receptor antagonists such as granisetron or ondansetron (Kytril® and Zofran®, respectively), or steroids are indicated. These can then be added as premedications prior to future doses of VELCADE to prophylax against these effects. All of these classes of antiemetics are active based on anecdotal experience, and there is also controlled data from the Phase II trial of VELCADE that documented an improvement in several disease- and therapy-related symptoms with the addition of dexamethasone, including nausea ***(17)***.

#### 3.4.3. Hypotension

Another event that can be noted is the development of hypotension, which was seen in 12% of patients on the Phase II trial of VELCADE, with orthostasis and syncope in a fraction of these ***(15)***. Vital signs should be monitored in patients receiving this agent, and it is also important to evaluate each person prior to initiation of therapy for factors that may predispose them to hypotension, such as the presence of baseline dehydration or an autonomic neuropathy. In our experience most episodes of hypotension have been generally brief and self-limited, and can be successfully treated through intravenous fluid resuscitation with saline if it is more prolonged. At some centers patients have been routinely given up to 500 mL of normal saline with each dose as a prophylactic measure, but at our institution we have not found this to be necessary. In those patients who have orthostasis that is not fluid responsive, prominent concerns in this population would include sepsis and adrenal insufficiency. If the former is ruled out, patients with persistent orthostasis should undergo testing of their adrenal axis, and may benefit from an empiric course of steroids with mineralocorticoid activity pending these results. For some patients, however, dose adjustments of concomitant antihypertensives have been necessary while they are receiving VELCADE, and in a few patients the orthostasis and hypotension did not resolve until the discontinuation of VELCADE altogether.

## 3.5. Management of Drug-Related Toxicities

### 3.5.1. Asthenia

The most common adverse event reported in Phase II clinical trials with VELCADE in patients with multiple myeloma has been asthenia ***(15)***, which has a tendency to be cumulative with continued therapy. For most patients the etiology of this presentation is multifactorial, and therefore its evaluation and therapy needs to be very individualized. Some themes can be more commonly seen, however, which include gastrointestinal complications, which are discussed in **Subheading 3.5.2.**, and anemia, whose management is discussed in **Subheading 3.5.3.** Anorexia can be a significant contributor, and such patients may benefit from appetite stimulants such as megestrol acetate, corticosteroids, and dronabinol (Marinol®). Depression should always be considered, and can be treated with any of the commonly used agents, including serotonin reuptake inhibitors such as paroxetine (Paxil®) and sertraline (Zoloft®). At our institution we have used mirtazapine (Remeron®) to good effect, in that in addition to its antidepressant actions it can also stimulate appetite in many patients. Consultation with a psychiatrist, especially with one who has special experience and/or interest in patients with oncologic histories, is also a worthy intervention. Patients may also benefit from additional services such as support groups. For a generalized decrease in alertness a gentle trial of methylphenidate (Ritalin®) may be of benefit, though this should always be preceded by a thorough evaluation to rule out other causes of mental status changes. Special concerns in this patient population would include a chronic infection or a hyperviscosity syndrome. Finally, patients may most benefit from a judiciously applied treatment delay, to allow for a greater time to recover from the morbidity of multiple myeloma as well as its therapy. In our experience, patients who have an excellent response to their first two cycles of therapy seem to continue to have responsive disease even after an extra week off treatment after cycle two.

### 3.5.2. Gastrointestinal Effects

Treatment of nausea and emesis that is associated with infusion of VELCADE was discussed in **Subheading 3.4.2.**, and similar interventions can be applied to patients who have delayed nausea and emesis, which are basically those used with most other chemotherapeutics. VELCADE is also associated with diarrhea in some, and constipation in others. In regards to the former, if pseudomembranous colitis or other infections are ruled out, standard therapy with agents such as loperamide (Imodium®) or, for more severe cases, deodorized tincture of opium, is generally successful. In patients refractory to this therapy, consideration should be given to octreotide (Sandostatin®). Constipation in patients treated with VELCADE is often multifactorial, and contributed to in particular by concurrent use of narcotic analgesics, as well as hypercalcemia and dehydration. Once the latter two are corrected, if present, a regimen is generally required to correct constipation consisting of stool softeners, such as docusate (Colace®), along with bulk agents containing psyllium, such as Metamucil®. Drugs that combine these two categories are also available, such as senna with docusate (Senokot-S®). Patients with more refractory constipation may require agents such as sorbitol or polyethylene glycol (MiraLax™), while others may benefit from the addition of metoclopramide (Reglan®). Ultimately, any agents and combinations that have proven to be effective in the individual practitioner's hands for therapy of constipation due to other agents, such as the vinca alkaloids, are likely to be effective for patients treated with VELCADE as well.

#### *3.5.3. Cytopenias*

Clinical trials with VELCADE in patients with hematologic malignancies have documented that this agent is associated with leukopenias including neutropenia, anemia, and thrombocytopenia, with the last of these being most prevalent. Despite the incidence of neutropenia few episodes of febrile neutropenia were reported in the Phase II study ***(15)***, and routine use of colony stimulating factors is not recommended. Use of either granulocyte colony stimulating factor (GCSF; Neupogen®) or granulocyte macrophage colony stimulating factor (GM-CSF; Leukine®) should be guided by standard criteria, such as those established by the American Society of Clinical Oncology ***(18)***. In those patients who have indications for the use of such agents, we have found that both GCSF and GM-CSF are well tolerated at standard doses.

Most patients with multiple myeloma who have been treated on clinical trials with VELCADE have been on supportive care for a preexisting anemia. However, patients may develop a new or worsening anemia during therapy with this proteasome inhibitor, and require either initiation of some type of erythropoietin, or an increased dose of the agent they are already receiving (including Procrit®, Epogen®, or Aranesp®). All of these agents in our experience seem to be well tolerated, and the choice of therapies should be left to the discretion of the treating physician. In some cases patients may also require either continuation of, or initiation of, packed red blood cell transfusions, but it is notable that patients who responded to VELCADE therapy in the Phase II trial had an improvement in their hemoglobin value over time ***(15)***. Thus, it is possible that a new or worsening anemia may indicate a failure of the patient's disease to respond to therapy with VELCADE, but this hypothesis has not yet been evaluated in a prospective fashion.

#### *3.5.4. Neuropathies*

Patients diagnosed with multiple myeloma not infrequently develop neurologic complications of their disease. Spinal cord compression leading to a radiculopathy, for example, occurs in about 10–15%, and is often due to either a vertebral compression fracture or a paraspinal mass. A variety of peripheral neuropathies are also seen at presentation, especially in association with amyloidosis. Many patients develop some worsening of their baseline neuropathy, or develop a new one, either from progression of their underlying disease, or from the effects of therapy with neurotoxic agents, most notably from the vinca alkaloid vincristine and related agents, or thalidomide (Thalomid®). VELCADE has also been noted to be able to induce a peripheral neuropathy, which is predominantly sensory, although mixed sensorimotor neuropathies have been described as well. Patients with a preexisting neuropathy may be more at risk for the neuropathic effects of VELCADE, and may suffer a worsening of their symptoms ***(12,13,15)***.

Clinical experience with the neuropathic effects of VELCADE is still accumulating, and some trials have incorporated formal and extensive neurological studies in an attempt to elucidate the mechanism by which this effect occurs. Such information may in the future help to guide prophylactic therapy that might decrease the incidence of this toxicity, but at this time only anecdotal information is available. In our study of VELCADE with pegylated liposomal doxorubicin, patients received pyridoxine because of reports suggesting vitamin B6 might decrease palmar plantar erythrodysesthesia (PPE). The incidence and severity of neuropathy seemed to be decreased in this study in comparison with our previous single agent VELCADE experience, but a larger trial of this intervention, preferably with randomization, would be needed before such a recommendation could be made with more confidence of benefit.

One of the most important aspects of managing patients receiving VELCADE is to carefully evaluate for signs and symptoms of neuropathy beforehand, so that an accurate assessment of their initial status is obtained. For those with a mild neuropathy at baseline, initiation of symptomatic care with agents used in the treatment of neuropathy may be appropriate, such as gabapentin (Neurontin®) or amitriptyline. These tend to be more effective for treating neuropathic pain than for sensory neuropathy presenting with numbness. Both of these agents may also be of benefit for patients who develop a new onset, mild (grade 1) neuropathy, which for sensory neuropathy is defined by the loss of deep tendon reflexes and/or development of paresthesias but without a functional impairment, while for neuropathic pain it is defined as mild pain, also without any functional impairment. Other agents that have anecdotally shown benefit include L-carnitine, L-glutamine, and α-lipoic acid (personal communication, Dr. Paul Richardson).

Some patients develop more severe neuropathies, including those that lead to interference in function, but not to the point of preventing activities of daily living. Dose adjustments of VELCADE are recommended under these conditions by 25–50%, and consultation with a neurologist, especially one who has expertise in the treatment of chemotherapy-related neuropathies, is helpful. With respect to the former recommendation, the 25% dose reduction in clinical trials has most commonly been from 1.30 to 1.00 mg/m$^2$, and the latter level of dosing has shown activity against multiple myeloma in another Phase II clinical trial of VELCADE ***(19)***. In some cases with a combined sensory neuropathy and neuropathic pain, VELCADE may need to be held prior to restarting this drug at a decreased dose. Additional agents that may prove beneficial at this point include narcotic analgesics such as the longer acting opioids, including sustained release morphine or oxycodone (MS Contin® and OxyContin®, respectively), or topical agents such as a fentanyl patch (Duragesic®). These may need to be combined with shorter acting agents for breakthrough pain, such as immediate release morphine sulfate or oxycodone, especially while the doses of the longer-acting agents are being titrated to the level appropriate for management of each individual's pain level and pattern. Patients whose therapy has resulted in a neuropathy that impairs their performance of activities of daily living should have their VELCADE discontinued. Details on the recommended dose modification criteria for patients with neuropathy can be found in the package insert for this agent.

### 3.6. Monitoring the Success of Therapy

Patients receiving therapy with VELCADE should be followed clinically, and also should have monitoring of their disease burden performed after completion of every two cycles by the parameters most applicable to the individual sites of disease. These could include serum protein electrophoresis with immunofixation and densitometric monoclonal protein quantitation, 24-h urine collection for light chain determination, bone marrow aspirate and biopsy, and/or surrogate marker studies such as $\beta_2$-microglobulin and C-reactive protein. This should be performed in addition to routine radiographic assessments of lytic bony disease, such as with skeletal surveys, and possibly computed tomography or magnetic resonance imaging for extramedullary disease. More frequent evaluations than every 6 wk of the serum and/or urine can be performed at the discretion of the patient and their health care provider. However, it should be noted that in clinical trials patients occasionally manifested an increase in their monoclonal protein component after cycle 1, but by the end of cycle 2 this would have declined to levels below the pretreatment baseline. Thus, such patients who have an early increase in their apparent disease burden without appearance of clear clinical signs of progression, such as a new hyperviscosity syndrome, should not be classified as refractory, and should be continued on therapy for one more cycle.

## 4. Notes

This chapter has provided an overview into the selection of patients with multiple myeloma who might be able to benefit from therapy with VELCADE, and guidelines for drug preparation, administration, and management of some of the more common early and late toxicities. A full description of these issues is beyond the scope of this chapter, however, and physicians contemplating the use of this agent in their patients are strongly encouraged to read the most up-to-date package insert, consult their Pharmacist, and also contact Millennium Pharmaceuticals, Inc. to discuss the use of VELCADE with health care providers that have experience with this agent.

For those readers who have an interest in using bortezomib as a laboratory tool, however, it should be noted that the commercially available preparation of VELCADE has a 10-fold weight/weight excess of the inert sugar mannitol. This preparation shows robust activity in in vivo xenograft models (Millennium Pharmaceuticals Inc.; personal communication). During tissue culture studies using cell lines, however, it is possible that mannitol might have an impact on in vitro experiments by changing the tonicity of the media, depending on the concentrations used. In using VELCADE in the laboratory, therefore, two controls should be prepared, both of which should contain vehicle, and one of which would contain a mannitol concentration comparable to that of commercially available VELCADE. Alternatively, other proteasome specific inhibitors that are commercially available can be substituted, including lactacystin ***(20)*** or epoxomicin ***(21)***.

## Acknowledgments

The author would like to acknowledge research support from the Leukemia and Lymphoma Society, R6206-02, and from the Department of Defense Breast Cancer Research Program, BC991049.

## References

1. Orlowski, M. and Wilk, S. (2000) Catalytic activities of the 20 S proteasome, a multicatalytic proteinase complex. *Arch. Biochem. Biophys.* **383,** 1–16.
2. Kisselev, A. F. and Goldberg, A. L. (2001) Proteasome inhibitors: from research tools to drug candidates. *Chem. Biol.* **8,** 739–758.
3. Orlowski, R. Z. (1999) The role of the ubiquitin-proteasome pathway in apoptosis. *Cell Death Differ.* **6,** 303–313.
4. Grimm, L. M. and Osborne, B. A. (1999) Apoptosis and the proteasome. *Results Probl. Cell Differ.* **23,** 209–228.
5. Orlowski, R. Z., Eswara, J. R., Lafond-Walker, A., Grever, M. R., Orlowski, M., and Dang, C. V. (1998) Tumor growth inhibition induced in a murine model of human Burkitt's lymphoma by a proteasome inhibitor. *Cancer Res.* **58,** 4342–4348.
6. Masdehors, P., Omura, S., Merle-Beral, H., et al. (1999) Increased sensitivity of CLL-derived lymphocytes to apoptotic death activation by the proteasome-specific inhibitor lactacystin. *Br. J. Haematol.* **105,** 752–757.
7. Hideshima, T., Richardson, P., Chauhan, D., et al. (2001) The proteasome inhibitor PS-341 inhibits growth, induces apoptosis, and overcomes drug resistance in human multiple myeloma cells. *Cancer Res.* **61,** 3071–3076.
8. Orlowski, R. Z. and Baldwin, A. S. (2002) NF-kappaB as a therapeutic target in cancer. *Trends Mol. Med.* **8,** 385–389.
9. Orlowski, R. Z., Small, G. W., and Shi, Y. Y. (2002) Evidence that inhibition of p44/42 mitogen-activated protein kinase signaling is a factor in proteasome inhibitor-mediated apoptosis. *J. Biol. Chem.* **277,** 27864–27871.
10. Adams, J., Palombella, V. J., Sausville, E. A., et al. (1999) Proteasome inhibitors: a novel class of potent and effective antitumor agents. *Cancer Res.* **59,** 2615–2622.
11. Voorhees, P. M., Dees, E. C., O'Neil, B., and Orlowski, R. Z. (2003) The proteasome as a target for cancer therapy. *Clin. Cancer Res.* **9,** 6316–6325.
12. Aghajanian, C., Soignet, S., Dizon, D. S., et al. (2002) A phase I trial of the novel proteasome inhibitor PS341 in advanced solid tumor malignancies. *Clin. Cancer Res.* **8,** 2505–2511.
13. Orlowski, R. Z., Stinchcombe, T. E., Mitchell, B. S., et al. (2002) Phase I trial of the proteasome inhibitor PS-341 in patients with refractory hematologic malignancies. *J. Clin. Oncol.* **20,** 4420–4427.

14. Mitsiades, N., Mitsiades, C. S., Richardson, P. G., et al. (2003) The proteasome inhibitor PS-341 potentiates sensitivity of multiple myeloma cells to conventional chemotherapeutic agents: therapeutic applications. *Blood* **101,** 2377–2380.
15. Richardson, P. G., Barlogie, B., Berenson, J., et al. (2003) A phase 2 study of bortezomib in relapsed, refractory myeloma. *N. Engl. J. Med.* **348,** 2609–2617.
16. Orlowski, R. Z. and Dees, E. C. (2002) Advances in applying drugs that impact the ubiquitin-proteasome pathway to the therapy of breast cancer. *Breast Cancer Res.* **5,** 1–7.
17. Jagannath, S., Richardson, P., Barlogie, B., et al. and the SUMMIT/CREST Investigators. (2003) Phase II trials of bortezomib in combination with dexamethasone in multiple myeloma: Assessment of additional benefits to combination in patients with sub-optimal responses to bortezomib alone. *Proc. Amer. Soc. Clin. Oncol.* **22,** 582 (Abstract 2341).
18. Byrne, J. L., Haynes, A. P., and Russell, N. H. (1997) Use of haemopoietic growth factors: commentary on the ASCO/ECOG guidelines. American Society of Clinical Oncology/ Eastern Co-operative Oncology Group. *Blood Rev.* **11,** 16–27.
19. Jagannath, S., Barlogie, B., Berenson, J., et al. (2002) A Phase II multicenter randomized study of the proteasome inhibitor bortezomib (VELCADE™, formerly PS-341) in multiple myeloma patients relapsed after front-line therapy. *Blood* **100,** 812a (Abstract 3207).
20. Fenteany, G. and Schreiber, S. L. (1998) Lactacystin, proteasome function, and cell fate. *J. Biol. Chem.* **273,** 8545–8548.
21. Sin, N., Kim, K. B., Elofsson, M., et al. (1999) Total synthesis of the potent proteasome inhibitor epoxomicin: a useful tool for understanding proteasome biology. *Bioorg. Med. Chem. Lett.* **9**, 2283–2288.
22. Orlowski, R. Z., Voorhees, P. M., Garcia, R. A., et al. (2003) Phase I study of the proteasome inhibitor bortezomib and pegylated, liposomal doxorubicin in patients with refractory hematologic malignancies. *Proc. Am. Soc. Clin. Oncol.* **22,** 200 (Abstract 801).

# 24

# Parkinson's Disease

## *Assays for the Ubiquitin Ligase Activity of Neural Parkin*

**Michael G. Schlossmacher and Hideki Shimura**

**Summary**

The identification of monogenic variants of Parkinson's disease (PD) has provided novel insights into its unknown pathogenesis. As the first protein linked to autosomal-recessive forms of PD, Parkin became a welcome tool to explain biochemical and neuropathological observations that had suggested involvement of the ubiquitin–proteasome system (UPS) in PD. Based on cellular expression studies and biochemical in vitro experiments, several researchers ascribed an E3-type, E2-dependent ubiquitin protein ligase activity to wild-type (but not mutant) Parkin proteins. Although the individual components of the proposed Parkin ubiquitin ligase complex in the normal human brain remain to be identified and the E3 ligase effect of Parkin function has not yet been confirmed in an animal model, the scientific exploration of a protein with several links to the UPS has provided many leads in PD research. This chapter describes assays that the authors have used to examine the cellular and in vitro effects of neural Parkin.

**Key Words:** Degradation; neurons; Parkinson's disease; proteasome; ubiquitin ligase.

## 1. Introduction

Parkinson's disease (PD) is the second most common neurodegenerative syndrome in men after Alzheimer's disease. Its main histopathological substrate is the progressive cell loss of select groups of neurons, in particular the population of dopamine-producing cells in the substantia nigra *pars* compacta of the human midbrain. This neurodegenerative process frequently involves intracellular inclusion formation comprising dystrophic neurites and aggresome-like Lewy bodies. Lewy inclusions are pathognomonic of PD and contain misfolded, fibril-forming polypeptides. The cause of PD remains essentially unknown, but three pathogenetic elements appear to play a role: excess oxidative stress linked to dopamine metabolism, mitochondrial dysfunction, as demonstrated by complex I deficiency, and impaired protein degradation ***(1–3)***. Although PD is encountered most frequently as a sporadic disease in persons older

From: *Methods in Molecular Biology, vol. 301, Ubiquitin–Proteasome Protocols*
Edited by: C. Patterson and D. M. Cyr © Humana Press Inc., Totowa, NJ

than 60 yr of age ("idiopathic PD"), by now 10 loci have been identified in pedigrees with familial phenotypes, highlighting the importance of hereditary factors. Among the few monogenic variants of heritable parkinsonism known to date, two genes stand out for historical, neuropathological and functional reasons. Very rare, autosomal-dominant mutations (including multiplications) in the *snca* gene, which encodes the abundant presynaptic protein α-synuclein (αS), result in a gain-of-function effect. In contrast, the more frequent, autosomal-recessive homozygous and compound heterozygous genotypes found in the *parkin* gene lead to a presumed loss-of-function effect ***(4–7)***.

Three seminal observations suggest that the ubiquitin–proteasome system (UPS) plays a role in the degeneration of postmitotic neurons in PD: First, the abundant presence of monomers as well as conjugates of ubiquitin (Ub) within the hallmark Lewy inclusions, where Ub is also the second most abundant, known constituent after phosphorylated αS ***(1,8–9)***; second, the discovery of a mutant genotype in a single PD family along with an infrequent, protective polymorphism in a Ub C-terminal hydrolase gene, *uch-L1*. *Uch-L1* encodes an enzyme with two distinct activities, namely weak Ub hydrolysis and surprisingly, Ub protein ligase function ***(10,11)***; and third, the discovery of *parkin*, the second largest human gene that encodes a protein with E3-type Ub protein ligase activity in vitro ***(6,12–14)***. Although mutations in both *parkin* alleles invariably lead to dopaminergic cell loss in humans, resulting in the onset of PD between the ages of 12 and 71 yr, inactivation of the *parkin* gene in adult mice does not ***(7,15–17)***. These unexpected findings led researchers to pursue three complementary goals: (1) to better characterize the cellular localization of parkin proteins and their individual E3 ligase activities in neural cells; (2) to dissect the modular units of the Skp1/Cullin/F-box (SCF)-like *p*arkin *Ub li*gase *c*omplex (referred to as E3$^{Public}$ in this text in light of the still emerging consensus on its constituents); and (3) to identify endogenous substrates of E3$^{Public}$ from neural tissue that may promote cell toxicity in dopamine-producing cells upon their accumulation in nonubiquitinated forms (***9,16–23***; reviewed in **refs.** ***2,3***).

Full-length human parkin is transcribed from 12 exons, which encode a 465-amino-acid-long polypeptide that migrates at 52–53 kDa under reducing, denaturing gel electrophoresis conditions. A total of four identifiable domains link parkin to the UPS: First, the N-terminus harbors a "type-2 Ub-like" domain that plays a role in protein–protein interactions and thus appears distinct from the protein-modifying effect of "type-1 Ub-like" polypeptides ***(6,9);*** second and third, the C-terminal two thirds of the protein contain two classical really-interesting-new gene (RING)-finger domains represented by a canonical $C_3H_1C_4$-motif; and fourth, an atypical RING-finger domain that is located between the latter two comprising a $C_6H_1C_1$-motif (dubbed "in-between-RING" domain; **Fig. 1**). As a likely member of the zinc-binding group of RING-positive proteins, parkin conveys E3 Ub ligase activity in vitro and in cellular expression studies following the recruitment of one of two principal E2 Ub conjugating enzymes, UbcH7 or UbcH8. Heterologous expression studies suggested that the Ub protein ligase activity assembled by parkin mediates polyubiquitination of its target substrates in preparation for their degradation by the 26S proteasome complex ***(12–14)***. To date,

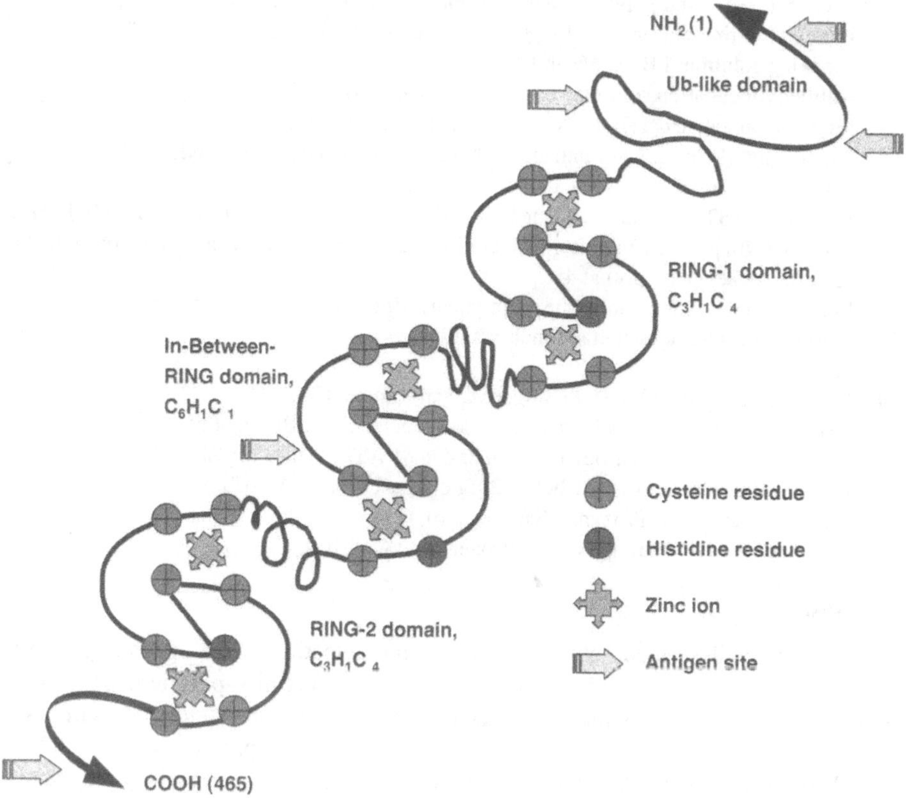


Fig. 1. Illustration of a hypothetical model of human parkin depicting its N-terminal, "type-2" ubiquitin-like domain, its two classical RING-finger motifs, and the noncanonical "In-Between-RING" domain.

several putative parkin binding partners, namely, Ubch7, UbcH8, Hsp70, the cochaperone/E4 protein CHIP, the F-box protein h-Sel10, the integral SCF component Cullin-1 and the proteasome subunit Rpn10 have been identified. In addition, eight putative substrates have been reported by as many investigators and include the septin family members CDC-rel1 and SEPT_v2, the αS-interactor synphilin-1, the endothelin-like Pael-receptor, αSp22, cyclin E, synaptotagmin XI, and the p38 subunit of the aminoacyl-tRNA synthetase complex as well as Parkin itself (*24*; reviewed in **refs. *2,3***).

This chapter is written to provide the interested reader with protocols that we have utilized in our research on the function of neural parkin.

## 2. Materials

1. Dot-blot apparatus and sodium dodecyl sulfate-polyacrylamide gel electrophoresis (SDS-PAGE) equipment.
2. 10–20% Tris glycine gradient gels.

3. Laemmli buffer and polyvinylidene difluoride membrane, 0.4 μ*M*.
4. Immunoblotting buffer (PBST): phosphate-buffered saline, 0.1% Tween-20
5. Blocking solution: PBST, 5% milk.
6. Antibody incubation solution: PBST, 1% bovine serum albumin.
7. Monoclonal anti-UbcH7, anti-α-synuclein and anti-Ub antibodies.
8. Supersignal (Pierce; IL) or enhanced chemiluminescence solution (NEN; MA.).
9. Glass homogenizer with both a loose and tight pestle.
10. Extraction (STEN) buffer: 50 m*M* Tris-HCl, pH 7.5, 0.14 *M* NaCl, 2 m*M* EDTA, 0.4% NP-40, protease inhibitor tablet (ethylenediaminetetraacetic acid-free; one tablet in 25 mL; Roche, Germany) at 4°C.
11. Immunofluorescence microscopy equipment with three-channel viewing.
12. Fluorescencein-conjugated secondary antibodies.
13. Microwave oven.
14. Citra antigen retrieval buffer (Biogenex, San Ramon, CA).
15. Vector immunohistochemistry staining kit (Vector Labs, Burlingame; CA).
16. Ubiquitination reaction buffer, 50 μL: 4 m*M* ATP, 2 m*M* $MgCl_2$, 50 m*M* Tris-HCl, pH 7.5, 100 ng E1, 2 μg of UbcH7, 2 μg of His-Ub or FLAG-Ub.
17. Deglycosylation kit (Prozyme; San Leandro, CA).
18. SDS-PAGE silver staining kit (Invitrogen Inc.; Carlsbad, CA).

## 3. Methods

The methods described here illustrate: (1) the detection of endogenous parkin in primary neuronal cultures and human brain using affinity-purified, polyclonal antibodies; (2) the demonstration of parkin-mediated Ub ligase function in vitro; and (3) the detection of one of neural parkin's reported binding partners within the UPS (UbcH7) and of a PD-linked putative substrate (αSp22) from human brain homogenates.

### *3.1. Affinity-Purified Antibodies to Parkin*

Topics covered under **Subheading 3.1.** include an outline for the generation of specific antibodies to parkin and their characterization using native conditions with dot blotting (nondenaturing, nonreducing) and conventional SDS/PAGE with Western blotting methods (denaturing and reducing). The development of these high-affinity anti-parkin antibodies and their detailed characterization allowed us to carry out experiments aimed at the investigation of endogenous parkin as well as epitope-tagged, heterologous proteins.

#### *3.1.1. Antibody Production*

Polyclonal rabbit antisera are routinely raised against synthetic peptides encompassing 10–40 amino acids of the target protein. The individual, to-be-synthesized peptide is selected based on combined performance in screening of its primary sequence to predict probabilities of three key parameters: immunogenicity, surface exposure and hydropathy *(25)*. We identified seven regions within the full-length sequence of human parkin (**Fig. 1**) with acceptable scores using all three criteria. Peptides of >95% purity (w/v; as assessed by high-performance liquid chromatography) were synthesized at our hospital's Biopolymer Facility *(26)*. Using a commercial vendor, ≤12 mg of sequence-verified peptide underwent covalent conjugation to a carrier

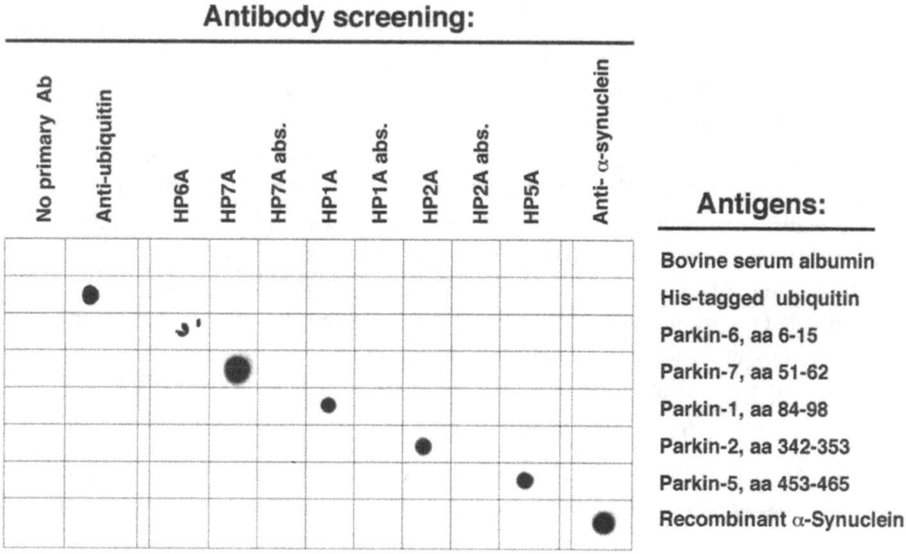


Fig. 2. Screening of affinity-purified parkin antibodies by dot blotting under nonreducing and nondenaturing conditions against synthetic peptides and recombinant proteins.

protein (keyhole limpet hemocyanin or rabbit serum albumin) and were injected into two rabbits per selected antigen over a 21-wk protocol that included an initial immunization step and five consecutive boosts. Serial phlebotomies with serum production and enzyme-linked immunosorbent assay (ELISA)-based calculation of antibody titres were carried out at the vendor's facility. Antisera with titres ranging from 1:10,000 to 1:120,000, as judged by ELISA, were pooled from sister rabbits (~200 mL) and processed over resin-immobilized peptide immunogen (~10 mg) by serial column chromatography at the same vendor facility *(27)*. Following the final elution by pH gradient of *a*ffinity-purified, polyclonal immunoglobulins to *h*uman *P*arkin (HP1A, HP2A, etc.; final volume, ~10 mL each; **Fig. 2**), antibodies were stored in sodium borate buffer. Individual aliquots (concentration, 0.12–1.2 μg/μL; 100 μL each) were lyophilized and parkin antibodies were stored at –20°C degrees. Of seven synthetic peptide-based immunization projects that were begun, five progressed to affinity-purification. Among those, all five met our criteria for the specific detection of endogenous and heterologous parkin proteins during subsequent comprehensive characterization, as outlined in the following.

### *3.1.2. Antibody Screening by Dot Blotting*

Synthetic peptides were thawed, diluted in PBS and loaded onto a 96-well dot blot-apparatus (100 ng and 10 ng/100 μL), which contained a mounted, prehydrated PVDF membrane. Loaded samples included aliquots of peptides that were used initially as antigens and of several recombinant proteins, among them full-length Ub, α synuclein and bovine serum albumin. For dot blotting, polypeptides were transferred onto the membrane by gravity and vacuum suction (30–45 min each). Individual membrane strips of peptide dots were probed with primary antibody preparations and subsequently developed by immunoblotting to demonstrate several characteristics: (1) the correct matchup of each affinity-purified antibody with its immunogen; (2) the degrees of sensitivity and specificity of each anti-parkin antibody for the cognate peptide; and (3) the lack of cross-reactivity with unrelated and PD-linked proteins. To this end, strips saw diluted affinity-purified anti-parkin antibodies (1:500–1:1000), anti-parkin that had been absorbed (by coincubation with the cognate antigen; 5 μg/μL of antibody), commercial monoclonal antibodies (e.g., to Ub and αS), as well as solutions that contained no primary antibody (**Fig. 2**). Anti-parkin antibodies had to specifically detect their cognate antigen by dot blotting and to have shown no cross-reactivity with other proteins before they were used in subsequent experiments (***9,18–20***).

### *3.1.3. Antibody Characterization by Western Blotting*

The next step in the characterization of anti-parkin antibodies was probing cell lysates and tissue homogenates by Western blotting. Lysates were obtained from dishes of confluent human SH-SY5Y neuroblastoma cells (that expressed no readily detectable endogenous parkin) after transfection with a vector that encodes N-terminally tagged wild-type parkin (*see* **Note 1**). An expression plasmid carrying the human *parkin* cDNA was generated by ligating the insert into a mammalian *p*cDNA3.1 expression vector utilizing restriction enzymes *Eco*RI and *Sal*I at the 5' and 3' ends, respectively. (A *p*GEM T-Easy cloning vector with the *parkin* cDNA had been kindly provided by Drs. Y. Mizuno and N. Hattori of Juntendo University in Tokyo, Japan.) For non-polymerase chain reaction (PCR)-based cloning of the *p*cDNA3.1*myc-parkin* plasmid and its stable transfection into SY5Y cells, we followed standard molecular biology protocols that will not be described here owing to space limitations (***9,12,19***). Extracts of human brain were prepared from frozen autopsy material (*see* **Note 2**) and loaded in parallel with lysates from myc~parkin-expressing SY5Y cells (for tissue, >10–30 μg/lane; for cells, 10 μg/lane). We employed Tris–glycine gradient gels for routine PAGE under reducing, SDS-denaturing conditions with subsequent Western blotting (**Fig. 3**). Six steps were carried out to obtain extracts from human brain:

1. Transfer <3.0 g wet tissue from a –80°C freezer onto ice (coronal sections of cortex, thickness 1 cm; collected within <15 h postmortem).
2. Remove macroscopically visible meninges, blood vessels, and large pieces of white matter.
3. During thawing, mince softening cortex into small cubes (~2 × 2 × 2 mm) by scalpel on ice during <30 min after the addition of STEN-lysis buffer (STEN-LB; 1 mL). The addition of 1–2% SDS will further enhance the yield of neural parkin from human (but not rodent) adult brain.
4. Transfer tissue cubes into a glass homogenizer, add STEN-LB (3X v/w; containing one tablet of protease inhibitors per 25 mL and NP-40, 0.2%), and begin homogenization with a loose glass pestle followed by a tight glass pestle (30–50 full strokes each).
5. Transfer total homogenate into a suitable ultracentrifuge tube and carry out ultracentrifugation (100,000 × rpm, 35 min, 4°C);
6. Take off supernatant, save aliquots to measure total protein concentration (100 μL, usually 6–9 μg/μL) and use without further delay or freezing step as "human brain extract."

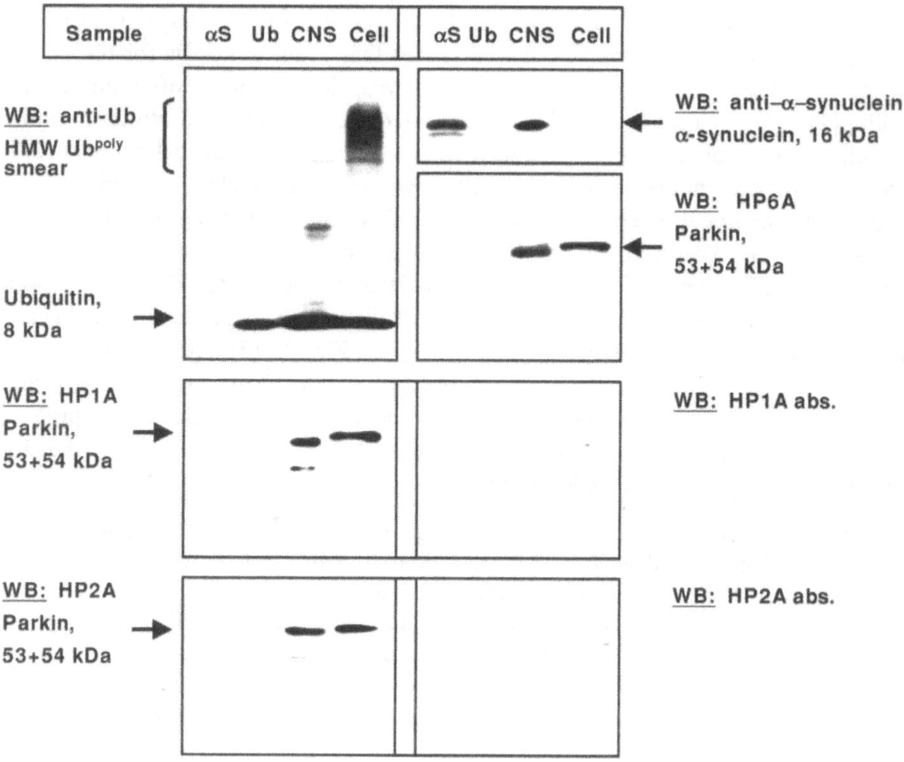


Fig. 3. Screening of affinity-purified antibodies to parkin by Western blotting under reducing and denaturing conditions by probing recombinant proteins, neural parkin from brain, and myc-tagged parkin expressed in transfected SH-SY5Y cells.

Following these experiments, anti-parkin antibodies that specifically detected both the 52- to 52.5-kDa neural parkin and the 52.5- to 53-kDa myc-parkin polypeptides by Western blotting, and that also did not cross-react with full-length αS or Ub under these conditions were then used for immunoprecipitation (IP) studies, indirect immunofluorescence microscopy, immuno-histochemical studies, immunoelectron microscopy, and in in vitro ubiquitination assays *(9,19,20)*.

### 3.2. Detection of Endogenous Parkin in Neurons

Topics covered under **Subheading 3.2.** include an outline for the specific detection of endogenous parkin in primary neuronal cultures by indirect immunofluorescence (IF) and the demonstration of E3 ubiquitin ligase-like activity in vitro, as mediated by IP-parkin from lysates of primary rat neurons.

#### 3.2.1. Detection of Parkin and UbcH7 in Neurons by Immunofluorescence

To characterize the staining specificity of our affinity-purified anti-parkin antibodies in untransfected vertebrate neurons, we first examined primary cortical cells derived from embryonic (E18) rat pups that were cultured for 7–10 days in vitro. Several authoritative chapters have been published that detail protocols for the microscopic dissection of Sprague–Dawley rat embryos, the precoating of culture dishes with polylysine as a substrate and for specific requirements of a neurotrophic culture medium to promote the development of these embryonic rat cortical neurons *(28)*; therefore, they are not reviewed here. Rather, we outline the technique used to detect endogenous neuronal proteins by indirect IF (**Fig. 4**). Six-well dishes were fitted with glass cover slips. Dissected cells were plated (age, embryonic d 18; density of $1 \times 10^6$ cells/well) and cultures were maintained for 7–10 d in vitro *(20,28)*. Following a brief fixation step, neurons were probed with three distinct antibody preparations: (1) anti-parkin antibodies (HP1A, HP2A; for epitopes, see **Fig. 2**); (2) anti-parkin competitively absorbed with its cognate antigen (antibody dilution, 1:50–1:100); and (3) specific, well-characterized antibodies to other antigens, including the E2 protein, UbcH7, and αS (**Fig. 4**). The following 11 steps should be carried out to probe cultured neurons by indirect IF microscopy:

1. Rinse cover slips in their wells with PBS at room temperature (RT; 3 × 1 min).
2. Use fresh paraformaldehyde (4% v/v in PBS) to fix cells during 15 min at RT.
3. Wash cover slips in their well with PBS at RT (3 × 1 min).
4. Permeabilize cells briefly with Triton-X100 detergent (0.25% v/v in PBS at RT; 5 min).
5. Wash cover slips in their well with PBS at RT (3 × 5 min).
6. Block cells with normal goat serum (10% in PBS; 30 min at RT).
7. Prepare antibodies in normal goat serum (1% in PBS; 200 µL of 1:100 primary antibody solution) and dispense onto an uncoated microscopic glass slide.
8. Carefully remove cover slips containing fixed neurons from their wells, place them face down onto corresponding antibody solution, and incubate for 60 min at RT (or at 4°C overnight).
9. Repeat washing steps in PBS, as before.
10. Prepare secondary antibodies in 1% normal goat serum (goat antimouse IgG, fluorescein isothiocyanate [FITC] dye-conjugated; goat antirabbit IgG, rhodamine dye-conjugated; dilution, 1:100 each) for 1 h at RT;
11. Repeat washing steps in PBS, as before, and mount cover slips in aqueous mounting medium.

## 3.3. Ubiquitin Ligase Activity of Neuronal Parkin

### 3.3.1. Neuronal Parkin Confers E3 Ubiquitin Ligase-Like Activity In Vitro

As three distinct anti-parkin antibodies, HP1A, HP2A and HP5A specifically and reproducibly generated the same staining pattern for endogenous parkin in paraformaldehyde-fixed neuronal cultures, we subsequently screened their ability to bind the full-length polypeptide under native conditions in IP experiments. There, anti-parkin HP5A (directed to the C-terminus) demonstrated both affinity and avidity for myc-parkin from transfected HEK293 cells. Anti-parkin HP1A (directed to an antigen near the N-terminus) demonstrated higher affinity for human parkin than for rat parkin in IP experiments using comparable concentrations. We decided to choose the third antibody, HP2A (directed to an antigen within the IBR-domain; **Fig. 1**), for IP of parkin from cultured cortical neurons and examine its in vitro Ub ligase activity

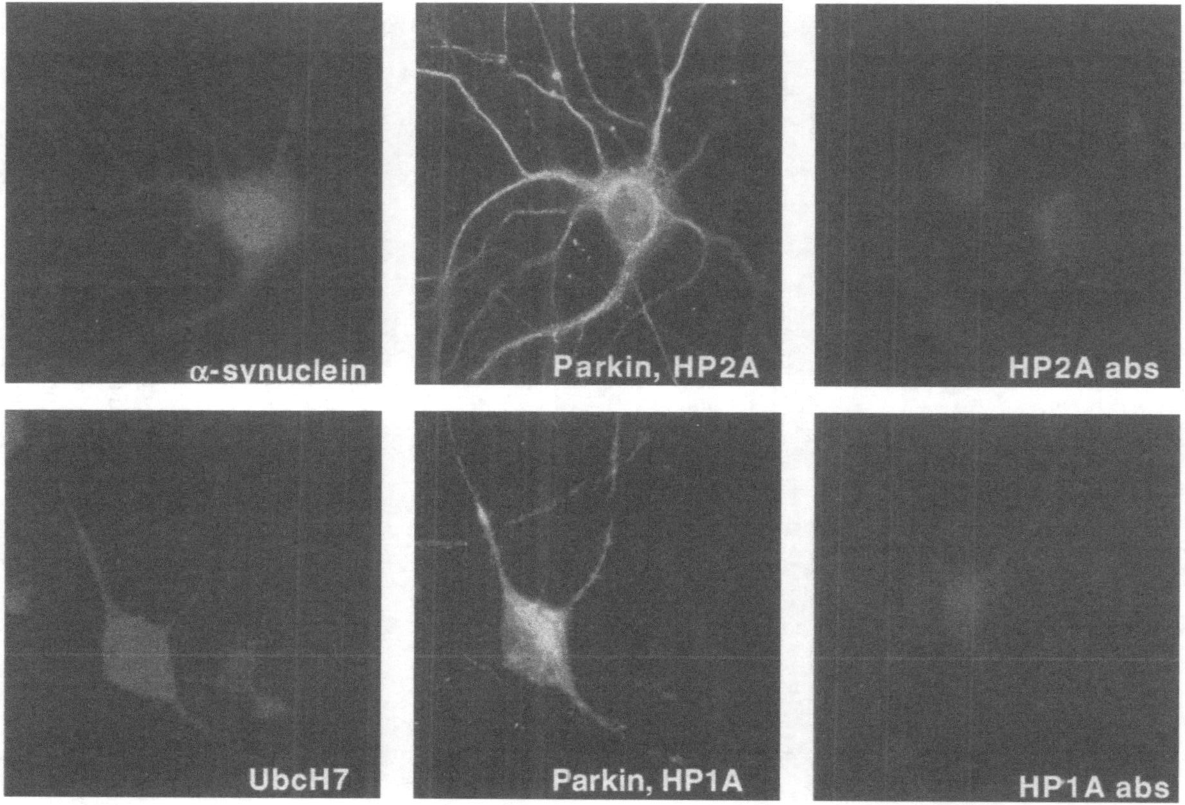


Fig. 4. Staining of endogenous proteins in paraformaldehyde-fixed primary neurons from embryonic rat cortex by indirect immunofluorescence. Antibodies used (left to right): anti-α-synuclein, anti-parkin (HP2A), HP2A absorbed **(top panel);** anti-UbcH7, anti-parkin (HP1A), and HP1A absorbed **(lower panel)**.

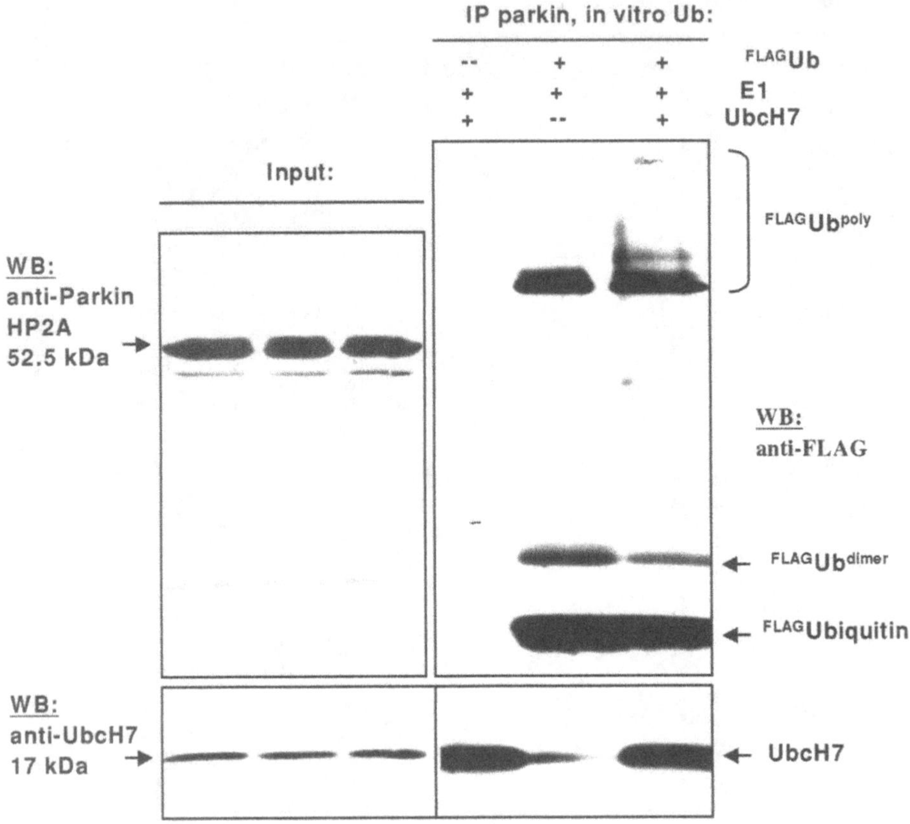


Fig. 5. Detection of in vitro ubiquitination using endogenous parkin from rat neurons **(upper left panel)** after its immunoprecipitation (IP). IP-rat parkin was coincubated with UbcH7 (both endogenous from co-IP and exogenous; *see* **bottom panels**), excess E1, FLAG-tagged ubiquitin, ATP, and $MgCl_2$. The in vitro reaction was monitored by Western blotting with anti-FLAG antibody **(upper right panel)**.

**(Fig. 5)**. The strategy behind this experiment entailed: first, to enrich endogenous parkin together with its constitutively assembled cofactors and substrates from lysates of primary neurons by anti-parkin IP; second, to facilitate Ub ligase activity of the enriched $E3^{Public}$ complex in vitro by the addition of excess E2 protein (UbcH7) and excess E1 under ATP-containing buffer conditions; and third, to demonstrate actual E3-type activity of parkin after adding excess amounts of recombinant, epitope-tagged Ub to the in vitro reaction mix, as followed by the detection of substrate modification by probing Ub-conjugates with a tag-directed, commercial antibody **(Fig. 5)**.

To facilitate in vitro ubiquitination reaction by IP-parkin, we followed these eight steps:

1. Lyse two dishes of 7-d-old primary rat cortical neurons (diameter, 10 cm; density, ~$2 \times 10^6$ neurons/dish) per IP arm in STEN-LB containing NP-40 (0.2%; volume, 500 L/dish).

2. Obtain and combine supernatants, as described in **Subheading 3.1.3.**, and prepare equal aliquots for several IP reactions (1200 μL/arm).
3. IP parkin by adding 10 μL of anti-parkin antibody (HP2A, 1:100) or HP2A absorbed (**Subheading 3.1.2.**) and Protein–A Sepharose beads (wet slurry, 50 μL) and rock at 4°C for 100 min.
4. Sediment IP beads (1000*g* at 4°C, 2 min) and wash pellets in cold PBS (two times, 5 min each time) (*see* **Note 4**).
5. Following the third centrifugation step, wash IP pellets in cold PBS buffer containing ATP (4 m*M*).
6. Remove final wash buffer and add 40 μL of in vitro ubiquitin reaction buffer to the beads on ice.
7. Transfer mix into water bath (37°C) and incubate for 60 min. Terminate reaction by the addition of 2X (or 4X) Lämmli buffer (25 μL) and load onto SDS-PAGE for subsequent Western blotting.
8. Probe membranes with monoclonal anti-UbcH7 and anti-FLAG antibodies (**Fig. 5**).

The above outlined experiment resulted in the covalent conjugation of FLAG-Ub onto unknown neuronal substrates in vitro. This reaction appeared more pronounced by the addition of exogenous human UbcH7 than with endogenous rat UbcH7 alone, a portion of which had co-IP'ed with rat parkin (**Fig. 5**). These results revealed that a pool of rat parkin-derived from freshly processed primary neurons—could be IP'ed with anti-parkin and induced to promote E3-type Ub ligase activity in vitro (*see* **Notes 3** and **4**).

## 3.4. Detection of Parkin in Human Brain

The topic covered under the **Subheading 3.4.** outlines the detection of parkin in human neurons by immunohistochemistry of sections from adult control brain.

### 3.4.1. Parkin Staining in Human Brain by Immunohistochemistry

To characterize the cellular distribution of parkin in the adult human nervous system, we probed brain sections from neurologically normal control cases by immunohistochemistry (**Fig. 6**). Tissue blocks (~1 × 1 × 0.5 cm) are routinely obtained by pathologists during autopsy from a variety of neuroanatomic sites for general histological and specific immunohistochemical purposes. There, specimens are first fixed in formalin, paraffin-embedded, serially cut (thickness, 10 μm) and mounted on microscopic glass carriers. In our immunohistochemical studies, we used polyclonal anti-parkin HP1A and HP2A and absorbed anti-parkin as well as monoclonal antibodies to tyrosine hydroxylase (the rate-limiting enzyme in the synthesis of dopamine) and the presynaptic protein, αS, to examine the staining pattern of dopaminergic neurons in the substantia nigra of the anterior midbrain, of cortical neurons in the neocortex, and of peripheral autonomic gangliocytes (**Fig. 6**). For staining purposes, we used the biotin–streptavidin enzyme immunoassay method and carried out the following 16-step protocol (adapted from **ref. *29***):

1. Deparaffinize and hydrate tissue sections by serial dipping (100 dips per dish) in the following solutions: Histoclear, Histoclear, Histoclear, and 100% ethanol (v/v, 50:50), 100% ethanol, 100% ethanol, 95% ethanol, 75% ethanol, 50% ethanol, and water.

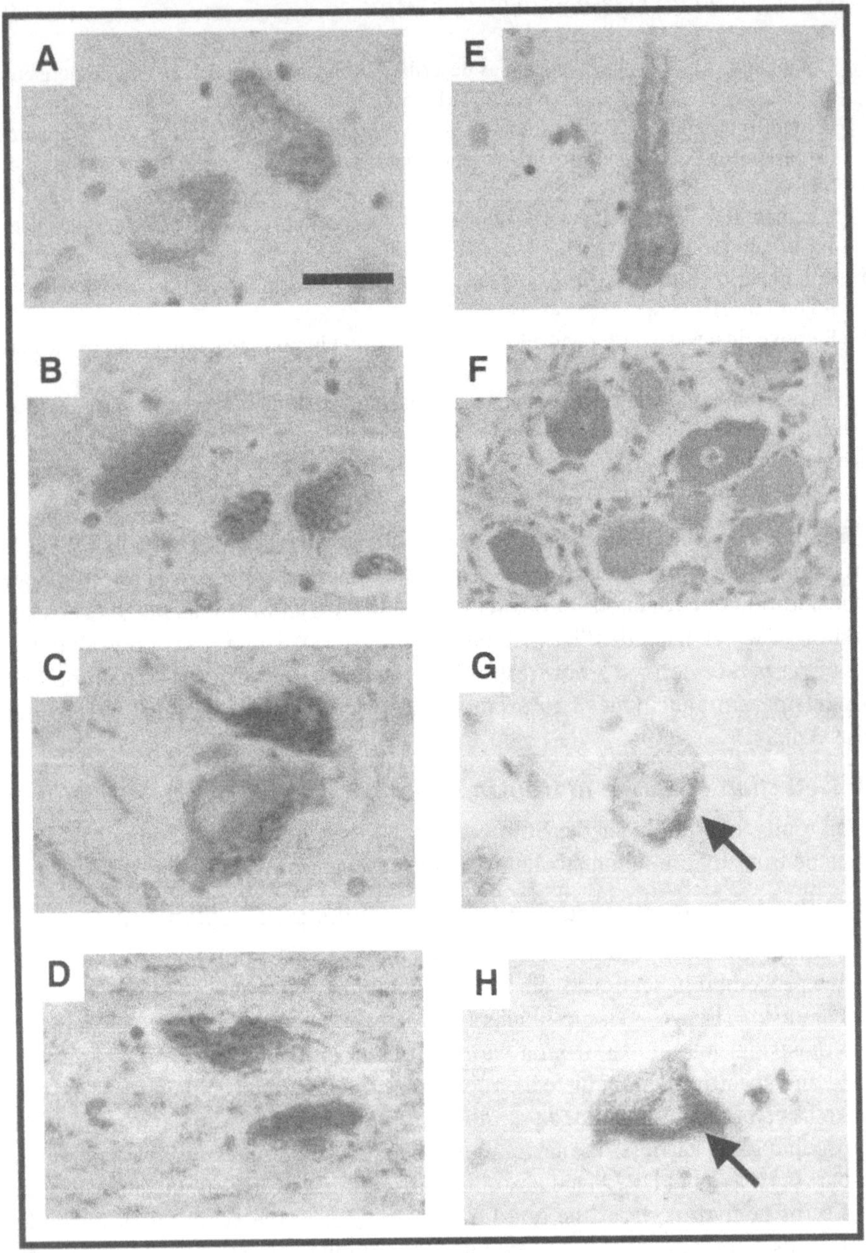


Fig. 6. Immunohistochemistry of parkin reactivity in human neurons of the central and peripheral nervous system. Staining with anti-parkin HP2A (**A,D**), HP1A (**B,C**), HP2A absorbed (**G**), and HP1A absorbed (**H**) is shown on sections of dopaminergic neurons in the substantia nigra pars compacta (**A–D,G,H**), of cortical neurons in the frontal cortex (**E**) and of peripheral autonomic gangliocytes (**F**). Note the autofluorescence of neuromelanin granules in the perinuclear space of dopaminergic neurons (*arrows* in **G,H**) and the staining of both vesicular structures in the cytoplasm and extracellular neuropil by anti-parkin. Bar length = 50 μm.

2. Incubate tissue sections in a glass container containing methanol (99 mL) and hydrogen peroxide (1 mL of 30% solution) for 10 min.
3. Rinse slides in deionized water (2 × 5 min).
4. Bring sections to 100°C by microwaving in Citra antigen retrieval buffer during four cyclical boiling (5 s) and cooling (10 s) steps. Transfer sections to RT.
5. Block sections in goat serum for 30 min at RT in Tris-buffered saline (TBS, 10%; 50 m*M* Tris-HCl, pH 7.5, 140 m*M* NaCl).
6. Add 300 μL of primary antibody (1% goat serum in TBS) onto glass slides and incubate overnight at 4°C (or for 1.5 h at RT; dilution of antibody, 1:50–1:200) in a wet chamber.
7. Wash glass slides in TBS (2 × 5 min).
8. Add 300 μL of secondary antibody by diluting 22 μL of biotinylated, affinity-purified antirabbit or antimouse antibody (provided with the Vector Laboratory Vectastain ABC kit) in 5 mL of blocking solution (as in **Subheading 3.4.1., step 5**) for 30 min at RT. At the same time, prepare fresh streptavidin–horseradish peroxidase (HRP)-containing complex by combining reagent A and B, as provided in the kit (45 μL each/5 mL of TBS).
9. Wash glass slides in TBS (2 × 5 min).
10. Add 300 μL of streptavidin–HRP-containing ABC complex and incubate at RT for 30 min).
11. Wash glass slides twice in Tris buffer (50 m*M*, 2 × 5 min).
12. Develop slides in glass container with freshly prepared substrate, 2,2'-azino-di-benzthiazoline-6-sulfonic acid (20 mg added to 50 m*M* Tris-HCl, pH 7.5, filtered through Whatman paper containing 33 μL of 30% hydrogen peroxide) and develop for <10 min at RT.
13. Stop reaction by putting glass slides into a deionized water bath at RT for >10 min.
14. Counterstain sections with filtered hematoxylin. Dip slides 16 times and return to water. Destain with single dip in acid–alcohol solution (99 mL of 80% ethanol, 1 mL of 1 *M* HCl) and quickly return to deionized water.
15. Dehydrate sections by running through serial dipping outlined in **step 1** in reverse order.
16. Cover slip sections with 25 × 50 mm-sized cover slips using nonaqueous mounting medium.

The above outlined experiments resulted in the strong and specific staining of neurons and their neuritic processes by anti-parkin in human midbrain, of groups of neurons in layer IV of the neocortex and of peripheral gangliocytes **(Fig. 6)** *(9)*.

### *3.5. Ubiquitin Ligase Activity by Parkin From Brain*

The topic covered under the **Subheading 3.5.** includes a protocol to detect the in vitro Ub ligation of neural substrate proteins from human brain. These experiments demonstrated in vitro Ub-ligase activity for the enriched, endogenous IP-parkin protein consistent with previous findings from cellular exogenous parkin overexpression experiments **(Fig. 7)** *(12–14)*.

#### *3.5.1. Functionally Active E3$^{Public}$ From Human Brain*

The rationale behind this experiment was summarized above under **Subheading 3.3.** In the following, it was applied to previously frozen specimens of human brain (rather than freshly lysed neurons), as carried out during the following six steps **(Fig. 7)**:

1. Obtain extracts from human brain, as detailed under **Subheading 3.1.3.**
2. Prepare several IP arms that see either: (a) anti-parkin HP1A, HP2A or HP5A with an aliquot of noncognate peptide (i.e., mock absorption); (b) antibody with its cognate peptide (i.e., absorption with the actual immunogen; 5 μg of peptide/μL of antibody);

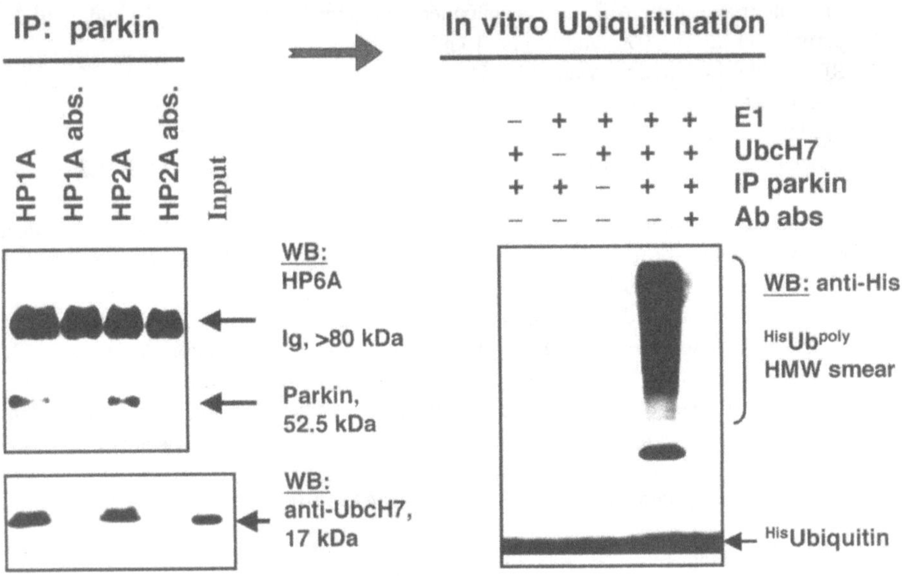


Fig. 7. In vitro ubiquitination using parkin from human brain extracts following its immunoprecipitation (IP; **left panels**). IP-parkin was incubated with exogenous UbcH7, excess E1, His-tagged ubiquitin, ATP, and $MgCl_2$. Reaction was monitored by Western blotting with anti-His antibody **(right panel)**.

(c) no antibody solution (i.e., negative IP control); and (d) an antibody to another RING-dependent (e.g., HHARi) or a non–RING-dependent E3 protein if available (e.g., E6-AP) (*19*).

3. Add each antibody solution to up to 2.0 mL of human brain extract (>4 mg) and Protein–A Sepharose beads, as outlined under **Subheading 3.3.1.** (*see* **Note 5**).
4. Wash IP pellets and prepare an in vitro ubiquitination reaction, as outlined under **Subheading 3.3.1.** using 2 µg of both exogenous UbcH7 and His (or FLAG)-tagged Ub.
5. Following incubation in a water bath (37°C, 1 h), add 4X Laemmli buffer at RT, and gently agitate beads and repellet them (1000 rpm; 30 s) to release all proteins from Protein A-Sepharose beads (*see* **Note 6**).
6. Analyze IP samples by loading equal aliquots onto Tris–glycine gradient gels (4–20% or 10–20%) for SDS-PAGE. Probe Western blots with antibodies to parkin, UbcH7, and to the chosen Ub tag (*see* **Note 7**).

## 3.6. Ubiquitination of Brain Substrates by Myc-Parkin

### 3.6.1. Detection of E3 Activity by IP-Myc-Parkin After Pull Downs

Ubiquitination of neural substrates from human brain extracts can also be observed using heterologous myc-parkin from transiently transfected HEK293 cells (*12,19*). This alternative method is useful when screening for substrates in parkin-deficient human brain or when analyzing naturally occurring, *parkin*-linked mutations for their loss-of-function effect (*19*). Such an assay works under the assumptions that IP-myc-parkin is (1) able to assemble essential co-factors of $E3^{Public}$ from lysates of HEK293 cells or from extracts of human brain (or both); (2) binds to parkin's natural substrates, which should be more abundantly present in the human brain than in nonneural

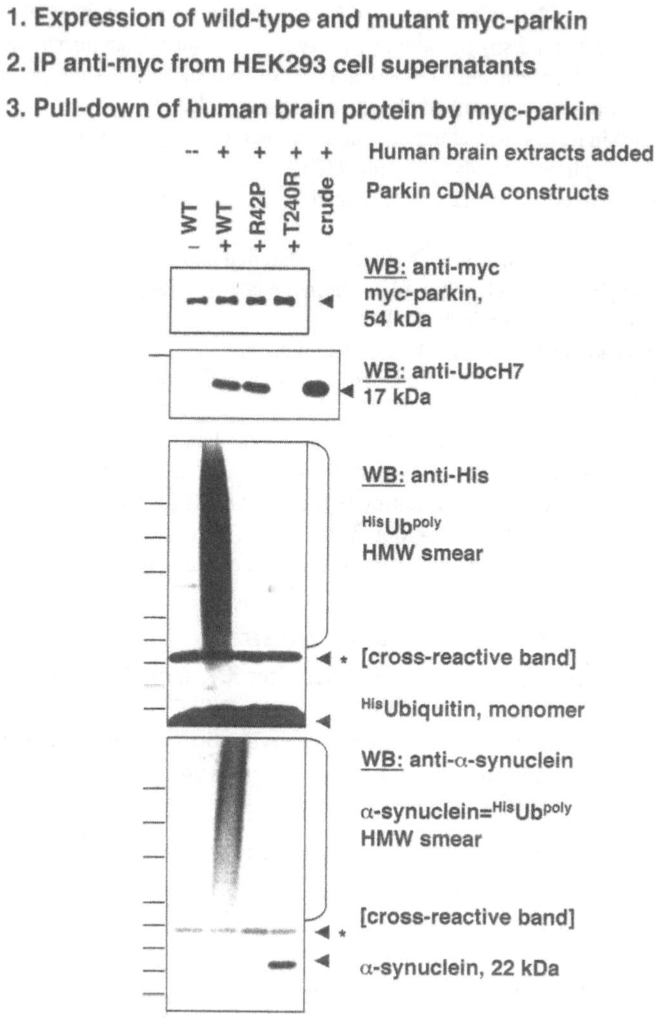


Fig. 8. The in vitro ubiquitination reaction using wild-type myc-tagged human parkin as E3 was carried out in three steps: (1) immunoprecipitation (IP) of myc-parkin by anti-myc from cell culture lysates following their transfection with wild-type or mutant human parkin cDNA constructs (top panel); (2) pull-downs of endogenous UbcH7 and neural substrates from human brain extracts by IP'ed myc-parkin (second from top); and (3) in vitro reaction using IP-myc-parkin, endogenous and exogenous UbcH7, excess E1, His-tagged ubiquitin, ATP, and $MgCl_2$, as assayed by Western blotting (bottom two panels).

HEK293 cells; and (3) is able to carry out E2-dependent, RING-mediated in vitro ubiquitination despite an approx 20 amino-acid-long myc-tag at the N-terminus ***(12,19)*** (*see* **Note 8**). We have used the following protocol to facilitate in vitro ubiquitination of neural substrates by myc-parkin protein **(Fig. 8)**:

1. Transfect HEK293 cell cultures (>90% confluence using) with *p*cDNA3.1 plasmid (10 µg DNA/10 cm dish) containing wild-type or mutant human *parkin* constructs.
2. Lyse cells and obtain cellular extracts from one or two dishes per IP (and thus ubiquitination) arm, as outlined under **Subheading 3.1.3.**
3. IP myc-parkin from supernatants by using commercial, polyclonal anti-myc antibody and Protein A– Sepharose beads (120 min, 4°C).
4. Wash beads in cold PBS (2 × 5 min each, 4°C) and pellet by centrifugation (2000*g*, 2 min).
5. Add extracts from human brain obtained as detailed under **Subheading 3.1.3.** to IP-myc parkin-carrying beads (1.5–2.0 mL of homogenate per ubiquitination arm; >6 mg) and perform pull-down (90 min each, 4°C).
6. Wash pelleted beads in cold PBS (2 × 5 min each, 4°C);
7. Following the third centrifugation step, wash IP pellets in cold PBS buffer containing ATP (4 m*M*).
8. Proceed with in vitro ubiquitination step, as outlined under **Subheading 3.5.1.**
9. Perform SDS/PAGE and Western blotting with monoclonal anti-myc antibody to monitor efficiency of IP, with anti-His antibody (or anti-FLAG for the chosen Ub tag) to monitor efficiency of in vitro Ub ligation, with antibodies to candidate E2 proteins (for co-IP effect) and with antibodies to neural proteins that may represent substrates of parkin (e.g., αS).

### 3.7. Identification of a Neural Parkin Substrate

The protocols covered under the **Subheading 3.7.** include the identification of a human brain-derived putative parkin substrate, αSp22. The latter was found to represent a rare variant of the abundant presynaptic protein, αS, in human brain homogenates, and shown to accumulate in brains from parkin-deficient human PD cases ***(19,30)***.

#### 3.7.1. Detection of αSp22 and Its Ubiquitination by Parkin From Human Brain

The detection of an αS isoform in the pellets of IP-parkin from human brain was hypothesis-driven ***(9,19)***. The observation of this co-IP effect was the result of specifically probing Western blot membranes of sister gels with commercially available, well-characterized monoclonal anti-α-synuclein (αS) antibodies (**Fig. 8**, lowest panel). Membranes were developed with anti-αS immediately after in vitro ubiquitination assays, as outlined under **Subheadings 3.5.1.** and **3.5.2.**

The contribution of neural parkin to the steady-state levels of α5 in the aging vertebrate brain is still a matter of scientific debates and ongoing investigations ***(2,3,16,19)*** (**Note 9**).

## 4. Notes

1. Confluent dishes of parkin-expressing cells should be rinsed with 10 mL of cold PBS after the removal of conditioned medium before lysis in STEN-LB. This step will greatly reduce the degree of a cell death-associated proteolytic cleavage event at the N-terminus of parkin, which results in the removal of the 10-11 kDa Ub-like domain and generation of an approx 43-kDa C-terminal fragment of parkin ***(9)***.
2. Institutional review board (IRB) approval for the collection of to-be-discarded human tissue has to be obtained in advance and special precautions have to be implemented in the prevention of infections when handling human specimens that may contain potential pathogens ***(9)***.

3. We have observed that anti-parkin HP2A and HP5A only immunoprecipitate a portion of the expressed neuronal parkin from primary rat neurons, in contrast to transfected cells that express myc-parkin and parkin from human brain. The most likely explanation is that under these native conditions, the single amino acid substitution in the corresponding rat parkin sequences (when compared with the human peptides that were used as immunogens) reduces affinity or avidity (or both). This is not the case for the detection of rat parkin by Western blotting with anti-parkin HP2A following SDS-PAGE ***(9)***.
4. A significant modification that we introduced in the IP protocol has allowed us to detect both parkin substrates from human brain more easily and to more efficiently facilitate $E3^{Public}$–mediated ligase activity in vitro. Instead of stringent washes of IP pellets with Triton X-100- and SDS-containing buffers, we performed our washing steps in cold PBS. Using affinity-purified anti-parkin in IPs, this change led to an increase in sensitivity but not at the expense of specificity in detecting in vitro Ub ligase activity ***(19)***.
5. IP of parkin, IP of myc-parkin and lectin-pull down experiments are best completed within a 2-h time frame after their start, as we have repeatedly observed either proteolytic degradation of parkin substrates or loss of $E3^{Public}$ activity (or both) beyond a processing time of 120 min in vitro ***(30)***. Likewise, we have been unable to utilize homogenized brain tissue that had undergone a freeze–thaw cycle in functional assays.
6. In contrast to boiling of IP or lectin pull-down pellets (which results in melting of the agarose resin), heat treatment of PVDF membranes in PBS at 60–65°C frequently enhances immunoreactivity for αS isoforms (but not for parkin proteins) in subsequent Western blots.
7. The detection of human parkin (52–53 kDa) after anti-parkin IP with polyclonal antibodies (Ig heavy chain, 50–52 kDa) is best accomplished by using one of three methods to reduce possible background staining of rabbit Ig: (a) anti-parkin HP6A (1:500–1:1,000) and a drastically increased dilution of the secondary antirabbit Ig antibody (1:20,000–30,000 rather than the routine 1:7,000); (b) using a monoclonal (mouse) anti-parkin for Western blotting ***(22)***; or (c) by conjugating anti-parkin HP6A covalently to biotin or FITC and completing immunoblotting with an avidin- or anti-FITC-immunoassay method, respectively.
8. We observed that the increasing length of an N-terminal epitope-tag created as a result of the initial cloning step of *parkin* cDNA (that is the number of residues generated by base pairs corresponding to the myc-tag and the multiple cloning site) negatively affects the observed co-IP effect of αSp22, which is recruited by the N-terminus of parkin ***(9,19)***.
9. Detailed characterization of the approx 6000-Da mass difference of αSp22 over αS's normal 16-KDa migration revealed the presence of sialic acid and N-acetylgalacto-samine sugars ***(19)***. Comprehensive mass spectrometry analyses confirmed full-length αS as the only protein constituent, but detected no glycoprotein-like modification ***(19)***. Rather, subsequent glycobiological and lipid characterization experiments identified a strong but noncovalent association of human α-synuclein with glycosphingolipids ***(30)***.

## Acknowledgments

The authors thank M. Baker for advice regarding antibody production, M. P. Frosch and J. A. Chan for human brain tissue collection, C. Lemere for sharing her immunohistochemical protocol, N. Hattori and Y. Mizuno for the human *parkin* cDNA, M. Medina for the generation of stably transfected SY5Y cells, and K. S. Rasakham and P. Xu for technical assistance. This work was supported by the Grass Foundation (Braintree, MA), the Lefler Foundation (Harvard Medical School), the M.S.A. Fund (Brigham &

Women's Hospital), and the NINDS-NIH (NS02127) to M. Schlossmacher and a Pergolide Fellowship (Eli Lilly, Japan K. K.) to H. Shimura.

## References

1. Jellinger, K. A. and Mizuno, Y. (2003) Parkinson's disease. In *Neurodegeneration: The Molecular Pathology of Dementia and Movement Disorders* (Dickson, D. W., ed.), ISN Neuropath Press, Basel, pp. 159–186.
2. Dauer, W. and Prezdborski, S. (2003) Parkinson's disease: mechanisms and models. *Neuron* **39,** 889–909.
3. Chiechanover, A. and Brundin, P. (2003) The ubiquitin proteasome system in neurodegenerative diseases: sometimes the chicken, sometimes the egg. *Neuron* **40,** 427–446.
4. Golbe, L. I. and Maroudian, M. M. (2004) Alpha-synuclein in Parkinson's disease: light from two new angles. *Annal. Neurol.* **55,** 153–155.
5. Singleton, A., Farrer, M., Johnson J., et al. (2003) Alpha-synuclein locus triplication causes Parkinson's disease. *Science* **302,** 841.
6. Kitada, T., Asakawa, S., Hattori, N., et al. (1998) Mutations in the parkin gene cause autosomal-recessive juvenile parkinsonism. *Nature* **392,** 605–608.
7. Oliveira, S. A., Scott, W. K., Martin, E. R., et al. (2003) Parkin mutations and susceptibility alleles in late-onset Parkinson's disease. *Ann. Neurol.* **53,** 624–629.
8. Hasegawa, M., Fujiwara, H., Nonaka, T., et al. (2002) Phosphorylated alpha-synuclein is ubiquitinated in alpha-synucleinopathy lesions. *J. Biol. Chem.* **277,** 49071–49076.
9. Schlossmacher, M. G., Frosch, M. P., Gai, W. P., et al. (2002) Parkin localizes to the Lewy bodies of Parkinson disease and dementia with Lewy bodies *Am. J. Pathol.* **160,** 1655–1667.
10. Leroy, E., Boyer, R., Auburger, G., et al. (1998) The ubiquitin pathway in Parkinson's disease. *Nature* **395,** 451–452.
11. Liu, Y., Fallon, L., Lashuel, H. A., Liu, Z., and Lansbury, P. T. (2002) The *UCH-L1* gene encodes two opposing enzymatic activities that affect alpha-synuclein degradation and Parkinson's disease susceptibility. *Cell* **111,** 209–218.
12. Shimura, H., Hattori, N., Kubo, S., et al. (2000) Familial Parkinson disease gene product, parkin, is a ubiquitin-protein ligase *Nat. Gen.* **25,** 302–305.
13. Zhang, Y., Gao, J., Chung, K. K., et al. (2000) Parkin functions as an E2-dependent ubiquitin-protein ligase and promotes the degradation of the synaptic vesicle-associated protein, CDCrel-1. *Proc. Nat. Acad. Sci. USA* **97,** 13354–13359.
14. Imai, Y., Soda, M., and Takahashi, R. (2000) Parkin suppresses unfolded protein stress-induced cell death through its E3 ubiquitin-protein ligase activity. *J. Biol. Chem.* **275,** 35661–35664.
15. Lücking, C. B., Durr, A., Bonifati, V., et al. (2000) Association between early-onset Parkinson's disease and mutations in the *parkin* gene. French Parkinson's disease genetics study group. *N. Engl. J. Med.* **342**, 1560–1567.
16. Goldberg, M. S., Fleming, S. M., Palacino, J. J., et al. (2003) Parkin-deficient mice exhibit nigrostriatal deficits but no loss of dopaminergic neurons. *J. Biol. Chem.* **278,** 43628–43635.
17. Itier, J. M., Ibanez, P., Mena, M. A., et al. (2003) Parkin gene inactivation alters behaviour and dopamine neurotransmission in the mouse. *Hum. Mol. Gen.* **12**, 2277–2291.
18. Shimura, H., Hattori, N., Kubo, S., et al. (1999) Immunohistochemical and subcellular localization of parkin protein: absence of protein in autosomal recessive juvenile parkinsonism patients. *Ann. Neurol.* **45,** 668–672.

19. Shimura, H., Schlossmacher, M. G., Hattori, N., et al. (2001) Ubiquitination of a new form of alpha-synuclein by parkin from human brain: implications for Parkinson's disease. *Science* **293,** 263–269.
20. Cookson, M. R., Lockhart, P. J., McLendon, C., O'Farrell, C., Schlossmacher, M., and Farrer, M. J. (2003) RING finger 1 mutations in parkin produce altered localization of the protein. *Hum. Mol. Gen.* **12,** 2957–2965.
21. Petrucelli, L., O'Farrell, C., Lockhart, P. J., et al. (2003) Parkin protects against the toxicity associated with mutant alpha-synuclein: proteasome dysfunction selectively affects catecholaminergic neurons. *Neuron* **36,** 1007–1019.
22. Staropoli, J. F., McDermott, C., Martinat, C., et al. (2003) Parkin is a component of an SCF-like ubiquitin ligase complex and protects postmitotic neurons from kainate toxicity. *Neuron* **37,** 735–749.
23. Yang, Y., Nishimura, I., Imai, Y., Takahashi, R., and Lu, B. (2003) Parkin suppresses dopaminergic neuron-selective neurotoxicity by Pael-R in Drosophila. *Neuron* **37,** 911–924.
24. Choi, P., Snyder, H., Petrucelli, H., et al. (2003) SEPT_v2 is a parkin-binding protein. *Brain Res. Mol. Brain Res.* **117,** 179–189.
25. Kyte, J. and Doolittle, R. F. (1982) A simple method for displaying the hydropathic character of a protein *J. Mol. Biol.* **157,** 105–132.
26. *http://research.bwh.harvard.edu/biopolymer.htm.*
27. *http://www.openbiosystems.com/custom_peptide_antibody_services.php.*
28. Martinez, M. C., Ochiishi, T., Majewski, M., and Kosik, S. K. (2003) Dual regulation of neuronal morphogenesis by a delta-catenin-cortactin complex and Rho. *J. Cell. Biol.* **162,** 99–111.
29. Stoltzner, S. E., Grenfell, T. J., Mori, C., Wisniewski, K. E., Selkoe, D. J., and Lemere, C. A. (2000) Temporal accrual of complement proteins in amyloid plaques in Down's syndrome with Alzheimer's disease. *Am. J. Pathol.* **156,** 489–499.
30. Schlossmacher, M. G., Cullen, V., and Müthing, J. (2005) Glucocerebrosidase, α-synuclein and Parkinson's disease. *New Engl. J. Med.*, in press.

# Index

FSC
www.fsc.org
MIX
Papier | Fördert gute Waldnutzung
FSC® C083411

Kolibri 360 GmbH
Ferdinand-Jühlke-Straße 7
99095 Erfurt, Deutschland
produktsicherheit@kolibri360.de